An Introduction to Multicomplex Spaces and Functions

PURE AND APPLIED MATHEMATICS

A Program of Monographs, Textbooks, and Lecture Notes

MONOGRAPHS AND TEXTBOOKS IN PURE AND APPLIED MATHEMATICS

1. *K. Yano,* Integral Formulas in Riemannian Geometry (1970) *(out of print)*
2. *S. Kobayashi,* Hyperbolic Manifolds and Holomorphic Mappings (1970) *(out of print)*
3. *V. S. Vladimirov,* Equations of Mathematical Physics (A. Jeffrey, editor; A. Littlewood, translator) (1970) *(out of print)*
4. *B. N. Pshenichnyi,* Necessary Conditions for an Extremum (L. Neustadt, translation editor; K. Makowski, translator) (1971)
5. *L. Narici, E. Beckenstein, and G. Bachman,* Functional Analysis and Valuation Theory (1971)
6. *D. S. Passman,* Infinite Group Rings (1971)
7. *L. Dornhoff,* Group Representation Theory (in two parts). Part A: Ordinary Representation Theory. Part B: Modular Representation Theory (1971, 1972)
8. *W. Boothby and G. L. Weiss (eds.),* Symmetric Spaces: Short Courses Presented at Washington University (1972)
9. *Y. Matsushima,* Differentiable Manifolds (E. T. Kobayashi, translator) (1972)
10. *L. E. Ward, Jr.,* Topology: An Outline for a First Course (1972) *(out of print)*
11. *A. Babakhanian,* Cohomological Methods in Group Theory (1972)
12. *R. Gilmer,* Multiplicative Ideal Theory (1972)
13. *J. Yeh,* Stochastic Processes and the Wiener Integral (1973) *(out of print)*
14. *J. Barros-Neto,* Introduction to the Theory of Distributions (1973) *(out of print)*
15. *R. Larsen,* Functional Analysis: An Introduction (1973) *(out of print)*
16. *K. Yano and S. Ishihara,* Tangent and Cotangent Bundles: Differential Geometry (1973) *(out of print)*
17. *C. Procesi,* Rings with Polynomial Identities (1973)
18. *R. Hermann,* Geometry, Physics, and Systems (1973)
19. *N. R. Wallach,* Harmonic Analysis on Homogeneous Spaces (1973) *(out of print)*
20. *J. Dieudonné,* Introduction to the Theory of Formal Groups (1973)
21. *I. Vaisman,* Cohomology and Differential Forms (1973)
22. *B. -Y. Chen,* Geometry of Submanifolds (1973)
23. *M. Marcus,* Finite Dimensional Multilinear Algebra (in two parts) (1973, 1975)
24. *R. Larsen,* Banach Algebras: An Introduction (1973)
25. *R. O. Kujala and A. L. Vitter (eds.),* Value Distribution Theory: Part A; Part B: Deficit and Bezout Estimates by Wilhelm Stoll (1973)
26. *K. B. Stolarsky,* Algebraic Numbers and Diophantine Approximation (1974)
27. *A. R. Magid,* The Separable Galois Theory of Commutative Rings (1974)
28. *B. R. McDonald,* Finite Rings with Identity (1974)
29. *J. Satake,* Linear Algebra (S. Koh, T. A. Akiba, and S. Ihara, translators) (1975)

30. *J. S. Golan,* Localization of Noncommutative Rings (1975)
31. *G. Klambauer,* Mathematical Analysis (1975)
32. *M. K. Agoston,* Algebraic Topology: A First Course (1976)
33. *K. R. Goodearl,* Ring Theory: Nonsingular Rings and Modules (1976)
34. *L. E. Mansfield,* Linear Algebra with Geometric Applications: Selected Topics (1976)
35. *N. J. Pullman,* Matrix Theory and Its Applications (1976)
36. *B. R. McDonald,* Geometric Algebra Over Local Rings (1976)
37. *C. W. Groetsch,* Generalized Inverses of Linear Operators: Representation and Approximation (1977)
38. *J. E. Kuczkowski and J. L. Gersting,* Abstract Algebra: A First Look (1977)
39. *C. O. Christenson and W. L. Voxman,* Aspects of Topology (1977)
40. *M. Nagata,* Field Theory (1977)
41. *R. L. Long,* Algebraic Number Theory (1977)
42. *W. F. Pfeffer,* Integrals and Measures (1977)
43. *R. L. Wheeden and A. Zygmund,* Measure and Integral: An Introduction to Real Analysis (1977)
44. *J. H. Curtiss,* Introduction to Functions of a Complex Variable (1978)
45. *K. Hrbacek and T. Jech,* Introduction to Set Theory (1978)
46. *W. S. Massey,* Homology and Cohomology Theory (1978)
47. *M. Marcus,* Introduction to Modern Algebra (1978)
48. *E. C. Young,* Vector and Tensor Analysis (1978)
49. *S. B. Nadler, Jr.,* Hyperspaces of Sets (1978)
50. *S. K. Segal,* Topics in Group Rings (1978)
51. *A. C. M. van Rooij,* Non-Archimedean Functional Analysis (1978)
52. *L. Corwin and R. Szczarba,* Calculus in Vector Spaces (1979)
53. *C. Sadosky,* Interpolation of Operators and Singular Integrals: An Introduction to Harmonic Analysis (1979)
54. *J. Cronin,* Differential Equations: Introduction and Quantitative Theory (1980)
55. *C. W. Groetsch,* Elements of Applicable Functional Analysis (1980)
56. *I. Vaisman,* Foundations of Three-Dimensional Euclidean Geometry (1980)
57. *H. I. Freedman,* Deterministic Mathematical Models in Population Ecology (1980)
58. *S. B. Chae,* Lebesgue Integration (1980)
59. *C. S. Rees, S. M. Shah, and C. V. Stanojević,* Theory and Applications of Fourier Analysis (1981)
60. *L. Nachbin,* Introduction to Functional Analysis: Banach Spaces and Differential Calculus (R. M. Aron, translator) (1981)
61. *G. Orzech and M. Orzech,* Plane Algebraic Curves: An Introduction Via Valuations (1981)
62. *R. Johnsonbaugh and W. E. Pfaffenberger,* Foundations of Mathematical Analysis (1981)
63. *W. L. Voxman and R. H. Goetschel,* Advanced Calculus: An Introduction to Modern Analysis (1981)
64. *L. J. Corwin and R. H. Szcarba,* Multivariable Calculus (1982)
65. *V. I. Istrătescu,* Introduction to Linear Operator Theory (1981)
66. *R. D. Järvinen,* Finite and Infinite Dimensional Linear Spaces: A Comparative Study in Algebraic and Analytic Settings (1981)

MONOGRAPHS AND TEXTBOOKS IN PURE AND APPLIED MATHEMATICS

1. *K. Yano,* Integral Formulas in Riemannian Geometry (1970) *(out of print)*
2. *S. Kobayashi,* Hyperbolic Manifolds and Holomorphic Mappings (1970) *(out of print)*
3. *V. S. Vladimirov,* Equations of Mathematical Physics (A. Jeffrey, editor; A. Littlewood, translator) (1970) *(out of print)*
4. *B. N. Pshenichnyi,* Necessary Conditions for an Extremum (L. Neustadt, translation editor; K. Makowski, translator) (1971)
5. *L. Narici, E. Beckenstein, and G. Bachman,* Functional Analysis and Valuation Theory (1971)
6. *D. S. Passman,* Infinite Group Rings (1971)
7. *L. Dornhoff,* Group Representation Theory (in two parts). Part A: Ordinary Representation Theory. Part B: Modular Representation Theory (1971, 1972)
8. *W. Boothby and G. L. Weiss (eds.),* Symmetric Spaces: Short Courses Presented at Washington University (1972)
9. *Y. Matsushima,* Differentiable Manifolds (E. T. Kobayashi, translator) (1972)
10. *L. E. Ward, Jr.,* Topology: An Outline for a First Course (1972) *(out of print)*
11. *A. Babakhanian,* Cohomological Methods in Group Theory (1972)
12. *R. Gilmer,* Multiplicative Ideal Theory (1972)
13. *J. Yeh,* Stochastic Processes and the Wiener Integral (1973) *(out of print)*
14. *J. Barros-Neto,* Introduction to the Theory of Distributions (1973) *(out of print)*
15. *R. Larsen,* Functional Analysis: An Introduction (1973) *(out of print)*
16. *K. Yano and S. Ishihara,* Tangent and Cotangent Bundles: Differential Geometry (1973) *(out of print)*
17. *C. Procesi,* Rings with Polynomial Identities (1973)
18. *R. Hermann,* Geometry, Physics, and Systems (1973)
19. *N. R. Wallach,* Harmonic Analysis on Homogeneous Spaces (1973) *(out of print)*
20. *J. Dieudonné,* Introduction to the Theory of Formal Groups (1973)
21. *I. Vaisman,* Cohomology and Differential Forms (1973)
22. *B.-Y. Chen,* Geometry of Submanifolds (1973)
23. *M. Marcus,* Finite Dimensional Multilinear Algebra (in two parts) (1973, 1975)
24. *R. Larsen,* Banach Algebras: An Introduction (1973)
25. *R. O. Kujala and A. L. Vitter (eds.),* Value Distribution Theory: Part A; Part B: Deficit and Bezout Estimates by Wilhelm Stoll (1973)
26. *K. B. Stolarsky,* Algebraic Numbers and Diophantine Approximation (1974)
27. *A. R. Magid,* The Separable Galois Theory of Commutative Rings (1974)
28. *B. R. McDonald,* Finite Rings with Identity (1974)
29. *J. Satake,* Linear Algebra (S. Koh, T. A. Akiba, and S. Ihara, translators) (1975)

30. *J. S. Golan*, Localization of Noncommutative Rings (1975)
31. *G. Klambauer*, Mathematical Analysis (1975)
32. *M. K. Agoston*, Algebraic Topology: A First Course (1976)
33. *K. R. Goodearl*, Ring Theory: Nonsingular Rings and Modules (1976)
34. *L. E. Mansfield*, Linear Algebra with Geometric Applications: Selected Topics (1976)
35. *N. J. Pullman*, Matrix Theory and Its Applications (1976)
36. *B. R. McDonald*, Geometric Algebra Over Local Rings (1976)
37. *C. W. Groetsch*, Generalized Inverses of Linear Operators: Representation and Approximation (1977)
38. *J. E. Kuczkowski and J. L. Gersting*, Abstract Algebra: A First Look (1977)
39. *C. O. Christenson and W. L. Voxman*, Aspects of Topology (1977)
40. *M. Nagata*, Field Theory (1977)
41. *R. L. Long*, Algebraic Number Theory (1977)
42. *W. F. Pfeffer*, Integrals and Measures (1977)
43. *R. L. Wheeden and A. Zygmund*, Measure and Integral: An Introduction to Real Analysis (1977)
44. *J. H. Curtiss*, Introduction to Functions of a Complex Variable (1978)
45. *K. Hrbacek and T. Jech*, Introduction to Set Theory (1978)
46. *W. S. Massey*, Homology and Cohomology Theory (1978)
47. *M. Marcus*, Introduction to Modern Algebra (1978)
48. *E. C. Young*, Vector and Tensor Analysis (1978)
49. *S. B. Nadler, Jr.*, Hyperspaces of Sets (1978)
50. *S. K. Segal*, Topics in Group Rings (1978)
51. *A. C. M. van Rooij*, Non-Archimedean Functional Analysis (1978)
52. *L. Corwin and R. Szczarba*, Calculus in Vector Spaces (1979)
53. *C. Sadosky*, Interpolation of Operators and Singular Integrals: An Introduction to Harmonic Analysis (1979)
54. *J. Cronin*, Differential Equations: Introduction and Quantitative Theory (1980)
55. *C. W. Groetsch*, Elements of Applicable Functional Analysis (1980)
56. *I. Vaisman*, Foundations of Three-Dimensional Euclidean Geometry (1980)
57. *H. I. Freedman*, Deterministic Mathematical Models in Population Ecology (1980)
58. *S. B. Chae*, Lebesgue Integration (1980)
59. *C. S. Rees, S. M. Shah, and C. V. Stanojević*, Theory and Applications of Fourier Analysis (1981)
60. *L. Nachbin*, Introduction to Functional Analysis: Banach Spaces and Differential Calculus (R. M. Aron, translator) (1981)
61. *G. Orzech and M. Orzech*, Plane Algebraic Curves: An Introduction Via Valuations (1981)
62. *R. Johnsonbaugh and W. E. Pfaffenberger*, Foundations of Mathematical Analysis (1981)
63. *W. L. Voxman and R. H. Goetschel*, Advanced Calculus: An Introduction to Modern Analysis (1981)
64. *L. J. Corwin and R. H. Szcarba*, Multivariable Calculus (1982)
65. *V. I. Istrătescu*, Introduction to Linear Operator Theory (1981)
66. *R. D. Järvinen*, Finite and Infinite Dimensional Linear Spaces: A Comparative Study in Algebraic and Analytic Settings (1981)

67. *J. K. Beem and P. E. Ehrlich,* Global Lorentzian Geometry (1981)
68. *D. L. Armacost,* The Structure of Locally Compact Abelian Groups (1981)
69. *J. W. Brewer and M. K. Smith, eds.,* Emmy Noether: A Tribute to Her Life and Work (1981)
70. *K. H. Kim,* Boolean Matrix Theory and Applications (1982)
71. *T. W. Wieting,* The Mathematical Theory of Chromatic Plane Ornaments (1982)
72. *D. B. Gauld,* Differential Topology: An Introduction (1982)
73. *R. L. Faber,* Foundations of Euclidean and Non-Euclidean Geometry (1983)
74. *M. Carmeli,* Statistical Theory and Random Matrices (1983)
75. *J. H. Carruth, J. A. Hildebrant, and R. J. Koch,* The Theory of Topological Semigroups (1983)
76. *R. L. Faber,* Differential Geometry and Relativity Theory: An Introduction (1983)
77. *S. Barnett,* Polynomials and Linear Control Systems (1983)
78. *G. Karpilovsky,* Commutative Group Algebras (1983)
79. *F. Van Oystaeyen and A. Verschoren,* Relative Invariants of Rings: The Commutative Theory (1983)
80. *I. Vaisman,* A First Course in Differential Geometry (1984)
81. *G. W. Swan,* Applications of Optimal Control Theory in Biomedicine (1984)
82. *T. Petrie and J. D. Randall,* Transformation Groups on Manifolds (1984)
83. *K. Goebel and S. Reich,* Uniform Convexity, Hyperbolic Geometry, and Nonexpansive Mappings (1984)
84. *T. Albu and C. Năstăsescu,* Relative Finiteness in Module Theory (1984)
85. *K. Hrbacek and T. Jech,* Introduction to Set Theory, Second Edition, Revised and Expanded (1984)
86. *F. Van Oystaeyen and A. Verschoren,* Relative Invariants of Rings: The Noncommutative Theory (1984)
87. *B. R. McDonald,* Linear Algebra Over Commutative Rings (1984)
88. *M. Namba,* Geometry of Projective Algebraic Curves (1984)
89. *G. F. Webb,* Theory of Nonlinear Age-Dependent Population Dynamics (1985)
90. *M. R. Bremner, R. V. Moody, and J. Patera,* Tables of Dominant Weight Multiplicities for Representations of Simple Lie Algebras (1985)
91. *A. E. Fekete,* Real Linear Algebra (1985)
92. *S. B. Chae,* Holomorphy and Calculus in Normed Spaces (1985)
93. *A. J. Jerri,* Introduction to Integral Equations with Applications (1985)
94. *G. Karpilovsky,* Projective Representations of Finite Groups (1985)
95. *L. Narici and E. Beckenstein,* Topological Vector Spaces (1985)
96. *J. Weeks,* The Shape of Space: How to Visualize Surfaces and Three-Dimensional Manifolds (1985)
97. *P. R. Gribik and K. O. Kortanek,* Extremal Methods of Operations Research (1985)
98. *J.-A. Chao and W. A. Woyczynski, eds.,* Probability Theory and Harmonic Analysis (1986)
99. *G. D. Crown, M. H. Fenrick, and R. J. Valenza,* Abstract Algebra (1986)
100. *J. H. Carruth, J. A. Hildebrant, and R. J. Koch,* The Theory of Topological Semigroups, Volume 2 (1986)

101. *R. S. Doran and V. A. Belfi,* Characterizations of C*-Algebras: The Gelfand-Naimark Theorems (1986)
102. *M. W. Jeter,* Mathematical Programming: An Introduction to Optimization (1986)
103. *M. Altman,* A Unified Theory of Nonlinear Operator and Evolution Equations with Applications: A New Approach to Nonlinear Partial Differential Equations (1986)
104. *A. Verschoren,* Relative Invariants of Sheaves (1987)
105. *R. A. Usmani,* Applied Linear Algebra (1987)
106. *P. Blass and J. Lang,* Zariski Surfaces and Differential Equations in Characteristic $p > 0$ (1987)
107. *J. A. Reneke, R. E. Fennell, and R. B. Minton.* Structured Hereditary Systems (1987)
108. *H. Busemann and B. B. Phadke,* Spaces with Distinguished Geodesics (1987)
109. *R. Harte,* Invertibility and Singularity for Bounded Linear Operators (1988).
110. *G. S. Ladde, V. Lakshmikantham, and B. G. Zhang,* Oscillation Theory of Differential Equations with Deviating Arguments (1987)
111. *L. Dudkin, I. Rabinovich, and I. Vakhutinsky,* Iterative Aggregation Theory: Mathematical Methods of Coordinating Detailed and Aggregate Problems in Large Control Systems (1987)
112. *T. Okubo,* Differential Geometry (1987)
113. *D. L. Stancl and M. L. Stancl,* Real Analysis with Point-Set Topology (1987)
114. *T. C. Gard,* Introduction to Stochastic Differential Equations (1988)
115. *S. S. Abhyankar,* Enumerative Combinatorics of Young Tableaux (1988)
116. *H. Strade and R. Farnsteiner,* Modular Lie Algebras and Their Representations (1988)
117. *J. A. Huckaba,* Commutative Rings with Zero Divisors (1988)
118. *W. D. Wallis,* Combinatorial Designs (1988)
119. *W. Więsław,* Topological Fields (1988)
120. *G. Karpilovsky,* Field Theory: Classical Foundations and Multiplicative Groups (1988)
121. *S. Caenepeel and F. Van Oystaeyen,* Brauer Groups and the Cohomology of Graded Rings (1989)
122. *W. Kozlowski,* Modular Function Spaces (1988)
123. *E. Lowen-Colebunders,* Function Classes of Cauchy Continuous Maps (1989)
124. *M. Pavel,* Fundamentals of Pattern Recognition (1989)
125. *V. Lakshmikantham, S. Leela, and A. A. Martynyuk,* Stability Analysis of Nonlinear Systems (1989)
126. *R. Sivaramakrishnan,* The Classical Theory of Arithmetic Functions (1989)
127. *N. A. Watson,* Parabolic Equations on an Infinite Strip (1989)
128. *K. J. Hastings,* Introduction to the Mathematics of Operations Research (1989)
129. *B. Fine,* Algebraic Theory of the Bianchi Groups (1989)
130. *D. N. Dikranjan, I. R. Prodanov, and L. N. Stoyanov,* Topological Groups: Characters, Dualities, and Minimal Group Topologies (1989)

131. *J. C. Morgan II,* Point Set Theory (1990)
132. *P. Biler and A. Witkowski,* Problems in Mathematical Analysis (1990)
133. *H. J. Sussmann,* Nonlinear Controllability and Optimal Control (1990)
134. *J.-P. Florens, M. Mouchart, and J. M. Rolin,* Elements of Bayesian Statistics (1990)
135. *N. Shell,* Topological Fields and Near Valuations (1990)
136. *B. F. Doolin and C. F. Martin,* Introduction to Differential Geometry for Engineers (1990)
137. *S. S. Holland, Jr.,* Applied Analysis by the Hilbert Space Method (1990)
138. *J. Okniński,* Semigroup Algebras (1990)
139. *K. Zhu,* Operator Theory in Function Spaces (1990)
140. *G. B. Price,* An Introduction to Multicomplex Spaces and Functions (1991)
141. *P. L. Sachdev,* Ordinary Differential Equations and Their Applications (1991)
142. *R. B. Darst,* Introduction to Linear Programming: Applications and Extensions (1991)

Other Volumes in Preparation

An Introduction to Multicomplex Spaces and Functions

G. BALEY PRICE

University of Kansas
Lawrence, Kansas

CRC Press is an imprint of the
Taylor & Francis Group, an **informa** business

CRC Press
Taylor & Francis Group
6000 Broken Sound Parkway NW, Suite 300
Boca Raton, FL 33487-2742

First issued in paperback 2019

ISBN-13: 978-0-8247-8345-7 (hbk)
ISBN-13: 978-0-367-40311-9 (pbk)

Library of Congress Cataloging-in-Publication Data

Price, G. Baley (Griffith Baley),
An introduction to multicomplex spaces and functions / G. Baley Price.
p. cm. — (Monographs and textbooks in pure and applied mathematics ; 140)
Includes bibliographical references and index.
ISBN 0-8247-8345-X
1. Banach algebras. 2. Holomorphic functions. I. Title. II. Title: Multicomplex spaces and functions. III. Series.
QA326.P75 1990
512′.55—dc20 90-41639
CIP

Visit the Taylor & Francis Web site at
http://www.taylorandfrancis.com

and the CRC Press Web site at
http://www.crcpress.com

Foreword

This book arose out of a doctoral thesis in the early 1950s which was influenced by Professor Price. I expect it will influence new students now. It is a book for analysts and algebraists as well.

The theory of functions of one complex variable, whether treated in the manner of Gauss, Cauchy, Riemann, or Weierstrass, is one of the enduring staples of every mathematician's education. It was natural to attempt extension to the case of several complex variables, to see where the natural extensions occur and where they do not. This study was started by Weierstrass, followed by Poincaré, Cousin, and Picard, who solved and formulated basic problems.

In another direction there is the theory of functions of a quaternion variable, associated with Fueter in Switzerland. The author studies another extension. It was Segre in Italy who had the most definite influence. The book develops the subject as nearly parallel as was possible to the theory of functions of a complex variable. This is in spite of the fact that there exist divisors of zero. "They only add interest."

The mathematical community will be grateful to Professor Price for returning to the subject and writing *An Introduction to Multicomplex Spaces and Functions.*

Olga Taussky Todd
California Institute of Technology
Pasadena, California

Preface

This book treats two subjects: first, a class of Banach algebras known collectively as multicomplex spaces, and second, the theory of holomorphic functions defined on these multicomplex spaces. These spaces and functions are closely related to the space of complex numbers and to the theory of holomorphic functions of a complex variable. A brief history of the development of the ideas that have led to this study will be helpful.

Although Karl Friedrich Gauss (1777–1855) had very early discovered some of the properties of functions of a complex variable, he published little, and from 1814 on Augustin Louis Cauchy (1789–1857) became the effective founder of the theory of functions of a complex variable. The foundations of the subject rest on the concept of an algebra as well as on ideas about the theory of functions. Originally the only algebra known to mathematics was the algebra of real numbers, but the intrusion and discovery of the complex numbers initiated a broadening of the concept. George Peacock (1791–1858) in 1830 published his *Treatise on Algebra* in an effort to give algebra a logical structure similar to that in Euclid's *Elements.* In 1833 Sir William Rowan Hamilton (1805–1865) presented a paper to the Irish Academy in which he developed a formal algebra of real numbers which is precisely the algebra of the complex numbers as usually understood today. In 1843 Hamilton discovered quaternions, and in 1844 Hermann Günther Grassmann (1809–1877) published his *Ausdehnungslehre.* Benjamin Peirce (1809–1880) in 1864 presented his paper entitled *Linear Associative Algebra* to the American

Association for the Advancement of Science, but it was not published until 1881. There had been only one algebra at the beginning of the century, but Benjamin Peirce worked out multiplication tables for 162 linear associative algebras. Special algebras were discovered and investigated in detail; several have been described already. In addition, Arthur Cayley (1821–1895), Benjamin Peirce, and Charles S. Peirce (1839–1914) developed matrix algebras; William Kingdon Clifford (1845–1879) developed the Clifford algebras, of which octonians, or biquaternions, are special cases; and the vector analysis of Josiah Williard Gibbs (1839–1903) appeared in 1881 and 1884.

Finally, in 1892, in the search for and development of special algebras, Corrado Segre (1860–1924) published a paper [12] (see bibliography at the end of the book) in which he treated an infinite set of algebras whose elements he called bicomplex numbers, tricomplex numbers, ..., n-complex numbers, ... A bicomplex number is an element of the form $(x_1+i_1x_2)+i_2(x_3+i_1x_4)$, where $x_1,\ldots,x_4$ are real numbers, $i_1^2=i_2^2=-1$, and $i_1i_2=i_2i_1$. The bicomplex numbers can be embedded in euclidean space of 2^2 dimensions. Alternatively, a bicomplex number is an element $z_1+i_2z_2$, where z_1 and z_2 are complex numbers, and the rules of operation are formally the same as for complex numbers. Segre used two bicomplex numbers to form a tricomplex number $[(x_1+i_1x_2)+i_2(x_3+i_1x_4)]+i_3[(x_5+i_1x_6)+i_2(x_7+i_1x_8)]$. The units are 1, i_1, i_2, i_3, i_1i_2, i_1i_3, i_2i_3, $i_1i_2i_3$; all multiplications are commutative, and $i_1^2=i_2^2=i_3^2=-1$. The tricomplex numbers are embedded in euclidean space of 2^3 dimensions. Segre showed that his construction could be iterated indefinitely to form n-complex numbers which are embedded in euclidean space of 2^n dimensions. Segre showed that the bicomplex numbers contain divisors of zero, and he showed that every bicomplex number $z_1+i_2z_2$ can be represented as the complex combination $(z_1-i_1z_2)[(1+i_1i_2)/2)]$ $+(z_i+i_1z_2)[(1-i_1i_2)/2]$ of the idempotent elements $(1+i_1i_2)/2$ and $(1-i_1i_2)/2$. In this book, the elements of the algebras introduced by Segre are called bicomplex numbers and, collectively, multicomplex numbers. With the addition of the euclidean norm of the space in which Segre's algebras are embedded, they become the Banach algebras which are the bicomplex space and the multicomplex spaces of this book.

The theory of functions of a complex variable is based on the algebra of complex numbers, and the discovery of linear associative algebras has led to many efforts to develop similar theories of functions in other algebras. As early as 1894 Scheffers [11] investigated the generalization of functions of a complex variable, and by 1940 the literature on the subject was enormous: the bibliography of James A. Ward's paper [18] entitled *Theory of Analytic Functions in Linear Associative Algebras* contained 81 references (see also some sections in Hille's book [4]).

Michiji Futagawa [3] in 1928 and 1932 seems to have been the first to consider the theory of functions of a bicomplex variable. Although his quaternary variable is equivalent to the bicomplex variable, Futagawa did not treat it as such. The hypercomplex system of Friedrich Ringleb [9] is more general than the bicomplex algebra; he showed in 1933 that Futagawa's system is a special case of his own. Ringleb's results included one of the fundamental theorems of the subject; he showed that every holomorphic function of a bicomplex variable (a function that has a derivative) can be represented by two holomorphic functions of a complex variable. There was considerable activity in the field for several years: Scorza Dragoni wrote a paper [2] on holomorphic functions of a bicomplex variable in 1934; Ugo Morin (1901–?) investigated in 1935 the algebra of the bicomplex numbers [6]; Spampinato wrote three papers (see [13], [14], [15]) on functions of a bicomplex variable in 1935 and 1936; and Tsurusaburo Takasu (1890–?) published a paper [16] on a generalized bicomplex variable in 1943.

In 1953 James D. Riley published a paper [8] entitled *Contributions to the Theory of Functions of a Bicomplex Variable.* The first page of the paper contains the following footnote:

> This paper was written as a Ph.D. thesis at the University of Kansas under the supervision of Prof. V. Wolontis and many of the problems and numerous changes have been suggested by him. The author wishes to express his appreciation. The project was originally proposed by Prof. G. B. Price, and a preliminary investigation was made by him.

I no longer remember how or where I first learned about functions of a bicomplex variable, and the only outcome of my "preliminary investigation" was Riley's thesis, supervised by Wolontis. Other activities and subjects claimed my attention, and I had no opportunity to make any further study of the field. And there the matter stood until 1984, when Olga Taussky Todd expressed regret that I had published nothing on the subject. The remark led me to undertake the writing of this book.

I decided to make a fresh start, and I have developed the subject as nearly like the theory of functions of a complex variable as I could. I have used $\mathbb{C}_0$ to denote the real numbers $\mathbb{R}$; I have used $\mathbb{C}_1$ to denote the complex numbers $\mathbb{C}$, $\mathbb{C}_2$ to denote the bicomplex numbers. More generally, $\mathbb{C}_n$ denotes, for $n \geqslant 2$, the n-complex numbers of Segre. The addition of the norm in $\mathbb{R}^{2^n}$, in which $\mathbb{C}_n$ has a natural embedding, converts the n-complex numbers into a Banach algebra, which also is denoted by $\mathbb{C}_n$. The first four chapters of this book contain a detailed treatment of $\mathbb{C}_2$ and of the differentiable functions on $\mathbb{C}_2$.

The fifth chapter treats $\mathbb{C}_n$, $n \geqslant 3$, and its differentiable functions, and the emphasis is on large values of n. Insofar as my knowledge goes, this is the first treatment of this part of the subject. There is a wealth of homomorphic representations of the space $\mathbb{C}_n$ and of the holomorphic functions in it. There is a matrix algebra homomorphic to $\mathbb{C}_n$; the matrices in it are called Cauchy–Riemann matrices. The determinants of Cauchy–Riemann matrices can be factored into determinants of matrices of lower order. Elementary methods suffice for the proof for small values of n, but new methods are needed for large n because of the size and complexity of the mass of details. For example, the real Cauchy–Riemann matrix in $\mathbb{C}_{10}$ is a $2^{10} \times 2^{10}$ matrix with more than a million elements. A holomorphic function in $\mathbb{C}_n$ maps a set in $\mathbb{C}_n$ into a set in $\mathbb{C}_n$; properties of the Cauchy–Riemann matrices can be used to prove that the jacobian of the mapping is nonnegative at every point, and that it is positive at every point at which the derivative is not zero or a divisor of zero. A holomorphic function in $\mathbb{C}_n$ can be represented by 2^n functions of 2^n real variables; these functions satisfy a system of Cauchy–Riemann differential equations. A holomorphic function in $\mathbb{C}_n$ can be represented also by 2^{n-1} functions of 2^{n-1} complex variables (variables in $\mathbb{C}_1$); these functions satisfy another system of Cauchy–Riemann differential equations. These two representations are the first two in a sequence; in the final one, the holomorphic function can be represented by a pair of holomorphic functions of two variables in $\mathbb{C}_{n-1}$, and there is a corresponding pair of Cauchy–Riemann differential equations. The section headings in the Contents give a more complete indication of the topics treated in the entire book.

Among other results in $\mathbb{C}_n$, $n \geqslant 3$, Chapter 5 proves the fundamental theorem of the integral calculus, Cauchy's integral theorem, and Cauchy's integral formula. These theorems are proved in $\mathbb{C}_n$ without any appeal to, or use of, functions of a complex variable in $\mathbb{C}_1$. The theory of holomorphic functions in $\mathbb{C}_n$ seems to be as complete and detailed as the theory of holomorphic functions in $\mathbb{C}_1$, but it is more interesting because of the vastly richer structure. Some have stated that a theory of functions in $\mathbb{C}_n$, $n \geqslant 2$, is impossible because of the presence of divisors of zero, but the presence of these singular elements does not hinder the development of the theory – they only add interest.

The theory of holomorphic functions in $\mathbb{C}_n$ is a natural continuation of the theory of holomorphic functions of a complex variable. The results given above suggest some of the reasons the subject is interesting. I hope that this book provides an introduction which will display some of the beauty and interest of the field, and that it will make the subject accessible to others. The Epilogue (Chapter 6) describes areas that have not been examined thus far. The investigation of m-dimensional multicomplex spaces and of functions of several multicomplex variables seems to be a natural next step.

Anyone who has had an introduction to the theory of functions of a complex variable should be able to read this book without difficulty. With respect to multicomplex spaces and functions, it is self-contained and complete. The explanations are full and complete, and there are exercises to provide examples and to complete the exposition.

I take this opportunity to acknowledge my indebtedness to Olga Taussky Todd and to thank her for encouraging the exploration of the multicomplex spaces and their holomorphic functions and for writing the Foreword for this book. Also, it is a pleasure to acknowledge the assistance of Sharon Gumm and to thank her for her careful typing of the manuscript.

G. Baley Price

Contents

1

The Bicomplex Space

1. INTRODUCTION

This chapter contains an introduction to the Banach algebra of bicomplex numbers. This space of bicomplex numbers is the first in an infinite sequence of multicomplex spaces which are generalizations of the space of complex numbers.

For convenience, the real numbers $\mathbb{R}$ are usually denoted in this book by $\mathbb{C}_0$, and the complex numbers $\mathbb{C}$ are denoted by $\mathbb{C}_1$. An element in $\mathbb{C}_1$ is a number of the form

(1) $\quad x_1 + i_1x_2, \qquad x_1, x_2 \text{ in } \mathbb{C}_0,\ i_1^2 = -1.$

An element in the space $\mathbb{C}_2$ of bicomplex numbers is a number of the form

(2) $\quad x_1 + i_1x_2 + i_2x_3 + i_1i_2x_4,$

$\qquad x_1, \ldots, x_4 \text{ in } \mathbb{C}_0,\ i_1^2 = -1,\ i_2^2 = -1,\ i_1i_2 = i_2i_1.$

Map the element (2) into the point $(x_1, \ldots, x_4)$ in $\mathbb{C}_0^4$; this mapping embeds $\mathbb{C}_2$ into $\mathbb{C}_0^4$. Addition of two elements in $\mathbb{C}_2$ and scalar multiplication of an element in $\mathbb{C}_2$ by a real number are defined in the usual way; $\mathbb{C}_2$ is a linear space with respect to addition and scalar multiplication. The norm of $x_1+i_1x_2+i_2x_3+i_1i_2x_4$ is defined to be the norm of $(x_1, \ldots, x_4)$ in $\mathbb{C}_0^4$; with this norm, $\mathbb{C}_2$ is a normed linear space. Since $\mathbb{C}_0^4$ is complete, $\mathbb{C}_2$ is complete and thus a Banach space. Two elements in $\mathbb{C}_2$ are multiplied as if they were polynomials; the assumptions in (2) show that $\mathbb{C}_2$ is closed under multiplication. Also, multiplication is associative and commutative, and $\mathbb{C}_2$ is a Banach algebra.

There is a second representation of elements in $\mathbb{C}_2$ which is important. From (2),

$$(3) \qquad x_1+i_1x_2+i_2x_3+i_1i_2x_4=(x_1+i_1x_2)+i_2(x_3+i_1x_4),$$

and (1) shows that $x_1+i_1x_2$ and $x_3+i_1x_4$ are complex numbers z_1 and z_2 in $\mathbb{C}_1$. Thus the element (2) can be represented as

$$(4) \qquad z_1+i_2z_2, \qquad z_1,\, z_2 \text{ in } \mathbb{C}_1,\ i_2^2=-1.$$

Map the element (4) into the point (z_1, z_2) in $\mathbb{C}_1^2$; this mapping embeds $\mathbb{C}_2$ into $\mathbb{C}_1^2$. Also, $\mathbb{C}_0$ is a subspace of $\mathbb{C}_1$, and $\mathbb{C}_1$ is a subspace of $\mathbb{C}_2$. The representation (4) is the reason why the elements (2) are called *bicomplex numbers*. In some cases the representation (4) is merely a matter of notational convenience; in others, it is essential to the development of the theory. Furthermore, some interesting problems arise from a comparison of results obtained from the two methods of representing elements in $\mathbb{C}_2$.

An element which is equal to its square is called an *idempotent* element. There are four idempotent elements in $\mathbb{C}_2$; they are

$$(5) \qquad 0,\ 1,\ \frac{1+i_1i_2}{2},\quad \frac{1-i_1i_2}{2}.$$

Furthermore, for every $z_1+i_2z_2$ in $\mathbb{C}_2$,

$$(6) \qquad z_1+i_2z_2=(z_1-i_1z_2)\left(\frac{1+i_1i_2}{2}\right)+(z_1+i_1z_2)\left(\frac{1-i_1i_2}{2}\right).$$

Also, if $\|\ \|$ denotes the norm of elements in $\mathbb{C}_2$, then

$$(7) \qquad \|z_1+i_2z_2\|=(|z_1|^2+|z_2|^2)^{1/2}=(x_1^2+x_2^2+x_3^2+x_4^2)^{1/2},$$

$$(8) \qquad \|z_1+i_2z_2\|=\left(\frac{|z_1-i_1z_2|^2+|z_1+i_1z_2|^2}{2}\right)^{1/2}.$$

There is an important difference between $\mathbb{C}_1$ and $\mathbb{C}_2$: the complex numbers form a field, but the bicomplex numbers do not since they contain divisors of zero. Thus

$$(9) \qquad \left(\frac{1+i_1i_2}{2}\right)\left(\frac{1-i_1i_2}{2}\right)=0, \qquad \left\|\frac{1+i_1i_2}{2}\right\|=\left\|\frac{1-i_1i_2}{2}\right\|=\frac{\sqrt{2}}{2},$$

$$(10) \qquad \left(\frac{1+i_1i_2}{2}\right)^2=\frac{1+i_1i_2}{2}, \qquad \left(\frac{1-i_1i_2}{2}\right)^2=\frac{1-i_1i_2}{2}.$$

The properties in (6), (9), and (10) show that the algebraic operations of addition, subtraction, multiplication, and division can be carried out on elements in $\mathbb{C}_2$ by performing the corresponding operations on the complex coefficients $z_1 - i_1 z_2$ and $z_1 + i_1 z_2$ in (6). The properties stated in (6)–(10), and the properties of the set of divisors of zero are the keys which unlock much of the theory. This chapter establishes the statements which have been made in this introduction and uses them to develop the fundamental properties of the bicomplex space $\mathbb{C}_2$.

2. $\mathbb{C}_2$: A LINEAR SPACE

The purpose of this section is to define the set $\mathbb{C}_2$ of bicomplex numbers, the addition $\oplus$ of bicomplex numbers, the scalar multiplication $\odot$ of a bicomplex number by a real scalar a, and to prove that the system $(\mathbb{C}_2, \oplus, \odot)$ is a linear space.

2.1 DEFINITION The set $\mathbb{C}_2$ and $\ominus$ (equals) are defined by the following statements:

(1) $$\mathbb{C}_2 = \{x_1 + i_1 x_2 + i_2 x_3 + i_1 i_2 x_4 \colon x_1, \ldots, x_4 \text{ in } \mathbb{C}_0,\ i_1^2 = i_2^2 = -1,\ i_1 i_2 = i_2 i_1\};$$

(2) $$(x_1 + i_1 x_2 + i_2 x_3 + i_1 i_2 x_4) \ominus (y_1 + i_1 y_2 + i_2 y_3 + i_1 i_2 y_4) \quad \text{if and only if } x_i = y_i,\ i = 1, \ldots, 4.$$

Addition is the operation on $\mathbb{C}_2$ defined by the function

(3) $$\oplus : \mathbb{C}_2 \times \mathbb{C}_2 \to \mathbb{C}_2,\quad (x_1 + i_1 x_2 + i_2 x_3 + i_1 i_2 x_4,\ y_1 + i_1 y_2 + i_2 y_3 + i_1 i_2 y_4) \mapsto (x_1 + y_1) + i_1(x_2 + y_2) + i_2(x_3 + y_3) + i_1 i_2 (x_4 + y_4).$$

Scalar multiplication is the operation on $\mathbb{C}_2$ defined by the function

(4) $$\odot : \mathbb{C}_0 \times \mathbb{C}_2 \to \mathbb{C}_2,\quad (a,\ x_1 + i_1 x_2 + i_2 x_3 + i_1 i_2 x_4) \mapsto ax_1 + i_1 a x_2 + i_2 a x_3 + i_1 i_2 a x_4.$$

Observe that equality, addition, and scalar multiplication in $\mathbb{C}_2$ are defined in terms of equality, addition, and multiplication of real numbers. This fact provides the basis for the proof of the following theorem.

2.2 THEOREM The system $(\mathbb{C}_2, \oplus, \odot)$ is a linear space.

Proof. The set $\mathbb{C}_2$ with the operation addition is a commutative group. First, $\mathbb{C}_2$ is closed under addition, and addition is associative. The identity for addition is $0+i_1 0+i_2 0+i_1 i_2 0$, which will be denoted hereafter by 0 and called zero. The inverse of $x_1+i_1x_2+i_2x_3+i_1i_2x_4$ is $(-x_1)+i_1(-x_2)+i_2(-x_3)+i_1i_2(-x_4)$, and addition is commutative. All of these statements are true because the corresponding properties are true for addition in $\mathbb{C}_0$. If 1 is the unit in $\mathbb{C}_0$, then $1 \odot (x_1+i_1x_2+i_2x_3+i_1i_2x_4) = x_1+i_1x_2+i_2x_3+i_1i_2x_4$. Finally, scalar multiplication has the properties of closure, associativity, and distributivity required to complete the proof that $(\mathbb{C}_2, \oplus, \odot)$ is a linear space. □

In the future, equals $\ominus$, addition $\oplus$, and scalar multiplication $\odot$ will be denoted by $=$, $+$, and juxtaposition, respectively.

The space $\mathbb{C}_2$ of bicomplex numbers is embedded in $\mathbb{C}_0^4$ by mapping the bicomplex number $x_1+i_1x_2+i_2x_3+i_1i_2x_4$ into the point (x_1, x_2, x_3, x_4) in $\mathbb{C}_0^4$. The bicomplex numbers $x_1+i_1 0+i_2 0+i_1 i_2 0$ are isomorphic to the real numbers $\mathbb{C}_0$, and for simplicity they are called real numbers; $0+i_1 0+i_2 0+i_1 i_2 0$ and $1+i_1 0+i_2 0+i_1 i_2 0$ are called zero and one and denoted by 0 and 1. The set $\{x_1+i_1x_2+i_2 0+i_1 i_2 0\colon x_1, x_2 \in \mathbb{C}_0\}$ and $\mathbb{C}_1$ are isomorphic under corresponding operations.

Exercises

2.1 The results in this section have been given for the bicomplex space $\mathbb{C}_2$ with elements in the real form $x_1+i_1x_2+i_2x_3+i_1i_2x_4$. Repeat the entire section for the bicomplex space with elements in the complex form $z_1+i_2z_2$ [see (4) in Section 1].

2.2 There are two forms of the bicomplex space $\mathbb{C}_2$, one with real elements $x_1+i_1x_2+i_2x_3+i_1i_2x_4$ and the other with complex elements $z_1+i_2z_2$. In each form of $\mathbb{C}_2$ there is an operation called addition (see Definition 2.1 and Exercise 2.1). Establish an isomorphism between the two forms of $\mathbb{C}_2$ with respect to these operations of addition and thus prove that, from the abstract point of view, the two forms of $\mathbb{C}_2$ are identical.

3. $\mathbb{C}_2$: A BANACH SPACE

This section contains the definition of a norm $\| \ \|$ on the linear space $(\mathbb{C}_2, \oplus, \odot)$ and shows that the system $(\mathbb{C}_2, \oplus, \odot, \| \ \|)$ is a Banach space.

3.1 DEFINITION Define the function $\| \ \| : \mathbb{C}_2 \to \mathbb{R}_{\geq 0}$ as follows: for every $x_1+i_1x_2+i_2x_3+i_1i_2x_4$ in $\mathbb{C}_2$,

(1) $$\|x_1 + i_1x_2 + i_2x_3 + i_1i_2x_4\| = (x_1^2 + x_2^2 + x_3^2 + x_4^2)^{1/2}.$$

3.2 THEOREM The function $\| \ \| : \mathbb{C}_2 \to \mathbb{R}_{\geq 0}$ is a norm on the linear space $(\mathbb{C}_2, \oplus, \odot)$.

Proof. The function $\| \ \|: \mathbb{C}_2 \to \mathbb{R}_{\geq 0}$ is known to have the following properties: for every $x_1+i_1x_2+i_2x_3+i_1i_2x_4$ and $y_1+i_1y_2+i_2y_3+i_1i_2y_4$ in $\mathbb{C}_2$ and a in $\mathbb{C}_0$,

(2) $\quad \|x_1 + i_1x_2 + i_2x_3 + i_1i_2x_4\| \geq 0,$

(3) $\quad \|x_1 + i_1x_2 + i_2x_3 + i_1i_2x_4\| = 0$ if and only if

$$x_1 + i_1x_2 + i_2x_3 + i_1i_2x_4 = 0,$$

(4) $\quad \|a(x_1 + i_1x_2 + i_2x_3 + i_1i_2x_4)\| = |a| \, \|x_1 + i_1x_2 + i_2x_3 + i_1i_2x_4\|,$

(5) $\quad \|(x_1 + i_1x_2 + i_2x_3 + i_1i_2x_4) + (y_1 + i_1y_2 + i_2y_3 + i_1i_2y_4)\|$

$$\leq \|x_1 + i_1x_2 + i_2x_3 + i_1i_2x_4\| + \|y_1 + i_1y_2 + i_2y_3 + i_1i_2y_4\|.$$

A function with these properties is called a norm; therefore $\| \ \|: \mathbb{C}_2 \to \mathbb{R}_{\geq 0}$ is a norm on $(\mathbb{C}_2, \oplus, \odot)$. □

The space $\mathbb{C}_0^4$ with the euclidean norm is known to be a complete space. Because $\mathbb{C}_2$ is embedded in $\mathbb{C}_0^4$ so that $x_1+i_1x_2+i_2x_3+i_1i_2x_4$ corresponds to (x_1, x_2, x_3, x_4), and because the norm on $\mathbb{C}_2$ is the same as the norm on $\mathbb{C}_0^4$, then the normed linear space $(\mathbb{C}_2, \oplus, \odot, \| \ \|)$ is a complete space. By definition, a space which is linear, normed, and complete is a Banach space. These statements prove the following theorem.

3.3 THEOREM The system $(\mathbb{C}_2, \oplus, \odot, \| \ \|)$ is a Banach space.

The norm $\| \ \|$ is defined in (1), but it has other representations. If $z_1 = x_1 + i_1x_2$ and $z_2 = x_3 + i_1x_4$, then

(6)
$$\begin{aligned} \|x_1 + i_1x_2 + i_2x_3 + i_1i_2x_4\| &= (x_1^2 + x_2^2 + x_3^2 + x_4^2)^{1/2} \\ &= [|z_1|^2 + |z_2|^2]^{1/2} \\ &= \|z_1 + i_2z_2\|. \end{aligned}$$

Exercises

3.1 Prove the statements in equations (2)–(5). [*Hint.* To prove (5), assume Schwarz's inequality.]

3.2 If ζ, ζ_1, and ζ_2 denote elements in $\mathbb{C}_2$, prove that
  (a) $\|-\zeta\| = \|\zeta\|$,
  (b) $|\|\zeta_1\| - \|\zeta_2\|| \leq \|\zeta_1 + \zeta_2\| \leq \|\zeta_1\| + \|\zeta_2\|$,
  (c) $|\|\zeta_1\| - \|\zeta_2\|| \leq \|\zeta_1 - \zeta_2\| \leq \|\zeta_1\| + \|\zeta_2\|$.

3.3 Let X be a set, and let $d: X \times X \to \mathbb{R}_{\geq 0}$, $(x, y) \mapsto d(x, y)$, be a function with the following properties: for every x, y, z in X,
  (i) $d(x, y) \geq 0$, $d(x, y) = 0$ if and only if $x = y$,
  (ii) $d(x, y) = d(y, x)$,
  (iii) $d(x, z) \leq d(x, y) + d(y, z)$ (triangle inequality).

A function d with these properties is called a *distance function* on X, and $d(x, y)$ is called the *distance* from x to y. The system (X, d) is called a *metric space.*

Define a function $d: \mathbb{C}_2 \times \mathbb{C}_2 \to \mathbb{R}_{\geqslant 0}$, $(\zeta_1, \zeta_2) \mapsto d(\zeta_1, \zeta_2)$, by setting $d(\zeta_1, \zeta_2) = \|\zeta_1 - \zeta_2\|$ for every ζ_1, ζ_2 in $\mathbb{C}_2$. Prove that this d is a distance function on $\mathbb{C}_2$ and thus that $(\mathbb{C}_2, d)$ is a metric space.

3.4 A Cauchy sequence in $\mathbb{C}_2$ is a function $s: \mathbb{N} \to \mathbb{C}_2$, $n \mapsto \zeta_n$, with the following property: to each $\varepsilon > 0$ there corresponds an n_0 in $\mathbb{N}$ for which $\|\zeta_n - \zeta_m\| < \varepsilon$ for all m, n such that $n \geqslant n_0$ and $m \geqslant n_0$.
   (a) Prove that a Cauchy sequence in $\mathbb{C}_2$ is bounded.
   (b) A normed linear space is said to be *complete* if every Cauchy sequence in the space has a limit in the space. Assume that the Bolzano–Weierstrass cluster point theorem is true in $\mathbb{C}_0^4$ and prove that $\mathbb{C}_2$ is a complete space and thus a Banach space.
   (c) Assume that the Bolzano–Weierstrass cluster point theorem is true in $\mathbb{C}_0$ and use this fact to prove that $\mathbb{C}_2$ is complete.

3.5 If z, z_1, z_2 are in $\mathbb{C}_1$, show that $z_1 + i_2 z_2$ and $(zz_1) + i_2(zz_2)$ are in $\mathbb{C}_2$. Show also that $\|(zz_1) + i_2(zz_2)\| = |z| \; \|z_1 + i_2 z_2\|$.

4. MULTIPLICATION

This section defines multiplication $\otimes$ in $\mathbb{C}_2$; it establishes the properties of this operation; and it proves an inequality for the product of two elements in $\mathbb{C}_2$. Thus the system $(\mathbb{C}_2, \oplus, \odot, \| \; \|, \otimes)$ is shown to be a Banach algebra. In the future, multiplication $\otimes$ will usually be denoted by juxtaposition, and $\mathbb{C}_2$ will usually denote the system as well as the set of elements. A nonsingular element has a multiplicative inverse; a singular element does not have an inverse. The final theorem in the section identifies all of the singular elements in $\mathbb{C}_2$.

4.1 DEFINITION The product $(x_1 + i_1 x_2 + i_2 x_3 + i_1 i_2 x_4) \otimes (y_1 + i_1 y_2 + i_2 y_3 + i_1 i_2 y_4)$ is the element in $\mathbb{C}_2$ obtained by multiplying $x_1 + i_1 x_2 + i_2 x_3 + i_1 i_2 x_4$ and $y_1 + i_1 y_2 + i_2 y_3 + i_1 i_2 y_4$ as if they were polynomials and then using the relations $i_1^2 = -1$, $i_2^2 = -1$, and $i_1 i_2 = i_2 i_1$ to simplify the result. The following display exhibits this product and the final result.

(1)

	y_1	$i_1 y_2$	$i_2 y_3$	$i_1 i_2 y_4$
x_1	$x_1 y_1$	$+\; i_1 x_1 y_2$	$+\; i_2 x_1 y_3$	$+\; i_1 i_2 x_1 y_4$
$i_1 x_2$	$+\; i_1 x_2 y_1$	$-\; x_2 y_2$	$+\; i_1 i_2 x_2 y_3$	$-\; i_2 x_2 y_4$
$i_2 x_3$	$+\; i_2 x_3 y_1$	$+\; i_1 i_2 x_3 y_2$	$-\; x_3 y_3$	$-\; i_1 x_3 y_4$
$i_1 i_2 x_4$	$+\; i_1 i_2 x_4 y_1$	$-\; i_2 x_4 y_2$	$-\; i_1 x_4 y_3$	$+\; x_4 y_4$

(2) $$(x_1 + i_1x_2 + i_2x_3 + i_1i_2x_4) \otimes (y_1 + i_1y_2 + i_2y_3 + i_1i_2y_4)$$
$$= (x_1y_1 - x_2y_2 - x_3y_3 + x_4y_4)$$
$$+ i_1(x_1y_2 + x_2y_1 - x_3y_4 - x_4y_3)$$
$$+ i_2(x_1y_3 - x_2y_4 + x_3y_1 - x_4y_2)$$
$$+ i_1i_2(x_1y_4 + x_2y_3 + x_3y_2 + x_4y_1).$$

4.2 EXAMPLE Definition 4.1 shows that

(3) $$(2 + i_14 - i_23 + i_1i_25)(6 - i_18 + i_23 - i_1i_27)$$
$$= (18 - i_128 + i_256 + i_1i_252).$$

If z_1, z_2, w_1, w_2 are elements in $\mathbb{C}_1$, then

(4) $$(z_1 + i_2z_2)(w_1 + i_2w_2) = (z_1w_1 - z_2w_2) + i_2(z_1w_2 + z_2w_1).$$

The formula in (4) emphasizes once more the formal similarities of complex and bicomplex numbers.

4.3 THEOREM The following statements describe properties of multiplication in $\mathbb{C}_2$.

(5) $\mathbb{C}_2$ is closed under multiplication.

(6) Multiplication is associative.

(7) Multiplication is distributive with respect to addition.

(8) Multiplication is commutative.

(9) There is a unit element for multiplication; it is $(1+i_10+i_20+i_1i_20)$, which is usually denoted by 1.

Proof. The statements in the theorem can be verified by straightforward calculation which employs the definitions of the operations and the properties of $\mathbb{C}_2$. □

4.4 THEOREM If z is in $\mathbb{C}_1$ and $z_1+i_2z_2$ and $w_1+i_2w_2$ are in $\mathbb{C}_2$, then

(10) $$\|z(z_1 + i_2z_2)\| = |z|\ \|z_1 + i_2z_2\|,$$

(11) $$\|(z_1 + i_2z_2)(w_1 + i_2w_2)\| \leqslant \sqrt{2}\,\|z_1 + i_2z_2\|\ \|w_1 + i_2w_2\|.$$

The inequality in (11) is the best possible.

Proof. Since z can be considered to be the bicomplex number $z+i_20$, the product $z(z_1+i_2z_2)$ is defined; equation (4) shows that

(12) $$z(z_1 + i_2z_2) = (zz_1) + i_2(zz_2).$$

Thus multiplication by z has the character of scalar multiplication. By (6) in Section 3,

$$\begin{aligned}(13)\qquad \|z(z_1 + i_2 z_2)\| &= (|zz_1|^2 + |zz_2|^2)^{1/2} = (|z|^2|z_1|^2 + |z|^2|z_2|^2)^{1/2}\\ &= |z|(|z_1|^2 + |z_2|^2)^{1/2} = |z|\ \|z_1 + i_2 z_2\|.\end{aligned}$$

Thus (10) is true. Next, since multiplication is distributive with respect to addition,

$$(14)\qquad (z_1 + i_2 z_2)(w_1 + i_2 w_2) = z_1(w_1 + i_2 w_2) + i_2 w_2(w_1 + i_2 w_2).$$

Since $\|z_1(w_1 + i_2 w_2)\| = |z_1|\ \|w_1 + i_2 w_2\|$ and $\|i_2 z_2(w_1 + i_2 w_2)\| = |z_2|\ \|w_1 + i_2 w_2\|$, the triangle inequality for the norm shows that

$$\begin{aligned}(15)\qquad \|(z_1 + i_2 z_2)(w_1 + i_2 w_2)\| &\leqslant |z_1|\ \|w_1 + i_2 w_2\| + |z_2|\ \|w_1 + i_2 w_2\|\\ &\leqslant (|z_1| + |z_2|)\|w_1 + i_2 w_2\|.\end{aligned}$$

Now Schwarz's inequality shows that

$$(16)\qquad (|z_1| + |z_2|) \leqslant \sqrt{2}(|z_1|^2 + |z_2|^2)^{1/2} = \sqrt{2}\|z_1 + i_2 z_2\|,$$

and (15) and (16) establish (11). To prove that the inequality in (11) is the best possible, observe that

$$(17)\qquad \left\|\frac{1 + i_1 i_2}{2}\right\| = \frac{\sqrt{2}}{2}, \qquad \left\|\frac{1 - i_1 i_2}{2}\right\| = \frac{\sqrt{2}}{2},$$

$$\left\|\frac{1 + i_1 i_2}{2}\frac{1 + i_1 i_2}{2}\right\| = \left\|\frac{1 + i_1 i_2}{2}\right\| = \frac{\sqrt{2}}{2} = \sqrt{2}\left\|\frac{1 + i_1 i_2}{2}\right\|\left\|\frac{1 + i_1 i_2}{2}\right\|,$$

$$\left\|\frac{1 - i_1 i_2}{2}\frac{1 - i_1 i_2}{2}\right\| = \left\|\frac{1 - i_1 i_2}{2}\right\| = \frac{\sqrt{2}}{2} = \sqrt{2}\left\|\frac{1 - i_1 i_2}{2}\right\|\left\|\frac{1 - i_1 i_2}{2}\right\|.$$

The proof of Theorem 4.4 is complete. □

4.5 EXAMPLE Although (17) shows that there are pairs of numbers $z_1 + i_2 z_2$ and $w_1 + i_2 w_2$ in $\mathbb{C}_2$ for which the equality holds in (11), there are many pairs for which the inequality holds. The extreme case of an inequality is the following:

$$(18)\qquad \left\|\frac{1 + i_1 i_2}{2}\frac{1 - i_1 i_2}{2}\right\| = \|0\| = 0 < \sqrt{2}\left\|\frac{1 + i_1 i_2}{2}\right\|\left\|\frac{1 - i_1 i_2}{2}\right\|.$$

In most cases, the norm of a product lies between the extremes shown in (17) and (18). For example,

$$(19)\qquad \|2 + i_1 4 - i_2 3 + i_1 i_2 5\| = \sqrt{54},$$
$$\|6 - i_1 8 + i_2 3 - i_1 i_2 7\| = \sqrt{158},$$

and the norm of the product is, by (3),

$$(20)\qquad \|18 - i_1 28 + i_2 56 + i_1 i_2 52\| = \sqrt{6948}.$$

Thus

$$(21)\qquad \|(2 + i_1 4 - i_2 3 + i_1 i_2 5)(6 - i_1 8 + i_2 3 - i_1 i_2 7)\|$$
$$= \frac{\sqrt{6948}}{\sqrt{54}\sqrt{158}} \|2 + i_1 4 - i_2 3 + i_1 i_2 5\| \; \|6 - i_1 8 + i_2 3 - i_1 i_2 7\|.$$

Since

$$(22)\qquad 0 < \frac{\sqrt{6948}}{\sqrt{54}\sqrt{158}} < 1,$$

the two elements in $\mathbb{C}_2$ in (19) satisfy the inequality in (11), but this case is quite different from the extreme situations shown in (17) and (18).

4.6 THEOREM The system $(\mathbb{C}_2, \oplus, \odot, \|\ \|, \otimes)$ is a Banach algebra.

This theorem follows from the definition of a Banach algebra and the properties of the system which have been established in preceding theorems. In the usual definition of a Banach algebra, the norm of the product of two elements is required to be equal to or less than the product of the norms of these elements. Thus, strictly speaking, $\mathbb{C}_2$ is a modified Banach algebra.

4.7 DEFINITION If $\zeta_1 : (z_1 + i_2 z_2)$ and $\zeta_2 : (w_1 + i_2 w_2)$ are two elements in $\mathbb{C}_2$ and $\zeta_1 \zeta_2 = 1$, then each of the elements ζ_1 and ζ_2 is said to be the (multiplicative) *inverse* of the other. An element which has an inverse is said to be *nonsingular*, and an element which does not have an inverse is said to be *singular*.

4.8 THEOREM An element $\zeta_1 : (z_1 + i_2 z_2)$ is nonsingular if and only if

$$(23)\qquad |z_1^2 + z_2^2| \neq 0,$$

and it is singular if and only if

$$(24)\qquad |z_1^2 + z_2^2| = 0.$$

Proof. By definition, ζ_1 is nonsingular if and only if there is a $\zeta_2 : (w_1 + i_2 w_2)$ such that $\zeta_1 \zeta_2 = 1$. By (4) this equation is equivalent to the following system:

$$(25) \qquad z_1 w_1 - z_2 w_2 = 1, \qquad z_2 w_1 + z_1 w_2 = 0.$$

This system has a (unique) solution for (w_1, w_2) if

$$(26) \qquad \det \begin{bmatrix} z_1 & -z_2 \\ z_2 & z_1 \end{bmatrix} = z_1^2 + z_2^2 \neq 0.$$

Thus ζ_1 is nonsingular if (23) is true. If (24) is true, the following considerations show that (25) has no solution. There are two possibilities: (a) $z_1 = z_2 = 0$, and (*b*) $z_2 = \pm i_1 z_1 \neq 0$. In case (a) the two equations in (25) are inconsistent, and the system has no solution. If $z_2 = i_1 z_1$, then (25) becomes

$$(27) \qquad z_1(w_1 - i_1 w_2) = 1, \qquad z_1(w_1 - i_1 w_2) = 0;$$

if $z_2 = -i_1 z_1$, then (25) can be simplified to

$$(28) \qquad z_1(w_1 + i_1 w_2) = 1, \qquad z_1(w_1 + i_1 w_2) = 0.$$

The systems in (27) and (28) are inconsistent, and thus (25) has no solution if (24) is true. Thus ζ_1 is nonsingular if and only if (23) is true, and it is singular if and only if (24) is true. □

4.9 COROLLARY The element $z_1 + i_2 z_2$ is nonsingular if and only if

$$(29) \qquad |z_1 - i_1 z_2| \, |z_1 + i_1 z_2| \neq 0,$$

and it is singular if and only if

$$(30) \qquad |z_1 - i_1 z_2| \, |z_1 + i_1 z_2| = 0.$$

4.10 THEOREM Let $\zeta_1 : (z_1 + i_2 z_2)$ and $\zeta_2 : (w_1 + i_2 w_2)$ be two elements in $\mathbb{C}_2$. Then $\zeta_1 \zeta_2$ is singular if and only if at least one of the elements ζ_1 and ζ_2 is singular.

Proof. By (4),

$$(31) \qquad (z_1 + i_2 z_2)(w_1 + i_2 w_2) = (z_1 w_1 - z_2 w_2) + i_2(z_1 w_2 + z_2 w_1).$$

Also

$$(32) \qquad \begin{aligned} &(z_1 w_1 - z_2 w_2)^2 + (z_1 w_2 + z_2 w_1)^2 = (z_1^2 + z_2^2)(w_1^2 + w_2^2), \\ &|(z_1 w_1 - z_2 w_2)^2 + (z_1 w_2 + z_2 w_1)^2| = |z_1^2 + z_2^2| \, |w_1^2 + w_2^2|. \end{aligned}$$

Then Theorem 4.8 shows that $\zeta_1 \zeta_2$ is singular if and only if at least one of the elements ζ_1 and ζ_2 is singular. □

4.11 EXAMPLE The elements

$$(33)\qquad \zeta_1:\left(\frac{1+i_1i_2}{2}\right),\qquad \zeta_2:\left(\frac{1-i_1i_2}{2}\right),$$

are nonzero elements, but since

$$(34)\qquad \left(\frac{1}{2}\right)^2+\left(\frac{i_1}{2}\right)^2=0,\qquad \left(\frac{1}{2}\right)^2+\left(\frac{-i_1}{2}\right)^2=0,$$

each of them is singular by Theorem 4.8. Then Theorem 4.10 shows that, for every element $z_1+i_2z_2$ in $\mathbb{C}_2$, the elements $\zeta_1(z_1+i_2z_2)$ and $\zeta_2(z_1+i_2z_2)$ are singular. Theorem 4.8 provides a second proof; since

$$(35)\qquad \left(\frac{1+i_1i_2}{2}\right)(z_1+i_2z_2)=\left(\frac{z_1}{2}-\frac{i_1z_2}{2}\right)+i_2\left[i_1\left(\frac{z_1}{2}-\frac{i_1z_2}{2}\right)\right],$$

then

$$(36)\qquad \left(\frac{z_1}{2}-\frac{i_1z_2}{2}\right)^2+\left[i_1\left(\frac{z_1}{2}-\frac{i_1z_2}{2}\right)\right]^2=0,$$

and $\zeta_1(z_1+i_2z_2)$ is singular by (24) in Theorem 4.8. Similar arguments show that $\zeta_2(z_1+i_2z_2)$ is singular.

4.12 DEFINITION Define $V:\mathbb{C}_2\to\mathbb{R}_{\geqslant 0}$ to be the function such that, if $\zeta:(x_1+i_1x_2+i_2x_3+i_1i_2x_4)$ is in $\mathbb{C}_2$, then

$$(37)\qquad V(\zeta)=|(x_1+i_1x_2)^2+(x_3+i_1x_4)^2|^2.$$

4.13 THEOREM If $\zeta_1:(x_1+i_1x_2+i_2x_3+i_1i_2x_4)$ and $\zeta_2:(y_1+i_1y_2+i_2y_3+i_1i_2y_4)$ are elements in $\mathbb{C}_2$, then

$$(38)\qquad V(\zeta_1)\geqslant 0,$$

$$(39)\qquad V(\zeta_1)=(x_1^2-x_2^2+x_3^2-x_4^2)^2+(2x_1x_2+2x_3x_4)^2,$$

$$(40)\qquad V(\zeta_1\zeta_2)=V(\zeta_1)V(\zeta_2).$$

Proof. The inequality (38) follows from the definition of $V(\zeta)$ in (37). To establish (39), calculate $V(\zeta_1)$ from the definition in (37). To prove (40), represent ζ_1 and ζ_2 as $z_1+i_2z_2$ and $w_1+i_2w_2$, respectively. Then (31) and (37) show that

$$(41)\qquad V(\zeta_1\zeta_2)=|(z_1w_1-z_2w_2)^2+(z_1w_2+z_2w_1)^2|^2.$$

Then since $V(\zeta_1)=|z_1^2+z_2^2|^2$ and $V(\zeta_2)=|w_1^2+w_2^2|^2$ by (37), equations (41) and (32) show that

$$(42)\qquad V(\zeta_1\zeta_2)=|z_1^2+z_2^2|^2\,|w_1^2+w_2^2|^2=V(\zeta_1)V(\zeta_2).$$

□

4.14 COROLLARY The element $\zeta:(z_1+i_2z_2)$ is nonsingular if and only if $V(\zeta)>0$, and is singular if and only if $V(\zeta)=0$.

Definition 4.12 and inequality (38) show that this corollary is merely a restatement of Theorem 4.8.

There is only one element in $\mathbb{C}_0$ which does not have a multiplicative inverse; it is the element 0. Denote the set whose only element is 0 by $\mathcal{O}_0$. There is only one element in $\mathbb{C}_1$ which does not have an inverse; it is $0+i_10$. Denote the set whose only element is $0+i_10$ by $\mathcal{O}_1$. Example 4.11 shows that there are many elements in $\mathbb{C}_2$ which do not have inverses; denote the set of these elements by $\mathcal{O}_2$. Since $\mathbb{C}_0$ is isomorphic to a subset of $\mathbb{C}_1$ and $\mathbb{C}_1$ is isomorphic to a subset of $\mathbb{C}_2$, it is customary to say simply that $\mathbb{C}_0$ *is* a subset of $\mathbb{C}_1$ and $\mathbb{C}_1$ *is* a subset of $\mathbb{C}_2$. Then $0+i_10+i_20+i_1i_20$ is the single element in $\mathcal{O}_0$ and $\mathcal{O}_1$, and it belongs also to $\mathcal{O}_2$. Thus

(43) $$\mathcal{O}_0=\mathcal{O}_1\subsetneqq\mathcal{O}_2.$$

Exercises

4.1 Verify each of the following statements:

(a) $$\left(\frac{1+i_1i_2}{2}\right)\left(\frac{1+i_1i_2}{2}\right)=\frac{1+i_1i_2}{2},\qquad \left(\frac{1-i_1i_2}{2}\right)\left(\frac{1-i_1i_2}{2}\right)=\frac{1-i_1i_2}{2},\qquad \left(\frac{1+i_1i_2}{2}\right)\left(\frac{1-i_1i_2}{2}\right)=0;$$

(b) $$\left\|\frac{1+i_1i_2}{2}\right\|=\frac{\sqrt{2}}{2},\qquad \left\|\frac{1-i_1i_2}{2}\right\|=\frac{\sqrt{2}}{2};$$

(c) $$V\left(\frac{1+i_1i_2}{2}\right)=0,\qquad V\left(\frac{1-i_1i_2}{2}\right)=0;$$

(d) $(1+i_1i_2)/2$ and $(1-i_1i_2)/2$ are singular elements.

(e) If $z_1+i_2z_2$ is an element in $\mathbb{C}_2$, then $(z_1+i_2z_2)[(1+i_1i_2)/2]$ and $(z_1+i_2z_2)[(1-i_1i_2)/2]$ are singular elements.

4.2 Prove the following theorem: if $z_1+i_2z_2$ and $w_1+i_2w_2$ are elements in $\mathbb{C}_2$, then

(a) $$z_1+i_2z_2=(z_1-i_1z_2)\left(\frac{1+i_1i_2}{2}\right)+(z_1+i_1z_2)\left(\frac{1-i_1i_2}{2}\right),$$

$$w_1+i_2w_2=(w_1-i_1w_2)\left(\frac{1+i_1i_2}{2}\right)+(w_1+i_1w_2)\left(\frac{1-i_1i_2}{2}\right);$$

(b) $(z_1+i_2z_2)+(w_1+i_2w_2)$

$$=[(z_1-i_1z_2)+(w_1-i_1w_2)]\left(\frac{1+i_1i_2}{2}\right)+[(z_1+i_1z_2)$$

$$+(w_1+i_1w_2)]\left(\frac{1-i_1i_2}{2}\right);$$

(c) $(z_1+i_2z_2)(w_1+i_2w_2)$

$$=(z_1-i_1z_2)(w_1-i_1w_2)\left(\frac{1+i_1i_2}{2}\right)$$

$$+(z_1+i_1z_2)(w_1+i_1w_2)\left(\frac{1-i_1i_2}{2}\right).$$

4.3 If $z_1+i_2z_2$ is an element in $\mathbb{C}_2$, show that

$$(z_1+i_2z_2)\left(\frac{1+i_1i_2}{2}\right)=(z_1-i_1z_2)\left(\frac{1+i_1i_2}{2}\right),$$

$$(z_1+i_2z_2)\left(\frac{1-i_1i_2}{2}\right)=(z_1+i_1z_2)\left(\frac{1-i_1i_2}{2}\right).$$

4.4 If $z_1+i_2z_2$ is a singular element, show that at least one of the following statements is true:

$$z_1+i_2z_2=(z_1-i_1z_2)\left(\frac{1+i_1i_2}{2}\right),$$

$$z_1+i_2z_2=(z_1+i_1z_2)\left(\frac{1-i_1i_2}{2}\right).$$

4.5 Prove the following statements:

(a) If c_1 and c_2 are in $\mathbb{C}_1$ and

$$c_1\left(\frac{1+i_1i_2}{2}\right)+c_2\left(\frac{1-i_1i_2}{2}\right)=0,$$

then $c_1=0$ and $c_2=0$.

(b) If both of the equations in Exercise 4.4 are true, then $z_1+i_2z_2=0$.

5. FRACTIONS AND QUOTIENTS

This section contains the definition of the quotient of two elements in $\mathbb{C}_2$, and it shows that a quotient (fraction) exists if and only if the divisor (denominator) is nonsingular. Next, the section establishes the standard cancellation laws and some necessary inequalities for quotients.

5.1 DEFINITION Let $\zeta_1:(z_1+i_2z_2)$ and $\zeta_2:(w_1+i_2w_2)$ be elements in $\mathbb{C}_2$. If there exists a unique element $\eta:(u_1+i_2u_2)$ in $\mathbb{C}_2$ such that $\zeta_1=\zeta_2\eta$, then the quotient (or fraction) ζ_1/ζ_2 exists and

$$\frac{\zeta_1}{\zeta_2}=\eta. \tag{1}$$

5.2 THEOREM The fraction ζ_1/ζ_2 is defined if and only if ζ_2 is nonsingular. If ζ_2 is nonsingular and ζ_2^{-1} denotes the inverse of ζ_2, then

$$\frac{\zeta_1}{\zeta_2}=\zeta_1\zeta_2^{-1}. \tag{2}$$

Proof. By (1), ζ_1/ζ_2 is a bicomplex number $\eta:(u_1+i_2u_2)$ such that $\zeta_1=\zeta_2\eta$. This equation is equivalent to the following system of equations in the unknowns u_1, u_2:

$$w_1u_1-w_2u_2=z_1, \qquad w_2u_1+w_1u_2=z_2. \tag{3}$$

This system of equations has a unique solution if and only if $w_1^2+w_2^2\neq 0$, that is, if and only if ζ_2 is nonsingular (see Theorem 4.8). Thus if ζ_2 is nonsingular, there is an element η in $\mathbb{C}_2$ such that $\zeta_1=\zeta_2\eta$. Since ζ_2 is nonsingular, it has an inverse ζ_2^{-1}, and $\zeta_1\zeta_2^{-1}=\zeta_2\eta\zeta_2^{-1}=\zeta_2\zeta_2^{-1}\eta=\eta$. Therefore, $\eta=\zeta_1\zeta_2^{-1}$, and the proof of (2) and of Theorem 5.2 is complete. □

5.3 COROLLARY If $\zeta_1:(z_1+i_2z_2)$ and $\zeta_2:(w_1+i_2w_2)$ are elements in $\mathbb{C}_2$ and $w_1^2+w_2^2\neq 0$, then the fraction ζ_1/ζ_2 is defined and

$$\begin{aligned}\frac{\zeta_1}{\zeta_2}&=\frac{z_1w_1+z_2w_2}{w_1^2+w_2^2}+i_2\left(\frac{z_2w_1-z_1w_2}{w_1^2+w_2^2}\right)\\&=\frac{(z_1+i_2z_2)(w_1-i_2w_2}{w_1^2+w_2^2}.\end{aligned} \tag{4}$$

Proof. The value of ζ_1/ζ_2 is $u_1+i_2u_2$, where u_1 and u_2 satisfy the system of equations in (3). The solution of these equations gives the result in (4). Formally, the value of the quotient is obtained as in the complex case:

$$\begin{aligned}\frac{z_1+i_2z_2}{w_1+i_2w_2}&=\frac{(z_1+i_2z_2)(w_1-i_2w_2)}{(w_1+i_2w_2)(w_1-i_2w_2)}\\&=\frac{(z_1+i_2z_2)(w_1-i_2w_2)}{w_1^2+w_2^2}.\end{aligned} \tag{5}$$

□

5.4 THEOREM (Cancellation Laws)
Let $\zeta_1, \zeta_2, \zeta_3$ be elements in $\mathbb{C}_2$.

(6) If ζ_3 is nonsingular and $\zeta_1\zeta_3=\zeta_2\zeta_3$, then $\zeta_1=\zeta_2$.

(7) If ζ_2 and ζ_3 are nonsingular, then $\dfrac{\zeta_1\zeta_3}{\zeta_2\zeta_3}=\dfrac{\zeta_1}{\zeta_2}$.

Proof. Since ζ_3 is nonsingular, it has an inverse ζ_3^{-1} by Definition 4.7. Since $\zeta_1\zeta_3=\zeta_2\zeta_3$ by hypothesis, then $(\zeta_1\zeta_3)\zeta_3^{-1}=(\zeta_2\zeta_3)\zeta_3^{-1}$. Then since multiplication is associative, $\zeta_1(\zeta_3\zeta_3^{-1})=\zeta_2(\zeta_3\zeta_3^{-1})$ or $\zeta_1=\zeta_2$. Thus (6) is true. Consider (7). Since ζ_2 and ζ_3 are nonsingular by hypothesis, then $\zeta_2\zeta_3$ is nonsingular by Theorem 4.10. Thus both of the fractions in (7) exist by Theorem 5.2. Let η denote the value of $\zeta_1\zeta_3/\zeta_2\zeta_3$. Then

(8) $$\zeta_1\zeta_3=(\zeta_2\zeta_3)\eta$$

and

(9) $$\zeta_1=\zeta_2\eta$$

by (6). Thus by (8) and (9),

(10) $$\frac{\zeta_1\zeta_3}{\zeta_2\zeta_3}=\eta=\frac{\zeta_1}{\zeta_2},$$

and (7) is true. The proof of Theorem 5.4 is complete. □

5.5 THEOREM If $w_1+i_2w_2$ is a nonsingular element in $\mathbb{C}_2$, then

(11) $$\frac{\sqrt{2}}{2}\frac{\|z_1+i_2z_2\|}{\|w_1+i_2w_2\|}\leqslant\left\|\frac{z_1+i_2z_2}{w_1+i_2w_2}\right\|$$
$$=\frac{\|(z_1+i_2z_2)(w_1-i_2w_2)\|}{|w_1^2+w_2^2|}\leqslant\frac{\sqrt{2}\|z_1+i_2z_2\|\,\|w_1+i_2w_2\|}{|w_1^2+w_2^2|}.$$

Proof. Since $w_1+i_2w_2$ is nonsingular, then $w_1^2+w_2^2\neq0$ and the fraction $(z_1+i_2z_2)/(w_1+i_2w_2)$ exists by Corollary 5.3; its value is an element $u_1+i_2u_2$ in $\mathbb{C}_2$ such that

(12) $$z_1+i_2z_2=(w_1+i_2w_2)(u_1+i_2u_2).$$

Then Theorem 4.4 (11) shows that

(13) $$\|z_1+i_2z_2\|\leqslant\sqrt{2}\|w_1+i_2w_2\|\,\|u_1+i_2u_2\|,$$
$$\frac{\sqrt{2}}{2}\frac{\|z_1+i_2z_2\|}{\|w_1+i_2w_2\|}\leqslant\left\|\frac{z_1+i_2z_2}{w_1+i_2w_2}\right\|,$$

and the inequality on the left in (11) is true. By (4) and Theorem 4.4,

(14) $$\left\|\frac{z_1+i_2z_2}{w_1+i_2w_2}\right\|=\left\|\frac{(z_1+i_2z_2)(w_1-i_2w_2)}{w_1^2+w_2^2}\right\|$$
$$=\frac{\|(z_1+i_2z_2)(w_1-i_2w_2)\|}{|w_1^2+w_2^2|}$$
$$\leqslant\frac{\sqrt{2}\|z_1+i_2z_2\|\,\|w_1+i_2w_2\|}{|w_1^2+w_2^2|}.$$

□

5.6 COROLLARY If $w_1 + i_2w_2$ is nonsingular, then

$$\frac{\sqrt{2}}{2}\frac{\|z_1 + i_2z_2\|}{\|w_1 + i_2w_2\|} \leqslant \left\|\frac{z_1 + i_2z_2}{w_1 + i_2w_2}\right\| \leqslant \frac{\sqrt{2}\|z_1 + i_2z_2\|\ \|w_1 + i_2w_2\|}{[V(w_1 + i_2w_2)]^{1/2}}. \tag{15}$$

Proof. The inequalities in (15) follow from (11) and the definition of $V(w_1+i_2w_2)$ in Definition 4.12. □

5.7 EXAMPLE In both $\mathbb{C}_1$ and $\mathbb{C}_2$, the norm of a fraction is large when the denominator is close to a singular element. At first glance, the situation in the two spaces seems different, but the apparent difference results from the fact that there is only one singular element (namely zero) in $\mathbb{C}_1$ but many singular elements in $\mathbb{C}_2$. In (15), $V(w_1+i_2w_2)$ can be considered to be a measure of how close $w_1+i_2w_2$ is to a singular element. If $w_2=i_1rw_1$, where $r\in\mathbb{C}_0$, then

$$[V(w_1 + i_2w_2)]^{1/2} = |w_1^2|\,|1 - r^2|, \tag{16}$$
$$\|w_1 + i_2w_2\| = |w_1|(1 + r^2)^{1/2},$$

$$\left\|\frac{z_1 + i_2z_2}{w_1 + i_2w_2}\right\| \leqslant \frac{\sqrt{2}\|z_1 + i_2z_2\|(1 + r^2)^{1/2}}{|w_1|\,|1 - r^2|}. \tag{17}$$

If $|1 - r^2|$ is close to zero, then $V(w_1+i_2w_2)$ is close to zero although $\|w_1+i_2w_2\|$ is not small. Thus the bound for $(z_1+i_2z_2)/(w_1+i_2w_2)$ on the right in (15) may be very large [see (17)] in spite of the fact that $\|w_1+i_2w_2\|$ is also large. The bound shown in (15) is necessarily large because the norm of the fraction $(z_1+i_2z_2)/(w_1+i_2w_2)$ is large in situations such as the one described in (16) and (17) [compare (11)].

Exercises

5.1 If $w_1^2+w_2^2\neq 0$, show that

$$\frac{z_1 + i_2z_2}{w_1 + i_2w_2} = (z_1 + i_2z_2)\left(\frac{1}{w_1 + i_2w_2}\right).$$

5.2 If $w_1^2+w_2^2\neq 0$, show that

$$\frac{z_1 + i_2z_2}{w_1 + i_2w_2} = \frac{(z_1 + i_2z_2)(w_1 - i_2w_2)}{(w_1 + i_2w_2)(w_1 - i_2w_2)},$$

and explain how this equation provides an easy way to calculate the value of the bicomplex fraction on the left.

5.3 (a) If $c \in \mathbb{C}_1$, the bicomplex numbers of the form c and $i_2 c$ can be called complex numbers and pure imaginary bicomplex numbers, respectively. If $c \neq 0$, show that division by c and $i_2 c$ is always possible.

(b) Assume that $c \neq 0$, and express each of the following fractions as a bicomplex number:

$$\frac{z_1 + i_2 z_2}{c}, \qquad \frac{z_1 + i_2 z_2}{i_2 c}.$$

5.4 Show that division by the following bicomplex numbers is impossible:

(a) $\dfrac{1 + i_1 i_2}{2}, \qquad \dfrac{1 - i_1 i_2}{2};$

(b) $(z_1 + i_2 z_2)\left(\dfrac{1 + i_1 i_2}{2}\right), \qquad (z_1 + i_2 z_2)\left(\dfrac{1 - i_1 i_2}{2}\right),$

$(z_1 + i_2 z_2) \in \mathbb{C}_2$.

5.5 Let $\zeta_1, \ldots, \zeta_4$ be elements in $\mathbb{C}_2$, and assume that ζ_2 and ζ_4 are nonsingular. Show that each of the fractions in the following equations is defined and that the equations hold.

$$\frac{\zeta_1}{\zeta_2}\frac{\zeta_3}{\zeta_4} = \frac{\zeta_1\zeta_3}{\zeta_2\zeta_4}, \qquad \frac{\zeta_1}{\zeta_2} + \frac{\zeta_3}{\zeta_2} = \frac{\zeta_1 + \zeta_3}{\zeta_2}, \qquad \frac{\zeta_1}{\zeta_2} + \frac{\zeta_3}{\zeta_4} = \frac{\zeta_1\zeta_4 + \zeta_2\zeta_3}{\zeta_2\zeta_4}.$$

5.6 (a) Let A be a 2 by 2 matrix $[a_{ij}]$ whose elements a_{ij} are bicomplex numbers in $\mathbb{C}_2$. Show that the definition of the determinant of a complex matrix can be applied to define a determinant of A which has all of the elementary properties of the determinant of a complex matrix.

(b) Consider the system of equations which, in matrix form, is

$$\begin{bmatrix} a_{11} & a_{12} \\ a_{21} & a_{22} \end{bmatrix}\begin{bmatrix} \zeta_1 \\ \zeta_2 \end{bmatrix} = \begin{bmatrix} c_1 \\ c_2 \end{bmatrix}, \qquad c_1, c_2 \text{ in } \mathbb{C}_2.$$

If $\det A$ is a nonsingular element in $\mathbb{C}_2$, show that the system of equations has a unique solution which is

$$\zeta_1 = \frac{\begin{vmatrix} c_1 & a_{12} \\ c_2 & a_{22} \end{vmatrix}}{\det A}, \qquad \zeta_2 = \frac{\begin{vmatrix} a_{11} & c_1 \\ a_{21} & c_2 \end{vmatrix}}{\det A}.$$

(c) Does Cramer's rule apply in this case? Is the solution ζ_1, ζ_2 obtained in (b) a pair of numbers in $\mathbb{C}_1$ or a pair of numbers in $\mathbb{C}_2$? If $c_1 = c_2 = 0$, what is the solution of the system of equations?

(d) Solve the same problem for a similar system of 3 equations in 3 unknowns, and for n equations in n unknowns.

6. THE IDEMPOTENT REPRESENTATION

This section defines idempotent elements and shows that there are four idempotent elements in $\mathbb{C}_2$. Two of these idempotent elements, namely $(1+i_1i_2)/2$ and $(1-i_1i_2)/2$, play an important role since every element in $\mathbb{C}_2$ has a unique representation as a linear combination of them. This section presents the properties of these idempotent elements, and it describes the properties of $\mathbb{C}_2$ and of its operations in terms of the idempotent representation of its elements.

6.1 DEFINITION Let ζ_1 and ζ_2 be elements in $\mathbb{C}_2$. If $\zeta_1^2=\zeta_1$, then ζ_1 is called an *idempotent* element. If $\zeta_1\neq 0$, $\zeta_2\neq 0$, and $\zeta_1\zeta_2=0$, then ζ_1 and ζ_2 are called *divisors of zero*.

6.2 THEOREM There are four and only four idempotent elements in $\mathbb{C}_2$, and they are

$$(1)\qquad 0, \qquad 1, \qquad \frac{1+i_1i_2}{2}, \qquad \frac{1-i_1i_2}{2}.$$

Proof. Let $z_1+i_2z_2$ be an element in $\mathbb{C}_2$. Then $z_1+i_2z_2$ is an idempotent element if and only if $(z_1+i_2z_2)^2=z_1+i_2z_2$. This equation is equivalent to the following two equations:

$$(2)\qquad z_1^2-z_2^2=z_1, \qquad 2z_1z_2=z_2.$$

The second of these equations is satisfied if $z_2=0$ or $z_1=\frac{1}{2}$. If $z_2=0$, the first equation is satisfied by $z_1=0$ or $z_1=1$. If $z_1=\frac{1}{2}$, the first equation is satisfied by $z_2=\pm i_1/2$. Thus the two equations in (2) have the four solutions

$$(3)\qquad \begin{array}{llll} z_1=0, & z_1=1, & z_1=1/2, & z_1=1/2, \\ z_2=0, & z_2=0, & z_2=i_1/2, & z_2=-i_2/2. \end{array}$$

The corresponding elements $z_1+i_2z_2$ are those in (1), and the proof of Theorem 6.2 is complete. □

The following notation will be used for the third and fourth idempotent elements in (1):

$$(4)\qquad e_1=\frac{1+i_1i_2}{2}, \qquad e_2=\frac{1-i_1i_2}{2}.$$

6.3 THEOREM The idempotent elements e_1, e_2 have the following properties:

$$(5)\qquad e_1^2=e_1, \qquad e_2^2=e_2, \qquad e_1e_2=0;$$

$$(6)\qquad \|e_1\|=\sqrt{2}/2, \qquad \|e_2\|=\sqrt{2}/2;$$

(7) $V(e_1) = 0$, $V(e_2) = 0$, and e_1, e_2 are singular elements in $\mathbb{C}_2$;

(8) e_1, e_2 are linearly independent with respect to complex constants of combination; that is, if c_1, c_2 are in $\mathbb{C}_1$ and $c_1e_1+c_2e_2=0$, then $c_1=c_2=0$.

Proof. The statements in (5) can be verified easily from the definitions in (4), and (6) follows from (4) and Definition 3.1. Statement (7) follows from (4), Definition 4.12, and Corollary 4.14. To prove (8), let c_1, c_2 be complex numbers in $\mathbb{C}_1$. Then the equation $c_1e_1+c_2e_2=0$ is

$$(9)\qquad c_1\left(\frac{1+i_1i_2}{2}\right)+c_2\left(\frac{1-i_1i_2}{2}\right)=0.$$

This equation is equivalent to the system

$$(10)\qquad c_1+c_2=0, \qquad i_1c_1-i_1c_2=0.$$

Since the determinant of the matrix of coefficients in this system is $-2i_1$, then $c_1=0$ and $c_2=0$. The proof of all parts of Theorem 6.3 is complete. □

6.4 THEOREM (Idempotent Representation) Every element $\zeta:(z_1+i_2z_2)$ in $\mathbb{C}_2$ has the following unique representation:

$$(11)\qquad \zeta=(z_1-i_1z_2)e_1+(z_1+i_1z_2)e_2.$$

Proof. Let c_1, c_2 be complex members in $\mathbb{C}_2$ such that $c_1e_1+c_2e_2=\zeta$. Then

$$(12)\qquad c_1\left(\frac{1+i_1i_2}{2}\right)+c_2\left(\frac{1-i_1i_2}{2}\right)=z_1+i_2z_2,$$

and this equation is equivalent to the following system of equations:

$$(13)\qquad \frac{c_1}{2}+\frac{c_2}{2}=z_1, \qquad \frac{i_1c_1}{2}-\frac{i_1c_2}{2}=z_2.$$

These equations have the following unique solution:

$$(14)\qquad c_1=z_1-i_1z_2, \qquad c_2=z_1+i_1z_2.$$

Thus ζ has the unique representation shown in (11), and the proof of Theorem 6.4 is complete. □

6.5 DEFINITION Equation (11) is the *idempotent representation* of the element $\zeta:(z_1+i_2z_2)$ in $\mathbb{C}_2$. Also, $z_1-i_1z_2$ and $z_1+i_1z_2$ are the *idempotent components* of $z_1+i_2z_2$.

6.6 THEOREM Let $z_1+i_2z_2$ and $w_1+i_2w_2$ be elements in $\mathbb{C}_2$. Then

$$(15)\qquad (z_1+i_2z_2)+(w_1+i_2w_2) = [(z_1-i_1z_2)+(w_1-i_1w_2)]e_1+[(z_1+i_1z_2)+(w_1+i_1w_2)]e_2,$$

$$(16)\quad (z_1 + i_2 z_2)(w_1 + i_2 w_2) = [(z_1 - i_1 z_2)(w_1 - i_1 w_2)]e_1 + [(z_1 + i_1 z_2)(w_1 + i_1 w_2)]e_2,$$

$$(17)\quad (z_1 + i_2 z_2)^n = (z_1 - i_1 z_2)^n e_1 + (z_1 + i_1 z_2)^n e_2,\ n = 0, 1, \ldots.$$

If $(w_1 - i_1 w_2) \neq 0$ and $(w_1 + i_1 w_2) \neq 0$, then

$$(18)\quad \frac{z_1 + i_2 z_2}{w_1 + i_2 w_2} = \left(\frac{z_1 - i_1 z_2}{w_1 - i_1 w_2}\right)e_1 + \left(\frac{z_1 + i_1 z_2}{w_1 + i_1 w_2}\right)e_2.$$

Proof. Equation (15) follows from the idempotent representation of elements in $\mathbb{C}_2$ and from properties of addition and multiplication. To prove (16), multiply the idempotent representations of $z_1 + i_2 z_2$ and $w_1 + i_2 w_2$ and then use the properties of e_1, e_2 in (5) and the properties of multiplication to simplify the result. Induction and (16) can be used to prove (17). Consider (18). Since $(w_1 - i_1 w_2) \neq 0$ and $(w_1 + i_1 w_2) \neq 0$ by hypothesis, $w_1 + i_2 w_2$ is nonsingular, and by Theorem 5.2 the fraction $(z_1 + i_2 w_2)/(w_1 + i_2 w_2)$ is defined; let $u_1 + i_2 u_2$ denote its value. Then by Definition 5.1,

$$(19)\quad z_1 + i_2 z_2 = (w_1 + i_2 w_2)(u_1 + i_2 u_2).$$

Since

$$(20)\quad \begin{aligned} z_1 + i_2 z_2 &= (z_1 - i_1 z_2)e_1 + (z_1 + i_1 z_2)e_2, \\ (w_1 + i_2 w_2)(u_1 + i_2 u_2) &= [(w_1 - i_1 w_2)(u_1 - i_1 u_2)]e_1 + [(w_1 + i_1 w_2)(u_1 + i_1 u_2)]e_2, \end{aligned}$$

equation (19) and the fact that the idempotent representation is unique show that

$$(21)\quad \begin{aligned} z_1 - i_1 z_2 &= (w_1 - i_1 w_2)(u_1 - i_1 u_2), \\ z_1 + i_1 z_2 &= (w_1 + i_1 w_2)(u_1 + i_1 u_2). \end{aligned}$$

These equations show that, since $(w_1 - i_1 w_2)(w_1 + i_1 w_2) \neq 0$,

$$(22)\quad u_1 - i_1 u_2 = \frac{z_1 - i_1 z_2}{w_1 - i_1 w_2}, \qquad u_1 + i_1 u_2 = \frac{z_1 + i_1 z_2}{w_1 + i_1 w_2}.$$

Thus

$$(23)\quad \begin{aligned} \frac{z_1 + i_2 z_2}{w_1 + i_2 w_2} &= u_1 + i_2 u_2 \\ &= (u_1 - i_1 u_2)e_1 + (u_1 + i_1 u_2)e_2, \\ &= \left(\frac{z_1 - i_1 z_2}{w_1 - i_1 w_2}\right)e_1 + \left(\frac{z_1 + i_1 z_2}{w_1 + i_1 w_2}\right)e_2. \end{aligned}$$

Therefore, (18) is true, and the proof of Theorem 6.6 is complete. □

6.7 COROLLARY If $z_1+i_2z_2$ and $w_1+i_2w_2$ are elements in $\mathbb{C}_2$ such that $z_1-i_1z_2=0$ and $w_1+i_1w_2=0$ (or $z_1+i_1z_2=0$ and $w_1-i_1w_2=0$), then $(z_1+i_2w_2)(w_1+i_2w_2)=0$. If $z_1+i_2z_2\neq 0$ and $w_1+i_2w_2\neq 0$, then these elements are divisors of zero.

Proof. If $z_1-i_1z_2=0$ and $w_1+i_1w_2=0$ (or $z_1+i_1z_2=0$ and $w_1-i_1w_2=0$), then Theorem 6.6 (16) shows that

(24) $$(z_1+i_2z_2)(w_1+i_2w_2)=0e_1+0e_2=0.$$

If, in addition, $z_1+i_2z_2\neq 0$ and $w_1+i_2w_2\neq 0$, then these elements are divisors of zero by Definition 6.1. □

6.8 THEOREM If $z_1+i_2z_2$ is an element in $\mathbb{C}_2$, then

(25) $$\left(\frac{|z_1-i_1z_2|^2+|z_1+i_1z_2|^2}{2}\right)^{1/2}=(|z_1|^2+|z_2|^2)^{1/2}=\|z_1+i_2z_2\|.$$

Proof. Let $z_1=x_1+i_1y_1$ and $z_2=x_2+i_1y_2$. Then

(26) $$z_1-i_1z_2=(x_1+i_1y_1)-i_1(x_2+i_1y_2)=(x_1+y_2)+i_1(y_1-x_2),$$
$$z_1+i_1z_2=(x_1+i_1y_1)+i_1(x_2+i_1y_2)=(x_1-y_2)+i_1(y_1+x_2).$$

Then

(27) $$\begin{aligned}|z_1-i_1z_2|^2+|z_1+i_1z_2|^2&=(x_1+y_2)^2+(y_1-x_2)^2\\&\quad+(x_1-x_2)^2+(y_1+x_2)^2\\&=2(x_1^2+y_1^2+x_2^2+y_2^2)\\&=2(|z_1|^2+|z_2|^2)\\&=2\|z_1+i_2z_2\|^2.\end{aligned}$$

The formula in (25) follows from these equations. □

Exercises

6.1 Show that an element $z_1+i_2z_2$ in $\mathbb{C}_2$ is zero if and only if both of its idempotent components are zero.

6.2 Define the i_2-conjugate bicomplex number of $z_1+i_2z_2$ to be $z_1-i_2z_2$. Show that $(z_1+i_2z_2)(z_1-i_2z_2)$ is a complex number in $\mathbb{C}_1$. Compare this result with the corresponding property of conjugate complex numbers in $\mathbb{C}_1$.

6.3 Define the i_1i_2-conjugate bicomplex number of $z_1+i_2z_2$ to be $\bar{z}_1-i_2\bar{z}_2$.
(a) Show that

$$(z_1+i_2z_2)(\bar{z}_1-i_2\bar{z}_2)=|z_1-i_1z_2|^2e_1+|z_1+i_1z_2|^2e_2.$$

(b) Show that $z_1+i_2z_2=0$ if and only if $(z_1+i_2z_2)(\bar{z}_1-i_2\bar{z}_2)=0$. Compare this result with a property of complex numbers in $\mathbb{C}_1$.

6.4 Find the bicomplex numbers which are nth roots of unity. [*Solution.* The problem is to find the solutions in $\mathbb{C}_2$ of the equation $(z_1+i_2z_2)^n=1$. Use the idempotent representation to show that this equation is

$$(z_1 - i_1z_2)^n e_1 + (z_1 + i_1z_2)^n = 1e_1 + 1e_2.$$

Show that this equation is satisfied if and only if

$$(z_1 - i_1z_2)^n = 1, \qquad (z_1 + i_1z_2)^n = 1.$$

Since these are polynomial equations in $\mathbb{C}_1$, each equation has n roots; they are the n nth roots of unity. Denote them by $\omega_1,\ldots,\omega_n$. Show that the equation $(z_1+i_2z_2)^n=1$ has exactly n^2 bicomplex roots, and that they are $\omega_ie_1+\omega_je_2$ for $i,j=1,\ldots,n$.]

6.5 Show that

$$\sum_{k=0}^{n} (a_k + i_2b_k)(z_1 + i_2z_2)^k = 0, \qquad a_k + i_2b_k \in \mathbb{C}_2,\ a_n^2 + b_n^2 \neq 0,$$

has n^2 roots (if roots are counted with their multiplicities), and explain how to find these roots. [*Hint.* Set $z_1+i_2z_2=(z_1-i_1z_2)e_1+(z_1+i_1z_2)e_2$, and use Theorem 6.6 to show that the given polynomial equation is

$$\left[\sum_{k=0}^{n} (a_k - i_1b_k)(z_1 - i_1z_2)^k\right]e_1 + \left[\sum_{k=0}^{n} (a_k + i_1b_k)(z_1 + i_1z_2)^k\right]e_2 = 0e_1 + 0e_2.$$

Show that the roots of the given equation can be found by solving the two equations

$$\sum_{k=0}^{n} (a_k - i_1b_k)(z_1 - i_1z_2)^k = 0, \qquad a_n - i_1b_n \neq 0,$$

$$\sum_{k=0}^{n} (a_k + i_1b_k)(z_1 + i_1z_2)^k = 0, \qquad a_n + i_1b_n \neq 0.]$$

6.6 Consider again the polynomial equation in Exercise 6.5. Assume that $a_n-i_1b_n\neq 0$, $a_n+i_1b_n=0$, and $a_{n-1}+i_1b_{n-1}\neq 0$. Show that the equation has $n(n-1)$ roots if the roots are counted with their multiplicities.

6.7 Let $z_1+i_2z_2$ and $w_1+i_2w_2$ be two elements in $\mathbb{C}_2$. Show that $z_1+i_2z_2=w_1+i_2w_2$ if and only if $z_1-i_1z_2=w_1-i_1w_2$ and $z_1+i_1z_2=w_1+i_1w_2$.

6.8 Let $[a_{ij}]$, $i, j = 1, \ldots, n$, be a matrix with elements a_{ij} in $\mathbb{C}_2$. As a matter of notation, set $a_{ij} = \alpha_{ij}e_1 + \beta_{ij}e_2$.

(a) Use Theorem 6.6 and the definition of the determinant to show that

$$\det[a_{ij}] = \det[\alpha_{ij}]e_1 + \det[\beta_{ij}]e_2.$$

(b) Show that $\det[a_{ij}]$ is nonsingular if and only if $\det[\alpha_{ij}] \neq 0$ and $\det[\beta_{ij}] \neq 0$.

6.9 Consider the following system of equations (compare Exercise 5.6):

$$\begin{bmatrix} a_{11} & \cdots & a_{1n} \\ \vdots & & \vdots \\ a_{n1} & \cdots & a_{nn} \end{bmatrix} \begin{bmatrix} \zeta_1 \\ \vdots \\ \zeta_n \end{bmatrix} = \begin{bmatrix} c_1 \\ \vdots \\ c_n \end{bmatrix}.$$

Here $[a_{ij}]$ is the matrix in Exercise 6.8. Also, $c_1, \ldots, c_n$ are given numbers in $\mathbb{C}_2$ and $\zeta_1, \ldots, \zeta_n$ are unknown numbers in $\mathbb{C}_2$. As a matter of notation, set

$$\zeta_i = z_i e_1 + w_i e_2, \qquad c_i = \gamma_i e_1 + \delta_i e_2, \qquad i = 1, \ldots, n.$$

(a) Show that the solution of the given system can be found by solving two systems

$$\begin{bmatrix} \alpha_{11} & \cdots & \alpha_{1n} \\ \vdots & & \vdots \\ \alpha_{n1} & \cdots & \alpha_{nn} \end{bmatrix} \begin{bmatrix} z_1 \\ \vdots \\ z_n \end{bmatrix} = \begin{bmatrix} \gamma_1 \\ \vdots \\ \gamma_n \end{bmatrix},$$

$$\begin{bmatrix} \beta_{11} & \cdots & \beta_{1n} \\ \vdots & & \vdots \\ \beta_{n1} & \cdots & \beta_{nn} \end{bmatrix} \begin{bmatrix} w_1 \\ \vdots \\ w_n \end{bmatrix} = \begin{bmatrix} \delta_1 \\ \vdots \\ \delta_n \end{bmatrix}.$$

(b) Assume that $\det[a_{ij}]$ is a nonsingular element in $\mathbb{C}_2$ and show that the given system of equations has a unique solution.

(c) Assume again that $\det[a_{ij}]$ is nonsingular. Use a second method to show that the given system of equations has a unique solution and to find this solution.

6.10 Consider again the system of equations in Exercise 6.9. Investigate the solution of the system under the assumption that

(a) $\det[\alpha_{ij}] \neq 0$, $\det[\beta_{ij}] = 0$,

(b) $\det[\alpha_{ij}] = 0$, $\det[\beta_{ij}] = 0$.

6.11 Use the formula in Theorem 6.8 (25) to prove the following properties of the norm $\|\ \|$: for every $z_1 + i_2 z_2$ and $w_1 + i_2 w_2$ in $\mathbb{C}_2$ and z in $\mathbb{C}_1$:

(a) $\|z_1 + i_2 z_2\| \geqslant 0$, $\|z_1 + i_2 z_2\| = 0$ if and only if $z_1 + i_2 z_2 = 0$;

(b) $\|z(z_1 + i_2 z_2)\| = |z|\ \|z_1 + i_2 z_2\|$;

(c) $\|(z_1+i_2z_2)+(w_1+i_2w_2)\| \leqslant \|z_1+i_2z_2\|+\|w_1+i_2w_2\|$;
(d) $\|(z_1+i_2z_2)(w_1+i_2w_2)\| \leqslant \sqrt{2}\,\|z_1+i_2z_2\|\,\|w_1+i_2w_2\|$.
Compare this exercise with Theorems 3.2 and 4.4.

6.12 If $z_1+i_2z_2$ is in $\mathbb{C}_2$, show that

$$\|(z_1 + i_2z_2)^n\| \leqslant 2^{(n-1)/2}\|z_1 + i_2z_2\|^n, \qquad n = 1, 2, \ldots.$$

6.13 (a) If

$$z_1 + i_2z_2 = (2 + 3i_1) + i_2(4 + 5i_1),$$
$$w_1 + i_2w_2 = (1 - 2i_1) + i_2(3 - 2i_1),$$

show that

$$\|z_1+i_2z_2\| = (54)^{1/2}, \qquad \|w_1 + i_2w_2\| = (18)^{1/2},$$
$$\|(z_1 + i_2z_2)(w_1 + i_2w_2)\| = (940)^{1/2},$$

and thus that

$$0 < \|(z_1 + i_2z_2)(w_1 + i_2w_2)\| < \|z_1 + i_2z_2\|\,\|w_1 + i_2w_2\|.$$

(b) Prove that

$$|z_1 - i_1z_2|^2\,|w_1 - i_1w_2|^2 + |z_1 + i_1z_2|^2\,|w_1 + i_1w_2|^2$$
$$\leqslant |z_1 - i_1z_2|^2\,|w_1 + i_1w_2|^2 + |z_1 + i_1z_2|^2\,|w_1 - i_1w_2|^2$$

is a necessary and sufficient condition that $z_1+i_2z_2$ and $w_1+i_2w_2$ satisfy the inequality

$$\|(z_1 + i_2z_2)(w_1 + i_2w_2)\| \leqslant \|z_1 + i_2z_2\|\,\|w_1 + i_2w_2\|.$$

(c) Verify that the numbers $z_1+i_2z_2$ and $w_1+i_2w_2$ in (a) satisfy the inequality in (a) by showing that they satisfy a sufficient condition obtained by the methods explained in (b).

6.14 Theorems 3.2 and 4.4 have shown that

$$0 \leqslant \|(z_1 + i_2z_2)(w_1 + i_2w_2)\| \leqslant \sqrt{2}\,\|z_1 + i_2z_2\|\,\|w_1 + i_2w_2\|.$$

Verify that there exist many pairs of nonzero numbers $z_1+i_2z_2$ and $w_1+i_2w_2$ in $\mathbb{C}_2$ for which the equality holds on the left, and many other pairs of nonzero numbers for which the equality holds on the right. (*Hint.* Theorem 6.3.)

6.15 Let z, z_1, z_2 and w_1, w_2 denote complex numbers in $\mathbb{C}_1$. Define a system $(\mathbb{A}, +, \cdot, \times, N)$ by the following statements:

Elements in $\mathbb{A}$: $(z_1 - i_1z_2,\ z_1 + i_1z_2)$
Equals (=): $(z_1-i_1z_2,\ z_1+i_1z_2)=(w_1-i_1w_2,\ w_1+i_1w_2)$ if and only if $z_1-i_1z_2=w_1-i_1w_2$ and $z_1+i_1z_2=w_1+i_1w_2$.

Addition $(+)$: $(z_1 - i_1z_2, z_1 + i_1z_2) + (w_1 - i_1w_2, w_1 + i_1w_2)$
$= [(z_1 - i_1z_2) + (w_1 - i_1w_2), (z_1 + i_1z_2) + (w_1 + i_1w_2)]$.
Scalar multiplication $(\cdot)$: $z \cdot (z_1 - i_1z_2, z_1 + i_1z_2) = [z(z_1 - i_1z_2), z(z_1 + i_1z_2)]$.
Multiplication $(\times)$: $(z_1 - i_1z_2, z_1 + i_1z_2) \times (w_1 - i_1w_2, w_1 + i_1w_2)$
$= [(z_1 - i_1z_2)(w_1 - i_1w_2), (z_1 + i_1z_2)(w_1 + i_1w_2)]$.
Norm N: $N(z_1 - i_1z_2, z_1 + i_1z_2) = \left(\dfrac{|z_1 - i_1z_2|^2 + |z_1 + i_1z_2|^2}{2}\right)^{1/2}$.

(a) Show that the system $(\mathbb{A}, +, \cdot, \times, N)$ is a Banach algebra.
(b) Show that, under the correspondence

$$z_1 + i_2z_2 \leftrightarrow (z_1 - i_1z_2, z_1 + i_1z_2),$$

the Banach algebra $(\mathbb{C}_2, \oplus, \odot, \otimes, \| \ \|)$ and $(\mathbb{A}, +, \cdot, \times, N)$ are isomorphic and that $\|z_1 + i_2z_2\| = N(z_1 - i_1z_2, z_1 + i_1z_2)$.

7. TWO PRINCIPAL IDEALS

The purpose of this section is to define a principal ideal in an algebra and then to describe the principal ideals in $\mathbb{C}_2$ determined by the idempotent elements e_1 and e_2. The section investigates the properties of these ideals and uses them to solve polynomial equations in $\mathbb{C}_2$.

7.1 DEFINITION An *ideal* I in an algebra A is a nonempty subset of A with the following properties:

(1) If α_1 and α_2 are in I, then $\alpha_1 - \alpha_2$ is in I;

(2) If α is in I and a is in A, then $a\alpha$ is in I.

The ideal determined by an element β in A is $\{a\beta : a \in A\}$, and it is called a *principal ideal*. The principal ideals in $\mathbb{C}_2$ determined by e_1 and e_2 are denoted by I_1 and I_2 respectively; thus

(3) $I_1 = \{(z_1 + i_2z_2)e_1 : (z_1 + i_2z_2) \in \mathbb{C}_2\}$,

(4) $I_2 = \{(z_1 + i_2z_2)e_2 : (z_1 + i_2z_2) \in \mathbb{C}_2\}$.

Since $z_1 + i_2z_2 = (z_1 - i_1z_2)e_1 + (z_1 + i_1z_2)e_2$, then

(5) $(z_1 + i_2z_2)e_1 = (z_1 - i_1z_2)e_1, \qquad (z_1 + i_2z_2)e_2 = (z_1 + i_1z_2)e_2,$

and the ideals I_1, I_2 in $\mathbb{C}_2$ have in addition the following descriptions:

(6) $I_1 = \{(z_1 - i_1z_2)e_1 : z_1, z_2 \in \mathbb{C}_2\}$,

(7) $I_2 = \{(z_1 + i_1z_2)e_2 : z_1, z_2 \in \mathbb{C}_2\}$.

7.2 THEOREM The ideals I_1, I_2 are linear subspaces in $\mathbb{C}_2$ which have the single element 0 in common:

(8) $\quad I_1 \cap I_2 = \{0\}.$

Proof. If all elements in $\mathbb{C}_2$ are multiplied by a fixed element ζ_0 in $\mathbb{C}_2$, the set $\mathbb{C}_2$ is transformed into a subset of $\mathbb{C}_2$. This transformation is a linear transformation since

(9) $\quad \zeta_0(z_1\zeta_1 + z_2\zeta_2) = z_1(\zeta_0\zeta_1) + z_2(\zeta_0\zeta_2).$

Then by the definitions of I_1 and I_2 in (3) and (4), these ideals are linear subspaces of $\mathbb{C}_2$. To prove (8), let $(z_1 - i_1z_2)e_1$ in I_1 equal $(w_1 + i_1w_2)e_2$ in I_2 [see (6) and (7)]. Thus

(10) $\quad (z_1 - i_1z_2)e_1 = (w_1 + i_1w_2)e_2.$

Since, by Theorem 6.3 (8), e_1 and e_2 are linearly independent with respect to complex constants of combinations, then $z_1 - i_1z_2 = 0$ and $w_1 + i_1w_2 = 0$. Therefore, 0 is the one and only element in $I_1 \cap I_2$, and the proof of Theorem 7.2 is complete. □

7.3 THEOREM The product of a number in I_1 and a number in I_2 is zero. Two elements in $\mathbb{C}_2$ are divisors of zero if and only if one is in $I_1 - \{0\}$ and the other is in $I_2 - \{0\}$.

Proof. Let $(z_1 - i_1z_2)e_1$ and $(w_1 + i_1w_2)e_2$ be elements in I_1 and I_2 respectively [see (6) and (7)]. Then

(11) $\quad (z_1 - i_1z_2)e_1(w_1 + i_1w_2)e_2 = (z_1 - i_1z_2)(w_1 + i_1w_2)e_1e_2 = 0,$

and the first statement in the theorem is true. If $(z_1 - i_1z_2)e_1 \neq 0$ and $(w_1 + i_1w_2)e_2 \neq 0$, then these two numbers are divisors of zero. To show that there are no other divisors of zero, let $z_1 + i_2z_2$ and $w_1 + i_2w_2$ be two elements in $\mathbb{C}_2$ such that

(12) $\quad z_1 + i_2z_2 \neq 0, \qquad w_1 + i_2w_2 \neq 0, \qquad (z_1 + i_2z_2)(w_1 + i_2w_2) = 0.$

By Theorem 6.6 (16),

(13) $$(z_1 + i_2z_2)(w_1 + i_2w_2)$$
$$= (z_1 - i_1z_2)(w_1 - i_1w_2)e_1 + (z_1 + i_1z_2)(w_1 + i_1w_2)e_2.$$

Since $(z_1 + i_2z_2)(w_1 + i_2w_2) = 0$, then

(14) $\quad (z_1 - i_1z_2)(w_1 - i_1w_2) = 0, \qquad (z_1 + i_1z_2)(w_1 + i_1w_2) = 0.$

Equations (14) and (12) show that one of the following two cases holds:

(15) $\quad (z_1 - i_1z_2) \neq 0, \qquad (z_1 + i_1z_2) = 0, \qquad (w_1 - i_1w_2) = 0,$

$(w_1 + i_1w_2) \neq 0;$

$$(16) \qquad (z_1 - i_1 z_2) = 0, \qquad (z_1 + i_1 z_2) \neq 0, \qquad (w_1 - i_1 w_2) \neq 0,$$
$$(w_1 + i_1 w_2) = 0.$$

If (15) holds, then $z_1 + i_2 z_2$ is in $I_1 - \{0\}$ by (6) since $z_1 + i_2 z_2 = (z_1 - i_1 z_2)e_1$, and $w_1 + i_2 w_2$ is in $I_2 - \{0\}$ by (7) since $w_1 + i_2 w_2 = (w_1 + i_1 w_2)e_2$. If (16) holds, similar considerations show that $z_1 + i_2 z_2$ is in $I_2 - \{0\}$ and $w_1 + i_2 w_2$ is in $I_1 - \{0\}$. In both cases, $z_1 + i_2 z_2$ and $w_1 + i_2 w_2$ are divisors of zero. The proof of Theorem 7.3 is complete. □

7.4 THEOREM An element $z_1 + i_2 z_2$ in $\mathbb{C}_2$ is singular if and only if $z_1 + i_2 z_2 \in I_1 \cup I_2$; it is nonsingular if and only if $z_1 + i_2 z_2 \notin I_1 \cup I_2$.

Proof. By Corollary 4.9, $z_1 + i_2 z_2$ is singular if and only if $|z_1 - i_1 z_2|$ $|z_1 + i_1 z_2| = 0$. Since $z_1 + i_2 z_2 = (z_1 - i_1 z_2)e_1 + (z_1 + i_1 z_2)e_2$, equations (6) and (7) show that $z_1 + i_2 z_2$ is singular if and only if $z_1 + i_2 z_2 \in I_1 \cup I_2$. Similarly, $z_1 + i_2 z_2$ is nonsingular if and only if $z_1 + i_2 z_2 \notin I_1 \cup I_2$. □

7.5 COROLLARY If $\mathcal{O}_2$ is the set of singular elements in $\mathbb{C}_2$ as defined at the end of Section 4, then

$$(17) \qquad \mathcal{O}_2 = I_1 \cup I_2.$$

An element in the complement of $\mathcal{O}_2$, that is, a nonsingular element, is often called a regular element in $\mathbb{C}_2$.

7.6 THEOREM If $\zeta : (x_1 + i_1 x_2 + i_2 x_3 + i_1 i_2 x_4)$ denotes an element in $\mathbb{C}_2$, then

$$(18) \qquad I_1 = \{x_1 + i_1 x_2 + i_2 x_3 + i_1 i_2 x_4 : x_1 - x_4 = 0 \text{ and } x_2 + x_3 = 0\},$$
$$(19) \qquad I_2 = \{x_1 + i_1 x_2 + i_2 x_3 + i_1 i_2 x_4 : x_1 + x_4 = 0 \text{ and } x_2 - x_3 = 0\}.$$

Proof. By the idempotent representation,

$$(20) \qquad (x_1 + i_1 x_2) + i_2(x_3 + i_1 x_4)$$
$$= [(x_1 + i_1 x_2) - i_1(x_3 + i_1 x_4)]e_1 + [(x_1 + i_1 x_2)$$
$$+ i_1(x_3 + i_1 x_4)]e_2.$$

Now (16) shows that ζ is in I_1 if and only if $(x_1 + i_1 x_2) + i_1(x_3 + i_1 x_4) = 0$, that is, if and only if $x_1 - x_4 = 0$ and $x_2 + x_3 = 0$. Thus (18) is true. Similarly, (7) shows that ζ is in I_2 if and only if $(x_1 + i_1 x_2) - i_1(x_3 + i_1 x_4) = 0$, that is, if and only if $x_1 + x_4 = 0$ and $x_2 - x_3 = 0$. Thus (19) is true, and the proof is complete. □

Theorem 7.6 provides an easy proof that $I_1 \cap I_2 = \{0\}$, for if ζ is in I_1 and in I_2, then

$$(21) \qquad x_1 - x_4 = 0, \quad x_2 + x_3 = 0, \qquad x_1 + x_4 = 0, \qquad x_2 - x_3 = 0.$$

The only solution of these four equations is $x_i = 0$, $i = 1, \ldots, 4$. Thus $\zeta = 0$, and (8) is true.

7.7 THEOREM The ideals I_1, I_2 are closed sets in $\mathbb{C}_2$; the set $\mathcal{O}_2$ of singular elements in $\mathbb{C}_2$ is closed in $\mathbb{C}_2$; and the set of regular elements (the complement of $\mathcal{O}_2$) is an open set in $\mathbb{C}_2$. Every point in $\mathcal{O}_2$ is a limit point of the set of regular elements.

Proof. Each of the planes $x_1 - x_4 = 0$ and $x_2 + x_3 = 0$ is a closed set in $\mathbb{C}_0^4$. Since the intersection of two closed sets is a closed set, (18) shows that I_1 is closed. Similar arguments show that I_2 is closed. Since the union of two closed sets is closed, $I_1 \cup I_2$ is closed, and $\mathcal{O}_2$ is closed by (17). Then the set of regular elements, the complement of $\mathcal{O}_2$, is an open set. Finally, if ζ is in $I_1 \cup I_2$, then every neighborhood of ζ contains points $x_1 + i_1 x_2 + i_2 x_3 + i_1 i_2 x_4$ which are not in $I_1 \cup I_2$ by (18) and (19); that is, every neighborhood of ζ contains regular elements, and ζ is a limit point of the set of regular elements in $\mathbb{C}_2$. The proof of Theorem 7.7 is complete. □

The existence of divisors of zero in $\mathbb{C}_2$ but not in $\mathbb{C}_1$ is one of the significant differences between the algebras $\mathbb{C}_1$ and $\mathbb{C}_2$, and the solution of polynomial equations emphasizes this difference. Exercise 6.5 outlines a proof that the equation

$$(22) \qquad \sum_{k=0}^{n} (a_k + i_2 b_k)(z_1 + i_2 z_2)^k = 0, \qquad a_n^2 + b_n^2 \neq 0,$$

has n^2 solutions. A further examination of this equation is instructive. Assume that (22) has the following n roots.

$$(23) \qquad r_k + i_2 s_k, \qquad k = 1, \ldots, n.$$

Then the remainder theorem and the factor theorem show that (22) can be given the form

$$(24) \qquad (a_n + i_2 b_n) \prod_{k=1}^{n} [(z_1 + i_2 z_2) - (r_k + i_2 s_k)] = 0.$$

Since $a_n^2 + b_n^2 \neq 0$ by (22), then $a_n + i_2 b_n$ is nonsingular, and (24) is equivalent to

$$(25) \qquad \prod_{k=1}^{n} [(z_1 + i_2 z_2) - (r_k + i_2 s_k)] = 0.$$

This form of the equation displays n of its roots, and it is not immediately apparent how the remaining n^2-n roots are to be found. Theorem 7.3 supplies the answer. There are exactly two ways in which a value for $z_1+i_2z_2$ causes the polynomial in (25) to vanish; they are the following: (a) the value of $z_1+i_2z_2$ makes one of the factors equal to zero; and (b) the value of $z_1+i_2z_2$ makes one factor equal to an element in I_1 and another factor equal to an element in I_2. These facts will now be used to find the n^2 roots of the polynomial equation $P(z_1+i_2z_2)=0$ in (25). Assume that this equation has n distinct roots as follows:

(26) $\quad r_k + i_2s_k = (r_k - i_1s_k)e_1 + (r_k + i_1s_k)e_2;$

(27) $\quad r_p - i_1s_p \neq r_q - i_1s_q, \qquad r_p + i_1s_p \neq r_q + i_1s_q,$

$\quad p \neq q,\ p, q = 1, \ldots, n.$

7.8 THEOREM If the equation $P(z_1+i_2z_2)=0$ in (25) has n distinct roots which satisfy (27), then it has n^2 roots; they are

(28) $\quad (r_p - i_1s_p)e_1 + (r_q + i_1s_q)e_2, \qquad p, q = 1, \ldots, n,$

and these n^2 roots are distinct.

Begin the proof of this theorem by first proving the following lemma.

7.9 LEMMA If $P(z_1+i_2z_2)=0$ in (25) has two roots

(29) $\quad r_p + i_2s_p, \qquad r_q + i_2s_q,$

which satisfy (27), then

(30) $\quad (r_p - i_1s_p)e_1 + (r_q + i_1s_q)e_2,$

(31) $\quad (r_q - i_1s_q)e_1 + (r_p + i_1s_q)e_2,$

are also two roots of $P(z_1+i_2z_2)=0$; they are distinct and distinct from the roots in (29).

Proof. To prove the lemma, show that there is a unique value of $z_1+i_2z_2$ such that

(32) $\quad [(z_1 + i_2z_2) - (r_p + i_2s_p)] \in I_1 - \{0\},$

$\quad [(z_1 + i_2z_2) - (r_q + i_2s_q)] \in I_2 - \{0\},$

and also a unique value for $z_1+i_2z_2$ such that

(33) $\quad (z_1 + i_2z_2) - (r_p + i_2s_p) \in I_2 - \{0\},$

$\quad (z_1 + i_2z_2) - (r_q + i_2s_q) \in I_1 - \{0\}.$

By (6) and (7), $z_1+i_2z_2$ satisfies (32) if and only if there exist numbers w_p and w_q in $\mathbb{C}_1$ such that

$$(z_1+i_2z_2)-(r_p+i_2s_p)=w_pe_1, \qquad w_p\neq 0, \tag{34}$$

$$(z_1+i_2z_2)-(r_q+i_2s_q)=w_qe_2, \qquad w_q\neq 0. \tag{35}$$

The unknown quantities in these equations are z_1, z_2, w_p, and w_q; equations (34) and (35) are equivalent to the following equations:

$$z_1-r_p=\frac{w_p}{2}, \qquad z_2-s_p=\frac{i_1w_p}{2}, \tag{36}$$

$$z_1-r_q=\frac{w_q}{2}, \qquad z_2-s_q=-\frac{i_1w_q}{2}.$$

These equations are linear in z_1, z_2, w_p, and w_q. Since the determinant of their matrix of coefficients is $i_1/2$, they have the following unique solution:

$$z_1=\frac{(r_p+r_q)+i_1(s_p-s_q)}{2}, \qquad z_2=\frac{-i_1(r_p-r_q)+(s_p+s_q)}{2}, \tag{37}$$

$$w_p=-(r_p-r_q)+i_1(s_p-s_q), \qquad w_q=(r_p-r_q)+i_1(s_p-s_q).$$

Because $r_p+i_2s_p$ and $r_q+i_2s_q$ are roots of (25) which satisfy (27), the last two equations in (37) show that $w_p\neq 0$ and $w_q\neq 0$. Thus there exists a unique value for $z_1+i_2z_2$ which satisfies (34) and (35); this value is a root of the polynomial equation (25) by Theorem 7.3. Since by (37), (34), and (35),

$$z_1+i_2z_2=(r_q-i_1s_q)e_1+(r_p+i_1s_p)e_2, \tag{38}$$

this root is different from the two roots in (29). In the same way, (6) and (7) show that $z_1+i_2z_2$ satisfies (33) if and only if there exist elements w_p and w_q in $\mathbb{C}_1$ such that

$$(z_1+i_2z_2)-(r_p+i_2s_p)=w_pe_2, \qquad w_p\neq 0, \tag{39}$$

$$(z_1+i_2z_2)-(r_q+i_2s_q)=w_qe_1, \qquad w_q\neq 0. \tag{40}$$

As before, the unknown quantities in these equations are z_1, z_2, w_p, and w_q; equations (39) and (40) are equivalent to the following equations:

$$z_1-r_p=\frac{w_p}{2}, \qquad z_2-s_p=-\frac{i_1w_p}{2}, \tag{41}$$

$$z_1-r_q=\frac{w_q}{2}, \qquad z_2-s_q=\frac{i_1w_q}{2}.$$

These equations are linear in z_1, z_2, w_p, and w_q; the determinant of their matrix of coefficients is $-i_1/2$; and their unique solution is

$$z_1=\frac{(r_p+r_q)-i_1(s_p-s_q)}{2}, \qquad z_2=\frac{i_1(r_p-r_q)+(s_p+s_q)}{2}, \tag{42}$$

$$w_p = (r_p - r_q) - i_1(s_p - s_q), \qquad w_q = -(r_p - r_q) - i_1(s_p - s_q).$$

Because $r_p + i_2 s_p$ and $r_q + i_2 s_q$ are roots of (25) which satisfy (27), the last two equations in (42) show that $w_p \neq 0$ and $w_q \neq 0$. Thus there exists a unique value for $z_1 + i_2 z_2$ which satisfies (39) and (40) and hence (33); this value is a root of the polynomial equation (25) by Theorem 7.3. Since by (42),

$$(43) \qquad z_1 + i_2 z_2 = (r_p - i_1 s_p)e_1 + (r_q + i_1 s_q)e_2,$$

and since the roots in (29) satisfy (27) by hypothesis, the root in (43) is distinct from the one in (38) and from those in (29). Thus the polynomial equation $P(z_1 + i_2 z_2) = 0$ has the two roots in (30) and (31) [see (38) and (43)] in addition to the two roots in (29), and the four roots are distinct. The proof of Lemma 7.9 is complete. □

Proof of Theorem 7.8. The equation $P(z_1 + i_2 z_2) = 0$ has, by hypothesis, the following n roots,

$$(44) \qquad (r_k - i_1 s_k)e_1 + (r_k + i_1 s_k)e_2, \qquad k = 1, \ldots, n;$$

they satisfy (27). Lemma 7.9 shows that, corresponding to each distinct pair $(r_p + i_2 s_p)$, $(r_q + i_2 s_q)$ of these roots, the equation $P(z_1 + i_2 z_2) = 0$ has two additional roots as follows:

$$(45) \qquad (r_p - i_1 s_p)e_1 + (r_q + i_1 s_q)e_2,$$

$$(46) \qquad (r_q - i_1 s_q)e_1 + (r_p + i_1 s_p)e_2, \qquad p \neq q.$$

Since a pair of roots can be selected from the n roots in (44) in $n(n-1)/2$ ways, and since two roots can be constructed from each pair, there are $n(n-1)$ roots of the form shown in (45) and (46). Thus the total number of roots is $n + n(n-1)$ or n^2. The roots in (44), (45), and (46) can be described as follows:

$$(47) \qquad (r_p - i_1 s_p)e_1 + (r_q + i_1 s_q)e_2, \qquad p, q = 1, \ldots, n.$$

Since the n roots in (26) satisfy (27) by hypothesis, the n^2 roots in (47) are distinct. The proof of Theorem 7.8 is complete. □

Exercises

7.1 Establish each of the following statements in two ways:

(a) $3 + 7i_1 - 7i_2 + 3i_1 i_2$ and $(a - i_1 b) + i_2(b + i_1 a)$, $a - i_1 b \neq 0$, are in $I_1 - \{0\}$.

(b) $7 + 4i_1 + 4i_2 - 7i_1 i_2$ and $(a + i_1 b) + i_2(b - i_1 a)$, $a + i_1 b \neq 0$, are in $I_2 - \{0\}$.

(c) The product of an element in (a) and an element in (b) is zero.

(d) $\|3 + 7i_1 - 7i_2 + 3i_1 i_2\| = 116^{1/2}$;
$\|(a + i_1 b) + i_2(b - i_1 a)\| = [2(a^2 + b^2)]^{1/2}$.

7.2 Show that the following equation is satisfied by every $z_1+i_2z_2$ in I_2:

$$\sum_{k=1}^{n}(a_k+i_2b_k)(z_1+i_2z_2)^k=0,$$

$(a_k+i_2b_k)\in I_1$ for $k=1,\ldots,n$.

7.3 Prove the following fundamental theorem of algebra: Every polynomial equation

$$\sum_{k=0}^{n}(a_k+i_2b_k)(z_1+i_2z_2)^k=0, \qquad a_n^2+b_n^2\neq 0, \quad n\geqslant 1,$$

has at least one root in $\mathbb{C}_2$. (*Hint.* Exercise 6.5.)

7.4 Prove the following Remainder Theorem: Let $P(z_1+i_2z_2)$ denote the polynomial

$$\sum_{k=0}^{n}(a_k+i_2b_k)(z_1+i_2z_2)^k=0, \qquad a_n^2+b_n^2\neq 0,\ n\geqslant 1,$$

and let $Q(z_1+i_2z_2)$ and R be the quotient and constant remainder obtained by dividing $P(z_1+i_2z_2)$ by $(z_1+i_2z_2)-(r_1+i_2r_2)$. Then

(a) $P(z_1+i_2z_2)=[(z_1+i_2z_2)-(r_1+i_2r_2)]\ Q(z_1+i_2z_2)+R$;

(b) $Q(z_1+i_2z_2)$ is a polynomial of degree $n-1$ whose leading coefficient is $a_n+i_2b_n$;

(c) $R=P(r_1+i_2r_2)$.

7.5 Prove the following factor theorem: If $P(z_1+i_2z_2)$ is the polynomial in Exercise 7.4, and if $P(r_1+i_2r_2)=0$, then $[(z_1+i_2z_2)-(r_1+i_2r_2)]$ is a factor of $P(z_1+i_2z_2)$. Thus if $P(r_1+i_2r_2)=0$, then

$$P(z_1+i_2z_2)=[(z_1+i_2z_2)-(r_1+i_2r_2)]\ Q(z_1+i_2z_2).$$

7.6 Let $P(z_1+i_2z_2)$ be the polynomial in Exercise 7.4. Prove that $P(z_1+i_2z_2)$ can be factored as follows:

$$P(z_1+i_2z_2)=(a_n+i_2b_n)\prod_{k=1}^{n}[(z_1+i_2z_2)-(r_k+i_2s_k)].$$

7.7 Let $P(z_1+i_2z_2)$ be the polynomial $(z_1+i_2z_2)^2-5(z_1+i_2z_2)+6$.

(a) Show that

$$2, \quad 3, \quad \frac{5-i_1i_2}{2}, \quad \frac{5+i_1i_2}{2},$$

are four roots of $P(z_1+i_2z_2)=0$.

(b) Use the factor theorem in Exercise 7.5 to show that $P(z_1+i_2z_2)$ can be factored into linear factors in two essentially different ways as follows:

$$P(z_1+i_2z_2)=[(z_1+i_2z_2)-2][(z_1+i_2z_2)-3];$$

$$P(z_1+i_2z_2)=\left[(z_1+i_2z_2)-\left(\frac{5-i_1i_2}{2}\right)\right]\left[(z_1+i_2z_2)-\left(\frac{5+i_1i_2}{2}\right)\right].$$

7.8 Let $P(z_1+i_2z_2)$ be the polynomial

$$\sum_{k=0}^{n}(a_k+i_2b_k)(z_1+i_2z_2)^k=0, \qquad a_n^2+b_n^2\neq 0,\ n\geqslant 1.$$

(a) Use the fundamental theorem of algebra, the factor theorem, and the methods used to prove Theorem 7.8 to show that the equation $P(z_1+i_2z_2)=0$ has n^2 roots (which may not all be distinct).

(b) Use the factor theorem to prove that $P(z_1+i_2z_2)$ can be factored into linear factors as follows:

$$P(z_1+i_2z_2)=(a_n+i_2b_n)\prod_{k=1}^{n}[(z_1+i_2z_2)-(r_k+i_2s_k)].$$

(c) Assume that no two of the roots of $P(z_1+i_2z_2)=0$ are equal. Prove that $P(z_1+i_2z_2)$ can be factored into linear factors in $n!$ essentially different ways. [*Hint.* The equation $P(z_1+i_2z_2)=0$ has a root $r_1+i_2s_1$ by the fundamental theorem of algebra. Then

$$P(z_1+i_2z_2)=[(z_1+i_2z_2)-(r_1+i_2s_1)]\ Q_1(z_1+i_2z_2).$$

The equation $Q_1(z_1+i_2z_2)=0$ has a root $r_2+i_2s_2$, and

$$\begin{aligned}&P(z_1+i_2z_2)\\&\quad=[(z_1+i_2z_2)-(r_1+i_2s_1)][(z_1+i_2z_2)\\&\qquad-(r_2+i_2s_2)]\ Q_2(z_1+i_2z_2).\end{aligned}$$

A continuation of this process shows that $P(z_1+i_2z_2)=0$ has n roots, and then the methods used in proving Theorem 7.8 show that it has n^2 roots. To factor $P(z_1+i_2z_2)$, use any one of the $(n-1)^2$ roots of $Q_1(z_1+i_2z_2)=0$ for the second factor, and so on. This process constructs $n^2(n-1)^2\cdots 2^21^2$, or $(n!)^2$, strings of factors. Since each set of n factors can be arranged in $n!$ different orders, there are $(n!)^2/n!$, or $n!$, essentially different ways to factor $P(z_1+i_2z_2)$. Compare Exercise 7.7.]

7.9 (a) Show that the following equation [a special case of (22)] has n roots in $\mathbb{C}_1$:

$$\sum_{k=0}^{n}a_k(z_1+i_2z_2)^k=0, \qquad a_k\in\mathbb{C}_1,\ k=0,1,\ldots,n,\ a_n\neq 0.$$

(b) Assume that no two of the complex roots in (a) are equal. Prove that the equation has $n(n-1)$ distinct bicomplex roots which are in $\mathbb{C}_2$ but not in $\mathbb{C}_1$.

(c) Prove that the n^2 distinct roots of the equation in (a) occur in i_2-conjugate bicomplex pairs. Compare Exercise 7.7. (*Hint.* Exercise 6.2. The i_2-conjugate bicomplex number of a number in $\mathbb{C}_1$ is the number itself.)

7.10 (a) Show that the following equation [a special case of (22)] has n roots in $\mathbb{C}_1$:

$$\sum_{k=0}^{n} a_k(z_1+i_2z_2)^k=0, \qquad a_k \text{ in } \mathbb{C}_0,\ k=0, 1, \ldots, n,\ a_n \neq 0.$$

(b) Show that, in (a), the complex roots in $\mathbb{C}_1$ occur in conjugate complex pairs. (*Hint.* The conjugate complex number of a number in $\mathbb{C}_0$ is the number itself.)

(c) Assume that the n complex roots of the equation in (a) are distinct numbers in $\mathbb{C}_1$. Prove that the equation has $n(n-1)$ distinct bicomplex roots which are not in $\mathbb{C}_1$.

(d) Prove that the roots of the equation in (a) occur in i_1i_2-conjugate bicomplex pairs. (*Hint.* Exercise 6.3.)

8. THE AUXILIARY COMPLEX SPACES

Define the complex spaces A_1, A_2 as follows:

(1) $$A_1 = \{z_1 - i_1z_2\colon z_1 \text{ and } z_2 \text{ in } \mathbb{C}_1\},$$
$$A_2 = \{z_1 + i_1z_2\colon z_1 \text{ and } z_2 \text{ in } \mathbb{C}_1\}.$$

Since each element in $\mathbb{C}_1$ can be represented in the form $z_1-i_1z_2$ and $z_1+i_1z_2$ (and in many ways), the elements in A_1 and A_2 are the same as the elements in $\mathbb{C}_1$. Nevertheless, because of the special representations $z_1-i_1z_2$ and $z_1+i_1z_2$, the special notation A_1 and A_2 is convenient. The idempotent representation $(z_1-i_1z_2)e_1+(z_1+i_1z_2)e_2$ associates with each point $z_1+i_2z_2$ in $\mathbb{C}_2$ the points $z_1-i_1z_2$ and $z_1+i_1z_2$ in A_1 and A_2 respectively, and to each pair of points $(z_1-i_1z_2, z_1+i_1z_2)$ in $A_1 \times A_2$ there corresponds a unique point in $\mathbb{C}_2$. Define functions $h_1: \mathbb{C}_2 \to A_1$, $h_2: \mathbb{C}_2 \to A_2$, and $H: A_1 \times A_2 \to \mathbb{C}_2$ as follows:

(2) $$h_1(z_1 + i_2z_2) = z_1 - i_1z_2, \qquad z_1 + i_2z_2 \text{ in } \mathbb{C}_2,\ z_1 - i_1z_2 \text{ in } A_1;$$
$$h_2(z_1 + i_2z_2) = z_1 + i_1z_2, \qquad z_1 + i_2z_2 \text{ in } \mathbb{C}_2,\ z_1 + i_1z_2 \text{ in } A_2;$$
$$H(z_1 - i_1z_2, z_1 + i_1z_2)$$
$$= (z_1 - i_1z_2)e_1 + (z_1 + i_1z_2)e_2, \qquad (z_1 - i_1z_2, z_1 + i_1z_2) \in A_1 \times A_2.$$

The purpose of this section is to establish the properties of these functions or mappings.

The functions h_1, h_2, restricted to a set X in $\mathbb{C}_2$, map X into sets X_1, X_2 in A_1, A_2 respectively; the function H, restricted to a set A in $A_1 \times A_2$, maps A into a set Y in $\mathbb{C}_2$. Thus

$$(3) \qquad \begin{aligned} &h_1(X) = X_1, \qquad X \subset \mathbb{C}_2,\ X_1 \subset A_1; \\ &h_2(X) = X_2, \qquad X \subset \mathbb{C}_2,\ X_2 \subset A_2; \\ &H(A) = Y, \qquad A \subset A_1 \times A_2,\ Y \subset \mathbb{C}_2. \end{aligned}$$

8.1 DEFINITION Let R and S be rings. A function $h: R \to S$, $u \mapsto h(u)$, is called a *homomorphism* if and only if

$$(4) \qquad h(u + v) = h(u) + h(v), \qquad u, v \text{ in } R,$$

$$(5) \qquad h(uv) = h(u)h(v).$$

8.2 THEOREM The mapping $h_1: \mathbb{C}_2 \to A_1$ is a homomorphism which maps I_2 into $\{0\}$ in A_1, and $h_2: \mathbb{C}_2 \to A_2$ is a homomorphism which maps I_1 into $\{0\}$ in A_2.

Proof. By Theorem 6.6,

$$(6) \qquad \begin{aligned} &(z_1 + i_2 z_2) + (w_1 + i_2 w_2) \\ &\quad = [(z_1 - i_1 z_2) + (w_1 - i_1 w_2)]e_1 \\ &\qquad + [(z_1 + i_1 z_2) + (w_1 + i_1 w_2)]e_2, \end{aligned}$$

$$(7) \qquad \begin{aligned} &(z_1 + i_2 z_2)(w_1 + i_2 w_2) \\ &\quad = [(z_1 - i_1 z_2)(w_1 - i_1 w_2)]e_1 \\ &\qquad + [(z_1 + i_1 z_2)(w_1 + i_1 w_2)]e_2. \end{aligned}$$

Then by (2),

$$(8) \qquad \begin{aligned} h_1[(z_1 + i_2 z_2) + (w_1 + i_2 w_2)] &= (z_1 - i_1 z_2) + (w_1 - i_1 w_2) \\ &= h_1(z_1 + i_2 z_2) + h_1(w_1 + i_2 w_2), \\ h_1[(z_1 + i_2 z_2)(w_1 + i_2 w_2)] &= (z_1 - i_1 z_2)(w_1 - i_1 w_2) \\ &= h_1(z_1 + i_2 z_2) h_1(w_1 + i_2 w_2). \end{aligned}$$

Thus h_1 is a homomorphism by Definition 8.1. In the same way, (6) and (7) and the definition of h_2 in (2) show that h_2 is a homomorphism. The proof will be completed by showing that

$$(9) \qquad h_1(I_2) = \{0\},$$

$$(10) \qquad h_2(I_1) = \{0\}.$$

If $z_1+i_2z_2$ is in I_2, then (7) in Section 7 shows that $z_1-i_1z_2=0$; then (2) shows that $h_1(z_1+i_2z_2)=0$, and (9) is true. In the same way, if $z_1+i_2z_2$ is in I_1, then $z_1+i_1z_2=0$ by (6) in Section 7, and $h_2(z_1+i_2z_2)=0$ by (2). Thus (10) is true, and the proof of Theorem 8.2 is complete. □

Theorem 8.2 emphasizes that the mappings $h_1:\mathbb{C}_2\to A_1$ and $h_2:\mathbb{C}_2\to A_2$ are many-to-one mappings. Nevertheless, h_1 and h_2 map each element $z_1+i_2z_2$ into a unique pair of elements $(z_1-i_1z_2, z_1+i_1z_2)$ in $A_1\times A_2$. Thus given a point $z_1+i_2z_2$ in $\mathbb{C}_2$, the equations

$$(11)\qquad z_1-i_1z_2=w_1,\qquad z_1+i_1z_2=w_2,$$

define a unique element (w_1,w_2) in $A_1\times A_2$ which corresponds to $z_1+i_2z_2$ in $\mathbb{C}_2$. Furthermore, given (w_1,w_2) in $A_1\times A_2$, the equations (11) have the unique solution

$$(12)\qquad z_1=\frac{w_1+w_2}{2},\qquad z_2=\frac{i_1(w_1-w_2)}{2},$$

which defines the unique point $z_1+i_2z_2$ in $\mathbb{C}_2$ which corresponds to the element (w_1,w_2) in $A_1\times A_2$.

Let X be a set in $\mathbb{C}_2$. The restrictions $h_1|_X:X\to A_1$ and $h_2|_X:X\to A_2$ map X into sets X_1 and X_2 [see (3)] as follows:

$$(13)\qquad X_1=\{w_1\in A_1: w_1=h_1(z_1+i_2z_2),\ z_1+i_2z_2\in X\},$$

$$(14)\qquad X_2=\{w_2\in A_2: w_2=h_2(z_1+i_2z_2),\ z_1+i_2z_2\in X\}.$$

An understanding of the relation between X and the pair X_1, X_2 is important for later work, and several examples will provide an introduction to the study of these sets.

8.3 EXAMPLE Let X be the set $\{(z_1^k+i_2z_2^k)$ in $\mathbb{C}_2$: $k=1,\dots,n\}$ such that

$$(15)\qquad z_1^k+i_2z_2^k=(z_1^k-i_1z_2^k)e_1+(z_1^k+i_1z_2^k)e_2,$$

$$(16)\qquad z_1^p-i_1z_2^p\neq z_1^q-i_1z_2^q,\qquad z_1^p+i_1z_2^p\neq z_1^q+i_1z_2^q,\qquad p\neq q,\ p,q=1,\dots,n.$$

Then

$$(17)\quad h_1(z_1^k+i_2z_2^k)=z_1^k-i_1z_2^k,\qquad h_2(z_1^k+i_2z_2^k)=z_1^k+i_1z_2^k,\qquad k=1,\dots,n,$$

$$(18)\quad X_1=\{(z_1^k-i_1z_2^k)\in A_1: k=1,\dots,n\},\quad X_2=\{(z_1^k+i_1z_2^k)\in A_2: k=1,\dots,n\}.$$

In this case $h_1|_X$ and $h_2|_X$ are one-to-one mappings of X into A_1 and A_2 respectively. The cartesian product of X_1 and X_2 is

$$(19)\qquad \{(z_1^p-i_1z_2^p,\ z_1^q+i_1z_2^q): p,q=1,\dots,n\}.$$

Now H maps $X_1 \times X_2$ into a set in $\mathbb{C}_2$, but this set is not X since

(20) $X = \{H(z_1^k - i_1 z_2^k, z_1^k + i_1 z_2^k) \text{ in } \mathbb{C}_2: k = 1, \ldots, n\}.$

It is necessary to know how the points in X_1 are paired with the points in X_2 in order to construct X. As (20) shows, X is the image under H of a proper subset of $X_1 \times X_2$.

8.4 EXAMPLE Let X be the following set of elements in $\mathbb{C}_2$:

(21) $(z_1^p - i_1 z_2^p)e_1 + (z_1^q + i_1 z_2^q)e_2, \qquad p, q = 1, \ldots, n,$

(22) $(z_1^p - i_1 z_2^p) \neq (z_1^q - i_1 z_2^q), \qquad (z_1^p + i_1 z_2^p) \neq (z_1^q + i_1 z_2^q), \qquad p, q = 1, \ldots, n.$

Then

(23) $h_1[(z_1^p - i_1 z_2^p)e_1 + (z_1^q + i_1 z_2^q)e_2] = z_1^p - i_1 z_2^p, \qquad p, q = 1, \ldots, n.$

(24) $h_2[(z_1^p - i_1 z_2^p)e_1 + (z_1^q + i_1 z_2^q)e_2] = z_1^q + i_1 z_2^q, \qquad p, q = 1, \ldots, n.$

In this case, $h_1|_X$ and $h_2|_X$ are n-to-1 mappings of X into A_1 and A_2, respectively. Because of the special nature of X, it can be reconstructed easily from X_1 and X_2. Since

(25) $$\begin{aligned} X &= \{(z_1^p - i_1 z_2^p)e_1 + (z_1^q + i_1 z_2^q)e_2: p, q = 1, \ldots, n\} \\ &= \{H(z_1^p - i_1 z_2^p, z_1^q + i_1 z_2^q): p, q = 1, \ldots, n\}, \end{aligned}$$

then X is the image under H of the cartesian product $X_1 \times X_2$.

8.5 EXAMPLE Let X_1, X_2 be given sets of elements w_1, w_2 in A_1, A_2, respectively. Set

(26) $X = \{z_1 + i_2 z_2 \text{ in } \mathbb{C}_2: z_1 + i_2 z_2 = w_1 e_1 + w_2 e_2, w_1 \in X_1, w_2 \in X_2\}.$

In this case, X is the image under H [see (2)] of the set

(27) $\{(w_1, w_2): (w_1, w_2) \in X_1 \times X_2\}.$

Thus X is the image under H of the cartesian set $X_1 \times X_2$.

8.6 THEOREM Let X_1, X_2 be sets in A_1, A_2 which have more than one point each, and let X be the set in $\mathbb{C}_2$ such that $X = H(X_1, X_2)$. Then

(28) each of the mappings $h_1|_X : X \to X_1$ and $h_2|_X : X \to X_2$ is a many-to-one mapping;

and

(29) there is a one-to-one correspondence between points $z_1 + i_2 z_2$ in X and pairs of points (w_1, w_2) in the cartesian product $X_1 \times X_2$.

Proof. To prove (28), let $a-i_1b$ be a fixed point in X_1, and let w_2 be a variable point in X_2. Set

$$z_1 + i_2z_2 = (a - i_1b)e_1 + w_2e_2. \tag{30}$$

Then $z_1+i_2z_2$ is in X, and $h_1(z_1+i_2z_2)=a-i_1b$. Thus corresponding to each point w_2 in X_2 there is a point $z_1+i_2z_2$ in X such that $h_1(z_1+i_2z_2)=a-i_1b$. Since X_2 has more than one point by hypothesis, $h_1|_X$ is a many-to-one mapping. Similar arguments show that $h_2|_X$ is a many-to-one mapping, and (28) is true. To prove (29), observe first that to each point $z_1+i_2z_2$ in X there corresponds a unique pair (w_1, w_2) in $X_1 \times X_2$ by (2). To complete the proof, show as follows that $(h_1|_X, h_2|_X): X \to X_1 \times X_2$ has an inverse. If $(w_1, w_2) \in X_1 \times X_2$, then

$$w_1 = z_1 - i_1z_2, \qquad w_2 = z_1 + i_1z_2. \tag{31}$$

Since these are the linear equations (11) which have the unique solution (12), there is a unique point $z_1+i_2z_2$ which corresponds to (w_1, w_2) in $X_1 \times X_2$; hence, (29) is true, and the proof is complete. □

8.7 THEOREM Let X be a set in $\mathbb{C}_2$, and let h_1 and h_2 map X into X_1 in A_1 and X_2 in A_2, respectively.

(32) If X is an open set in $\mathbb{C}_2$, then X_1 and X_2 are open sets in A_1 and A_2.

(33) If X is a convex set in $\mathbb{C}_2$, then X_1 and X_2 are convex sets in A_1 and A_2.

(34) If X is star-shaped with respect to $a+i_2b$ in $\mathbb{C}_2$, then X_1 and X_2 are star-shaped with respect to $a-i_1b$ and $a+i_1b$, respectively.

Proof. To prove (32), show that each point in X_1 has a neighborhood in X_1 and that each point in X_2 has a neighborhood in X_2. Let w_1^0 be a point in X_1; then there is some point $a+i_2b$ in X such that $w_1^0=h_1(a+i_2b)=a-i_1b$. Also, $h_2(a+i_2b)$ is a point $w_2^0=a+i_1b$ in X_2. Since X is open, there is a neighborhood $N(a+i_2b, \varepsilon)$ which is contained in X. We shall show that $N(a-i_1b, \varepsilon) \subset X_1$ and $N(a+i_1b, \varepsilon) \subset X_2$. Choose arbitrary points w_1 and w_2 such that

$$w_1 \in N(a - i_1b, \varepsilon), \qquad w_2 \in N(a + i_1b, \varepsilon). \tag{35}$$

Proof is still required to show that $w_1 \in X_1$ and $w_2 \in X_2$. Because of (35),

$$|w_1 - (a - i_1b)| < \varepsilon, \qquad |w_2 - (a + i_1b)| < \varepsilon. \tag{36}$$

Let $z_1+i_2z_2$ be the unique point in $\mathbb{C}_2$ which corresponds to (w_1, w_2) by (12); then

$$w_1=h_1(z_1+i_2z_2)=z_1-i_1z_2, \qquad w_2=h_2(z_1+i_2z_2)=z_1+i_1z_2. \tag{37}$$

The proof will show that $z_1+i_2z_2$ is not only in $\mathbb{C}_2$ but also in X. By (36), (37), and Theorem 6.8,

$$\begin{aligned}(38)\qquad &\|(z_1+i_2z_2)-(a+i_2b)\|\\ &=\left[\frac{|(z_1-i_1z_2)-(a-i_1b)|^2+|(z_1+i_1z_2)-(a+i_1b)|^2}{2}\right]^{1/2}\\ &<\left(\frac{\varepsilon^2+\varepsilon^2}{2}\right)^{1/2}=\varepsilon.\end{aligned}$$

Then $(z_1+i_2z_2)\in N(a+i_2b,\varepsilon)$, and therefore $w_1\in X_1$ and $w_2\in X_2$ by (37). To summarize, $a-i_1b$ is an arbitrary point in X_1; and w_1, which was chosen as an arbitrary point in $N(a-i_1b,\varepsilon)$, is in X_1. Therefore, $N(a-i_1b,\varepsilon)\subset X_1$ and X_1 is an open set in A_1. A similar proof can be given to show that a neighborhood $N(c+i_1d,\varepsilon)$ of an arbitrary point $c+i_1d$ in X_2 is in X_2. Thus the proof is complete that X_1 and X_2 are open sets in A_1 and A_2, respectively.

To prove (33), let w_1^1 and w_1^2 be two points in X_1. Then there are points $a+i_2b$ and $c+i_2d$, not necessarily unique, such that

$$(39)\qquad w_1^1=h_1(a+i_2b)=a-i_1b,\qquad w_1^2=h_1(c+i_2d)=c-i_1d.$$

Also, there are points w_2^1 and w_2^2 in X_2 such that

$$(40)\qquad w_2^1=h_2(a+i_2b)=a+i_1b,\qquad w_2^2=h_2(c+i_2d)=c+i_1d.$$

Then $a-i_1b$ and $c-i_1d$ are in X_1, and $a+i_1b$ and $c+i_1d$ are in X_2. Since X is convex by hypothesis, then $t(a+i_2b)+(1-t)(c+i_2d)$, $0\leqslant t\leqslant 1$, is in X and therefore

$$\begin{aligned}(41)\qquad &t(a-i_1b)+(1-t)(c-i_1d)\in X_1,\\ &t(a+i_1b)+(1-t)(c+i_1d)\in X_2,\qquad 0\leqslant t\leqslant 1.\end{aligned}$$

The first of these equations shows that X_1 is convex. To show that X_2 is convex, choose two arbitrary points in X_2 and use a similar argument to show that the segment connecting them is in X_2. The proof of (33) is complete.

To prove (34), assume that X is star-shaped with respect to $a+i_2b$. Then for every $z_1+i_2z_2$ in X the points

$$(42)\qquad t(z_1+i_2z_2)+(1-t)(a+i_2b),\qquad 0\leqslant t\leqslant 1,$$

are also in X, and

$$\begin{aligned}(43)\qquad &t(z_1-i_1z_2)+(1-t)(a-i_1b)\in X_1,\\ &t(z_1+i_1z_2)+(1-t)(a+i_1b)\in X_2,\qquad 0\leqslant t\leqslant 1.\end{aligned}$$

If w_1^0 is an arbitrary point in X_1, then there is at least one point $z_1+i_2z_2$ in X such that $w_1^0=h_1(z_1+i_2z_2)=z_1-i_1z_2$. For this point, $z_1+i_1z_2$ is in X_2. Also,

(42) holds for this point $z_1+i_2z_2$ since X is star-shaped with respect to $a+i_1b$; therefore, (43) is true for every w_1^0 in X_1. Thus X_1 is star-shaped with respect to $a-i_1b$. To prove that X_2 is star-shaped, let w_2^0 be an arbitrary point in X_2; choose $z_1+i_2z_2$ to a point in X such that $w_2^0=h_2(z_1+i_2z_2)=z_1+i_1z_2$, and then proceed as before. The proof of (34) and of all parts of Theorem 8.7 is complete. □

8.8 DEFINITION A set X in $\mathbb{C}_2$ is called a *cartesian set* if and only if there exist sets X_1 in A_1 and X_2 in A_2 such that

(44) $$X=\{z_1+i_2z_2 \text{ in } \mathbb{C}_2\colon z_1+i_2z_2=w_1e_1+w_2e_2, (w_1,w_2)\in X_1\times X_2\}.$$

If X satisfies (44), then X is the cartesian set *determined by X_1 and X_2*.

The set X in Example 8.3 is not a cartesian set, but the sets X in Examples 8.4 and 8.5 are cartesian sets.

8.9 THEOREM Let X be the cartesian set $\mathbb{C}_2$ determined by X_1 in A_1 and X_2 in A_2.

(45) If X_1 and X_2 are open sets in A_1 and A_2, respectively, then X is open in $\mathbb{C}_2$.

(46) If X_1 and X_2 are convex sets in A_1 and A_2, respectively, then X is convex in $\mathbb{C}_2$.

(47) If X_1 and X_2 are star-shaped with respect to $a-i_1b$ and $a+i_1b$, respectively, then X is star-shaped with respect to $a+i_2b$.

Proof. To prove (45), show that each point $a+i_2b$ in X has a neighborhood in X. Corresponding to $a+i_2b$ in X there are unique points $a-i_1b$ in X_1 and $a+i_1b$ in X_2:

(48) $$h_1(a+i_2b)=a-i_1b, \qquad h_2(a+i_2b)=a+i_1b.$$

Since X_1 and X_2 are open by hypothesis, there exists an $\varepsilon>0$ such that $N(a-i_1b,\varepsilon)\subset X_1$ and $N(a+i_1b,\varepsilon)\subset X_2$. The proof of (45) consists of showing that

(49) $$N(a+i_2b, \varepsilon/\sqrt{2})\subset X.$$

To prove this statement, let $z_1+i_2z_2$ be a point in $\mathbb{C}_2$ such that

(50) $$\|(z_1+i_2z_2)-(a+i_2b)\|<\varepsilon/\sqrt{2}.$$

To prove (49), it is necessary and sufficient to prove that $z_1+i_2z_2$ is in X. Now.

(51) $$z_1-i_1z_2=h_1(z_1+i_2z_2), \qquad z_1+i_1z_1=h_2(z_1+i_2z_2),$$

and $z_1 - i_1 z_2$ and $z_1 + i_1 z_2$ are in A_1 and A_2, respectively. Also, by Theorem 6.8 and the definition of distance in $\mathbb{C}_2$,

$$\begin{aligned}(52)\qquad &\|(z_1 + i_2 z_2) - (a + i_2 b)\| \\ &= \left[\frac{|(z_1 - i_1 z_2) - (a - i_1 b)|^2 + |(z_1 + i_1 z_2) - (a + i_1 b)|^2}{2}\right]^{1/2}.\end{aligned}$$

From (52) it follows that

$$(53)\qquad \|(z_1 + i_2 z_2) - (a + i_2 b)\| \geqslant \begin{cases} \dfrac{|(z_1 - i_1 z_2) - (a - i_1 b)|}{\sqrt{2}}, \\[2ex] \dfrac{|(z_1 + i_1 z_2) - (a + i_1 b)|}{\sqrt{2}}. \end{cases}$$

If either $|(z_1 - i_1 z_2) - (a - i_1 b)| \geqslant \varepsilon$ or $|(z_1 + i_1 z_2) - (a + i_1 b)| \geqslant \varepsilon$, then (53) contradicts (50); therefore,

$$(54)\qquad |(z_1 - i_1 z_2) - (a - i_1 b)| < \varepsilon, \qquad |(z_1 + i_1 z_2) - (a + i_1 b)| < \varepsilon,$$

and $(z_1 - i_1 z_2) \in N(a - i_1 b, \varepsilon)$ and $(z_1 + i_1 z_2) \in N(a + i_1 b, \varepsilon)$. Since these neighborhoods are in X_1 and X_2, respectively, then $z_1 - i_1 z_2 \in X_1$ and $z_1 + i_1 z_2 \in X_2$. Now X is a cartesian set by hypothesis; therefore, $z_1 + i_2 z_2$ is in X since

$$(55)\qquad z_1 + i_2 z_2 = (z_1 - i_1 z_2)e_1 + (z_1 + i_1 z_2)e_2.$$

The proof has shown that every point $z_1 + i_2 z_2$ in $N(a + i_2 b, \varepsilon/\sqrt{2}) \subset X$. Since $a + i_2 b$ is an arbitrary point in X, this set is open, and the proof of (45) is complete.

To prove (46), begin by letting $a + i_2 b$ and $c + i_2 d$ be two arbitrary points in X. Then $a - i_1 b$ and $c - i_1 d$ are in X_1, and $a + i_1 b$ and $c + i_1 d$ are in X_2. Since X_1 and X_2 are convex by hypothesis,

$$(56)\qquad \begin{aligned} &t(a - i_1 b) + (1 - t)(c - i_1 d) \in X_1, \\ &t(a + i_1 b) + (1 - t)(c + i_1 d) \in X_2, \end{aligned} \qquad 0 \leqslant t \leqslant 1.$$

Since X is a cartesian set by hypothesis, the point $z_1 + i_2 z_2$ such that

$$\begin{aligned}(57)\qquad z_1 + i_2 z_2 = {}& [t(a - i_1 b) + (1 - t)(c - i_1 d)]e_1 \\ &+ [t(a + i_1 b) + (1 - t)(c + i_1 d)]e_2\end{aligned}$$

is a point in X for $0 \leqslant t \leqslant 1$. Simplification of (57) shows that

$$(58)\qquad z_1 + i_2 z_2 = t(a + i_2 b) + (1 - t)(c + i_2 d), \qquad 0 \leqslant t \leqslant 1,$$

is a point in X on the segment which joins $a + i_2 b$ and $c + i_2 d$. Thus a segment (58) in X joins each two points $a + i_2 b$ and $c + i_2 d$ in X, and X is convex by definition. The proof of (46) is complete.

To begin the proof of (47), let $c+i_2d$ be a point in X. Then $a-i_1b$ and $c-i_1d$ are in X_1, and $a+i_1b$ and $c+i_1d$ are in X_2. Since X_1 and X_2 are star-shaped with respect to $a-i_1b$ and $a+i_1b$, respectively,

$$(59)\qquad \begin{aligned} &t(a-i_1b)+(1-t)(c-i_1d)\in X_1,\\ &t(a+i_1b)+(1-t)(c+i_1d)\in X_2, \end{aligned}\qquad 0\leqslant t\leqslant 1.$$

Then since X is the cartesian set determined by X_1 and X_2, the point $z_1+i_2z_2$ such that

$$(60)\qquad \begin{aligned} z_1+i_2z_2=[&t(a-i_1b)+(1-t)(c-i_1d)]e_1\\ &+[t(a+i_1b)+(1-t)(c+i_1d)]e_2 \end{aligned}$$

is a point in X for $0\leqslant t\leqslant 1$. Simplification of (60) shows that

$$(61)\qquad z_1+i_2z_2=t(a+i_2b)+(1-t)(c+i_2d),\qquad 0\leqslant t\leqslant 1,$$

is a point in X on the segment which joins $a+i_2b$ and $c+i_2d$. Thus a segment (61) in X joins $a+i_2b$ to the arbitrary point $c+i_2d$ in X, and X is star-shaped with respect to $a+i_2b$ by definition. The proof of (47) and of all parts of Theorem 8.9 is complete. □

8.10 THEOREM Let X be a cartesian set in $\mathbb{C}_2$ which is determined by X_1 in A_1 and X_2 in A_2. Let X be a proper subset of X' in $\mathbb{C}_2$. If h_1 and h_2 map X' into X'_1 and X'_2, then either X_1 is a proper subset of X'_1 or X_2 is a proper subset of X'_2, or both statements are true.

Proof. Since X is a proper subset of X' by hypothesis, there exists a point $z_1+i_2z_2$ such that $z_1+i_2z_2\in X'$ and $z_1+i_2z_2\notin X$. Since $z_1-i_1z_2=h_1(z_1+i_2z_2)$ and $z_1+i_1z_2=h_2(z_1+i_2z_2)$, then $z_1-i_1z_2\in X'_1$ and $z_1+i_1z_2\in X'_2$. If $z_1-i_1z_2\in X_1$, and $z_1+i_1z_2\in X_2$, then the hypothesis that X is a cartesian set determined by X_1 and X_2 shows that $z_1+i_2z_2\in X$. But this conclusion contradicts the fact that $z_1+i_2z_2\notin X$. Then either $z_1-i_1z_2\notin X_1$ or $z_1+i_1z_2\notin X_2$, or both statements are true. Therefore, X_1 is a proper subset of X'_1, or X_2 is a proper subset of X'_2, or both statements are true. The proof is complete. □

Recall that, by definition, a set is (arcwise) connected if and only if, for each two points in the set, there is a polygonal curve, in the set, which connects the two points.

8.11 THEOREM Let X be a set in $\mathbb{C}_2$, and let h_1 and h_2 map X into X_1 in A_1 and X_2 in A_2, respectively. If X is a connected set, then X_1 and X_2 are connected sets.

Proof. Let $a-i_1b$ and $c-i_1d$ be two points in X_1. Then there are points $a+i_2b$ and $c+i_2d$ in X such that

$$(62)\qquad h_1(a+i_2b)=a-i_1b,\qquad h_1(c+i_2d)=c-i_1d.$$

Because X is connected by hypothesis, there exists a polygonal curve which connects $a+i_2b$ and $c+i_2d$. The proof of (33) in Theorem 8.7 has shown that h_1 maps a segment of a line in X into a segment of a line in X_1; therefore, h_1 maps the polygonal curve which connects $a+i_2b$ and $c+i_2d$ into a polygonal curve in X_1 which connects $a-i_1b$ and $c-i_1d$. Then X_1 is a connected set. A similar proof shows that X_2 is a connected set. □

8.12 THEOREM Let X be the cartesian set in $\mathbb{C}_2$ determined by X_1 in A_1 and X_2 in A_2. If X_1 and X_2 are connected sets, then X is a connected set.

Proof. Let $a+i_2b$ and $c+i_2d$ be two points in X. Then $a-i_1b$ and $c-i_1d$ are points in X_1, and $a+i_1b$ and $c+i_1d$ are points in X_2. Since X_1 and X_2 are connected sets by hypothesis, there are polygonal curves which connect $a-i_1b$ and $c-i_1d$ in X_1, and also $a+i_1b$ and $c+i_1d$ in X_2. Assume, as a first case, that each of these polygonal curves contains a single segment. Thus

$$\begin{aligned} &t(a-i_1b)+(1-t)(c-i_1d)\in X_1, \\ &t(a+i_1b)+(1-t)(c+i_1d)\in X_2, \end{aligned} \qquad 0\leqslant t\leqslant 1. \tag{63}$$

Since X is a cartesian set determined by X_1 and X_2, then

$$[t(a-i_1b)+(1-t)(c-i_1d)]e_1+[t(a+i_1b)+(1-t)(c+i_1d)]e_2 \tag{64}$$

is in X for $0\leqslant t\leqslant 1$. Simplification of (64) shows that

$$[t(a+i_2b)+(1-t)(c+i_2d)]\in X, \qquad 0\leqslant t\leqslant 1. \tag{65}$$

Thus (65) defines a polygonal curve (consisting of a single segment) in X which connects $a+i_2b$ and $c+i_2d$, and X is a connected set in this case. In the general case, $a-i_1b$ and $c-i_1d$ are connected by a polygonal curve with m segments, and $a+i_1b$ and $c+i_1d$, by a polygonal curve with n segments. By suitable subdivisions of some of the segments in the two curves, it is possible to construct polygonal curves which connect the points as before but which now have the same number of segments. Now apply the argument in the first case to each pair of corresponding segments in the two polygonal curves. The result is a polygonal curve in X which connects $a+i_2b$ and $c+i_2d$. The proof is complete in all cases. □

Exercises

8.1 Add the following hypothesis to those in Theorem 8.7: the set X is the cartesian set determined by X_1 and X_2. Show that, with this additional hypothesis, simpler proofs of (32), (33), and (34) can be given than those in the proof of Theorem 8.7.

8.2 Show that each of the following sets is unbounded:

$$\{(z_1 + i_2z_2) \text{ in } \mathbb{C}_2: |h_1(z_1 + i_2z_2)| \leqslant r\},$$
$$\{(z_1 + i_2z_2) \text{ in } \mathbb{C}_2: |h_2(z_1 + i_2z_2)| \leqslant r\}.$$

Nevertheless, show that the set

$$\{(z_1+i_2z_2) \text{ in } \mathbb{C}_2: |h_1(z_1+i_2z_2)| \leqslant r \text{ and } |h_2(z_1+i_2z_2)| \leqslant r\}$$

is bounded, and find the radius of the smallest sphere which contains it.

8.3 Prove the following theorem. Let X be a cartesian set in $\mathbb{C}_2$ which is determined by X_1 and X_2. If $a+i_2b$ is a boundary point of X, then either $a-i_1b$ is a boundary point of X_1 or $a+i_1b$ is a boundary point of X_2, or both. (*Hint.* A point p is a boundary point of a set S if and only if every neighborhood of p contains points in S and also points in the complement of S.)

8.4 Prove the following theorem. Let X be a cartesian set in $\mathbb{C}_2$ which is determined by X_1 and X_2. If w_1^0 is a boundary point of X_1, then every point in the set

$$\{(z_1 + i_2z_2) \text{ in } \mathbb{C}_2: z_1 + i_2z_2 = w_1^0e_1 + w_2e_2, w_2 \in X_2\}$$

is a boundary point of X. Likewise, if w_2^0 is a boundary point of X_2, then every point in the set

$$\{(z_1 + i_2z_2) \text{ in } \mathbb{C}_2: z_1 + i_2z_2 = w_1e_1 + w_2^0e_2, w_1 \in X_2\}$$

is a boundary point of X.

8.5 Use Exercise 8.4 to construct an example of a set X in $\mathbb{C}_2$ every point of which is a boundary point of X.

8.6 Use Theorem 8.9 to construct an example of a set X in $\mathbb{C}_2$ every point of which is an interior point of the set.

8.7 Prove the following theorem. If X_1 and X_2 are closed sets in A_1 and A_2 respectively, then the cartesian set X in $\mathbb{C}_2$ determined by X_1 and X_2 is closed in $\mathbb{C}_2$.

9. THE DISCUS

This section defines a special cartesian set in $\mathbb{C}_2$ which is called a discus, and it establishes some of the properties of this set.

Let $a:(a_1+i_1a_2+i_2a_3+i_1i_2a_4)$ be a fixed point in $\mathbb{C}_2$. Set $\alpha=a_1+i_1a_2$ and $\beta=a_3+i_1a_4$; then

(1) $\quad a = a_1 + i_1a_2 + i_2a_3 + i_1i_2a_4 = \alpha + i_2\beta.$

Let r, r_1, r_2 denote numbers in $\mathbb{C}_0$ such that

(2) $\quad r > 0, \qquad r_1 > 0, \qquad r_2 > 0.$

Next, let w_1 and w_2 denote numbers in A_1 and A_2, respectively; observe that w_1 and w_2 are in fact complex numbers in $\mathbb{C}_1$. Recall that the open ball $B(a,r)$ and the closed ball $\bar{B}(a,r)$ with center a and radius r are defined as follows:

(3) $B(a,r) = \{z_1 + i_2z_2 \text{ in } \mathbb{C}_2\colon \|(z_1 + i_2z_2) - (\alpha + i_2\beta)\| < r\}$,

(4) $\bar{B}(a,r) = \{z_1 + i_2z_2 \text{ in } \mathbb{C}_2\colon \|(z_1 + i_2z_2) - (\alpha + i_2\beta)\| \leqslant r\}$.

As stated in Definition 8.8, a cartesian set X in $\mathbb{C}_2$ is determined by sets X_1, X_2, in A_1, A_2 respectively as follows:

(5) $X = \{z_1 + i_2z_2 \text{ in } \mathbb{C}_2\colon z_1 + i_2z_2 = w_1e_1 + w_2e_2,\ (w_1, w_2) \in X_1 \times X_2\}$.

9.1 DEFINITION If

(6) $X_1 = \{w_1 \in A_1\colon |w_1 - (\alpha - i_1\beta)| < r_1\}$,

$X_2 = \{w_2 \in A_2\colon |w_2 - (\alpha + i_1\beta)| < r_2\}$,

then X in (5) is called the *open discus with center a and radii r_1 and r_2* and denoted by $D(a; r_1, r_2)$. If

(7) $X_1 = \{w_1 \in A_1\colon |w_1 - (\alpha - i_1\beta)| \leqslant r_1\}$,

$X_2 = \{w_2 \in A_2\colon |w_2 - (\alpha + i_1\beta)| \leqslant r_2\}$,

then X in (5) is called the *closed discus with center a and radii r_1 and r_2* and denoted by $\bar{D}(a; r_1, r_2)$. Thus

(8) $$D(a; r_1, r_2) = \{z_1 + i_2z_2 \text{ in } \mathbb{C}_2\colon \\ z_1 + i_2z_2 = w_1e_1 + w_2e_2, \\ |w_1 - (\alpha - i_1\beta| < r_1,\ |w_2 - (\alpha + i_1\beta)| < r_2\},$$

(9) $$\bar{D}(a; r_1, r_2) = \{z_1 + i_2z_2 \text{ in } \mathbb{C}_2\colon \\ z_1 + i_2z_2 = w_1e_1 + w_2e_2, \\ |w_1 - (\alpha - i_1\beta| \leqslant r_1,\ |w_2 - (\alpha + i_1\beta)| \leqslant r_2\}.$$

9.2 THEOREM If $0 < r_1 \leqslant r_2$, then

(10) $B(a, r_1/\sqrt{2}) \subsetneqq D(a; r_1, r_2) \subsetneqq B(a, [(r_1^2 + r_2^2)/2]^{1/2})$,

(11) $\bar{B}(a, r_1/\sqrt{2}) \subsetneqq \bar{D}(a; r_1, r_2) \subsetneqq \bar{B}(a, [(r_1^2 + r_2^2)/2]^{1/2})$.

Proof. By Theorem 6.8,

(12) $$\|(z_1 + i_2z_2) - (\alpha + i_2\beta)\| \\ = \left[\frac{|w_1 - (\alpha - i_1\beta)|^2 + |w_2 - (\alpha + i_1\beta)|^2}{2}\right]^{1/2}.$$

If $z_1+i_2z_2$ is in $D(a; r_1,r_2)$, then by (5), (6), and (8),

$$(13)\qquad |w_1-(\alpha-i_1\beta)|<r_1, \qquad |w_2-(\alpha+i_1\beta)|<r_2;$$

hence, (12) shows that

$$(14)\qquad \|(z_1+i_2z_2)-(\alpha+i_2\beta)\|<\left[\frac{r_1^2+r_2^2}{2}\right]^{1/2},$$

and $z_1+i_2z_2$ is in $B(a, [(r_1^2+r_2^2)/2]^{1/2})$ by (3). Therefore,

$$(15)\qquad D(a; r_1,r_2)\subset B(a, [(r_1^2+r_2^2)/2]^{1/2}).$$

To prove that these two sets are not equal, let s be a number such that

$$(16)\qquad r_1<s<r_1\sqrt{2}.$$

Choose (w_1,w_2) so that

$$(17)\qquad |w_1-(\alpha-i_1\beta)|=s>r_1, \qquad |w_2-(\alpha+i_1\beta)|=0.$$

If $z_1+i_2z_2=w_1e_1+w_2e_2$, then $(z_1+i_2z_2)\notin D(a; r_1,r_2)$ (see (5) and (6)), but $z_1+i_2z_2$ is in $B(a, [(r_1^2+r_2^2)/2]^{1/2})$ since

$$(18)\qquad \|(z_1+i_2z_2)-(\alpha+i_2\beta)\|=\frac{s}{\sqrt{2}}<r_1\leqslant\left(\frac{r_1^2+r_2^2}{2}\right)^{1/2}.$$

Therefore, (15) and these statements about $z_1+i_2z_2$ show that

$$(19)\qquad D(a; r_1, r_2)\subsetneqq B(a, [(r_1^2+r_2^2)/2]^{1/2}),$$

and the statement on the right in (10) has been proved.

Next, consider the two sets on the left in (10). If $z_1+i_2z_2\in B(a,r_1/\sqrt{2})$, then

$$(20)\qquad \|(z_1+i_2z_2)-(\alpha+i_2\beta)\|<\frac{r_1}{\sqrt{2}}.$$

If $w_1=z_1-i_1z_2$ and $w_2=z_1+i_1z_2$, then the formula in (12) shows that

$$(21)\qquad \|(z_1+i_2z_2)-(\alpha+i_2\beta)\|\geqslant\begin{cases}|w_1-(\alpha-i_1\beta)|/\sqrt{2},\\ |w_2-(\alpha+i_1\beta)|/\sqrt{2}.\end{cases}$$

Recall that $r_1\leqslant r_2$. If either $|w_1-(\alpha-i_1\beta)|\geqslant r_1$, or $|w_2-(\alpha+i_1\beta)|\geqslant r_2$, then (21) contradicts (20). Therefore,

$$(22)\qquad |w_1-(\alpha-i_1\beta)|<r_1, \qquad |w_2-(\alpha+i_1\beta)|<r_2,$$

$z_1+i_2z_2$ is in $D(a; r_1,r_2)$ by (8), and

$$(23)\qquad B(a,r_1/\sqrt{2})\subset D(a; r_1,r_2).$$

To show that the set on the left is a proper subset of the one on the right, let t be a number such that $1/\sqrt{2}<t<1$. Choose (w_1, w_2) so that

(24) $\quad |w_1 - (\alpha - i_1\beta)| = tr_1, \qquad |w_2 - (\alpha + i_1\beta)| = tr_1.$

If $z_1+i_2z_2=w_1e_1+w_2e_2$, then (24) and (8) show that $z_1+i_2z_2$ is in $D(a; r_1, r_2)$, and (12) shows that $z_1+i_2z_2$ is not in $B(a, r_1/\sqrt{2})$ since

(25) $$\|(z_1 + i_2z_2) - (\alpha + i_2\beta)\| = \left(\frac{(tr_1)^2 + (tr_1)^2}{2}\right)^{1/2} = tr_1 > \frac{r_1}{\sqrt{2}}.$$

Therefore,

(26) $\quad B(a, r_1/\sqrt{2}) \subsetneqq D(a; r_1, r_2),$

and the proof of (10) is complete.

The proof of (11) is similar to the proof of (10). To begin, let $z_1+i_2z_2$ be a point in $\bar{D}(a; r_1, r_2)$. Then $|w_1-(\alpha-i_1\beta)|\leqslant r_1$, $|w_2-(\alpha+i_1\beta)|\leqslant r_2$ by (9), and

(27) $$\|(z_1 + i_2z_2) - (\alpha + i_2\beta)\| \leqslant \left(\frac{r_1^2 + r_2^2}{2}\right)^{1/2}$$

by (12). Thus $z_1+i_2z_2$ is in $\bar{B}(a, [(r_1^2+r_2^2)/2]^{1/2})$ by (4); therefore

(28) $\quad \bar{D}(a; r_1, r_2) \subset \bar{B}(a, [(r_1^2 + r_2^2)/2]^{1/2}).$

To prove that the two sets in (28) are not equal, choose s as in (16) and (w_1, w_2) as in (17). If $z_1+i_2z_2=w_1e_1+w_2e_2$, then $z_1+i_2z_2$ is not in $\bar{D}(a; r_1, r_2)$ by (9), but (16), (17), and (12) show that $z_1+i_2z_2$ is in $\bar{B}(a, [(r_1^2+r_2^2)/2]^{1/2})$ since (18) holds as before. Thus the set on the right in (28) contains an element not in the set on the left; therefore,

(29) $\quad \bar{D}(a; r_1, r_2) \subsetneqq \bar{B}(a, [(r_1^2 + r_2^2)/2]^{1/2}).$

Next, consider the two sets on the left in (11). If $z_1 + i_2z_2 \in \bar{B}(a, r_1/2)$, then

(30) $$\|(z_1 + i_2z_2) - (\alpha + i_2\beta)\| \leqslant \frac{r_1}{\sqrt{2}}.$$

If $w_1=z_1-i_1z_2$ and $w_2=z_1+i_1z_2$, then by (12),

(31) $$\|(z_1 + i_2z_2) - (\alpha + i_2\beta)\| \geqslant \begin{cases} |w_1 - (\alpha - i_1\beta)|/\sqrt{2}, \\ |w_2 - (\alpha + i_1\beta)|/\sqrt{2}. \end{cases}$$

Recall again that $r_1\leqslant r_2$. If either $|w_1-(\alpha-i_1\beta)|>r_1$, or $|w_2-(\alpha+i_1\beta)|>r_2$, then (31) contradicts (30). Therefore,

(32) $\quad |w_1 - (\alpha - i_1\beta)| \leqslant r_1, \qquad |w_2 - (\alpha + i_1\beta)| \leqslant r_2,$

$z_1+i_2z_2$ is in $\bar{D}(a;r_1,r_2)$ by (9), and

$$\bar{B}(a,r_1/\sqrt{2}) \subset \bar{D}(a;r_1,r_2). \tag{33}$$

To show that the set on the left is a proper subset of the one on the right, let t be a number such that $1/\sqrt{2}<t<1$. Choose (w_1,w_2) so that (24) is satisfied, and set $z_1+i_2z_2=w_1e_1+w_2e_2$. Then (24) shows that $z_1+i_2z_2$ is in $\bar{D}(a;r_1,r_2)$, and (25) shows that $z_1+i_2z_2$ is not in $\bar{B}(a,r_1/\sqrt{2})$. Therefore,

$$\bar{B}(a,r_1/\sqrt{2}) \subsetneqq \bar{D}(a;r_1,r_2). \tag{34}$$

The proof of (11) and of all parts of Theorem 9.2 is now complete. □

The next theorem uses the results in Exercises 8.3 and 8.4 to obtain a description of the boundary of the discus $\bar{D}(a;r_1,r_2)$. The open discus and the closed discus have the same boundary.

9.3 THEOREM Let $0<r_1\leqslant r_2$. The boundary *of* $\bar{D}(a;r_1,r_2)$ is the union of the following three disjoint sets:

$$\{z_1+i_2z_2 \text{ in } \mathbb{C}_2\colon |(z_1-i_1z_2)-(\alpha-i_1\beta)|=r_1,\quad |(z_1+i_1z_2)-(\alpha+i_1\beta)|=r_2\}, \tag{35}$$

$$\{z_1+i_2z_2 \text{ in } \mathbb{C}_2\colon |(z_1-i_1z_2)-(\alpha-i_1\beta)|=r_1,\quad |(z_1+i_1z_2)-(\alpha+i_1\beta)|<r_2\}, \tag{36}$$

$$\{z_1+i_2z_2 \text{ in } \mathbb{C}_2\colon |(z_1-i_1z_2)-(\alpha-i_1\beta)|<r_1,\quad |(z_1+i_1z_2)-(\alpha+i_1\beta)|=r_2\}. \tag{37}$$

The set in (35) is in the boundary of the ball $\bar{B}(a,\ [(r_1^2+r_2^2)/2]^{1/2})$, and the sets in (36) and (37) are in the interior of this ball. The minimum distance from $\alpha+i_2\beta$ to a point in the boundary of $\bar{D}(a;r_1,r_2)$ is $r_1/\sqrt{2}$.

Proof. The definitions of the sets in (35), (36), (37) show that they are disjoint. Since $\bar{D}(a;r_1,r_2)$ is a cartesian set, Exercises 8.3 and 8.4 show that each of these sets consists of boundary points of $\bar{D}(a;r_1,r_2)$. If $z_1+i_2z_2$ is in (35), then

$$\begin{aligned}\|(z_1+i_2z_2)-(\alpha+i_2\beta)\| &= \left[\frac{|(z_1-i_1z_2)-(\alpha-i_1\beta)|^2+|(z_1+i_1z_2)-(\alpha+i_1\beta)|^2}{2}\right]^{1/2}\\ &= \left(\frac{r_1^2+r_2^2}{2}\right)^{1/2},\end{aligned} \tag{38}$$

and $z_1+i_2z_2$ is in the boundary of $\bar{B}(a, [(r_1^2+r_2^2)/2]^{1/2})$. Thus (35) is contained in the boundary of this ball. Finally, since $0<r_1\leqslant r_2$, the points in the boundary at minimum distance from $\alpha+i_2\beta$ are the points $z_1+i_2z_2$ such that

$$(39)\qquad |(z_1-i_1z_2)-(\alpha-i_1\beta)|=r_1,\qquad |(z_1+i_1z_2)-(\alpha+i_1\beta)|=0.$$

For these points, $\|(z_1+i_2z_2)-(\alpha+i_2\beta)\|=r_1/\sqrt{2}$. The proof of Theorem 9.3 is complete. □

Theorems 9.2 and 9.3 explain why the name *discus* has been chosen for the set $\bar{D}(a; r_1, r_2)$. The set (35) lies in the boundary of the ball $\bar{B}(a, [(r_1^2+r_2^2)/2]^{1/2})$ and separates the remainder of the boundary into two sets (36) and (37) which lie in the interior of the ball. The discus is a thick set because Theorem 9.2 shows that $\bar{D}(a; r_1, r_2)$ contains $\bar{B}(a, r_1/\sqrt{2})$.

In a sense to be explained now, the points in the discuss $D(a; r_1, r_2)$ are "inside" certain curves which lie in its boundary (35). Let

$$(40)\qquad \begin{aligned} &g_1:[0,2\pi]\to A_1, &&\theta\mapsto g_1(\theta), &&g_1(2\pi)=g_1(0),\\ &g_2:[0,2\pi]\to A_2, &&\theta\mapsto g_2(\theta), &&g_2(2\pi)=g_2(0),\end{aligned}$$

be two continuous functions such that the equations

$$(41)\qquad w_1=g_1(\theta),\qquad w_2=g_2(\theta),\qquad 0\leqslant\theta\leqslant 2\pi,$$

define two simple closed curves, in A_1 and A_2, respectively, whose traces are the circles C_1:$\{w_1$ in A_1: $|w_1-(\alpha-i_1\beta)|=r_1\}$ and C_2:$\{w_2$ in A_2: $|w_2-(\alpha+i_1\beta)|=r_2\}$. Then the equation

$$(42)\qquad z_1+i_2z_2=g_1(\theta)e_1+g_2(\theta)e_2,\qquad 0\leqslant\theta\leqslant 2\pi,$$

describes a simple closed curve C whose trace is in the set (35) in the boundary of $\bar{D}(a; r_1, r_2)$; this trace also lies in the boundary of $\bar{B}(a, [(r_1^2+r_2^2)/2]^{1/2})$. Let $\gamma+i_2\delta$ be a point in $\mathbb{C}_2$ such that

$$(43)\qquad |(\gamma-i_1\delta)-(\alpha-i_1\beta)|<r_1,\qquad |(\gamma+i_1\delta)-(\alpha+i_1\beta)|<r_2.$$

Then $\gamma-i_1\delta$ and $\gamma+i_1\delta$ are inside C_1 and C_2, respectively, and $\gamma+i_2\delta$ is in $D(a; r_1, r_2)$. Furthermore, (42) shows that

$$(44)\qquad (z_1+i_2z_2)-(\gamma+i_2\delta)=[g_1(\theta)-(\gamma-i_1\delta)]e_1+[g_2(\theta)-(\gamma+i_1\delta)]e_2.$$

Since $\gamma-i_1\delta$ and $\gamma+i_1\delta$ are inside C_1 and C_2, then

$$(45)\qquad |g_1(\theta)-(\gamma-i_1\delta)|>0,\qquad |g_2(\theta)-(\gamma+i_1\delta)|>0,\qquad 0\leqslant\theta\leqslant 2\pi,$$

and (44) shows that

$$(46)\qquad [(z_1+i_2z_2)-(\gamma+i_2\delta)]\notin\mathcal{O}_2,\qquad 0\leqslant\theta\leqslant 2\pi.$$

Furthermore, if $\gamma + i_2\delta$ is a point in $D(a; r_1, r_2)$, then (43)–(46) are true. The points $\gamma + i_2\delta$ in $D(a; r_1, r_2)$ can be thought of as points "inside" the curve C. There are many curves C, but (46) is true for each such C and point $\gamma + i_2\delta$ in $D(a; r_1, r_2)$.

Exercises

9.1 Prove the following theorem (compare Theorem 9.2). If $0 < r_2 \leqslant r_1$, then

$$B(a, r_2/\sqrt{2}) \subsetneqq D(a; r_1, r_2) \subsetneqq B(a, [(r_1^2 + r_2^2)/2]^{1/2}),$$
$$\bar{B}(a, r_2/\sqrt{2}) \subsetneqq \bar{D}(a; r_1, r_2) \subsetneqq \bar{B}(a, [(r_1^2 + r_2^2)/2]^{1/2}).$$

9.2 Prove the following theorem (compare Theorem 9.3). Let $0 < r_2 \leqslant r_1$. Then the boundary of $\bar{D}(a; r_1, r_2)$ is the union of the following three disjoint sets:

(a) $\{z_1 + i_2z_2 \text{ in } \mathbb{C}_2: |(z_1 - i_1z_2) - (\alpha - i_1\beta)| = r_1,$
$|(z_1 + i_1z_2) - (\alpha + i_1\beta)| = r_2\},$

(b) $\{z_1 + i_2z_2 \text{ in } \mathbb{C}_2: |(z_1 - i_1z_2) - (\alpha - i_1\beta)| = r_1,$
$|(z_1 + i_1z_2) - (\alpha + i_1\beta)| < r_2\},$

(c) $\{z_1 + i_2z_2 \text{ in } \mathbb{C}_2: |(z_1 - i_1z_2) - (\alpha - i_1\beta)| < r_1,$
$|(z_1 + i_1z_2) - (\alpha + i_1\beta)| = r_2\}.$

The set in (a) is in the boundary of the ball $\bar{B}(a, [(r_1^2 + r_2^2)/2]^{1/2})$, and the sets in (b) and (c) are in the interior of the ball. The minimum distance from $\alpha + i_2\beta$ to a point in the boundary of $\bar{D}(a; r_1, r_2)$ is $r_2/\sqrt{2}$.

9.3 Define the functions g_1 and g_2 in (40) as follows:

$$\begin{aligned} g_1(\theta) &= (\alpha - i_1\beta) + r_1(\cos\theta + i_1\sin\theta), \\ g_2(\theta) &= (\alpha + i_1\beta) + r_2(\cos\theta + i_1\sin\theta), \end{aligned} \qquad 0 \leqslant \theta \leqslant 2\pi.$$

(a) Show that $w_1 = g_1(\theta)$ describes a circle in A_1 with center $\alpha - i_1\beta$ and radius r_1, and that $w_2 = g_2(\theta)$ describes a circle in A_2 with center $\alpha + i_1\beta$ and radius r_2.

(b) Describe the curve C defined by

$$z_1 + i_2z_2 = g_1(\theta)e_1 + g_2(\theta)e_2, \qquad 0 \leqslant \theta \leqslant 2\pi.$$

Show that the trace of C is contained in the boundary of $\bar{D}(a; r_1, r_2)$ and also in the boundary of $\bar{B}(a, [(r_1^2 + r_2^2)/2]^{1/2})$.

(c) Which points in $\mathbb{C}_2$ are described as being "inside" C? What special property do these points have with respect to points in the trace of C?

9.4 Repeat Exercise 9.3 for the functions defined by

$$\begin{aligned} g_1(\theta) &= (\alpha - i_1\beta) + r_1(\cos\theta - i_1\sin\theta), \\ g_2(\theta) &= (\alpha + i_1\beta) + r_2(\cos\theta + i_1\sin\theta), \end{aligned} \qquad 0 \leqslant \theta \leqslant 2\pi.$$

Show that the trace of the curve C defined by these functions is different from the trace of C in Exercise 9.3, but that the points in $D(a; r_1, r_2)$ are "inside" both curves C.

9.5 Repeat Exercise 9.3 for the functions g_1, g_2 defined by

$$\begin{aligned} g_1(\theta) &= (\alpha - i_1\beta) + r_1(\cos\theta - i_1\sin\theta), \\ g_2(\theta) &= (\alpha + i_1\beta) + r_2(\cos\theta - i_1\sin\theta), \end{aligned} \qquad 0 \leqslant \theta \leqslant 2\pi.$$

Show that the trace of C in this case is the same as the trace of C in Exercise 9.3. How do the two curves differ?

9.6 Use the fact, established in Theorem 9.3, that the minimum distance from $\alpha + i_2\beta$ to a point in the boundary of $\bar{D}(a; r_1, r_2)$ is $r_1/\sqrt{2}$ to show that $\bar{B}(a, r/\sqrt{2}) \subset \bar{D}(a; r_1, r_2)$.

9.7 If $0 < r_1 = r_2 = r$, show that

$$\bar{B}(a, r/\sqrt{2}) \subsetneqq \bar{D}(a; r, r) \subsetneqq \bar{B}(a, r).$$

Show that, in this special case, each of the boundary sets in (36) and (37) contains points in $\bar{B}(a, r/\sqrt{2})$.

9.8 Let X_1 be the set defined in (7), and let X_2 be the complex space A_2. The cartesian set

$$\{z_1 + i_2 z_2 \text{ in } \mathbb{C}_2\colon z_1 + i_2 z_2 = w_1 e_1 + w_2 e_2, (w_1, w_2) \in X_1 \times X_2\}$$

can be interpreted as a discus with $r_2 = \infty$. Show that $\bar{D}(a; r_1, \infty)$ is an unbounded set, and that (11) in this case is replaced by

$$\bar{B}(a, r_1/\sqrt{2}) \subsetneqq \bar{D}(a; r_1, \infty) \subsetneqq \mathbb{C}_2.$$

9.9 If $r_1 = 0$ and $r_2 > 0$, the set

$$\{z_1 + i_2 z_2 \text{ in } \mathbb{C}_2\colon z_1 + i_2 z_2 = w_1 e_1 + w_2 e_2, \\ |w_1 - (\alpha - i_1\beta)| = r_1, |w_2 - (\alpha + i_1\beta)| < r_2\}$$

can be considered a degenerate discus and denoted by $D(a; 0, r_2)$. If $B(a, 0)$ denotes the set with the single element $\alpha + i_2\beta$, show that

$$B(a, 0) \subsetneqq D(a; 0, r_2) \subsetneqq B(a, r_2).$$

Show also that every point in $D(a; 0, r_2)$ is a boundary point of this set.

9.10 Establish the following properties of the discus:

(a) $D(a; r_1, r_2)$ is an open set and $\bar{D}(a; r_1, r_2)$ is a closed set.

(b) $D(a; r_1, r_2)$ and $\bar{D}(a; r_1, r_2)$ are convex sets.

(c) $D(a; r_1, r_2)$ and $\bar{D}(a; r_1, r_2)$ are star-shaped with respect to each point in each of these sets.

(d) Each two points in $D(a; r_1, r_2)$, and also in $\bar{D}(a; r_1, r_2)$, can be connected by a polygonal curve in the set.

2

Functions Defined by Bicomplex Power Series

10. INTRODUCTION

This chapter begins the study of holomorphic functions of a bicomplex variable. These functions have many striking similarities to holomorphic functions of a complex variable; for example, holomorphic functions of both complex and bicomplex variables can be defined either as functions which can be represented locally by power series or as functions which have a derivative. The power series definition seems to provide the easier introduction to the theory of holomorphic functions of a bicomplex variable; accordingly, this chapter is devoted to a study of infinite series, and especially to a study of functions defined by power series in $\mathbb{C}_2$.

There are three different representations for an element in $\mathbb{C}_2$. In addition to $x_1+i_1x_2+i_2x_3+i_1i_2x_4$, they are $z_1+i_2z_2$ and $(z_1-i_1z_2)e_1+(z_1+i_1z_2)e_2$. Corresponding to these three representations of an element there are three representations for the norm of an element. This multiplicity of representations adds to the richness of the theory.

A summary of some of the results in the chapter will provide a helpful introduction for the reader. Let X be a domain in $\mathbb{C}_2$, and let $f:X\to\mathbb{C}_2$, $z_1+i_2z_2\mapsto f(z_1+i_2z_2)$ be a function defined in X. Then

$$(1)\qquad f(z_1+i_2z_2)=u(z_1,z_2)+i_2v(z_1,z_2),\qquad z_1+i_2z_2\in X,$$
$$u(z_1,z_2)\in\mathbb{C}_1,\ v(z_1,z_2)\in\mathbb{C}_1.$$

Since $f(z_1+i_2z_2)\in\mathbb{C}_2$, then $f(z_1+i_2z_2)$ has the following idempotent representation:

$$(2)\qquad f(z_1+i_2z_2)=[u(z_1,z_2)-i_1v(z_1,z_2)]e_1 + [u(z_1,z_2)+i_1v(z_1,z_2)]e_2.$$

There is no surprise in this statement, but what is surprising is the following: if $f: X\to\mathbb{C}_2$ is a holomorphic function (a function represented locally by power series), then $u(z_1,z_2)-i_1v(z_1,z_2)$ and $u(z_1,z_2)+i_1v(z_1,z_2)$ are values of holomorphic functions f_1, f_2 of the complex variables $z_1-i_1z_2$ and $z_1+i_1z_2$, respectively. Thus a holomorphic function of a bicomplex variable is represented by a pair of holomorphic functions f_1, f_2 of complex variables as follows:

$$(3)\qquad f(z_1+i_2z_2)=f_1(z_1-i_1z_2)e_1+f_2(z_1+i_1z_2)e_2,$$
$$f_1(z_1-i_1z_2)\in\mathbb{C}_1,\ f_2(z_1+i_1z_2)\in\mathbb{C}_1.$$

The power series representation of f provides a quick and easy demonstration of this representation. Furthermore, the construction can be reversed: a pair f_1, f_2 of holomorphic functions of the complex variables $z_1-i_1z_2$ and $z_1+i_1z_2$ can be used with (3) to define a holomorphic function f of the bicomplex variable $z_1+i_2z_2$. Many of the properties of f can be derived from properties of the pair of functions f_1, f_2 of complex variables. If f is represented by a power series, then power series in $\mathbb{C}_1$ which represent f_1 and f_2 can be obtained from (2). This chapter proves that the power series for f converges if and only if the power series for f_1 and f_2 converge. Since the power series for f_1 and f_2, in the variables $z_1-i_1z_2$ and $z_1+i_1z_2$, converge in circles in A_1 and A_2, it follows that the power series for f, in the variable $z_1+i_2z_2$, converges in a discus in $\mathbb{C}_2$.

After this chapter has established the fundamental properties of functions defined by power series in $\mathbb{C}_2$, it defines the elementary functions of a bicomplex variable and investigates some of their properties. The exponential function $e^{z_1+i_2z_2}$, the trigonometric functions $\cos(z_1+i_2z_2)$ and $\sin(z_1+i_2z_2)$, and the hyperbolic functions $\cosh(z_1+i_2z_2)$ and $\sinh(z_1+i_2z_2)$ are defined by power series obtained by replacing the complex variable z by the bicomplex variable $z_1+i_2z_2$ in the familiar power series for these functions. These new functions of a bicomplex variable are doubly periodic entire functions.

One of the theorems established in the chapter provides a method for finding the inverse of a holomorphic function of a bicomplex variable. An application of this theorem calculates the inverse of the exponential function; the result is an explicit formula for the logarithm function. The theory and the resulting formula show that $\log(z_1+i_2z_2)$ is defined for every $z_1+i_2z_2$ not in $\mathcal{O}_2$; and, corresponding to the fact that $e^{z_1+i_2z_2}$ is doubly periodic, $\log(z_1+i_2z_2)$ is a doubly infinitely multiple valued function.

11. LIMITS OF SEQUENCES

This section treats sequences in $\mathbb{C}_2$ and their limits. Since $\mathbb{C}_2$ is a Banach space, the standard definitions and theorems apply. Thus a sequence in $\mathbb{C}_2$ is a function $s:\mathbb{N}\to\mathbb{C}_2$, $n\mapsto s_n$, $s_n\in\mathbb{C}_2$; this sequence converges to a limit s^* if and only if to each $\varepsilon>0$ there corresponds an $n(\varepsilon)$ such that

(1) $\quad \|s_n - s^*\| < \varepsilon, \qquad \forall n \geqslant n(\varepsilon).$

The sequence s is a Cauchy sequence if and only if to each $\varepsilon>0$ there corresponds an $n(\varepsilon)$ such that

(2) $\quad \|s_n - s_m\| < \varepsilon, \qquad \forall n \geqslant n(\varepsilon), \forall m \geqslant n(\varepsilon).$

Finally, s converges to a limit in $\mathbb{C}_2$ if and only if it is a Cauchy sequence. These statements would seem to dispose of the subject of sequences in $\mathbb{C}_2$, but there is more to be said because of the multiple representations of an element and the norm in $\mathbb{C}_2$.

If ζ is an element in $\mathbb{C}_2$, then ζ and its norm have three representations as follows:

(3) $\quad \zeta = x_1 + i_1x_2 + i_2x_3 + i_1i_2x_4, \qquad \|\zeta\| = [x_1^2 + \cdots + x_4^2]^{1/2},$
$\quad x_i\in\mathbb{C}_0,\ i = 1,\ldots,4.$

(4) $\quad \zeta = z_1 + i_2z_2, \qquad \|\zeta\| = [|z_1|^2 + |z_2|^2]^{1/2}, \qquad z_1 = x_1 + i_1x_2,$
$\quad z_2 = x_3 + i_1x_4.$

(5) $\quad \zeta = (z_1 - i_1z_2)e_1 + (z_1 + i_1z_2)e_2,$

$$\|\zeta\| = \left[\frac{|z_1 - i_1z_2|^2 + |z_1 + i_1z_2|^2}{2}\right]^{1/2}.$$

Let $s:\mathbb{N}\to\mathbb{C}_2$, $n\mapsto s_n$, be a sequence such that

(6) $\quad s_n = x_{1n} + i_1x_{2n} + i_2x_{3n} + i_1i_2x_{4n}, \qquad n = 0, 1, \ldots.$

The representations of ζ in (3)–(5) show that the following significant sequences are associated with s:

(7) Four sequences in $\mathbb{C}_0$: x_{kn}, $n = 0, 1, \ldots$, $k = 1, \ldots, 4$;

(8) Two sequences in $\mathbb{C}_1$: z_{kn}, $n = 0, 1, \ldots$, $k = 1, 2$;

(9) Two sequences in A_1 and A_2: $z_{1n} - i_1z_{2n}$ and $z_{1n}+i_1z_{2n}$, $n = 0, 1, \ldots.$

11.1 THEOREM If $s:\mathbb{N}\to\mathbb{C}_2$, $s_n=x_{1n}+i_1x_{2n}+i_2x_{3n}+i_1i_2x_{4n}$, is a sequence and

(10) $\quad \lim_{n\to\infty} s_n = x_1^* + i_1x_2^* + i_2x_3^* + i_1i_2x_4^* = \zeta^*,$

then the following limits exist and have the values shown:

(11) $\lim_{n\to\infty} x_{kn} = x_k^*, \qquad k = 1,\ldots,4;$

(12) $\lim_{n\to\infty} z_{1n} = z_1^* = x_1^* + i_1 x_2^*, \qquad \lim_{n\to\infty} z_{2n} = z_2^* = x_3^* + i_1 x_4^*;$

(13) $\lim_{n\to\infty} (z_{1n} - i_1 z_{2n}) = z_1^* - i_1 z_2^*, \qquad \lim_{n\to\infty} (z_{1n} + i_1 z_{2n}) = z_1^* + i_1 z_2^*.$

Furthermore, if the limits exist as indicated in any one of the statements (11)–(13), then $\lim_{n\to\infty} s_n$ exists as stated in (10), and all of the statements (11)–(13) are true.

Proof. The three representations for the norm in (3)–(5) show that the following inequalities are valid:

(14) $|x_{kn}-x_k^*| \leqslant \|s_n-\zeta^*\| \leqslant |x_{1n}-x_1^*| + \cdots + |x_{4n}-x_4^*|, \qquad k=1,\ldots,4;$

(15) $|z_{kn}-z_k^*| \leqslant \|s_n-\zeta^*\| \leqslant |z_{1n}-z_1^*| + |z_{2n}-z_2^*|, \qquad k=1, 2;$

(16)
$$\left.\begin{array}{l}|(z_{1n}-i_1 z_{2n})-(z_1^*-i_1 z_2^*)|\\ |(z_{1n}+i_1 z_{2n})-(z_1^*+i_1 z_2^*)|\end{array}\right\} \leqslant \sqrt{2}\,\|s_n-\zeta^*\| \leqslant |(z_{1n}-i_1 z_{2n})-(z_1^*-i_1 z_2^*)| + |(z_{1n}+i_1 z_{2n})-(z_1^*+i_1 z_2^*)|.$$

If $\lim_{n\to\infty} s_n=\zeta^*$, then the inequalities on the left show that the statements in (11)–(13) are true. Furthermore, the inequalities on the right show that, if any one of the statements (11)–(13) is true, then $\lim_{n\to\infty} s_n=\zeta^*$, and the first part of the theorem shows that all of the statements in (11)–(13) are true. The proof of Theorem 11.1 is complete. □

Exercises

11.1 Let $s:\mathbb{N}\to\mathbb{C}_2$ be a sequence such that

$$s_n = \left(\frac{z_1 - i_1 z_2}{n}\right)e_1 + \left(\frac{s_1 + i_1 s_2}{n}\right)e_2.$$

Show in two ways that $\lim_{n\to\infty} s_n=0$.

11.2 Let ζ be an element in $\mathbb{C}_2$, and let $s_n:\mathbb{N}\to\mathbb{C}_2$ be the sequence such that $s_n=\zeta^n$.

(a) If $\zeta=e_1$ or if $\zeta=e_2$, show that $\|\zeta\|=\sqrt{2}/2$ and that $\lim_{n\to\infty} s_n$ exists; find the value of this limit.

(b) If $\zeta=\sqrt{2}e_1$ or if $\zeta=\sqrt{2}e_2$, show that $\|\zeta\|=1$ but that s_n diverges to infinity.

(c) If ζ is in $\mathbb{C}_2$ and $\|\zeta\|<1/\sqrt{2}$, show that $\lim_{n\to\infty} s_n=0$.

(d) Let ζ be a complex number $x_1+i_1x_2$ in $\mathbb{C}_2$ such that $1/\sqrt{2}<|x_1+i_1x_2|<1$. Show that $\lim_{n\to\infty} s_n=0$.

(e) Let $\zeta=\omega_1e_1+\omega_2e_2$, an rth root of unity (see Exercise 6.4). Show that s is a bounded sequence, and that $\|s_n\|\leqslant 1$. Show also that this sequence does not converge.

11.3 Let $s:\mathbb{N}\to\mathbb{C}_2$ be a sequence such that

$$s_n=\left(\cos\frac{2n\pi}{3}-i_1\sin\frac{2n\pi}{3}\right)e_1+\left(\cos\frac{2n\pi}{3}+i_1\sin\frac{2n\pi}{3}\right)e_2.$$

Show that this sequence is bounded but that it does not have a limit.

11.4 Let $\Sigma_{k=0}^{\infty} a_kz^k$ and $\Sigma_{k=0}^{\infty} b_kz^k$ be two complex infinite series which converge for $|z|<r$. Define the sequence $s:\mathbb{N}\to\mathbb{C}_2$ as follows:

$$s_n(z)=\sum_{k=0}^{n} a_kz^k+i_2\sum_{k=0}^{n} b_kz^k.$$

Show that $\lim_{n\to\infty} s_n(z)$ exists for $|z|<r$, and find the value of this limit.

11.5 Consider the infinite series

$$\sum_{k=0}^{\infty}\frac{\zeta^k}{k!},\qquad \zeta\in\mathbb{C}_2.$$

(a) Show that, if $\zeta=z_1+i_2z_2$, then

$$\sum_{k=0}^{n}\frac{\zeta^k}{k!}=\sum_{k=0}^{n}\frac{(z_1-i_1z_2)^k}{k!}e_1+\sum_{k=0}^{n}\frac{(z_1+i_1z_2)^k}{k!}e_2.$$

(b) Define the sequence $s:\mathbb{N}\to\mathbb{C}_2$ so that

$$s_n(\zeta)=\sum_{k=0}^{n}\frac{\zeta^k}{k!}.$$

Show that $s(\zeta)$ converges for every value of ζ, and that

$$\lim_{n\to\infty} s_n(\zeta)=e^{z_1-i_1z_2}e_1+e^{z_1+i_1z_2}e_2.$$

(c) Show that for every ζ in $\mathbb{C}_2$,

$$\lim_{n\to\infty} s_n(\zeta)\notin\mathcal{O}_2.$$

12. INFINITE SERIES

This section treats infinite series whose terms are constants in $\mathbb{C}_2$. Corresponding to each of the representations of an element and its norm in $\mathbb{C}_2$, there is a form for the infinite series and conditions for its convergence. Each representation has its own special importance.

The expression

$$(1)\qquad \sum_{k=0}^{\infty} \zeta_k, \qquad \forall k,\ \zeta_k \in \mathbb{C}_2,$$

is called an infinite series in $\mathbb{C}_2$. Define the sequence $s: \mathbb{N} \to \mathbb{C}_2$, $n \mapsto s_n$, by setting

$$(2)\qquad s_n = \sum_{k=0}^{n} \zeta_k, \qquad \forall n \in \mathbb{N}.$$

12.1 DEFINITION The infinite series (1) *converges* if and only if

$$(3)\qquad \lim_{n\to\infty} s_n$$

exists; if the limit (3) does not exist, the series *diverges*. If

$$(4)\qquad \lim_{n\to\infty} s_n = \zeta^*,$$

then ζ^* is called the *sum* of the series, and we write

$$(5)\qquad \sum_{k=0}^{\infty} \zeta_k = \zeta^*.$$

12.2 THEOREM In (1) let $\zeta_k = x_{1k} + i_1 x_{2k} + i_2 x_{3k} + i_1 i_2 x_{4k}$. Then the infinite series (1) converges and has the sum $\zeta^* = x_1^* + i_1 x_2^* + i_2 x_3^* + i_1 i_2 x_4^*$ if and only if the following infinite series converge and have the sums shown:

$$(6)\qquad \sum_{k=0}^{\infty} x_{1k} = x_1^*, \ldots, \sum_{k=0}^{\infty} x_{4k} = x_4^*.$$

Proof. Now

$$(7)\qquad s_n - \zeta^* = \left(\sum_{k=0}^{n} x_{1k} - x_1^*\right) + i_1\left(\sum_{k=0}^{n} x_{2k} - x_2^*\right) + i_2\left(\sum_{k=0}^{n} x_{3k} - x_3^*\right) + i_1 i_2\left(\sum_{k=0}^{n} x_{4k} - x_4^*\right).$$

Then [compare (14) in Section 11]

$$(8)\qquad \left|\sum_{k=0}^{n} x_{1k} - x_1^*\right| \leqslant \|s_n - \zeta^*\|, \ldots, \left|\sum_{k=0}^{n} x_{4k} - x_4^*\right| \leqslant \|s_n - \zeta^*\|.$$

If $\lim_{n\to\infty} s_n = \zeta^*$, then the inequalities in (8) show that the infinite series in (6) converge and have the sums indicated. Next, assume that the statements in (6) are true. Then

$$(9)\qquad \|s_n - \zeta^*\| \leqslant \left|\sum_{k=0}^{n} x_{1k} - x_1^*\right| + \cdots + \left|\sum_{k=0}^{n} x_{4k} - x_4^*\right|,$$

and this inequality shows that $\lim_{n\to\infty} s_n = \zeta^*$. The proof of Theorem 12.2 is complete. □

Set

(10) $z_{1k} = x_{1k} + i_1 x_{2k}, \qquad z_{2k} = x_{3k} + i_1 x_{4k}.$

Then

(11) $\zeta_k = z_{1k} + i_2 z_{2k}, \qquad \forall k \in \mathbb{N},$

and the infinite series in (1) can be written in the following additional form:

(12) $$\sum_{k=0}^{\infty} (z_{1k} + i_2 z_{2k}).$$

12.3 THEOREM The infinite series (1) converges and has the sum ζ^* if and only if the following infinite series converge and have the sums shown:

(13) $$\sum_{k=1}^{\infty} z_{1k} = z_1^* = x_1^* + i_1 x_2^*, \qquad \sum_{k=0}^{\infty} z_{2k} = z_2^* = x_3^* + i_1 x_4^*.$$

Proof. This theorem follows from the inequalities in (15) in Section 11 just as Theorem 12.2 follows from (14) in that section. □

By the idempotent representation of elements in $\mathbb{C}_2$,

(14) $z_{1k} + i_2 z_{2k} = (z_{1k} - i_1 z_{2k})e_1 + (z_{1k} + i_1 z_{2k})e_2.$

The properties of the idempotent representation show that the infinite series in (12) can be written as follows:

(15) $$\sum_{k=0}^{\infty} (z_{1k}+i_2 z_{2k}) = \left[\sum_{k=0}^{\infty} (z_{1k}-i_1 z_{2k})\right]e_2 + \left[\sum_{k=0}^{\infty} (z_{1k}+i_1 z_{2k})\right]e_2.$$

12.4 THEOREM The infinite series (1) [see also (12) and (15)] converges and has the sum $\zeta^* = z_1^* + i_2 z_2^*$ if and only if the following infinite series converge and have the sums shown:

(16) $$\sum_{k=0}^{\infty} (z_{1k} - i_1 z_{2k}) = z_1^* - i_1 z_2^*, \qquad \sum_{k=0}^{\infty} (z_{1k} + i_1 z_{2k}) = z_1^* + i_1 z_2^*.$$

Proof. The proof of this theorem depends on properties of the norm in the idempotent representation [see (5) and (16) in Section 11]. Now

(17) $$\begin{aligned} \sqrt{2}&\left\|\sum_{k=0}^{n} (z_{1k} + i_2 z_{2k}) - (z_1^* + i_2 z_2^*)\right\| \\ &\leqslant \left|\sum_{k=0}^{n} (z_{1k} - i_1 z_{2k}) - (z_1^* - i_1 z_2^*)\right| \\ &\quad + \left|\sum_{k=0}^{n} (z_{1k} + i_1 z_{2k}) - (z_1^* + i_1 z_2^*)\right|. \end{aligned}$$

This inequality shows that, if (16) is true, then the series in (12) [which is the series in (1)] converges and has the sum $z_1^* + i_2 z_2^*$. Each of the two absolute value terms on the right in (17) is equal to or less than the term on the left, and these inequalities show that the two series in (16) converge to the sums shown if (1) converges to $z_1^* + i_2 z_2^*$. The proof of Theorem 12.4 is complete. □

12.5 EXAMPLE Each of the theorems in this section can be used to construct convergent bicomplex infinite series from convergent real or complex series. The following example illustrates one method. Let

$$(18) \qquad \sum_{k=0}^{\infty} a_k = a^*, \qquad \sum_{k=0}^{\infty} b_k = b^*, \qquad a_k, b_k \in \mathbb{C}_1, \ \forall k \in \mathbb{N},$$

be two convergent complex infinite series, with the sums as shown, such that, for an infinite number of values of k in $\mathbb{N}$, it is true that $a_k \neq b_k$. Define $z_{1k} + i_2 z_{2k}$ as follows:

$$(19) \qquad z_{1k} + i_2 z_{2k} = a_k e_1 + b_k e_2 = \left(\frac{a_k + b_k}{2}\right) + i_2 \left[\frac{i_1(a_k - b_k)}{2}\right], \qquad \forall k \in \mathbb{N}.$$

Then

$$(20) \qquad z_{1k} = \frac{a_k + b_k}{2}, \qquad z_{2k} = \frac{i_1(a_k - b_k)}{2},$$

and

$$(21) \qquad \sum_{k=0}^{\infty} (z_{1k} + i_2 z_{2k})$$

is a bicomplex infinite series in $\mathbb{C}_2$ and not a complex infinite series in $\mathbb{C}_1$. Furthermore, the infinite series $\Sigma_{k=0}^{\infty}(z_{1k} - i_1 z_{2k})$ and $\Sigma_{k=0}^{\infty}(z_{1k} + i_1 z_{2k})$ are the given series in (18) since

$$(22) \qquad \begin{aligned} \left(\frac{a_k + b_k}{2}\right) - i_1 \left[\frac{i_1(a_k - b_k)}{2}\right] &= a_k, \\ \left(\frac{a_k + b_k}{2}\right) + i_1 \left[\frac{i_1(a_k - b_k)}{2}\right] &= b_k, \end{aligned} \qquad \forall k \in \mathbb{N}.$$

Thus the bicomplex infinite series (21) converges if the complex infinite series in (18) converge, and it diverges in all other cases.

Exercises

12.1 Give two proofs of the following theorem. If $\Sigma_{k=0}^{\infty}(z_{1k} + i_2 z_{2k})$ converges, then $\lim_{k\to\infty}(z_{1k} + i_2 z_{2k}) = 0$.

12.2 By definition, the series $\Sigma_{k=0}^{\infty}(z_{1k} + i_2 z_{2k})$ converges absolutely if and only if $\Sigma_{k=0}^{\infty} \|z_{1k} + i_2 z_{2k}\|$ converges. Prove that a series converges if it converges absolutely.

12.3 Give an example to show that absolute convergence is not a necessary condition for convergence.

12.4 Prove the following theorem. If $\Sigma_{k=0}^{\infty}(c_k+i_2d_k)$ is a series in $\mathbb{C}_2$ which converges absolutely, and if $\Sigma_{k=0}^{\infty}(a_k+i_2b_k)$ is a series in $\mathbb{C}_2$ such that

$$\|a_k + i_2b_k\| \leqslant \|c_k + i_kd_k\|, \qquad \forall k \in \mathbb{N},$$

then $\Sigma_{k=\infty}^{\infty}(a_k+i_2b_k)$ converges absolutely and

$$\left\|\sum_{k=0}^{\infty}(a_k + i_2b_k)\right\| \leqslant \sum_{k=0}^{\infty}\|c_k + i_2d_k\|.$$

12.5 Prove that $\Sigma_{k=0}^{\infty}(a_k+i_2b_k)$ converges absolutely if and only if $\Sigma_{k=0}^{\infty}(a_k-i_1b_k)$ and $\Sigma_{k=0}^{\infty}(a_k+i_1b_k)$ converge absolutely.

13. POWER SERIES

This section is devoted to an investigation of the convergence of power series in the bicomplex variable $z_1+i_2z_2$; the next section begins the study of the properties of functions represented by power series. The section begins by calling attention to some of the differences between power series in $\mathbb{C}_1$ and in $\mathbb{C}_2$. As Section 12 suggests (see Theorem 12.4), the idempotent series play a major role in determining the convergence of power series. The major results in this section are: (a) a power series converges if and only if the idempotent component power series converge; (b) the region of convergence of a bicomplex power series is the discus (including degenerate types) rather than the ball (see Section 9); and (c) the idempotent components of a bicomplex power series are power series in complex variables, namely, the idempotent components $z_1-i_1z_2$ and $z_1+i_1z_2$ of the bicomplex variable $z_1+i_2z_2$.

13.1 DEFINITION Let α_k, ζ, and ζ_0 denote elements in $\mathbb{C}_2$. A power series in the bicomplex variable ζ about the point ζ_0 is an infinite series of the form

$$(1) \qquad \sum_{k=0}^{\infty}\alpha_k(\zeta - \zeta_0)^k.$$

The series (1) converges if $\zeta=\zeta_0$; if $\|\alpha_k\|$ increases too rapidly with k, the series does not converge for any other value of ζ. The interesting series, however, are those which converge for values of ζ different from ζ_0.

13.2 THEOREM (Weierstrass) Let a_k, z, and z_0 denote complex numbers in $\mathbb{C}_1$. If there exists a $z_0\neq 0$ such that the terms of the power series $\Sigma_{k=0}^{\infty}a_kz^k$ are bounded for $z=z_0$:

$$(2) \qquad |a_kz_0^k| \leqslant M, \qquad \forall k \in \mathbb{N},$$

then the series converges absolutely for every z such that $|z|<|z_0|$, and uniformly with respect to z in $|z|\leqslant(1-\varepsilon)|z_0|$, $0<\varepsilon<1$.

Proof. By elementary properties of complex numbers,

$$(3)\qquad |a_k z^k| = \left|a_k z_0^k \frac{z^k}{z_0^k}\right| = |a_k z_0^k| \frac{|z^k|}{|z_0^k|} \leqslant M\left(\frac{|z|}{|z_0|}\right)^k.$$

If $|z|<|z_0|$, then $\Sigma_{k=0}^{\infty}(M|z|^k/|z_0|^k)$ converges by comparison with the geometric series, and $\Sigma_{k=0}^{\infty} a_k z^k$ converges absolutely by the comparison test. Also, $|a_k z^k|\leqslant M(1-\varepsilon)^k$ for all z in $|z|\leqslant(1-\varepsilon)|z_0|$, and $\Sigma_{k=0}^{\infty} a_k z^k$ converges uniformly in $|z|\leqslant(1-\varepsilon)|z_0|$ by the Weierstrass M-test. □

This theorem leads to the following conclusion. If $\Sigma_{k=0}^{\infty} a_k z^k$ converges for $z_0\neq 0$, then either it converges for all z or there is a circle in $\mathbb{C}_1$ in which it converges.

Weierstrass' theorem suggests the following conjecture for bicomplex power series.

13.3 CONJECTURE If the bicomplex power series

$$(4)\qquad \sum_{k=0}^{\infty} \alpha_k \zeta^k$$

converges at the fixed point

$$(5)\qquad \eta \neq 0, \qquad \eta\in\mathbb{C}_2,$$

then the series (4) converges absolutely for all ζ such that

$$(6)\qquad \|\zeta\| < \|\eta\|.$$

It might be thought that (3) would be a model for a proof of this conjecture, but even the first step

$$(7)\qquad \|\alpha_k \zeta^k\| = \left\|\alpha_k \eta^k \frac{\zeta^k}{\eta^k}\right\|$$

is not valid unless the hypothesis (5) is strengthened to

$$(8)\qquad \eta\notin\mathcal{O}_2.$$

Theorem 4.4 (11) shows that the next step in (3), in the present case, becomes

$$(9)\qquad \left\|\alpha_k \eta^k \frac{\zeta^k}{\eta^k}\right\| \leqslant \sqrt{2}\,\|\alpha_k \eta^k\| \left\|\frac{\zeta^k}{\eta^k}\right\|,$$

and Theorem 5.5 indicates that there is no useful way to relate

$$(10) \qquad \left\|\frac{\zeta^k}{\eta^k}\right\| \text{ and } \frac{\|\zeta^k\|}{\|\eta^k\|}.$$

Thus this effort to prove the conjecture fails.

13.4 THEOREM Let α_k and ζ denote bicomplex numbers in $\mathbb{C}_2$. If there exists a z_0 in $\mathbb{C}_1$, $z_0 \neq 0$, such that the terms of the power series $\Sigma_{k=0}^{\infty} \alpha_k \zeta^k$ are bounded for $\zeta = z_0$,

$$(11) \qquad \|\alpha_k z_0^k\| \leqslant M, \qquad \forall k \in \mathbb{N},$$

then the series $\Sigma_{k=0}^{\infty} \alpha_k \zeta^k$ converges absolutely for every ζ in $\mathbb{C}_2$ such that

$$(12) \qquad \|\zeta\| < \frac{|z_0|}{\sqrt{2}},$$

and uniformly with respect to ζ for every ζ such that

$$(13) \qquad \|\zeta\| \leqslant \frac{(1-\varepsilon)|z_0|}{\sqrt{2}}, \qquad 0 < \varepsilon < 1.$$

Proof. Since $z_0 \in \mathbb{C}_1$ and $z_0 \neq 0$, Theorems 5.4 and 4.4 (11), (10) show that

$$\begin{aligned}(14) \qquad \|\alpha_k \zeta^k\| &= \left\|\alpha_k z_0^k \frac{\zeta^k}{z_0^k}\right\| \\ &\leqslant \sqrt{2}\,\|\alpha_k z_0^k\| \left\|\frac{\zeta^k}{z_0^k}\right\| \\ &\leqslant \sqrt{2}M(\sqrt{2})^{k-1}\frac{\|\zeta\|^k}{|z_0|^k} \\ &\leqslant M\left[\frac{\sqrt{2}\|\zeta\|}{|z_0|}\right]^k.\end{aligned}$$

Thus $\Sigma_{k=0}^{\infty} \alpha_k \zeta^k$ converges absolutely for every ζ such that

$$(15) \qquad \frac{\sqrt{2}\|\zeta\|}{|z_0|} < 1 \qquad \text{or} \qquad \|\zeta\| < \frac{|z_0|}{\sqrt{2}}.$$

Also, (14) shows that $\|\alpha_k \zeta^k\| \leqslant M(1-\varepsilon)^k$ for all ζ such that $\|\zeta\| \leqslant (1-\varepsilon)|z_0|/\sqrt{2}$, and thus that $\Sigma_{k=0}^{\infty} \alpha_k \zeta^k$ converges uniformly in (13) by the Weierstrass M-test. The proof of Theorem 13.4 is complete. □

Theorem 13.4 gives some information about the convergence of $\Sigma_{k=0}^{\infty} \alpha_k \zeta^k$, but it is very incomplete. There may be no $z_0 \neq 0$ in $\mathbb{C}_1$ for which the series converges; in this case the theorem gives no information about the convergence of the series. Furthermore, Theorem 13.4 gives no information in any case about the divergence of the series. Thus a complete determination of the convergence and divergence of bicomplex power series is still lacking. Section 12 suggests that necessary and sufficient conditions for convergence can be found by studying the idempotent component series. In order to begin this study, it is necessary to replace the notation in (1) by one showing more detail. If $\zeta_0=0$ and $\alpha_k=a_k+i_2b_k$ and $\zeta=z_1+i_2z_2$, then the power series (1) is

$$\sum_{k=0}^{\infty}(a_k+i_2b_k)(z_1+i_2z_2)^k, \qquad \forall k\in\mathbb{N},\ a_k,\ b_k,\ z_1,\ z_2\in\mathbb{C}_1. \tag{16}$$

13.5 THEOREM The idempotent component series of the bicomplex power series (16) are the complex power series

$$\sum_{k=0}^{\infty}(a_k-i_1b_k)(z_1-i_1z_2)^k, \tag{17}$$

$$\sum_{k=0}^{\infty}(a_k+i_1b_k)(z_1+i_1z_2)^k; \tag{18}$$

that is,

$$\begin{aligned}\sum_{k=0}^{\infty}(a_k+i_2b_k)(z_1+i_2z_2)^k &= \left[\sum_{k=0}^{\infty}(a_k-i_1b_k)(z_1-i_2z_2)^k\right]e_1\\ &\quad+\left[\sum_{k=0}^{\infty}(a_k+i_1b_k)(z_1+i_1z_2)^k\right]e_2.\end{aligned} \tag{19}$$

Proof. Theorem 6.6 shows that

$$a_k+i_2b_k=(a_k-i_1b_k)e_1+(a_k+i_1b_k)e_2, \tag{20}$$

$$z_1+i_2z_2=(z_1-i_1z_2)e_1+(z_1+i_1z_2)e_2, \tag{21}$$

$$(z_1+i_2z_2)^k=(z_1-i_1z_2)^ke_1+(z_1+i_1z_2)^ke_2, \tag{22}$$

$$\begin{aligned}(a_k+i_2b_k)(z_1+i_2z_2)^k &= (a_k-i_1b_k)(z_1-i_1z_2)^ke_1\\ &\quad+(a_k+i_1b_k)(z_1+i_1z_2)^ke_2,\end{aligned} \tag{23}$$

$$\begin{aligned}&\sum_{k=0}^{\infty}(a_k+i_2b_k)(z_1+i_2z_2)^k\\ &\quad=\sum_{k=0}^{\infty}(a_k-i_1b_k)(z_1-i_1z_2)^ke_1+\sum_{k=0}^{\infty}(a_k+i_1b_k)(z_1+i_1z_2)^ke_2.\end{aligned} \tag{24}$$

The formula in (24) shows that the idempotent component series of (16) are the series in (17) and (18), and that the statement in (19) is true at least formally. Matters of convergence are treated in the next theorem. □

13.6 THEOREM If the bicomplex power series (16) converges at $z_1+i_2z_2$, then the complex power series (17) and (18) converge at $z_1-i_1z_2$ and $z_1+i_1z_2$, respectively. Conversely, if (17) and (18) converge at $z_1-i_1z_2$ and $z_1+i_1z_2$, respectively, then (16) converges at $z_1+i_2z_2$.

Proof. This theorem follows from Theorems 13.5 and 12.4. □

Some further explanation of Theorem 13.6 may be helpful. The series in (17) and (18) are the following complex power series:

$$(25) \qquad \sum_{k=0}^{\infty} (a_k - i_1 b_k)w^k, \qquad w \in \mathbb{C}_1,$$

$$(26) \qquad \sum_{k=0}^{\infty} (a_k + i_1 b_k)w^k, \qquad w \in \mathbb{C}_1.$$

Theorem 13.6 asserts that, if the bicomplex series (16) converges at $z_1+i_2z_2$, then (25) converges for $w=z_1-i_1z_2$ and (26) converges for $w=z_1+i_1z_2$. Also if (25) and (26) converge at w_1 and w_2, respectively, then (16) converges at the corresponding point $z_1+i_2z_2$ such that

$$(27) \qquad z_1 + i_2 z_2 = w_1 e_1 + w_2 e_2 = \left(\frac{w_1 + w_2}{2}\right) + i_2\left[\frac{i_1(w_1 - w_2)}{2}\right].$$

Thus Theorem 13.6 shows that the bicomplex power series (16) converges if and only if the complex power series (25) and (26) [or (17) and (18)] converge at the corresponding points. Since (25) and (26) are complex power series, and since known theorems give complete information (except on the circle of convergence) about the convergence and divergence of complex power series, it is possible to determine the convergence and divergence of bicomplex power series as stated in the next theorem.

13.7 THEOREM Let r_1 and r_2 denote the radii of the circles of convergence of (25) and (26), respectively. Then $\Sigma_{k=0}^{\infty}(a_k+i_2b_k)(z_1+i_2z_2)^k$ converges absolutely at every $z_1+i_2z_2$ in the discus $D(0; r_1, r_2)$, and it diverges at every $z_1+i_2z_2$ in the complement of $\bar{D}(0; r_1, r_2)$. The radii of convergence r_1 and r_2 may have the values 0 and ∞.

Proof. A complex power series converges absolutely in its circle of convergence. Thus the series (25) and (26) converge absolutely in the sets

$$(28) \qquad X_1 = \{w_1 \in \mathbb{C}_1 \colon |w_1| < r_1\},\ X_2 = \{w_2 \in \mathbb{C}_1 \colon |w_2| < r_2\}.$$

Then since

$$(29)\qquad \|(a_k + i_2 b_k)(z_1 + i_2 z_2)^k\| \leqslant |(a_k - i_1 b_k)(z_1 - i_1 z_2)^k|\ \|e_1\| + |(a_k + i_1 b_k)(z_1 + i_1 z_2)^k|\ \|e_2\|, \qquad \forall k \in \mathbb{N},$$

the series

$$(30)\qquad \sum_{k=0}^{\infty} (a_k + i_2 b_k)(z_1 + i_2 z_2)^k$$

converges absolutely (compare Exercise 12.4) in

$$(31)\qquad X = \{(z_1 + i_2 z_2) \in \mathbb{C}_2 \colon z_1 + i_2 z_2 = w_1 e_1 + w_2 e_2,\ (w_1, w_2) \in X_1 \times X_2\},$$

which, by Definition 9.1, is the discus $D(0; r_1, r_2)$. Thus the first conclusion has been proved. Next, let $z_1 + i_2 z_2$ be a point in the complement of $\bar{D}(0; r_1, r_2)$. If

$$(32)\qquad z_1 + i_2 z_2 = w_1 e_1 + w_2 e_2,$$

then either $|w_1| > r_1$ or $|w_2| > r_2$ (or both) because otherwise $z_1 + i_2 z_2$ would be a point in $\bar{D}(0; r_1, r_2)$ by Definition 9.1. Now (25) diverges for $|w| > r_1$, and (26) diverges for $|w| > r_2$. Thus if $z_1 + i_2 z_2$ is in the complement of $\bar{D}(0; r_1, r_2)$, at least one of the series

$$(33)\qquad \sum_{k=0}^{\infty} (a_k - i_1 b_k) w_1^k, \qquad \sum_{k=0}^{\infty} (a_k + i_1 b_k) w_2^k,$$

diverges; therefore, by Theorem 13.6, the bicomplex series (30) diverges at $z_1 + i_2 z_2$. The proof of Theorem 13.7 is complete. □

The region of convergence is called a discus $D(0; r_1, r_2)$ in all cases, but nine different types can be identified corresponding to zero, positive, and infinite values for r_1 and r_2. These types are the following (r_1 and r_2 denote positive numbers in this array):

$$(34)\qquad \begin{array}{lll} D(0; 0, 0) & D(0; 0, r_2) & D(0; 0, \infty) \\ D(0; r_1, 0) & D(0; r_1, r_2) & D(0; r_1, \infty) \\ D(0; \infty, 0) & D(0; \infty, r_2) & D(0; \infty, \infty). \end{array}$$

A description of $D(0; r_1, r_2)$ in some representative cases follows.

$$(35)\qquad r_1 = 0, \qquad r_2 = 0.$$

In this case, each of the series (25) and (26) converges only at $w = 0$, and the bicomplex power series (30) converges only for $z_1 + i_2 z_2 = 0$. The discus

$D(0; 0, 0)$ contains the single point $0+i_2 0$.

(36) $\quad r_1 = 0, \qquad r_2 > 0.$

In this case, the series (25) converges only at $w=0$, but (26) has a positive radius of convergence $r_2>0$. The bicomplex series (30) converges in the discus $D(0; 0, r_2)$, which is the set

$$\{z_1 + i_2 z_2 \in \mathbb{C}_2: z_1 + i_2 z_2 = w_2 e_2,\ w_2 \in \mathbb{C}_1,\ |w_2| < r_2\}.$$

Equation (7) in Section 7 shows that $D(0; 0, r_2) \subset I_2$. Since I_2 contains no complex number w such that $|w|>0$, this case provides an example of a bicomplex power series to which Theorem 13.4 cannot be applied: there is no z_0 in $\mathbb{C}_1$, $|z_0|>0$, for which $\|(a_k+i_2 b_k)z_0^k\| \leqslant M$ for $k=0, 1, \ldots$ [compare (11)]. If $r_1>0$ and $r_2=0$, the bicomplex series (30) converges in $D(0; r_1, 0)$ and this discus is contained in I_1; the details are similar to those in the case described.

(37) $\quad r_1 > 0, \qquad r_2 > 0.$

The bicomplex power series (30) converges in the nondegenerate discus $D(0; r_1, r_2)$ and diverges in the complement of $\bar{D}(0; r_1, r_2)$. If $0<r_1<r_2$, Theorem 9.2 shows that

$$B(0, r_1/\sqrt{2}) \subsetneqq D(0; r_1, r_2) \subsetneqq B\left(0, \left(\frac{r_1^2 + r_2^2}{2}\right)^{1/2}\right).$$

The case $0<r_2<r_1$ is similar.

(38) $\quad r_1 > 0, \qquad r_2 = \infty.$

In this case, (25) has a finite radius of convergence, but (26) converges for all w in $\mathbb{C}_1$ and represents an entire function. The discus $D(0; r_1, \infty)$ contains the ball $B(0; r_1/\sqrt{2})$, and it is an unbounded set (see Exercise 9.8). The case $r_1=\infty$, $r_2>0$ is similar to the one described.

(39) $\quad r_1 = \infty, \qquad r_2 = \infty.$

In this case, each of the series (25) and (26) converges for all w in $\mathbb{C}_1$, and represents an entire function. Then $D(0; \infty, \infty)=\mathbb{C}_2$, the bicomplex power series (30) converges for every $z_1+i_2 z_2$ in $\mathbb{C}_2$, and the function which it represents is called an entire function of a bicomplex variable.

Equation (19) in Theorem 13.5 shows that the bicomplex power series

(40) $$\sum_{k=0}^{\infty} (a_k + i_2 b_k)(z_1 + i_2 z_2)^k$$

determines the following two complex power series [see (25) and (26)]:

(41) $$\sum_{k=0}^{\infty} (a_k - i_1 b_k) w^k, \qquad \sum_{k=0}^{\infty} (a_k + i_1 b_k) w^k, \qquad w \in \mathbb{C}_1.$$

This construction can be reversed. The next theorem shows that each two complex power series determine a unique bicomplex power series. This result suggests the great variety and complexity of functions represented by bicomplex series.

13.8 THEOREM If

$$(42) \qquad \sum_{k=0}^{\infty} \alpha_k w^k, \qquad \sum_{k=0}^{\infty} \beta_k w^k, \qquad \alpha_k, \beta_k, w \text{ in } \mathbb{C}_1,$$

are two arbitrary complex power series, then there exists a unique bicomplex series

$$(43) \qquad \sum_{k=0}^{\infty} (a_k + i_2 b_k)(z_1 + i_2 z_2)^k$$

such that

$$(44) \qquad \sum_{k=0}^{\infty} (a_k + i_2 b_k)(z_1 + i_2 z_2)^k = \left[\sum_{k=0}^{\infty} \alpha_k (z_1 - i_1 z_2)^k\right] e_1 + \left[\sum_{k=0}^{\infty} \beta_k (z_1 + i_1 z_2)^k\right] e_2.$$

Proof. Theorem 13.5 shows that (44) is true if and only if

$$(45) \qquad \begin{aligned} a_k - i_1 b_k &= \alpha_k, \\ a_k + i_1 b_k &= \beta_k, \end{aligned} \qquad \forall k \in \mathbb{N}.$$

These equations have the unique solution

$$(46) \qquad a_k = \frac{\alpha_k + \beta_k}{2}, \qquad b_k = \frac{i_1(\alpha_k - \beta_k)}{2}, \qquad \forall k \in \mathbb{N}.$$

If the coefficients in (43) have the values given in (46), then (45) and Theorem 13.5 show that (44) is true. Theorem 13.6 shows that the bicomplex power series (43) converges at $z_1 + i_2 z_2$ if and only if the complex power series on the right in (44) converge at $z_1 - i_1 z_2$ and $z_1 + i_1 z_2$, respectively. If the radii of convergence of the series in (42) are r_1 and r_2, respectively, then Theorem 13.7 shows that the bicomplex power series (43) converges at each $z_1 + i_2 z_2$ in $D(0; r_1, r_2)$. □

13.9 EXAMPLE Let $\Sigma_{k=0}^{\infty} \alpha_k w^k$ be a complex power series which converges only for $w = 0$, and let $\Sigma_{k=0}^{\infty} \beta_k w^k$ be a complex power series which converges for every value of w [see (42)]. Then the bicomplex power series $\Sigma_{k=0}^{\infty} (a_k + i_2 b_k)(z_1 + i_2 z_2)^k$ can be constructed from these power series as explained in the proof of Theorem 13.8, and (44) states the relation between the complex and the bicomplex series. For what values of $z_1+i_2z_2$ does the bicomplex series converge? The idempotent representation shows that

$$(47)\qquad z_1 + i_2z_2 = (z_1 - i_1z_2)e_1 + (z_1 + i_1z_2)e_2.$$

Then (44) shows that $\Sigma_{k=0}^{\infty}(a_k+i_2b_k)(z_1+i_2z_2)^k$ converges for every value of $z_1+i_2z_2$ such that $z_1-i_1z_2=0$ but only for these values. The points $z_1+i_2z_2$ in $\mathbb{C}_2$ such that $z_1-i_1z_2=0$ are the points in I_2 and also in $D(0;0,\infty)$; therefore, $D(0;0,\infty)=I_2$. Since no point in I_2 has a neighborhood which is in I_2 (see Theorem 7.7), the set of points of convergence of the bicomplex series contains no neighborhood of a point.

13.10 EXAMPLE The terms of the series

$$(48)\qquad \sum_{k=0}^{\infty}(x_1 + i_1x_2 + i_2x_3 + i_1i_2x_4)^k, \qquad x_i\in\mathbb{C}_0,\ i = 1,\dots,4,$$

are bounded for $x_1+i_1x_2+i_2x_3+i_1i_2x_4=1$, and Theorem 13.4 (12) shows that (48) converges for all $x_1+i_1x_2+i_2x_3+i_1i_2x_4$ such that

$$(49)\qquad \|x_1 + i_1x_2 + i_2x_3 + i_1i_2x_4\| < \frac{1}{\sqrt{2}}.$$

But (48) converges for still more values of its bicomplex variable. Theorem 13.5 shows that the idempotent component series of (48) are

$$(50)\qquad \sum_{k=0}^{\infty}[(x_1 + i_1x_2) - i_1(x_3 + i_1x_4)]^k,$$

$$(51)\qquad \sum_{k=0}^{\infty}[(x_1 + i_1x_2) + i_1(x_3 + i_1x_4)]^k.$$

These are complex power series, and the radius of convergence of each one is 1; that is, $r_1=1$ and $r_2=1$. Then Theorem 13.7 shows that (48) converges in $D(0;1,1)$ and diverges in the complement of $\bar{D}(0;1,1)$. Theorem 9.2 states that $D(0;1,1)$ contains the ball $B(0,1/\sqrt{2})$, and this result agrees with that already established in (49).

13.11 EXAMPLE The idempotent component series of the bicomplex series

$$(52)\qquad \sum_{k=0}^{\infty}(a_{1k}+i_1a_{2k}+i_2a_{3k}+i_1i_2a_{4k})(x_1+i_1x_2+i_2x_3+i_1i_2x_4)^k$$

are the complex power series

$$(53)\qquad \sum_{k=0}^{\infty}[(a_{1k}+i_1a_{2k})-i_1(a_{3k}+i_1a_{4k})][(x_1+i_1x_2)-i_1(x_3+i_1x_4)]^k,$$

(54) $$\sum_{k=0}^{\infty} [(a_{1k}+i_1a_{2k})+i_1(a_{3k}+i_1a_{4k})][(x_1+i_1x_2)+i_1(x_3+i_1x_4)]^k.$$

Assume that the radii of convergence of (53) and (54) are r_1 and r_2, respectively, and that $r_1 \leqslant r_2$. Theorem 13.6 shows that (52) converges at every $x_1+i_1x_2+i_2x_3+i_1i_2x_4$ such that both of the following inequalities are satisfied:

(55) $$|(x_1 + i_1x_2) - i_1(x_3 + i_1x_4)| < r_1,$$

(56) $$|(x_1 + i_1x_2) + i_1(x_3 + i_1x_4)| < r_2.$$

Thus (52) converges in $D(0; r_1, r_2)$. The inequalities (55), (56) are

(57) $$|(x_1 + x_4) + i_1(x_2 - x_3)| < r_1,$$

(58) $$|(x_1 - x_4) + i_1(x_2 + x_3)| < r_2,$$

and these inequalities are equivalent to the following:

(59) $$x_1^2 + x_2^2 + x_3^2 + x_4^2 + 2x_1x_4 - 2x_2x_3 < r_1^2,$$

(60) $$x_1^2 + x_2^2 + x_3^2 + x_4^2 - 2x_1x_4 + 2x_2x_3 < r_2^2.$$

If $\zeta:(x_1+i_1x_2+i_2x_3+i_1i_2x_4)$ satisfies both of these inequalities, then their addition shows that

(61) $$\|\zeta\| < \left(\frac{r_1^2 + r_2^2}{2}\right)^{1/2}.$$

Thus, $D(0; r_1, r_2) \subset B(0, [(r_1^2+r_2^2)/2]^{1/2})$. Next, show that $B(0, r_1/\sqrt{2}) \subset D(0; r_1, r_2)$. Assume that

(62) $$x_1^2 + x_2^2 + x_3^2 + x_4^2 < \frac{r_1^2}{2}.$$

Then since $|2x_1x_4| \leqslant x_1^2+x_4^2$ and $|2x_2x_3| \leqslant x_2^2+x_3^2$,

(63) $$\begin{aligned} &|x_1^2 + x_2^2 + x_3^2 + x_4^2 \pm 2x_1x_4 \mp 2x_2x_3| \\ &\leqslant |x_1^2 + x_2^2 + x_3^2 + x_4^2| + |2x_1x_4| + |2x_2x_3| \\ &\leqslant 2(x_1^2 + x_2^2 + x_3^2 + x_4^2) \\ &< r_1^2 \leqslant r_2^2. \end{aligned}$$

Thus if x satisfies (62), it also satisfies (59) and (60), and $B(0, r_1/\sqrt{2}) \subset D(0; r_1, r_2)$. In an effort to understand better the nature of $D(0; r_1, r_2)$, we shall investigate the intersection of this set with each of the six two-dimensional coordinate planes in $\mathbb{C}_0^4$.

(64) $$x_3 = 0, \qquad x_4 = 0.$$

Here (57) and (58) are $|x_1+i_1x_2|<r_1$ and $|x_1+i_1x_2|<r_2$. Since $r_1\leqslant r_2$, both inequalities are satisfied in the circle $x_1^2+x_2^2<r_1^2$.

(65) $\quad x_2=0, \qquad x_4=0.$

Here (57) and (58) are $|x_1-i_1x_3|<r_1$ and $|x_1+i_1x_3|<r_2$, and both are satisfied in the circle $x_1^2+x_3^2<r_1^2$.

(66) $\quad x_2=0, \qquad x_3=0.$

Here (57) and (58) are $|x_1+x_4|<r_1$ and $|x_1-x_4|<r_2$, and both inequalities are satisfied at all points (x_1,x_4) in the rectangle

$$-r_1<x_1+x_4<r_1,\ -r_2<x_1-x_4<r_2.$$

(67) $\quad x_1=0, \qquad x_4=0.$

Here (57) and (58) are $|i_1(x_2-x_3)|<r_1$ and $|i_1(x_2+x_3)|<r_2$, and both inequalities are satisfied in the rectangle

$$-r_1<x_2-x_3<r_1,\ -r_2<x_2+x_3<r_2.$$

(68) $\quad x_1=0, \qquad x_3=0.$

Here (57) and (58) are $|x_4+i_1x_2|<r_1$ and $|-x_4+i_1x_2|<r_2$, and both are satisfied in the circle $x_2^2+x_4^2<r_1^2$.

(69) $\quad x_1=0, \qquad x_2=0.$

Here (57) and (58) are $|x_4-i_1x_3|<r_1$ and $|-x_4+i_1x_3|<r_2$, and both are satisfied in the circle $x_3^2+x_4^2<r_1^2$.

Finally, (64)–(69) contain many examples to show that $B(0,r_1/\sqrt{2})$ is a proper subset of $D(0;r_1,r_2)$, and that this discus is a proper subset of $B(0,[(r_1^2+r_2^2)/2]^{1/2})$.

Exercises

13.1 Find the discus in which each of the following power series converges:

(a) $\sum_{k=0}^{\infty} k^p(z_1+i_2z_2)^k$,

(b) $\sum_{k=1}^{\infty} \frac{(z_1+i_2z_2)^k}{k}$,

(c) $\sum_{k=0}^{\infty} \frac{(z_1+i_2z_2)^k}{k!}$.

13.2 Find the bicomplex infinite series whose idempotent representations are the following; find the discus in which each series converges.

(a) $$\sum_{k=0}^{\infty}\left(\cos\frac{2k\pi}{3}-i_1\sin\frac{2k\pi}{3}\right)(z_1-i_1z_2)^k e_1 + \sum_{k=0}^{\infty}\left(\cos\frac{2k\pi}{3}+i_1\sin\frac{2k\pi}{3}\right)(z_1+i_1z_2)^k e_2.$$

(b) $$\sum_{k=0}^{\infty} k^k(z_1-i_1z_2)^k e_1+\sum_{k=0}^{\infty}(z_1+i_1z_2)^k e_2.$$

13.3 Show that

$$\sum_{k=0}^{\infty}\frac{(z_1+i_2z_2)^k}{k!}=\frac{e^{z_1+i_1z_2}+e^{z_1-i_1z_2}}{2}+i_2\frac{e^{z_1+i_1z_2}-e^{z_1-i_1z_2}}{2i_1},$$

and find the values of $z_1+i_2z_2$ for which this statement is true.

13.4 Construct an example to show that Conjecture 13.3 is false.

13.5 Show that the coefficient $a_k-i_1b_k$ in the series in (25) is equal to the coefficient $a_k+i_1b_k$ in (26) for all k if and only if the coefficients in $\Sigma_{k=0}^{\infty}(a_k+i_2b_k)(z_1+i_2z_2)^k$ are in $\mathbb{C}_1$ for all k. Show that $r_1=r_2$ if the coefficients in the infinite series are in $\mathbb{C}_1$.

14. FUNCTIONS REPRESENTED BY POWER SERIES

This section begins the study of functions which are represented by power series in the bicomplex variable $x_1+i_1x_2+i_2x_3+i_1i_2x_4$. As shown in the last section, every series $\Sigma_{k=0}^{\infty}(a_k+i_2b_k)(z_1+i_2z_2)^k$ has associated with it the two series

$$\sum_{k=0}^{\infty}(a_k-i_1b_k)(z_1-i_1z_2)^k,\qquad \sum_{k=0}^{\infty}(a_k+i_1b_k)(z_1+i_1z_2)^k.$$

If the first of these series converges only for $z_1-i_1z_2=0$, or if the second series converges only for $z_1+i_1z_2=0$, then an early theorem in this section shows that $\Sigma_{k=0}^{\infty}(a_k+i_2b_k)(z_1+i_2z_2)^k$ represents a degenerate type of function: (a) the power series converges in no ball $B(a+i_2b,\varepsilon)$, and (b) the power series represents a function which fails to have a unique power series representation. Lacking interest, these functions are excluded from further study, and henceforth all power series are assumed to converge in a discus with $r_1>0$ and $r_2>0$. Functions included in this class may or may not have isolated zeros. This section contains theorems concerning the representations and properties of such functions, theorems concerning their zeros, and uniqueness theorems.

14.1 THEOREM Let f be a function which is defined by a power series in $z_1+i_2z_2$ as follows:

(1) $$f(z_1+i_2z_2)=\sum_{k=0}^{\infty}(a_k+i_2b_k)(z_1+i_2z_2)^k.$$

If

(2) $$\sum_{k=0}^{\infty}(a_k-i_1b_k)(z_1-i_1z_2)^k$$

converges only for $z_1-i_1z_2=0$, or if

(3) $$\sum_{k=0}^{\infty}(a_k+i_1b_k)(z_1+i_1z_2)^k$$

converges only for $z_1+i_1z_2=0$, then there is no ball $B(a+i_2b,\varepsilon)$ in which f is defined. Furthermore, the power series (1) in $z_1+i_2z_2$ which represents f in its region of convergence is not unique.

Proof. Consider the case in which (2) converges only for $z_1-i_1z_2=0$, and in which (3) converges in $\{z_1+i_1z_2\in\mathbb{C}_1\colon\ |z_1+i_1z_2|<r_2\}$. The points $x_1+i_1x_2+i_2x_3+i_1i_2x_4$ in $\mathbb{C}_2$ which satisfy these two conditions are the points such that

(4) $$x_1+x_4=0,\qquad x_2-x_3=0,$$

(5) $$(x_1-x_4)^2+(x_2+x_3)^2<r_2^2.$$

The points $(x_1,\ldots,x_4)$ which satisfy (4) form a two-dimensional linear space in $\mathbb{C}_0^4$, and (4) and (5) together describe a bounded subset of this linear subspace. No point in this convergence set of (1) is the center $a+i_2b$ of a ball $B(a+i_2b,\varepsilon)$ which is contained in the set. In the same way, it can be shown that (1) converges in no ball if (3) converges only for $z_1+i_1z_2=0$.

The proof of Theorem 14.1 will be completed by showing that there are many power series in $z_1+i_2z_2$ which represent f in the set in which (1) converges if (2) converges only for $z_1-i_1z_2=0$, or if (3) converges only for $z_1+i_1z_2=0$. Consider the case in which (2) converges only for $z_1-i_1z_2=0$; the other case is similar. Then

(6) $$\sum_{k=0}^{\infty}(a_k-i_1b_k)(z_1-i_1z_2)^k$$

diverges for all $z_1-i_1z_2\neq 0$. Let $a_1'+i_1b_1'$ be a bicomplex number such that

(7) $$a_1'-i_1b_1'\neq a_1-i_1b_1,\qquad a_1'+i_1b_1'=a_1+i_1b_1.$$

There are many such $a_1'+i_1b_1'$; one example is $(a_1+1)+i_2(b_1+i_1)$. We shall now show that

$$(8)\qquad (a_0+i_2b_0)+(a_1'+i_2b_1')(z_1+i_2z_2)+\sum_{k=2}^{\infty}(a_k+i_2b_k)(z_1+i_2z_2)^k,$$

which is different from the power series in (1) because of (7), nevertheless represents f in the set where these series converge. The idempotent component series for (8) are

$$(9)\qquad (a_0-i_1b_0)+(a_1'-i_1b_1')(z_1-i_1z_2)+\sum_{k=2}^{\infty}(a_k-i_1b_k)(z_1-i_1z_2)^k,$$

$$(10)\qquad (a_0+i_1b_0)+(a_1'+i_1b_1')(z_1+i_1z_2)+\sum_{k=2}^{\infty}(a_k+i_1b_k)(z_1+i_1z_2)^k.$$

The series (9), as shown by (2), converges only for $z_1-i_1z_2=0$, and its value when it converges is $a_0-i_1b_0$. Thus the value of (9) where it converges equals the value of the corresponding series for (1) where it converges. Because of (7), the series (10) is exactly the same as the corresponding series for (1). Thus (8) represents a function whose values are

$$(11)\qquad (a_0-i_1b_0)e_1+\sum_{k=0}^{\infty}(a_k+i_1b_k)(z_1+i_1z_2)^ke_2$$

for $z_1-i_1z_2=0$ and $|z_1+i_1z_2|<r_2$. This function is exactly the function f represented by the series in (1). Thus the two distinct power series (1) and (8) represent the same function in the set in which both converge. The construction shows that there is a large class of power series which represent the function f in (1). The proof of Theorem 14.1 is complete. □

Theorem 14.1 shows that the class of functions represented by power series

$$(12)\qquad \sum_{k=0}^{\infty}(a_k+i_2b_k)(z_1+i_2z_2)^k,$$

such that $\Sigma_{k=0}^{\infty}(a_k-i_1b_k)(z_1-i_1z_2)^k$ converges only for $z_1-i_1z_2=0$, or $\Sigma_{k=0}^{\infty}(a_k+i_1b_k)(z_1+i_1z_2)^k$ converges only for $z_1+i_1z_2=0$, is not a very interesting class of functions. These facts lead us to make the following hypothesis.

14.2 HYPOTHESIS The assumption will be made henceforth, without further mention, that all power series (12) converge in a nondegenerate discus $D(0;r_1,r_2)$ with $r_1>0$ and $r_2>0$. More generally, the power series

$$(13)\qquad \sum_{k=0}^{\infty}(a_k+i_2b_k)[(z_1+i_2z_2)-(c_1+i_2c_2)]^k$$

about the point $c_1+i_2c_2$ in $\mathbb{C}_2$, will be assumed to converge in a discus $D(c_1+i_2c_2; r_1,r_2)$ with $r_1>0$ and $r_2>0$.

Let f be a function defined by a power series as follows:

$$(14)\qquad f(z_1+i_2z_2)=\sum_{k=0}^{\infty}(a_k+i_2b_k)(z_1+i_2z_2)^k, \qquad z_1+i_2z_2\in D(0; r_1,r_2).$$

Define two complex-valued functions f_1, f_2 on the circles of convergence of the idempotent component power series as follows:

$$(15)\qquad f_1(z_1-i_1z_2)=\sum_{k=0}^{\infty}(a_k-i_1b_k)(z_1-i_1z_2)^k, \qquad |z_1-i_1z_2|<r_1,$$

$$(16)\qquad f_2(z_1+i_1z_2)=\sum_{k=0}^{\infty}(a_k+i_1b_k)(z_1+i_1z_2)^k, \qquad |z_1+i_1z_2|<r_2.$$

Then (14), (15), (16), and Section 13 show that

$$(17)\qquad f(z_1+i_2z_2)=f_1(z_1-i_1z_2)e_1+f_2(z_1+i_1z_2)e_2,$$
$$\forall z_1+i_2z_2\in D(0; r_1,r_2).$$

Define two complex-valued functions u and v as follows:

$$(18)\qquad u(z_1,z_2)=\frac{f_1(z_1-i_1z_2)+f_2(z_1+i_1z_2)}{2},$$

$$(19)\qquad v(z_1,z_2)=\frac{i_1[f_1(z_1-i_1z_2)-f_2(z_1+i_1z_2)]}{2}, \quad \forall z_1+i_2z_2\in D(0; r_1,r_2).$$

Equations (18) and (19) show that

$$(20)\qquad f_1(z_1-i_1z_2)=u(z_1,z_2)-i_1v(z_1,z_2),$$
$$(21)\qquad f_2(z_1+i_1z_2)=u(z_1,z_2)+i_1v(z_1,z_2), \qquad \forall z_1+i_2z_2\in D(0; r_1,r_2).$$

Then (17), (20), and (21) show that

$$(22)\qquad f(z_1+i_2z_2)=u(z_1,z_2)+i_2v(z_1,z_2), \qquad \forall z_1+i_2z_2\in D(0; r_1,r_2).$$

The functions u and v are defined by (18) and (19), and (14) and (22) show that

$$(23)\qquad u(z_1,z_2)+i_2v(z_1,z_2)=\sum_{k=0}^{\infty}(a_k+i_2b_k)(z_1+i_2z_2)^k,$$
$$\forall z_1+i_2z_2\in D(0; r_1,r_2).$$

This relation will now be used to obtain power series representations for u and v. By the binomial theorem,

$$(24)\qquad (z_1+i_2z_2)^k=\sum_{r=0}^{k}\binom{k}{r}z_1^{k-r}(i_2z_2)^r=\varphi_k(z_1,z_2)+i_2\psi_k(z_1,z_2).$$

Here φ_k and ψ_k are polynomials in z_1, z_2 of degree k. Then

$$(25)\qquad (a_k+i_2b_k)(z_1+i_2z_2)^k$$
$$=[a_k\varphi_k(z_1,z_2)-b_k\psi_k(z_1,z_2)]+i_2[a_k\psi_k(z_1,z_2)+b_k\varphi_k(z_1,z_2)].$$

Define the complex-valued functions P_k and Q_k by the following equations:

$$(26)\qquad \begin{aligned} P_k(z_1,z_2)&=a_k\varphi_k(z_1,z_2)-b_k\psi_k(z_1,z_2), &\quad \forall z_1+i_2z_2\in D(0;r_1,r_2),\\ Q_k(z_1,z_2)&=a_k\psi_k(z_1,z_2)+b_k\varphi_k(z_1,z_2), &\quad k=0,1,\ldots. \end{aligned}$$

Here P_k and Q_k are polynomials of degree k in z_1 and z_2, and (25) and (26) show that

$$(27)\qquad \sum_{k=0}^{n}(a_k+i_2b_k)(z_1+i_2z_2)^k=\sum_{k=0}^{n}P_k(z_1,z_2)+i_2\sum_{k=0}^{n}Q_k(z_1,z_2).$$

14.3 THEOREM If the series in (23) converges in $D(0;r_1,r_2)$ and diverges in the complement of $\bar{D}(0;r_1,r_2)$, then the series

$$(28)\qquad \sum_{k=0}^{\infty}P_k(z_1,z_2),\qquad \sum_{k=0}^{\infty}Q_k(z_1,z_2),$$

which are power series in z_1 and z_2, converge for all $z_1+i_2z_2$ in $D(0;r_1,r_2)$ and

$$(29)\qquad u(z_1,z_2)=\sum_{k=0}^{\infty}P_k(z_1,z_2),\qquad v(z_1,z_2)=\sum_{k=0}^{\infty}Q_k(z_1,z_2).$$

Furthermore, if $z_1+i_2z_2$ is in the complement of $\bar{D}(0;r_1,r_2)$, then at least one of the series in (28) diverges at (z_1,z_2).

Proof. By (23) and (27),

$$(30)\qquad \begin{aligned} u(z_1,z_2)+i_2v(z_1,z_2)&=\lim_{n\to\infty}\sum_{k=0}^{n}(a_k+i_2b_k)(z_1+i_2z_2)^k\\ &=\lim_{n\to\infty}\left[\sum_{k=0}^{n}P_k(z_1,z_2)+i_2\sum_{k=0}^{n}Q_k(z_1,z_2)\right]. \end{aligned}$$

Now

$$(31)\qquad \left|u(z_1,z_2)-\sum_{k=0}^{n}P_k(z_1,z_2)\right|,\qquad \left|v(z_1,z_2)-\sum_{k=0}^{n}Q_k(z_1,z_2)\right|,$$

are each equal to or less than

$$(32)\qquad \left\|[u(z_1,z_2)+i_2v(z_1,z_2)]-\left[\sum_{k=0}^{n}P_k(z_1,z_2)+i_2\sum_{k=0}^{n}Q_k(z_1,z_2)\right]\right\|$$

by the definition of the norm in $\mathbb{C}_2$. Then because of (32) and (30), each of the series in (28) converges to the value stated in (29) for each (z_1, z_2) such that $z_1 + i_2 z_2$ is in $D(0; r_1, r_2)$. Next, let $z_1 + i_2 z_2$ be a point in the complement of $\bar{D}(0; r_1, r_2)$. If the two series in (28) were to converge at the corresponding point (z_1, z_2), then the series in (23) would converge at $z_1 + i_2 z_2$ because (27) shows that

$$(33) \qquad \sum_{k=0}^{\infty} (a_k + i_2 b_k)(z_1 + i_2 z_2)^k = \sum_{k=0}^{\infty} P_k(z_1, z_2) + i_2 \sum_{k=0}^{\infty} Q_k(z_1, z_2).$$

But the series on the left diverges since $z_1 + i_2 z_2$ is in the complement of $\bar{D}(0; r_1, r_2)$ by hypothesis. This contradiction shows that at least one of the series in (28) diverges at each (z_1, z_2) such that $z_1 + i_2 z_2$ is in the complement of $\bar{D}(0; r_1, r_2)$. The proof of Theorem 14.3 is complete. □

14.4 THEOREM If the power series (14) converges in $D(0; r_1, r_2)$, then u and v [see (18) and (19)] have an infinite number of continuous derivatives with respect to z_1 and z_2 at each (z_1, z_2) for which $z_1 + i_2 z_2$ is in $D(0; r_1, r_2)$, and

$$(34) \qquad \frac{\partial u}{\partial z_1} = \frac{\partial v}{\partial z_2}, \qquad \frac{\partial u}{\partial z_2} = -\frac{\partial v}{\partial z_1},$$

$$(35) \qquad \frac{\partial^2 u}{\partial z_1^2} + \frac{\partial^2 u}{\partial z_2^2} = 0, \qquad \frac{\partial^2 v}{\partial z_1^2} + \frac{\partial^2 v}{\partial z_2^2} = 0.$$

Proof. Now

$$(36) \qquad f_1(z_1 - i_1 z_2) = f_1(w_1) \qquad \text{where } w_1 = z_1 - i_1 z_2,$$

$$(37) \qquad f_2(z_1 + i_1 z_2) = f_2(w_2) \qquad \text{where } w_2 = z_1 + i_1 z_2,$$

and (15) and (16) show that f_1 and f_2 are holomorphic functions of the complex variables w_1 and w_2 for $|w_1| < r_1$ and $|w_2| < r_2$. Then f_1 and f_2 have an infinite number of derivatives in $|w_1| < r_1$ and $|w_2| < r_2$, respectively. Furthermore, by (18) and (19),

$$(38) \qquad \frac{\partial u}{\partial z_1} = \frac{1}{2}\left(\frac{\partial f_1}{\partial w_1}\frac{\partial w_1}{\partial z_1} + \frac{\partial f_2}{\partial w_2}\frac{\partial w_2}{\partial z_1}\right) = \frac{1}{2}\left(\frac{\partial f_1}{\partial w_1} + \frac{\partial f_2}{\partial w_2}\right),$$

$$\frac{\partial u}{\partial z_2} = \frac{1}{2}\left(\frac{\partial f_1}{\partial w_1}\frac{\partial w_1}{\partial z_2} + \frac{\partial f_2}{\partial w_2}\frac{\partial w_2}{\partial z_2}\right) = \frac{1}{2}\left(-i_1\frac{\partial f_1}{\partial w_1} + i_1\frac{\partial f_2}{\partial w_2}\right),$$

$$\frac{\partial v}{\partial z_1} = \frac{i_1}{2}\left(\frac{\partial f_1}{\partial w_1}\frac{\partial w_1}{\partial z_1} - \frac{\partial f_2}{\partial w_2}\frac{\partial w_2}{\partial z_1}\right) = \frac{i_1}{2}\left(\frac{\partial f_1}{\partial w_1} - \frac{\partial f_2}{\partial w_2}\right),$$

$$\frac{\partial v}{\partial z_2} = \frac{i_1}{2}\left(\frac{\partial f_1}{\partial w_1}\frac{\partial w_1}{\partial z_2} - \frac{\partial f_2}{\partial w_2}\frac{\partial w_2}{\partial z_2}\right) = \frac{i_1}{2}\left(-i_1\frac{\partial f_1}{\partial w_1} - i_1\frac{\partial f_2}{\partial w_2}\right).$$

These equations show that u and v satisfy the Cauchy–Riemann differential equations in (34). The second partial derivatives are calculated from (38) as follows:

$$(39)\qquad \frac{\partial^2 u}{\partial z_1^2}=\frac{1}{2}\left(\frac{\partial^2 f_1}{\partial w_1^2}\frac{\partial w_1}{\partial z_1}+\frac{\partial^2 f_2}{\partial w_2^2}\frac{\partial w_2}{\partial z_1}\right)=\frac{1}{2}\left(\frac{\partial^2 f_1}{\partial w_1^2}+\frac{\partial^2 f_2}{\partial w_2^2}\right),$$

$$\frac{\partial^2 u}{\partial z_2^2}=\frac{1}{2}\left(-i_1\frac{\partial^2 f_1}{\partial w_1^2}\frac{\partial w_1}{\partial z_2}+i_1\frac{\partial^2 f_2}{\partial w_2^2}\frac{\partial w_2}{\partial z_2}\right)=-\frac{1}{2}\left(\frac{\partial^2 f_1}{\partial w_1^2}+\frac{\partial^2 f_2}{\partial w_2^2}\right),$$

$$\frac{\partial^2 v}{\partial z_1^2}=\frac{i_1}{2}\left(\frac{\partial^2 f_1}{\partial w_1^2}\frac{\partial w_1}{\partial z_1}-\frac{\partial^2 f_2}{\partial w_2^2}\frac{\partial w_2}{\partial z_1}\right)=\frac{i_1}{2}\left(\frac{\partial^2 f_1}{\partial w_1^2}-\frac{\partial^2 f_2}{\partial w_2^2}\right),$$

$$\frac{\partial^2 v}{\partial z_2^2}=\frac{i_1}{2}\left(\frac{\partial^2 f_1}{\partial w_1^2}\frac{\partial w_1}{\partial z_2}+\frac{\partial^2 f_2}{\partial w_2^2}\frac{\partial w_2}{\partial z_2}\right)=-\frac{i_1}{2}\left(\frac{\partial^2 f_1}{\partial w_1^2}-\frac{\partial^2 f_2}{\partial w_2^2}\right).$$

These equations show that u and v satisfy Laplace's equation as stated in (35). Thus u and v satisfy the Cauchy–Riemann equations in (34), and they are harmonic functions of the complex variables z_1 and z_2. □

The next theorem begins the investigation of the relation of the zeros of f_1 and f_2 to the zeros of f.

14.5 THEOREM Let f, f_1, f_2 be the functions defined by the following power series:

$$(40)\qquad f(z_1+i_2z_2)=\sum_{k=0}^{\infty}(a_k+i_2b_k)(z_1+i_2z_2)^k,\qquad z_1+i_2z_2\in D(0;r_1,r_2),$$

$$(41)\qquad f_1(z_1-i_1z_2)=\sum_{k=0}^{\infty}(a_k-i_1b_k)(z_1-i_1z_2)^k,$$

$$(42)\qquad f_2(z_1+i_1z_2)=\sum_{k=0}^{\infty}(a_k+i_1b_k)(z_1+i_1z_2)^k.$$

If (40) converges at $z_1+i_2z_2$ and $f(z_1+i_2z_2)=0$, then (41) and (42) converge at $z_1-i_1z_2$ and $z_1+i_1z_2$, respectively, and $f_1(z_1-i_1z_2)=0$ and $f_2(z_1+i_1z_2)=0$. Conversely, if (41), (42) converge at $z_1-i_1z_2$, $z_1+i_1z_2$ and $f_1(z_1-i_1z_2)=0$, $f_2(z_1+i_1z_2)=0$, then (40) converges at $z_1+i_2z_2$ and $f(z_1+i_2z_2)=0$.

Proof. If (40) converges at $z_1+i_2z_2$, then (41), (42) converge at $z_1-i_1z_2$, $z_1+i_1z_2$ by Theorem 13.6. Also by (17),

$$(43)\qquad f(z_1+i_2z_2)=f_1(z_1-i_1z_2)e_1+f_2(z_1+i_1z_2)e_2.$$

If $f(z_1+i_2z_2)=0$, then (43) shows that

$$(44)\qquad f_1(z_1-i_1z_2)+f_2(z_1+i_1z_2)=0,$$

$$f_1(z_1-i_1z_2)-f_2(z_1+i_1z_2)=0.$$

These equations show that $f_1(z_1-i_1z_2)=0$ and $f_2(z_1+i_1z_2)=0$. Conversely, (40) converges if (41) and (42) converge (see Theorem 13.6), and $f(z_1+i_2z_2)=0$ if $f_1(z_1-i_1z_2)=0$ and $f_2(z_1+i_1z_2)=0$ [see (43)]. The proof of the entire theorem is complete. □

14.6 COROLLARY If $f_1(z_1-i_1z_2)=0$ and $f_2(z_1+i_1z_2)\neq 0$, then $f(z_1+i_2z_2)\in I_2$. If $f_1(z_1-i_1z_2)\neq 0$ and $f_2(z_1+i_1z_2)=0$, then $f(z_1+i_2z_2)\in I_1$.

Proof. This corollary follows from (43) above and the definitions of I_1 and I_2 in Definition 7.1 (3) and (4). □

The next theorem states necessary and sufficient conditions for one of the functions f_1, f_2 to vanish identically.

14.7 THEOREM Let f, f_1, f_2 be the functions in (40), (41), (42). A necessary and sufficient condition that

(45) $\qquad f_1(z_1 - i_1z_2) = 0, \qquad \forall z_1 + i_2z_2 \in D(0; r_1, r_2),$

is that $a_k - i_1b_k = 0$ for $k = 0, 1, \ldots$, or equivalently, that

(46) $\qquad a_k + i_2b_k \in I_2, \qquad \forall k \in \mathbb{N}.$

Also, a necessary and sufficient condition that

(47) $\qquad f_2(z_1 + i_1z_2) = 0, \qquad \forall z_1 + i_2z_2 \in D(0; r_1, r_2),$

is that $a_k+i_1b_k=0$ for $k=0, 1, \ldots$, or equivalently, that

(48) $\qquad a_k + i_2b_k \in I_1, \qquad \forall k \in \mathbb{N}.$

Proof. The function f_1 in (41) vanishes identically if and only if all coefficients $a_k - i_1b_k$ in its power series representation are zero. Since

(49) $\qquad a_k + i_2b_k = (a_k - i_1b_k)e_1 + (a_k + i_1b_k)e_2,$

then $a_k-i_1b_k=0$ if and only if $a_k+i_2b_k$ is in I_2. The proof of the remainder of Theorem 14.7 is similar. □

A function f defined by a power series (40) may or may not have zeros; if f has zeros, they may or may not be isolated. The next theorem describes the situation.

14.8 THEOREM Let f, f_1, f_2 be the functions in (40), (41), (42). If neither f_1 nor f_2 vanishes identically, then the zeros, if any, of f are isolated. If either f_1 or f_2 vanishes identically, then the zeros, if any, of f are not isolated. A necessary and sufficient condition that f vanish identically in $D(0; r_1, r_2)$ is that both f_1 and f_2 vanish identically.

Proof. If neither f_1 nor f_2 vanishes identically, then they are holomorphic functions of a complex variable whose zeros are isolated. By Theorem 14.5, $f(z_1+i_2z_2)=0$ if and only if $f_1(z_1-i_1z_2)=0$ and $f_2(z_1+i_1z_2)=0$. If f_1 and f_2 have zeros at $z_1-i_1z_2$ and $z_1+i_1z_2$, then, since their zeros are isolated, there are circles about $z_1-i_1z_2$ and $z_1+i_1z_2$, with radii ε_1 and ε_2 in which f_1 and f_2 have no other zero. Then $f(z_1+i_2z_2)=0$, but f has no other zero in $D(z_1+i_2z_2;\, \varepsilon_1,\varepsilon_2)$, and the zeros of f are isolated in this case. If f_2 vanishes identically, then

$$(50) \qquad f(z_1+i_2z_2)=f_1(z_1-i_1z_2)e_1.$$

If f_1 has no zero in $|z_1-i_1z_2|<r_1$, then f has no zero in $D(0;r_1,r_2)$. If $f_1(z_1^0-i_1z_2^0)=0$, then $f(z_1+i_2z_2)=0$ for every $z_1+i_2z_2$ such that

$$(51) \qquad z_1+i_2z_2=(z_1^0-i_1z_2^0)e_1+(z_1+i_1z_2)e_2, \qquad |z_1+i_1z_2|<r_2.$$

Thus f vanishes on a large connected set. The conclusions are similar if f_1 vanishes identically. The equation

$$(52) \qquad f(z_1+i_2z_2)=f_1(z_1-i_1z_2)e_1+f_2(z_1+i_1z_2)e_2, \quad \forall z_1+i_2z_2\in D(0;r_1,r_2),$$

shows that a necessary and sufficient condition that f vanish identically is that f_1 and f_2 vanish identically. The proof of Theorem 14.8 is complete. □

The next theorem establishes the uniqueness of the power series representation of a function of a bicomplex variable in $D(0;r_1,r_2)$.

14.9 THEOREM If

$$(53) \qquad \sum_{k=0}^{\infty}(a_k+i_2b_k)(z_1+i_2z_2)^k=\sum_{k=0}^{\infty}(c_k+i_2d_k)(z_1+i_2z_2)^k, \quad z_1+i_2z_2 \text{ in } D(0;r_1,r_2),$$

then

$$(54) \qquad a_k+i_2b_k=c_k+i_2d_k, \qquad \forall k\in\mathbb{N}.$$

Furthermore, if

$$(55) \qquad \sum_{k=0}^{\infty}(a_k-i_1b_k)(z_1-i_1z_2)^ke_1+\sum_{k=0}^{\infty}(a_k+i_1b_k)(z_1+i_1z_2)^ke_2$$
$$=\sum_{k=0}^{\infty}(c_k-i_1d_k)(z_1-i_1z_2)^ke_1+\sum_{k=0}^{\infty}(c_k+i_1d_k)(z_1+i_1z_2)^ke_2, \quad \forall z_1+i_2z_2\in D(0;r_1,r_2),$$

then

(56) $\quad a_k - i_1 b_k = c_k - i_1 d_k, \qquad a_k + i_1 b_k = c_k + i_1 d_k, \qquad \forall k \in \mathbb{N}.$

Proof. To begin the proof of (54), define the function $g : D(0; r_1, r_2) \to \mathbb{C}_2$ as follows:

(57) $$g(z_1 + i_2 z_2) = \sum_{k=0}^{\infty} [(a_k + i_2 b_k) - (c_k + i_2 d_k)](z_1 + i_2 z_2)^k.$$

Then

(58) $$g_1(z_1 - i_1 z_2) = \sum_{k=0}^{\infty} [(a_k - i_1 b_k) - (c_k - i_1 d_k)](z_1 - i_1 z_2)^k,$$
$$|z_1 - i_1 z_2| < r_1,$$
$$g_2(z_1 + i_1 z_2) = \sum_{k=0}^{\infty} [(a_k + i_1 b_k) - (c_k + i_1 d_k)](z_1 + i_1 z_2)^k,$$
$$|z_1 + i_1 z_2| < r_2.$$

The hypothesis in (53) shows that g is identically zero in $D(0; r_1, r_2)$; therefore, by Theorem 14.8, g_1 and g_2 vanish identically and Theorem 14.7 shows that

(59) $$\begin{aligned} (a_k - i_1 b_k) - (c_k - i_1 d_k) &= 0, \\ (a_k + i_1 b_k) - (c_k + i_1 d_k) &= 0, \end{aligned} \qquad \forall k \in \mathbb{N}.$$

These equations imply that $a_k = c_k$ and $b_k = d_k$, $k = 0, 1, \ldots$; therefore $a_k + i_2 b_k = c_k + i_2 d_k$ for all k and (54) is true. To prove that (55) implies (56), observe first that

(60) $$\sum_{k=0}^{\infty} [(a_k - i_1 b_k) - (c_k - i_1 d_k)](z_1 - i_1 z_2)^k e_1$$
$$+ \sum_{k=0}^{\infty} [(a_k + i_1 b_k) - (c_k + i_1 d_k)](z_1 + i_1 z_2)^k e_2 = 0$$

for all $z_1 + i_2 z_2$ in $D(0; r_1, r_2)$. This equation shows that the left side is the idempotent representation of a function which vanishes identically; thus, by Theorem 14.8,

(61) $$\sum_{k=\infty}^{\infty} [(a_k - i_1 b_k) - (c_k - i_1 d_k)](z_1 - i_1 z_2)^k = 0, \qquad |z_1 - i_1 z_2| < r_1,$$

(62) $$\sum_{k=0}^{\infty} [(a_k + i_1 b_k) - (c_k + i_1 d_k)](z_1 + i_1 z_2)^k = 0, \qquad |z_1 + i_1 z_2| < r_2.$$

Then Theorem 14.7 shows that all coefficients in (61) and (62) are zero, and (56) follows from this statement. The proof of Theorem 14.9 is complete. □

Exercises

14.1 (a) Show that the series

$$\sum_{k=0}^{\infty} (z_1 + i_2 z_2)^k$$

converges and defines a function f in the discus $D(0; 1, 1)$.

(b) Show that

$$\sum_{k=0}^{\infty} (z_1 + i_2 z_2)^k = \frac{1}{1 - (z_1 + i_2 z_2)}, \qquad \forall z_1 + i_2 z_2 \text{ in } D(0; 1, 1).$$

(c) Show that f has no zeros in $D(0; 1, 1)$.

14.2 (a) Let f be defined as follows:

$$f(z_1 + i_2 z_2) = \sum_{k=1}^{\infty} \left[\frac{(a_k - i_1 b_k) + i_2(i_1 a_k + b_k)}{2}\right](z_1 + i_2 z_2)^k,$$

$$a_k, b_k \in \mathbb{C}_1.$$

(b) Show that

$$f_1(z_1 - i_1 z_2) = \sum_{k=1}^{\infty} (a_k - i_1 b_k)(z_1 - i_1 z_2)^k,$$

$$f_2(z_1 + i_1 z_2) = 0, \qquad \forall z_1 + i_1 z_2 \text{ in } \mathbb{C}_1.$$

Assume that the series for f_1 converges for $|z_1 - i_1 z_2| < r_1$.

(c) Show that $f_1(0) = 0$. Show also that $f(z_1 + i_2 z_2) = 0$ for all $z_1 + i_2 z_2$ such that $z_1 + i_2 z_2 = 0e_1 + (z_1 + i_1 z_2)e_2$. Thus show that $f(z_1 + i_2 z_2) = 0$ for every $z_1 + i_2 z_2$ in I_2.

(d) Show that

$$f(z_1 + i_2 z_2) = \sum_{k=1}^{\infty} (a_k - i_1 b_k)e_1(z_1 + i_2 z_2)^k.$$

Substitute $(z_1 + i_1 z_2)e_2$ for $z_1 + i_2 z_2$ in this series and thus show that f equals zero at all points in I_2. Show also that all coefficients in the series for f are elements in I_1 and that therefore all coefficients in the series for f_2 are zero (compare Theorem 14.7).

14.3 Find a polynomial of degree n in $z_1 + i_2 z_2$ which has the value zero at every point in I_2.

14.4 (a) The exponential function $e^{z_1 + i_2 z_2}$ of the bicomplex variable $z_1 + i_2 z_2$ is defined by the exponential series as follows:

$$e^{z_1 + i_2 z_2} = \sum_{k=0}^{\infty} \frac{(z_1 + i_2 z_2)^k}{k!}.$$

(b) Show that the exponential series converges for all values of $z_1+i_2z_2$ in $\mathbb{C}_2$.

(c) Show that

$$e^{z_1+i_2z_2}=e^{z_1-i_1z_2}e_1+e^{z_1+i_1z_2}e_2, \quad \forall z_1+i_2z_2 \text{ in } \mathbb{C}_2.$$

Use this relation to show that the exponential function has no zero in $\mathbb{C}_2$. More generally, show that

$$e^{z_1+i_2z_2}\notin \mathcal{O}_2, \quad \forall z_1+i_2z_2 \text{ in } \mathbb{C}_2.$$

(d) Use the formula in (c) to show that

$$e^{z_1+i_2z_2}=\frac{e^{z_1+i_1z_2}+e^{z_1-i_1z_2}}{2} + i_2\frac{e^{z_1+i_1z_2}-e^{z_1-i_1z_2}}{2i_1}, \quad \forall z_1+i_2z_2 \text{ in } \mathbb{C}_2.$$

14.5 (a) Use the infinite series for the sine function to define the sine function for bicomplex values as follows:

$$\sin(z_1+i_2z_2)=\frac{z_1+i_2z_2}{1!}-\frac{(z_1+i_2z_2)^3}{3!}+\frac{(z_1+i_2z_2)^5}{5!}-\cdots.$$

(b) Show that the infinite series in (a) converges for all $z_1+i_2z_2$ in $\mathbb{C}_2$ and that

$$\sin(z_1+i_2z_2)=e_1\sin(z_1-i_1z_2)+e_2\sin(z_1+i_1z_2).$$

(c) Show that the zeros of the sine function of a bicomplex variable are the elements $z_1+i_2z_2$ such that

$$z_1+i_2z_2=\pi\left[\frac{m+n}{2}+\frac{i_2i_1(m-n)}{2}\right], \quad m,n=0, \pm 1, \pm 2,\ldots.$$

(d) The sine function of the bicomplex variable is, for $z_2=0$, the sine function of the complex variable z_1. Show that the zeros of the sine function of the complex variable are also zeros of the sine function of the bicomplex variable.

15. HOLOMORPHIC FUNCTIONS OF A BICOMPLEX VARIABLE

This section defines a class of functions to be known as holomorphic functions of a bicomplex variable. After establishing the existence and some of the properties of these functions, it proves that a domain in $\mathbb{C}_2$ is a domain of holomorphism if and only if it is a cartesian set. This result generalizes a theorem of Mittag–Leffler.

A set X in $\mathbb{C}_2$ is *open* if and only if for each point $z_1+i_2z_2$ in X there is a ball $B(z_1+i_2z_2, \varepsilon)$ in X, or, alternatively, a discus $D(z_1+i_2z_2; r_1, r_2)$, with $r_1>0$ and $r_2>0$, in X. A set X is (arcwise) *connected* if and only if each two points in X can be connected by a polygonal curve in X. A set X in $\mathbb{C}_2$ is called a *domain* if and only if it is both open and connected. Also, a set Z in $\mathbb{C}_1$ is a domain if and only if it is open and connected.

Let Z be a domain in $\mathbb{C}_1$, and let $g: Z \to \mathbb{C}_1$ be a function of the complex variable z in Z. If g has a derivative at each point z_0 in Z, then g is said to be *holomorphic* in Z. The set of holomorphic functions $g: Z \to \mathbb{C}_1$ is denoted by $H(Z)$. The following theorem states a well-known property of a holomorphic function $g: Z \to \mathbb{C}_1$.

15.1 THEOREM If $g \in H(Z)$, then for each point z_0 in Z there is a disk $|z-z_0|<r$ in Z and a power series $\Sigma_{k=0}^{\infty} b_k(z-z_0)^k$ such that

(1) $$g(z) = \sum_{k=0}^{\infty} b_k(z - z_0)^k, \qquad \forall z \text{ in } |z - z_0| < r.$$

This theorem and the results in Sections 13 and 14 suggest the following definition of a holomorphic function of a bicomplex variable.

15.2 DEFINITION A *holomorphic* function f of a bicomplex variable $z_1+i_2z_2$ is a function with the following properties: (a) f is defined on a domain X in $\mathbb{C}_2$, and $f(z_1+i_2z_2) \in \mathbb{C}_2$ for each $z_1+i_2z_2$ in X; and (b) for each $a_1+i_2a_2$ in X there exist a discus $D(a_1+i_2a_2; r_1, r_2)$, with $r_1>0$ and $r_2>0$, and a power series such that

(2) $$f(z_1 + i_2z_2) = \sum_{k=0}^{\infty} (a_k + i_2z_2)[(z_1 + i_2z_2) - (a_1 + i_2a_2)]^k$$

for all $z_1+i_2z_2$ in $D(a_1+i_2a_2; r_1, r_2)$. The set of holomorphic functions on X is denoted by $H(X)$.

A theorem is needed to establish the existence of holomorphic functions of a bicomplex variable. Sections 13 and 14 are devoted to a study of functions represented by a single power series in a discus $D(a_1+i_2a_2; r_1, r_2)$, but it has not been shown that these are holomorphic functions according to Definition 15.2. At least three methods are available for establishing the existence of holomorphic functions.

The first of these methods is based on the theory of power series. This theory can be used to show that a function represented by a power series on $D(a_1+i_2a_2; r_1, r_2)$ is holomorphic on this discus and in many cases on a larger set X. This method shows, trivially, that all polynomial functions

$$(3) \qquad f(z_1 + i_2 z_2) = \sum_{k=0}^{n} (a_k + i_2 b_k)(z_1 + i_2 z_2)^k$$

are holomorphic functions. Observe that

$$(4) \qquad z_1 + i_2 z_2 = [(z_1 + i_2 z_2) - (a_1 + i_2 a_2)] + (a_1 + i_2 a_2).$$

Replace $z_1 + i_2 z_2$ in (3) by its value in (4), and then rearrange the result in ascending powers of $[(z_1 + i_2 z_2) - (a_1 + i_2 a_2)]$. Thus f is represented by a terminating power series about the point $a_1 + i_2 a_2$, and the polynomial in (3) is a holomorphic function in $\mathbb{C}_2$ by Definition 15.2. Although interesting and important in the total theory, this method will not be developed further in this book.

A second method for establishing the existence of holomorphic functions of a bicomplex variable is based on the theory of differentiation and integration, and this method will be developed in detail in Chapter 3.

A third method uses the existence of holomorphic functions of a complex variable to establish the existence of holomorphic functions of a bicomplex variable and to develop their properties. The details of this method are presented in this section.

Let X be a domain in $\mathbb{C}_2$. Then h_1 and h_2 (see Section 8) map X into sets X_1 and X_2 in A_1 and A_2, respectively. Thus X_1 and X_2 are sets in the complex plane, and they are domains (see Theorems 8.7 and 8.11). Let $f_1 : X_1 \to \mathbb{C}_1$ and $f_2 : X_2 \to \mathbb{C}_1$ be holomorphic functions of the complex variables $z_1 - i_1 z_2$ and $z_1 + i_1 z_2$; such functions exist by the theory of functions of a complex variable. Define the function $f : X \to \mathbb{C}_2$ as follows:

$$(5) \qquad f(z_1 + i_2 z_2) = f_1(z_1 - i_1 z_2)e_1 + f_2(z_1 + i_1 z_2)e_2, \quad \forall z_1 + i_2 z_2 \text{ in } X.$$

15.3 THEOREM If $f_1 : X_1 \to \mathbb{C}_1$ and $f_2 : X_2 \to \mathbb{C}_1$ are holomorphic functions in $H(X_1)$ and $H(X_2)$, respectively, then the function $f : X \to \mathbb{C}_2$ defined in (5) is a holomorphic function in $H(X)$.

Proof. By the definition of X_1 and X_2, the points $z_1 - i_1 z_2$ and $z_1 + i_1 z_2$ are in X_1 and X_2, respectively, and thus (5) is a meaningful statement. Let $a_1 + i_2 a_2$ be a point in X; then since X is a domain, there is an $r > 0$ such that $N(a_1 + i_2 a_2, r) \subset X$. The proof of Theorem 8.7 shows that

$$(6) \qquad N(a_1 - i_1 a_2, r) \subset X_1, \qquad N(a_1 + i_1 a_2, r) \subset X_2,$$

$$(7) \qquad D(a_1 + i_2 a_2; r, r) \subset X.$$

Since $f_1 \in H(X_1)$ and $f_2 \in H(X_2)$, there are complex power series such that

$$(8) \qquad f_1(z_1 - i_1 z_2) = \sum_{k=0}^{\infty} \alpha_k[(z_1 - i_1 z_2) - (a_1 - i_1 a_2)]^k,$$
$$|(z_1 - i_1 z_2) - (a_1 - i_1 a_2)| < r,$$

$$(9) \qquad f_2(z_1 + i_1 z_2) = \sum_{k=0}^{\infty} \beta_k[(z_1 + i_1 z_2) - (a_1 + i_1 a_2)]^k,$$
$$|(z_1 + i_1 z_2) - (a_1 + i_1 a_2)| < r.$$

Then (5) shows that

$$(10) \qquad f(z_1 + i_2 z_2) = \sum_{k=0}^{\infty} \alpha_k[(z_1 - i_1 z_2) - (a_1 - i_1 a_2)]^k e_1 + \sum_{k=0}^{\infty} \beta_k[(z_1 + i_1 z_2) - (a_1 + i_1 a_2)]^k e_2$$

for every $z_1 + i_2 z_2$ in $D(a_1 + i_2 a_2; r, r)$. Now by a slight generalization of Theorem 13.8, there is a unique bicomplex power series such that

$$(11) \qquad \sum_{k=0}^{\infty} (a_k + i_2 b_k)[(z_1 + i_2 z_2) - (a_1 + i_2 a_2)]^k = \sum_{k=0}^{\infty} \alpha_k[(z_1 - i_1 z_2) - (a_1 - i_1 a_2)]^k e_1 + \sum_{k=0}^{\infty} \beta_k[(z_1 + i_1 z_2) - (a_1 + i_1 a_2)]^k e_2$$

for $|(z_1 - i_1 z_2) - (a_1 - i_1 a_2)| < r$ and $|(z_1 + i_1 z_2) - (a_1 + i_1 a_2)| < r$. Equations (10) and (11) show that

$$(12) \qquad f(z_1 + i_2 z_2) = \sum_{k=0}^{\infty} (a_k + i_2 b_k)[(z_1 + i_2 z_2) - (a_1 + i_2 a_2)]^k,$$
$$\forall z_1 + i_2 z_2 \in D(a_1 + i_2 a_2; r, r).$$

Thus for every point $a_1 + i_2 a_2$ in X there is a discus $D(a_1 + i_2 a_2; r, r)$ and a power series which converges and represents f in $D(a_1 + i_2 a_2; r, r)$. By Definition 15.2, $f: X \to \mathbb{C}_2$ is a holomorphic function in $H(X)$. The proof of Theorem 15.3 is complete. □

15.4 COROLLARY Let X, X_1, and X_2 be the sets described above, and let X' be the cartesian domain determined by X_1 and X_2. If X is not a cartesian set, then X is a proper subset of X', and there exists a holomorphic function $f': X' \to \mathbb{C}_2$ which is the holomorphic continuation of f from X into X':

$$(13) \qquad f'(z_1 + i_2 z_2) = f(z_1 + i_2 z_2), \qquad \forall z_1 + i_2 z_2 \in X.$$

Proof. In Theorem 15.3, X is assumed to be a domain, but it is not assumed to be a cartesian set. Thus X is the set of points $z_1+i_2z_2$ such that

$$(14)\qquad z_1 + i_2z_2 = (z_1 - i_1z_2)e_1 + (z_1 + i_1z_2)e_2, \qquad \forall z_1 + i_2z_2 \in X.$$

As $z_1+i_2z_2$ varies over X, the points $z_1-i_1z_2$ and $z_1+i_1z_2$ describe the sets X_1 and X_2. But the set of points $z_1+i_2z_2$ such that

$$(15)\qquad z_1 + i_2z_2 = (z_1 - i_1z_2)e_1 + (z_1 + i_1z_2)e_2,$$
$$\forall z_1 - i_1z_2 \in X_1,\ \forall z_1 + i_1z_2 \in X_2,$$

is a set X' which contains X; it is the cartesian set determined by X_1 and X_2. If X is a cartesian set, then $X'=X$, but X is a proper subset of X' if X is not a cartesian set. The proof of Theorem 15.3 can be changed as follows to construct a holomorphic function $f':X'\to\mathbb{C}_2$. Since f_1 and f_2 are defined in X_1 and X_2, respectively, replace (5) by

$$(16)\qquad f'(z_1 + i_2z_2) = f_1(z_1 - i_1z_2)e_1 + f_2(z_1 + i_1z_2)e_2,$$
$$\forall z_1 - i_1z_2 \in X_1,\ \forall z_1 + i_1z_2 \in X_2.$$

If $a_1-i_1a_2$ and $a_1+i_1a_2$ are points in X_1 and X_2, respectively, then there is an $r>0$ and power series such that (8) and (9) are satisfied. Then as shown in the proof of Theorem 15.3, there is a power series at $a_1+i_2a_2$ such that

$$(17)\qquad f'(z_1 + i_2z_2) = \sum_{k=0}^{\infty} (a_k + i_2b_k)[(z_1 + i_2z_2) - (a_1 + i_2a_2)]^k,$$
$$\forall z_1 + i_2z_2 \in D(a_1 + i_2a_2;\, r, r).$$

Thus f' is defined in X', and at each $a_1+i_2a_2$ in X' it is represented by a power series. Therefore, by Definition 15.2, f' is a holomorphic function in X', and (13) is true. Thus f' is the holomorphic continuation of f from X into the larger set X'. The proof of Corollary 15.4 is complete. □

Theorem 15.3 has shown that holomorphic functions $f:X\to\mathbb{C}_2$ can be constructed from holomorphic functions $f_1:X_1\to\mathbb{C}_1$ and $f_2:X_2\to\mathbb{C}_1$ of complex variables. The next theorem shows that all holomorphic functions $f:X\to\mathbb{C}_2$ in $H(X)$ are obtained in this manner.

15.5 THEOREM Let X be a domain in $\mathbb{C}_2$, and let $f:X\to\mathbb{C}_2$ be a holomorphic function in $H(X)$. Then there exist holomorphic functions $f_1:X_1\to\mathbb{C}_1$ in $H(X_1)$ and $f_2:X_2\to\mathbb{C}_1$ in $H(X_2)$ such that

$$(18)\qquad f(z_1 + i_2z_2) = f_1(z_1 - i_1z_2)e_1 + f_2(z_1 + i_1z_2)e_2,$$
$$\forall z_1 + i_2z_2 \text{ in } X.$$

Proof. There are two steps in the proof as follows: (a) there exist functions $f_1 : X_1 \to \mathbb{C}_1$ and $f_2 : X_2 \to \mathbb{C}_1$ which satisfy (18); and (b) the functions f_1 and f_2 are holomorphic functions in $H(X_1)$ and $H(X_2)$, respectively. Observe first that $f(z_1 + i_2 z_2)$ is a bicomplex number in $\mathbb{C}_2$ for each $z_1 + i_2 z_2$ in X. Also, each bicomplex number has an idempotent representation. Use this representation to define the functions $f_1 : X_1 \to \mathbb{C}_1$ and $f_2 : X_2 \to \mathbb{C}_1$ so that f_1 and f_2 are holomorphic functions. Let $a_1 + i_2 a_2$ be a point in X. Then, as shown in the proof of Theorem 15.3, there is an $\varepsilon > 0$ such that $N(a_1 + i_2 a_2, \varepsilon) \subset X$, $N(a_1 - i_1 a_2, \varepsilon) \subset X_1$, and $N(a_1 + i_1 a_2, \varepsilon) \subset X_2$. Also, there is an r, $0 < r \leqslant \varepsilon$, and a power series such that

$$\text{(19)} \qquad f(z_1 + i_2 z_2) = \sum_{k=0}^{\infty} (a_k + i_2 b_k)[(z_1 + i_2 z_2) - (a_1 + i_2 a_2)]^k,$$

$$\forall z_1 + i_2 z_2 \in D(a_1 + i_2 a_2; r, r).$$

Thus, as shown in Theorems 13.5 and 13.6,

$$\text{(20)} \qquad \sum_{k=0}^{\infty} (a_k + i_2 b_k)[(z_1 + i_2 z_2) - (a_1 + i_2 a_2)]^k$$

$$= \sum_{k=0}^{\infty} (a_k - i_1 b_k)[(z_1 - i_1 z_2) - (a_1 - i_1 a_2)]^k e_1$$

$$+ \sum_{k=0}^{\infty} (a_k + i_1 b_k)[(z_1 + i_1 z_2) - (a_1 + i_1 a_2)]^k e_2$$

for all $z_1 + i_2 z_2$ in $D(a_1 + i_2 a_2; r, r)$, or, equivalently, all $z_1 - i_1 z_2$ in $|(z_1 - i_1 z_2) - (a_1 - i_1 b_2)| < r$ and all $z_1 + i_1 z_2$ in $|(z_1 + i_1 z_2) - (a_1 + i_1 b_2)| < r$. Since the idempotent representation of a bicomplex number is unique, (18) and (20) show that

$$\text{(21)} \qquad f_1(z_1 - i_1 z_2) = \sum_{k=0}^{\infty} (a_k - i_1 b_k)[(z_1 - i_1 z_2) - (a_1 - i_1 a_2)]^k,$$

$$|(z_1 - i_1 z_2) - (a_1 - i_1 b_2)| < r.$$

$$f_2(z_1 + i_1 z_2) = \sum_{k=0}^{\infty} (a_k + i_1 b_k)[(z_1 + i_1 z_2) - (a_1 + i_1 a_2)]^k,$$

$$|(z_1 + i_1 z_2) - (a_1 + i_1 b_2)| < r.$$

Thus f_1 and f_2 are represented by power series in neighborhoods of $a_1 - i_1 a_2$ and $a_1 + i_1 a_2$, respectively. By the definition of the sets X_1 and X_2, every point in X_1 is a point $a_1 - i_1 a_2$ for some $a_1 + i_2 a_2$ in X. Thus f_1 is represented by a power series in a neighborhood of every point in X_1, and f_2 is represented by a power series in a neighborhood of every point in X_2. By known results in the theory of functions of a complex variable, f_1 and f_2 are holomorphic functions in X_1 and X_2, respectively. Thus the proof has shown that there exist holomorphic functions f_1 in $H(X_1)$ and f_2 in $H(X_2)$ such that

$$(22)\qquad f(z_1 + i_2 z_2) = f_1(z_1 - i_1 z_2)e_1 + f_2(z_1 + i_1 z_2)e_2, \quad \forall z_1 + i_2 z_2 \in X,$$

and the proof of Theorem 15.5 is complete. □

Theorems 15.3 and 15.5 show that $f: X \to \mathbb{C}_2$ is a holomorphic function of the bicomplex variable $z_1 + i_2 z_2$ if and only if there exist holomorphic functions $f_1: X_1 \to \mathbb{C}_1$ and $f_2: X_2 \to \mathbb{C}_1$ such that

$$(23)\qquad f(z_1 + i_2 z_2) = f_1(z_1 - i_1 z_2)e_1 + f_2(z_1 + i_1 z_2)e_2, \quad \forall z_1 + i_2 z_2 \in X.$$

Define functions u and v as follows [compare (18) and (19) in Section 14]:

$$(24)\qquad \begin{aligned} u(z_1, z_2) &= \frac{f_1(z_1 - i_1 z_2) + f_2(z_1 + i_1 z_2)}{2}, \\ v(z_1, z_2) &= \frac{i_1[f_1(z_1 - i_1 z_2) - f_2(z_1 + i_1 z_2)]}{2}, \end{aligned} \qquad \forall z_1 + i_2 z_2 \in X.$$

Then

$$(25)\qquad f_1(z_1 - i_1 z_2) = u(z_1, z_2) - i_1 v(z_1, z_2),$$

$$(26)\qquad f_2(z_1 + i_1 z_2) = u(z_1, z_2) + i_1 v(z_1, z_2),$$

$$(27)\qquad f(z_1 + i_2 z_2) = u(z_1, z_2) + i_2 v(z_1, z_2).$$

As in Section 14, u and v satisfy the Cauchy–Riemann differential equations, and each of these functions is a solution of Laplace's equation:

$$\begin{aligned} &(28)\qquad \frac{\partial u}{\partial z_1} = \frac{\partial v}{\partial z_2}, \qquad \frac{\partial u}{\partial z_2} = -\frac{\partial v}{\partial z_1}; \\ &(29)\qquad \frac{\partial^2 u}{\partial z_1^2} + \frac{\partial^2 u}{\partial z_2^2} = 0, \qquad \frac{\partial^2 v}{\partial z_1^2} + \frac{\partial^2 v}{\partial z_2^2} = 0, \end{aligned} \qquad \forall z_1 + i_2 z_2 \in X.$$

15.6 THEOREM Let $f: X \to \mathbb{C}_2$ be a holomorphic function. Then f has a zero at $z_1^0 + i_2 z_2^0$ if and only if the following conditions are satisfied:

$$(30)\qquad f_1(z_1^0 - i_1 z_2^0) = 0, \qquad f_2(z_1^0 + i_1 z_2^0) = 0;$$

$$(31)\qquad u(z_1^0, z_2^0) = 0, \qquad v(z_1^0, z_2^0) = 0.$$

If f_1 is not identically zero in X_1 and f_2 is not identically zero in X_2, then the zeros of f, if any, are isolated. If f_1 is identically zero in X_1, or if f_2 is identically zero in X_2, then the zeros of f, if any, are not isolated.

Proof. If $f(z_1^0+i_2z_2^0)=0$, then (23)–(27) show that (30) and (31) are true. Conversely, if (30) or (31) is satisfied, then (23) or (27) shows that $f(z_1^0+i_2z_2^0)=0$. Thus the first conclusion in the theorem has been established. Assume next that $f(z_1^0+i_2z_2^0)=0$ and that neither f_1 nor f_2 is identically zero. Then $f_1(z_1^0-i_1z_2^0)=0$ and $f_2(z_1^0+i_1z_2^0)=0$. Also, since f_1 and f_2 are holomorphic functions of a complex variable, their zeros are isolated. Thus there exist r_1 and r_2 such that f_1 has no other zero in $|(z_1-i_1z_2)-(z_1^0-i_1z_2^0)|<r_1$ and f_2 has no other zero in $|(z_1+i_1z_2)-(z_1^0+i_1z_2^0)|<r_2$. Thus f has a zero at $z_1^0+i_2z_2^0$ but no other zero in $D(z_1^0+i_2z_2^0;\ r_1,r_2)$, and the zeros of f are isolated. Assume finally that $f_1(w_1)=0$ for every w_1 in X_1. If f_2 has no zeros, then f has no zeros by the first part of the theorem. If $f_2(z_1^0+i_1z_2^0)=0$, then f has a zero at every point $z_1+i_2z_2$ such that

$$(32)\qquad z_1+i_2z_2=w_1e_1+(z_1^0+i_1z_2^0)e_2,\qquad \forall w_1\in X_1,$$

since

$$(33)\qquad f(z_1+i_2z_2)=f_1(w_1)e_1+f_2(z_1^0+i_1z_2^0)e_2=0.$$

The set of zeros in (32) is not an isolated set. Similar results can be established if f_2 is identically zero in X_2. The proof of all parts of Theorem 15.6 is complete. □

15.7 DEFINITION Let $f:X\to\mathbb{C}_2$ be a holomorphic function. A point $z_1^0+i_2z_2^0$ such that

$$(34)\qquad f(z_1^0+i_2z_2^0)\in I_1-\{0\}$$

is called an I_1*-point* of f, and a point $z_1^0+i_2z_2^0$ such that

$$(35)\qquad f(z_1^0+i_2z_2^0)\in I_2-\{0\}$$

is called an I_2*-point* of f.

15.8 THEOREM Let $f:X\to\mathbb{C}_2$ be a holomorphic function. Then $z_1^0+i_2z_2^0$ is an I_1-point of f if and only if

$$(36)\qquad f_1(z_1^0-i_1z_2^0)\neq 0,\qquad f_2(z_1^0+i_1z_2^0)=0,$$

and it is an I_2-point if and only if

$$(37)\qquad f_1(z_1^0-i_1z_2^0)=0,\qquad f_2(z_1^0+i_1z_2^0)\neq 0.$$

If neither f_1 nor f_2 has a zero, then f has no I_1-point and no I_2-point. Finally, I_1-points and I_2-points are never isolated.

Proof. By (3) and (4) in Section 7,

$$(38)\qquad I_1=\{(a+i_2b)e_1:\ (a+i_2b)\in\mathbb{C}_2\},$$

$$(39)\qquad I_2=\{(a+i_2b)e_2:\ (a+i_2b)\in\mathbb{C}_2\}.$$

By (23),

(40) $\quad f(z_1+i_2z_2)=f_1(z_1-i_1z_2)e_1+f_2(z_1+i_1z_2)e_2, \quad \forall z_1+i_2z_2\in X.$

Thus, by Definition 15.7, $z_1^0+i_2z_2^0$ is an I_1-point if and only if (36) is true, and it is an I_2-point if and only if (37) is true. If f_1 and f_2 have no zero, it is impossible to satisfy the conditions in (36) and (37). Finally, let f have an I_1-point at $z_1^0+i_2z_2^0$. Then

(41) $\quad f_1(z_1^0-i_1z_2^0)\neq 0, \qquad f_2(z_1^0+i_1z_2^0)=0.$

Since f_1 is continuous at $z_1^0-i_1z_2^0$, there is an $\varepsilon>0$ such that

(42) $\quad f_1(w_1)\neq 0, \qquad \forall w_1$ such that $|w_1-(z_1^0-i_1z_2^0)|<\varepsilon.$

Thus

(43) $\quad f_1(w_1)\neq 0, \qquad f_2(z_1^0+i_1z_2^0)=0,$

and f has an I_1-point at every point $z_1+i_2z_2$ such that

(44) $\quad z_1+i_2z_2=w_1e_1+(z_1^0+i_1z_2^0)e_2, \qquad |w_1-(z_1^0+i_1z_2^0)|<\varepsilon;$

therefore, the I_1-points of f are not isolated. The proof that the I_2-points of f are not isolated is similar. The proof of Theorem 15.8 is complete. □

15.9 COROLLARY Let $f:X\to\mathbb{C}_2$ be a holomorphic function. If f_1 is identically zero in X_1, then every $z_1+i_2z_2$ in X is either a zero or an I_2-point of f; if f_2 is identically zero in X_2, then every $z_1+i_2z_2$ in X is either a zero or an I_1-point of f.

Proof. If f_1 is identically zero, then

(45) $\quad f(z_1+i_2z_2)=f_2(z_1+i_1z_2)e_2, \qquad \forall z_1+i_2z_2\in X.$

If $f_2(z_1+i_1z_2)=0$, then $z_1+i_2z_2$ is a zero of f; if $f_2(z_1+i_1z_2)\neq 0$, then $z_1+i_2z_2$ is an I_2-point of f. If f_2 is identically zero,

(46) $\quad f(z_1+i_2z_2)=f_1(z_1-i_1z_2)e_1,$

and every $z_1+i_2z_2$ in X is either a zero or an I_1-point of f. □

A domain X is called a *domain of holomorphism* if there exists a holomorphic function in X which cannot be continued into a larger domain. This definition applies to holomorphic functions of complex and also bicomplex variables.

15.10 EXAMPLE This example shows that there exist domains of holomorphism for functions of a bicomplex variable. It is shown in the theory of functions of a complex variable that each of the power series

$$(47) \qquad \sum_{k=0}^{\infty} \frac{1}{k!} z^{2^k}, \qquad \sum_{k=0}^{\infty} z^{k!},$$

converges for $|z|<1$. However, each point of the circle $|z|=1$ is a singular point for each series; the unit circle is thus a *natural boundary* and each series is noncontinuable. Set

$$(48) \qquad f(z_1 + i_2 z_2) = \sum_{k=0}^{\infty} \frac{1}{k!} (z_1 - i_1 z_2)^{2^k} e_1 + \sum_{k=0}^{\infty} (z_1 + i_1 z_2)^{k!} e_2.$$

Then f is a holomorphic function of the bicomplex variable $z_1 + i_2 z_2$ in the discus $D(0; 1, 1)$. Since the power series in (47) are noncontinuable, the power series which represents $f : D(0; 1, 1) \to \mathbb{C}_2$ is noncontinuable. Thus $D(0; 1, 1)$ is a domain of holomorphism. It is clear from this example that every discus $D(a_1 + i_2 a_2; r_1, r_2)$ with $r_1 > 0$ and $r_2 > 0$ is a domain of holomorphism. But there are others, as we shall show.

Mittag-Leffler showed that every domain in $\mathbb{C}_1$ is a domain of holomorphism for functions f of a complex variable; in 1884 he proved the following theorem: given any domain D in the complex plane, there exists a function $f : D \to \mathbb{C}_1$ which has D as its domain of holomorphism. Corollary 15.4 has shown already that some domains X in $\mathbb{C}_2$ are not a domain of holomorphism for a function of a bicomplex variable. The following theorem determines all domains of holomorphism in $\mathbb{C}_2$.

15.11 THEOREM A domain X in $\mathbb{C}_2$ is a domain of holomorphism for functions of a bicomplex variable if and only if X is a cartesian set.

Proof. First, the condition is necessary. To prove this statement, assume that X in $\mathbb{C}_2$ is a domain of holomorphism but not a cartesian set. Then there is a function $f : X \to \mathbb{C}_2$ which cannot be continued into a larger domain. But Corollary 15.4 shows that X is a proper subset of a domain X' in which there is a holomorphic function $f' : X' \to \mathbb{C}_2$ such that $f'(z_1 + i_2 z_2) = f(z_1 + i_2 z_2)$ for all $z_1 + i_2 z_2$ in X. Thus the assumption that X is not a cartesian set has led to a contradiction, and the condition is necessary. Next, the condition is sufficient. Assume that X is a cartesian domain. Then the sets X_1 and X_2 are domains in $\mathbb{C}_1$, and Mittag-Leffler's theorem shows that they are domains of holomorphism. Thus there exist holomorphic functions $f_1 : X_1 \to \mathbb{C}_1$ and $f_2 : X_2 \to \mathbb{C}_1$ which cannot be continued into larger domains. Then the function $f : X \to \mathbb{C}_2$ such that

$$(49) \qquad f(z_1 + i_2 z_2) = f_1(z_1 - i_1 z_2) e_1 + f_2(z_1 + i_1 z_2) e_2, \quad \forall (z_1 - i_1 z_2, z_1 + i_1 z_2) \text{ in } X_1 \times X_2,$$

is holomorphic in X, and we shall now show that it cannot be continued into a larger domain. To prove this statement, assume that f can be continued into a domain X' which contains X as a proper subset. Then there are sets X'_1 and X'_2 such that $X_1 \subset X'_1$ and $X_2 \subset X'_2$; furthermore, f_1 and f_2 can be continued into X'_1 and X'_2, respectively, by Theorem 15.5. Since X is a cartesian set by hypothesis, and since X is a proper subset of X', Theorem 8.10 shows that at least one of the sets X_1, X_2 is a proper subset of the corresponding set X'_1, X'_2. For example, suppose that X_1 is a proper subset of X'_1; the procedure is similar if X_2 is a proper subset of X'_2. Then f_1 can be continued from X_1 into the larger set X'_1. This statement contradicts the assumption that f_1 cannot be continued from X_1 into a larger domain. This contradiction has resulted from the hypothesis that f can be continued from X into a larger domain X'. Therefore, $f: X \to \mathbb{C}_2$ is a function which cannot be continued from X into a larger domain, and the cartesian domain X is a domain of holomorphism. The proof of Theorem 15.11 is complete. □

Exercises

15.1 Show that each function f defined by the following statements is a holomorphic function in $\mathbb{C}_2$:

(a) $f(z_1 + i_2 z_2) = a$, a constant;

(b) $f(z_1 + i_2 z_2) = (z_1 + i_2 z_2)^n$, $\forall n \in \mathbb{N}$;

(c) $f(z_1 + i_2 z_2) =$ a polynomial in $z_1 + i_2 z_2$.

15.2 Let $\exp: \mathbb{C}_2 \to \mathbb{C}_2$ be a function such that

$$\exp(z_1 + i_2 z_2) = \sum_{k=0}^{\infty} \frac{(z_1 + i_2 z_2)^k}{k!}.$$

(a) Show that exp is a holomorphic function of $z_1 + i_2 z_2$.

(b) Show that $\exp[(z_1 + i_2 z_2) + (a_1 + i_2 a_2)] = \exp(z_1 + i_2 z_2) \exp(a_1 + i_2 a_2)$.

(c) Find the functions u and v in (24) for exp, and show that these functions satisfy the Cauchy–Riemann equations and Laplace's equation [compare Exercise 14.4(d)].

(d) Show that exp has no zeros and no I_1-points and no I_2-points.

15.3 If the power series $\Sigma_{k=0}^{\infty} a_k z^k$ in the complex variable z converges only for $z = 0$, it does not represent a holomorphic function. If the power series $\Sigma_{k=0}^{\infty} (a_k + i_2 b_k)(z_1 + i_2 z_2)^k$ converges only on a set in I_1, or only on a set in I_2, show that it does not represent a holomorphic function of the bicomplex variable $z_1 + i_2 z_2$.

15.4 The function $f: \mathbb{C}_2 \to \mathbb{C}_2$, $z_1 + i_2 z_2 \mapsto (z_1 + i_2 z_2)^2$, is a holomorphic function with an isolated zero at $z_1 + i_2 z_2 = 0$.

(a) Verify that the conditions in (30) and (31) for a zero at $z_1+i_2z_2=0$ are satisfied.
(b) Show by direct calculation that the conditions in (30) and (31) are equivalent for the function f.
(c) Use Theorem 15.6 to show that the zeros of f are isolated. How many solutions does the equation $f(z_1+i_2z_2)=0$ have?

15.5 The following equations define functions $f:\mathbb{C}_2\to\mathbb{C}_2$. Show that each of these functions is a holomorphic function in $\mathbb{C}_2$ and find all of its zeros.
(a) $f(z_1+i_2z_2)=\sin(z_1-i_1z_2)e_1+\sin(z_1+i_1z_2)e_2$.
(b) $f(z_1+i_2z_2)=\sin(z_1-i_1z_2)e_1+\cos(z_1+i_1z_2)e_2$.

15.6 Let $g:\mathbb{C}_1\to\mathbb{C}_1$ be a holomorphic function of the complex variable z. Define $f:\mathbb{C}_2\to\mathbb{C}_2$, $z_1+i_2z_2\mapsto f(z_1+i_2z_2)$, as follows:

$$f(z_1+i_2z_2)=g(z_1-i_1z_2)e_1+g(z_1+i_1z_2)e_2.$$

(a) Show that f is holomorphic in $\mathbb{C}_2$.
(b) Show that f has complex values when $z_1+i_2z_2$ has complex values; that is, show that $f(z_1+i_2z_2)=g(z_1)$, where $g(z_1)\in\mathbb{C}_1$, if $z_2=0$.

15.7 Let X be a domain in $\mathbb{C}_2$; let a be a constant in $\mathbb{C}_2$; and let $f:X\to\mathbb{C}_2$ and $g:X\to\mathbb{C}_2$ be holomorphic functions. Define functions af, $f+g$, and fg as follows:

$$\begin{aligned}(af)(z_1+i_2z_2)&=af(z_1+i_2z_2),\\ (f+g)(z_1+i_2z_2)&=f(z_1+i_2z_2)+g(z_1+i_2z_2), \qquad \forall z_1+i_2z_2\in X.\\ (fg)(z_1+i_2z_2)&=f(z_1+i_2z_2)g(z_1+i_2z_2),\end{aligned}$$

Prove that af, $f+g$, and fg are holomorphic functions in X.

16. ALGEBRAS OF HOLOMORPHIC FUNCTIONS

Let $BH(Z)$ denote the set of bounded holomorphic functions on a domain Z in $\mathbb{C}_1$. The system consisting of $BH(Z)$ and the operations of addition, multiplication, and scalar multiplication is an algebra. This algebra becomes a normed algebra with the addition of the sup norm. Furthermore, every sequence in $BH(Z)$ which is a Cauchy sequence in the sup norm converges to a function in $BH(Z)$, and thus the space of bounded holomorphic functions on Z is an algebra and a linear, normed, and complete space. This section reviews these properties of the bounded holomorphic functions $BH(Z)$ of a complex variable on a domain Z in $\mathbb{C}_1$ and then extends them to construct the Banach algebra of bounded holomorphic functions $BH(X)$ of a bicomplex variable on a domain X in $\mathbb{C}_2$.

Let Z be a domain in $\mathbb{C}_1$, and let $H(Z)$ denote the set of functions $g: Z \to \mathbb{C}_1$ which are holomorphic on Z. Let a be a constant in $\mathbb{C}_1$ and let $g: Z \to \mathbb{C}_1$ and $h: Z \to \mathbb{C}_1$ be functions in $H(Z)$. The algebraic operations of scalar multiplication, addition, and multiplication in $H(Z)$ are defined as follows:

$$(1) \qquad \begin{aligned} &(ag)(z) = ag(z), \\ &(g+h)(z) = g(z) + h(z), \qquad \forall z \text{ in } Z. \\ &(gh)(z) = g(z)h(z), \end{aligned}$$

The algebraic properties of scalar multiplication, addition, and multiplication follow from the algebraic properties of the complex numbers. If g and h are in $H(Z)$, then ag, $g+h$, and gh are in $H(Z)$; if g and h are in $BH(Z)$, then ag, $g+h$, and gh are in $BH(Z)$. If $g \in BH(Z)$, then

$$(2) \qquad \sup\{|g(z)|: z \in Z\}$$

is defined and is a number in $\mathbb{R}_{\geqslant 0}$.

16.1 DEFINITION The function $|\ |: BH(Z) \to \mathbb{R}_{\geqslant 0}$ such that

$$(3) \qquad |g| = \sup\{|g(z)|: z \in Z\}, \qquad \forall g \text{ in } BH(Z),$$

is called the *sup norm* on $BH(Z)$, and $|g|$ is called the *sup norm of g*.

The sup norm on $BH(Z)$ has the following properties of a norm: for all a in $\mathbb{C}_1$ and all g and h in $BH(Z)$,

$$(4) \qquad |g| \geqslant 0; \ |g| = 0 \text{ if and only if } g \text{ is identically zero on } Z;$$

$$(5) \qquad |ag| = |a|\,|g|;$$

$$(6) \qquad |g+h| \leqslant |g| + |h|;$$

$$(7) \qquad |gh| \leqslant |g|\,|h|.$$

A Cauchy sequence in $BH(Z)$ is a sequence g^n, $n \in \mathbb{N}$, with the following property: for each $\varepsilon > 0$ there exists an $n(\varepsilon)$ such that

$$(8) \qquad |g^n - g^m| < \varepsilon, \qquad \forall n \geqslant n(\varepsilon),\ \forall m \geqslant n(\varepsilon).$$

16.2 THEOREM The algebra of bounded holomorphic functions $BH(Z)$ with the sup norm is a Banach algebra.

Proof. The system consisting of the functions $BH(Z)$ and the operations defined in (1) is an algebra. Add the sup norm in (3) to this system; equations (4)–(7) show that the resulting system is a normed algebra. The proof of Theorem 16.2 can be completed by showing that this normed algebra is complete. Since for each z in Z,

$$(9) \qquad |g^n(z) - g^m(z)| \leqslant |g^n - g^m|,$$

each sequence $\{g^n(z)\colon n\in\mathbb{N}\}$ is a Cauchy sequence in $\mathbb{C}_1$. Since $\mathbb{C}_1$ is a complete Banach space, then

(10) $\quad \lim_{n\to\infty} g^n(z)$

exists for each z in Z and defines a function $g\colon Z\to\mathbb{C}_1$. By (8) and (9), for each z in Z,

(11) $\quad |g^n(z) - g^m(z)| < \varepsilon, \qquad \forall n \geqslant n(\varepsilon), \forall m \geqslant n(\varepsilon).$

Let $m\to\infty$; then

(12) $\quad |g^n(z) - g(z)| \leqslant \varepsilon, \qquad \forall n \geqslant n(\varepsilon), \forall z \text{ in } Z.$

Since g^n is a bounded function, (12) shows that g is bounded on Z. Since convergence in the sup norm is uniform convergence, a familiar proof shows that g is a continuous function. Then standard theorems in the theory of functions of a complex variable show that g is holomorphic in Z. Thus the normed algebra of bounded holomorphic functions $BH(Z)$ is complete in the sup norm, and it is a Banach algebra. □

There is a theorem similar to Theorem 16.2 for the bounded holomorphic functions $BH(X)$ of a bicomplex variable $z_1+i_2z_2$ on a domain X in $\mathbb{C}_2$, and we turn now to its proof.

The algebraic operations of scalar multiplication, addition, and multiplication in $H(X)$ are defined as follows:

(13)
$$(af)(z_1+i_2z_2)=af(z_1+i_2z_2), \qquad a\in\mathbb{C}_2,$$
$$(f+g)(z_1+i_2z_2)=f(z_1+i_2z_2)+g(z_1+i_2z_2), \qquad f, g\in H(X),$$
$$(fg)(z_1+i_2z_2)=f(z_1+i_2z_2)g(z_1+i_2z_2), \qquad \forall z_1+i_2z_2\in X.$$

The algebraic properties of scalar multiplication, addition, and multiplication follow from the algebraic properties of the bicomplex numbers. If f and g are in $H(X)$, then af, $f+g$, and fg are in $H(X)$ (see Exercise 15.7); if f and g are in $BH(X)$, then af, $f+g$, and fg are in $BH(X)$. If $f\in BH(X)$, then

(14) $\quad \sup\{\|f(z_1 + i_2z_2)\|\colon z_1 + i_2z_2\in X\}$

is defined and is a number in $\mathbb{R}_{\geqslant 0}$.

16.3 DEFINITION The function $\|\ \|\colon BH(X)\to\mathbb{R}_{\geqslant 0}$ such that

(15) $\quad \|f\| = \sup\{\|f(z_1 + i_2z_2)\|\colon z_1 + i_2z_2\in X\}, \qquad \forall f\in BH(X),$

is called the *sup norm* on $BH(X)$, and $\|f\|$ is called the *sup norm of f*.

The sup norm on $BH(X)$ has the following properties of a norm: for all a in $\mathbb{C}_2$ and all f and g in $BH(X)$,

(16) $\quad \|f\| \geqslant 0$; $\|f\| = 0$ if and only if f is identically zero on X;

(17) $\quad \|af\| = |a|\,\|f\|, \qquad a \in \mathbb{C}_1,\ f \in BH(X);$

(18) $\quad \|af\| \leqslant \sqrt{2}\,\|a\|\,\|f\|, \qquad a \in \mathbb{C}_2,\ f \in BH(X);$

(19) $\quad \|f + g\| \leqslant \|f\| + \|g\|, \qquad f, g \in BH(X);$

(20) $\quad \|fg\| \leqslant \sqrt{2}\,\|f\|\,\|g\|.$

A Cauchy sequence in $BH(X)$ is a sequence $\{f^n : n \in \mathbb{N}\}$ with the following property: for each $\varepsilon > 0$ there exists an $n(\varepsilon)$ such that

(21) $\quad \|f^n - f^m\| < \varepsilon, \qquad \forall n \geqslant n(\varepsilon),\ \forall m \geqslant n(\varepsilon).$

A lemma is required for the proof of the next theorem. First, recall certain facts and notation. If $f : X \to \mathbb{C}_2$ is a holomorphic function on the domain X in $\mathbb{C}_2$, then there are domains X_1, X_2, in $\mathbb{C}_1$ and holomorphic functions $f_1 : X_1 \to \mathbb{C}_1$ and $f_2 : X_2 \to \mathbb{C}_1$ such that

(22) $$f(z_1 + i_2 z_2) = f_1(z_1 - i_1 z_2)e_1 + f_2(z_1 + i_1 z_2)e_2, \quad \forall z_1 + i_2 z_2 \text{ in } X.$$

16.4 LEMMA The function $f : X \to \mathbb{C}_2$ is in $BH(X)$ if and only if $f_1 : X_1 \to \mathbb{C}_1$ is in $BH(X_1)$ and $f_2 : X_2 \to \mathbb{C}_1$ is in $BH(X_2)$. Also, $\{f^n : n \in \mathbb{N}\}$ is a Cauchy sequence in $BH(X)$ if and only if $\{f_1^n : n \in \mathbb{N}\}$ is a Cauchy sequence in $BH(X_1)$ and $\{f_2^n : n \in \mathbb{N}\}$ is a Cauchy sequence in $BH(X_2)$.

Proof. Theorems 15.3 and 15.5 have shown that $f : X \to \mathbb{C}_2$ is holomorphic if and only if $f_1 : X_1 \to \mathbb{C}_1$ and $f_2 : X_2 \to \mathbb{C}_1$ are holomorphic. To prove that f is bounded if and only if f_1 and f_2 are bounded, we shall show that

(23) $$\left.\begin{matrix} |f_1| \\ |f_2| \end{matrix}\right\} \leqslant \sqrt{2}\,\|f\| \leqslant |f_1| + |f_2|.$$

By (22),

(24) $$\sqrt{2}\,\|f(z_1 + i_2 z_2)\| = [|f_1(z_1 - i_1 z_2)|^2 + |f_2(z_1 + i_1 z_2)|^2]^{1/2}.$$

Then

(25) $$\begin{aligned} \sqrt{2}\,\|f(z_1 + i_2 z_2)\| &\leqslant |f_1(z_1 - i_1 z_2)| + |f_2(z_1 + i_1 z_2)|, \\ &\leqslant \sup_{X_1} |f_1(z_1 - i_1 z_2)| + \sup_{X_2} |f_2(z_1 + i_1 z_2)|, \\ &\leqslant |f_1| + |f_2|; \\ \sqrt{2}\,\sup_{X} \|f(z_1 + i_2 z_2)\| &\leqslant |f_1| + |f_2|, \\ &\leqslant |f_1| + |f_2|. \end{aligned}$$

Thus the inequality on the right in (23) has been established. To prove the one on the left, observe that, by (24),

$$(26)\qquad \left.\begin{matrix}|f_1(z_1 - i_1z_2)| \\ |f_2(z_1 + i_1z_2)|\end{matrix}\right\} \leqslant \sqrt{2}\,\|f(z_1 + i_2z_2)\|.$$

Inequality (26) is true for each $z_1 + i_2z_2$ and corresponding $z_1 - i_1z_2$ and $z_1 + i_1z_2$. Fix $z_1 - i_1z_2$ and $z_1 + i_1z_2$ at one set of corresponding values and observe that

$$(27)\qquad \left.\begin{matrix}|f_1(z_1 - i_1z_2)| \\ |f_2(z_1 + i_1z_2)|\end{matrix}\right\} \leqslant \sqrt{2}\sup_X \|f(z_1 + i_2z_2)\| = \sqrt{2}\,\|f\|.$$

Since these inequalities hold for each $z_1 - i_1z_2$ in X_1 and $z_1 + i_1z_2$ in X_2, then

$$(28)\qquad \sup_{X_1}|f_1(z_1 - i_1z_2)| \leqslant \sqrt{2}\,\|f\|, \qquad \sup_{X_2}|f_2(z_1 + i_1z_2)| \leqslant \sqrt{2}\,\|f\|,$$

$$\left.\begin{matrix}|f_1| \\ |f_2|\end{matrix}\right\} \leqslant \sqrt{2}\,\|f\|,$$

and the proof of (23) is complete. Now (23) shows that f is bounded if and only if f_1 and f_2 are bounded, and the proof of the first statement in the lemma is complete. Consider the second. Equation (22) shows that, corresponding to a sequence $\{f^n : n \in \mathbb{N}\}$ in $BH(X)$, there are sequences $\{f_1^n : n \in \mathbb{N}\}$ and $\{f_2^n : n \in \mathbb{N}\}$ in $BH(X_1)$ and $BH(X_2)$, respectively. Then the inequalities in (23) show that

$$(29)\qquad \left.\begin{matrix}|f_1^n - f_1^m| \\ |f_2^n - f_2^m|\end{matrix}\right\} \leqslant \sqrt{2}\,\|f^n - f^m\| \leqslant |f_1^n - f_1^m| + |f_2^n - f_2^m|.$$

If $\{f^n : n \in \mathbb{N}\}$ is a Cauchy sequence, the inequalities on the left show that $\{f_1^n : n \in \mathbb{N}\}$ and $\{f_2^n : n \in \mathbb{N}\}$ are Cauchy sequences. Also, if $\{f_1^n : n \in \mathbb{N}\}$ and $\{f_2^n : n \in \mathbb{N}\}$ are Cauchy sequences, the inequality on the right shows that $\{f^n : n \in \mathbb{N}\}$ is a Cauchy sequence. Thus $\{f^n : n \in \mathbb{N}\}$ is a Cauchy sequence if and only if $\{f_1^n : n \in \mathbb{N}\}$ and $\{f_2^n : n \in \mathbb{N}\}$ are Cauchy sequences, and the proof of Lemma 16.4 is complete. □

16.5 THEOREM The algebra of bounded holomorphic functions $BH(X)$ is a Banach algebra.

Proof. The system consisting of the functions $BH(X)$ and the operations defined in (13) is an algebra. Add the sup norm in (15) to this system; equations (16)–(20) show that the resulting system is a normed algebra. The proof of Theorem 16.5 can be completed by showing that this normed algebra is complete. Since for each $z_1 + i_2z_2$ in X,

$$(30)\qquad \|f^n(z_1 + i_2z_2) - f^m(z_1 + i_2z_2)\| \leqslant \|f^n - f^m\|,$$

each sequence $\{f^n(z_1+i_2z_2): n\in\mathbb{N}\}$ is a Cauchy sequence in $\mathbb{C}_2$ if $\{f^n: n\in\mathbb{N}\}$ is a Cauchy sequence in $BH(X)$. Since $\mathbb{C}_2$ is a (complete) Banach space, then

$$(31)\qquad \lim_{n\to\infty} f^n(z_1+i_2z_2)$$

exists for each $z_1+i_2z_2$ in X and defines a function $f: X\to\mathbb{C}_2$. Since $\{f^n: n\in\mathbb{N}\}$ is a Cauchy sequence, (30) shows that for each $\varepsilon>0$ there is an $n(\varepsilon)$ such that

$$(32)\qquad \|f^n(z_1+i_2z_2)-f^m(z_1+i_2z_2)\|<\varepsilon,\qquad \forall n\geqslant n(\varepsilon).$$

Let $m\to\infty$; then

$$(33)\qquad \|f^n(z_1+i_2z_2)-f(z_1+i_2z_2)\|\leqslant\varepsilon,\qquad \forall n\geqslant n(\varepsilon),\ \forall z_1+i_2z_2 \text{ in } X.$$

Since f^n is a bounded function, (33) shows that f is bounded on X. Since convergence in the sup norm is uniform convergence, a familiar proof shows that f is a continuous function. To complete the proof of Theorem 16.5, it is necessary to show that f is a holomorphic function in $H(X)$. In Section 42 of Chapter 4, methods will be developed which can be used to show directly that f is in $H(X)$, but at this point an appeal to Theorem 16.2 must be made.

Let $\{f^n: n\in\mathbb{N}\}$ be a Cauchy sequence in $BH(X)$. Then Lemma 16.4 shows that $\{f_1^n: n\in\mathbb{N}\}$ and $\{f_2^n: n\in\mathbb{N}\}$ are Cauchy sequences in $BH(X_1)$ and $BH(X_2)$, respectively. Then Theorem 16.2 shows that

$$(34)\qquad \lim_{n\to\infty} f_1^n(z_1-i_1z_2)=f_1(z_1-i_1z_2),\qquad \forall z_1-i_1z_2 \text{ in } X_1,$$

$$\lim_{n\to\infty} f_2^n(z_1+i_1z_2)=f_2(z_1+i_1z_2),\qquad \forall z_1+i_1z_2 \text{ in } X_2.$$

Here f_1 and f_2 are bounded holomorphic functions in $BH(X_1)$ and $BH(X_2)$, respectively. Define the function $f: X\to\mathbb{C}_2$ as follows:

$$(35)\qquad f(z_1+i_2z_2)=f_1(z_1-i_1z_2)e_1+f_2(z_1+i_1z_2)e_2,\qquad \forall z_1+i_2z_2 \text{ in } X.$$

Then f is a bounded function by (23), and it is a holomorphic function by Theorem 15.3. To complete the proof it is necessary to show that

$$(36)\qquad \lim_{n\to\infty} f^n=f.$$

By (23),

$$(37)\qquad \sqrt{2}\|f^n-f\|\leqslant|f_1^n-f_1|+|f_2^n-f_2|.$$

Since $\{f_1^n: n\in\mathbb{N}\}$ and $\{f_2^n: n\in\mathbb{N}\}$ are Cauchy sequences in $BH(X_1)$ and $BH(X_2)$, respectively, then

$$(38)\qquad \lim_{n\to\infty} f_1^n=f_1,\qquad \lim_{n\to\infty} f_2^n=f_2,$$

and (37) shows that

$$(39)\qquad \lim_{n\to\infty} f^n = f.$$

Thus every Cauchy sequence in $BH(X)$ has a limit in $BH(X)$, and the algebra of bounded holomorphic functions on X is a Banach algebra. The proof of Theorem 16.5 is complete. □

Let X be a domain in $\mathbb{C}_2$, and let A denote the Banach algebra of bounded holomorphic functions $BH(X)$. Then A has many interesting subsets. For example, the subset of A consisting of the constant functions on X with values in $\mathbb{C}_0$, denoted by $\mathbb{C}_0(X)$, is itself a Banach algebra since it is closed under the algebraic operations and complete in the sup norm on A. If $c \in \mathbb{C}_0(X)$ and $f \in A$, then it is not always true that $cf \in \mathbb{C}_0(X)$; therefore, $\mathbb{C}_0(X)$ is not an ideal in A (see Definition 7.1). Likewise, the subsets of A consisting of the constant functions on X with values in $\mathbb{C}_1$ or in $\mathbb{C}_2$, denoted by $\mathbb{C}_1(X)$ and $\mathbb{C}_2(X)$, respectively, are Banach algebras but not ideals in A.

The subset $P(X)$ of A consisting of polynomials in $z_1 + i_2 z_2$ is a ring and a normed algebra, but $P(X)$ is not a Banach algebra since it is not complete (the limit of a Cauchy sequence of polynomials need not be a polynomial). Also, $P(X)$ is not an ideal since the product of a polynomial and an arbitrary element in A need not be a polynomial.

16.6 DEFINITION Define the subsets $I_1(X)$ and $I_2(X)$ of A by the following equations:

$$(40)\qquad I_1(X) = \{e_1 g: g \in A\},$$

$$(41)\qquad I_2(X) = \{e_2 g: g \in A\}.$$

If f is in A, then

$$(42)\qquad f(z_1 + i_2 z_2) = f_1(z_1 - i_1 z_2)e_1 + f_2(z_1 + i_1 z_2)e_2, \quad \forall z_1 + i_2 z_2 \text{ in } X.$$

This representation of f provides new descriptions of $I_1(X)$ and $I_2(X)$ as stated in the next theorem.

16.7 THEOREM In the algebra A,

$$(43)\qquad I_1(X) = \{f \in A: f = e_1 f_1\} = \{f \in A: f_2(z_1 + i_1 z_2) = 0 \text{ in } X_2\},$$

$$(44)\qquad I_2(X) = \{f \in A: f = e_2 f_2\} = \{f \in A: f_1(z_1 - i_1 z_2) = 0 \text{ in } X_1\}.$$

Proof. Prove first that

$$(45)\qquad \{e_1 g: g \in A\} = \{f \in A: f = e_1 f_1\}.$$

Let f be an element in $\{e_1g: g\in A\}$. Then there is a g in A such that $f=e_1g$. Then $e_1f=e_1e_1g=e_1g$, and therefore

$$(46)\qquad f = e_1g = e_1f = e_1(e_1f_1 + e_2f_2) = e_1f_1.$$

Thus $f\in\{f\in A: f=e_1f_1\}$, and the proof is complete that

$$(47)\qquad \{e_1g: g\in A\} \subset \{f\in A: f = e_1f_1\}.$$

Next, let f be an element in $\{f\in A: f=e_1f_1\}$. Then

$$(48)\qquad f = e_1f_1, \qquad e_1f = e_1e_1f_1 = e_1f_1 = f.$$

Since $f=e_1f$, then $f\in\{e_1g: g\in A\}$, and the proof is complete that

$$(49)\qquad \{f\in A: f = e_1f_1\} \subset \{e_1g: g\in A\}.$$

Since each of the sets in (45) is contained in the other by (47) and (49), they are equal. Then (43) follows from (40) and (45). The proof of (44) is similar. The proof of Theorem 16.7 is complete. □

16.8 THEOREM The sets $I_1(X)$ and $I_2(X)$ are principal ideals in A which are also Banach algebras.

Proof. Since e_1 and e_2 can be considered constant functions in A, then (40), (41), and Definition 7.1 show that $I_1(X)$ and $I_2(X)$ are principal ideals in A. It is clear from (43) and (44) that $I_1(X)$ and $I_2(X)$ are each closed under the algebraic operations (scalar multiplication by a constant in $\mathbb{C}_2$, addition, and multiplication) and normed by the sup norm in A. The proof that $I_1(X)$ and $I_2(X)$ are Banach algebras can be completed by showing that they are complete. The proof follows.

Let $\{f^n: n\in\mathbb{N}\}$ be a sequence in $I_1(X)$. Then by (43)

$$(50)\qquad f^n = f_1^n e_1 + 0^n e_2,$$

and Lemma 16.4 shows that $\{f_1^n: n\in\mathbb{N}\}$ is a Cauchy sequence in $BH(X_1)$ and $\{0^n: n\in\mathbb{N}\}$ is a Cauchy sequence in $BH(X_2)$. Then Theorems 16.2 and 16.5 show that

$$(51)\qquad \lim_{n\to\infty} f_1^n = f_1, \qquad \lim_{n\to\infty} 0^n = 0,$$

$$(52)\qquad \lim_{n\to\infty} f^n = f, \qquad f = f_1e_1 + 0e_2.$$

Now f is a bounded holomorphic function in A by Theorem 16.5, and f is in $I_1(X)$ by (43) since $f = f_1e_1$. Thus every Cauchy sequence in $I_1(X)$ has a limit in $I_1(X)$, and the proof that $I_1(X)$ is a Banach algebra is complete. The proof that $I_2(X)$ is a Banach algebra is similar to the one just given for $I_1(X)$. The proof of Theorem 16.8 is complete. □

16.9 THEOREM Let A be the algebra of bounded holomorphic functions on X. If $f \in A$ but $f \notin I_1(X) \cup I_2(X)$, then the zeros of f, if any, are isolated. If $f \in I_1(X) \cup I_2(X)$, then the zeros of f, if any, are not isolated.

Proof. If f is in A but not in $I_1(X) \cup I_2(X)$, Theorem 16.7 shows that neither f_1 nor f_2 vanishes identically. Then Theorem 15.6 shows that the zeros of f are isolated. If $f \in I_1(X) \cup I_2(X)$, then at least one of the functions f_1, f_2 vanishes identically, and Theorem 15.6 shows that the zeros of f, if any, are not isolated. The proof is complete. □

16.10 THEOREM Let A be the algebra of bounded holomorphic functions on X. If $f \in I_1(X)$, then

$$(53) \qquad f(z_1 + i_2 z_2) \in I_1, \qquad \forall z_1 + i_2 z_2 \text{ in } X.$$

Furthermore, the power series representation of f about each point $a_1 + i_2 a_2$ in X has the form

$$(54) \qquad f(z_1 + i_2 z_2) = \sum_{k=0}^{\infty} (a_k + i_2 b_k)[(z_1 + i_2 z_2) - (a_1 + i_2 a_2)]^k,$$
$$(a_k + i_2 b_k) \in I_1, \ \forall k \text{ in } \mathbb{N}.$$

Also, if $f \in I_2(X)$, then

$$(55) \qquad f(z_1 + i_2 z_2) \in I_2, \qquad \forall z_1 + i_2 z_2 \text{ in } X,$$

and the power series representation of f about $a_1 + i_2 a_2$ in X has the form

$$(56) \qquad f(z_1 + i_2 z_2) = \sum_{k=0}^{\infty} (a_k + i_2 b_k)[(z_1 + i_2 z_2) - (a_1 + i_2 a_2)]^k,$$
$$(a_k + i_2 b_k) \in I_2, \ \forall k \text{ in } \mathbb{N}.$$

Proof. If f is in $I_1(X)$, then all values of f are in I_1 by (43); if $f \in I_2(X)$, then all values of f are in I_2 by (44). Thus the statements in (53) and (55) are true. If f is in A, then by Theorem 13.5(19) and equation (20) in Section 15,

$$(57) \qquad f(z_1 + i_2 z_2) = \sum_{k=0}^{\infty} (a_k - i_1 b_k)[(z_1 - i_1 z_2) - (a_1 - i_1 a_2)]^k e_1$$
$$+ \sum_{k=0}^{\infty} (a_k + i_1 b_k)[(z_1 + i_1 z_2) - (a_1 + i_1 a_2)]^k e_2,$$

and Theorem 13.8 shows that

$$(58) \qquad f(z_1 + i_2 z_2) = \sum_{k=0}^{\infty} (a_k + i_2 b_k)[(z_1 + i_2 z_2) - (a_1 + i_2 a_2)]^k.$$

If f is in $I_1(X)$, then $f_2 = 0$ by (43) and $a_k + i_1 b_k = 0$ for all k in $\mathbb{N}$. Then $a_k + i_2 b_k$ is in I_1 for all k in $\mathbb{N}$, and (54) is true. Also, if f is in $I_2(X)$, then $f_1 = 0$ by (44) and $a_k - i_1 b_k = 0$ for all k in $\mathbb{N}$. Then $a_k + i_2 b_k$ is in I_2 for all k in $\mathbb{N}$, and (56) is true. The proof of all parts of Theorem 16.10 is complete. □

Exercises

16.1 Let A denote the Banach algebra of bounded holomorphic functions $BH(X)$ on X, and let A_1 and A_2 denote the Banach algebras of bounded holomorphic functions $BH(X_1)$ and $BH(X_2)$ on X_1 and X_2, respectively. Let E be a subset of A, and let E_1 and E_2 be the sets of idempotent components f_1 and f_2 of the functions in E. Thus if f is in E, then $f=f_1e_1+f_2e_2$, and $f_1\in E_1$ and $f_2\in E_2$. Also, $E_1\subset A_1$ and $E_2\subset A_2$. Prove the following statements.

(a) If E is an ideal in A, then E_1 and E_2 are ideals in A_1 and A_2, respectively.

(b) If E is a principal ideal in A, then E_1 and E_2 are principal ideals in A_1 and A_2.

(c) If E is a Banach algebra, then E_1 and E_2 are Banach algebras.

16.2 Let E_1 and E_2 be subsets of A_1 and A_2, respectively; define E to be a subset of A as follows:

$$E=\{f\in A\colon f=f_1e_1+f_2e_2,\ f_1\in E_1 \text{ and } f_2\in E_2\}.$$

(a) If E_1 and E_2 are ideals in A_1 and A_2, respectively, show that E is an ideal in A.

(b) If E_1 and E_2 are principal ideals in A_1 and A_2, show that E is a principal ideal in A.

(c) If E_1 and E_2 are Banach algebras, show that E is a Banach algebra.

16.3 (a) Since $f=1\cdot f$ for every f in A, show that A is a principal ideal in A and the principal ideal determined by the constant function whose value is 1 at each $z_1+i_2z_2$ in X.

(b) Also, show that A_1 is a principal ideal in A_1 and the principal ideal determined by the constant function with value 1 in X_1.

(c) Finally, show that A_2 is a principal ideal in A_2 and the principal ideal determined by the constant function with value 1 in X_2.

16.4 Define sets O, O_1, O_2 as follows:

$$O=\{f\in A\colon f(z_1+i_2z_2)=0,\ \forall z_1+i_2z_2 \text{ in } X\};$$
$$O_1=\{f_1\in A_1\colon f_1(z_1-i_1z_2)=0,\ \forall z_1-i_1z_2 \text{ in } X_1\};$$
$$O_2=\{f_2\in A_2\colon f_2(z_1+i_1z_2)=0,\ \forall z_1+i_1z_2 \text{ in } X_2\}.$$

Show that the sets O, O_1, O_2 are principal ideals in A, A_1, A_2, respectively, and that in each case the principal ideal is determined by the zero function.

16.5 In Exercise 16.2, a set E in A is formed from sets E_1 in A_1 and E_2 in A_2. Prove the following:

(a) If $E_1=O_1$ and $E_2=O_2$, then $E=O$.

(b) If $E_1=A_1$ and $E_2=A_2$, then $E=A$.

(c) If $E_1 = A_1$ and $E_2 = O_2$, then $E = I_1(X)$.
(d) If $E_1 = O_1$ and $E_2 = A_2$, then $E = I_2(X)$.

16.6 Let $z_1^0 + i_2 z_2^0$ be a point in X, and define the subset B of A as follows:

$$B = \{f \text{ in } A\text{: } f(z_1^0 + i_2 z_2^0) = 0\}.$$

Show that B is a proper subset of A, an ideal in A, and a Banach algebra. Is B a principal ideal? Why?

17. ELEMENTARY FUNCTIONS

The purpose of this section is to define the sine and cosine, the hyperbolic sine and cosine, and the exponential functions of the bicomplex variable $z_1 + i_2 z_2$ and to establish their elementary properties. The account traces the development of these functions beginning with their definition as functions of a real variable, continuing with their extension as functions of a complex variable, and ending with a final extension to functions of a bicomplex variable. The emphasis is on insight and on proofs by the most elementary methods possible. As functions of a real variable, the trigonometric functions seem unrelated to the exponential function, but as functions of complex and bicomplex variables, the three are closely related. Furthermore, the exponential function, defined by the familiar infinite series, is the dominant function, and the functional equation $\exp(x+y) = \exp(x)\exp(y)$ seems to account for most of the properties of the trigonometric, the hyperbolic, and the exponential functions.

In the most elementary beginning, $\cos x$ and $\sin x$ are defined as the coordinates of a point, on the unit circle, determined by the central angle x.

17.1 THEOREM If the cosine and sine are the elementary functions just described, then

(1) $$\cos(x + 2\pi) = \cos x, \qquad \sin(x + 2\pi) = \sin x;$$

(2) $$\frac{d\cos x}{dx} = -\sin x, \qquad \frac{d\sin x}{dx} = \cos x;$$

(3) $$\cos x = 1 - \frac{x^2}{2!} + \frac{x^4}{4!} - \frac{x^6}{6!} + \cdots;$$

(4) $$\sin x = \frac{x}{1!} - \frac{x^3}{3!} + \frac{x^5}{5!} - \frac{x^7}{7!} + \cdots.$$

Proof. The fact that the cosine and sine functions are periodic with period 2π follows from the definition of the values of these functions as coordinates of a point on the unit circle. There are several elementary proofs of the formulas in (2) for the derivatives of the cosine and sine; the formulas show that these functions have an infinite number of derivatives. The infinite series in (3) and (4) follow from Taylor's theorem and the formulas for the derivatives in (2). □

17.2 DEFINITION The *exponential function* $\exp: \mathbb{C}_0 \to \mathbb{C}_0$, $x \mapsto \exp(x)$, is defined as follows:

$$\text{(5)} \qquad \exp(x) = \sum_{k=0}^{\infty} \frac{x^k}{k!}.$$

Elementary tests show that the infinite series in (5) converges for every x in $\mathbb{C}_0$.

17.3 THEOREM The exponential function has the following property: for every x and y in $\mathbb{C}_0$,

$$\text{(6)} \qquad \exp(x + y) = \exp(x)\exp(y).$$

Proof. The formula in (6) is usually proved by means of Taylor's theorem, but an application of this theorem requires that all of the derivatives of the function be known. The proof of (6) presented here uses only the definition in (5), the binomial theorem, and one theorem about absolutely convergent double series. The proof of (6) will be given first in the special case in which $x > 0$ and $y > 0$; this restriction will be removed later.

Assume then that $x > 0$ and $y > 0$, and observe that all terms in the series (5) for $\exp(x)$ and $\exp(y)$ are positive. Set

$$\text{(7)} \qquad s_n(x) = \sum_{k=0}^{n} \frac{x^k}{k!}, \qquad s_n(y) = \sum_{k=0}^{n} \frac{y^k}{k!}, \qquad s_n(x + y) = \sum_{k=0}^{n} \frac{(x + y)^k}{k!}.$$

$$\text{(8)} \qquad \begin{array}{c|cccccc}
 & 1 & \frac{y}{1!} & \frac{y^2}{2!} & \frac{y^3}{3!} & \cdots & \frac{y^n}{n!} \\
\hline
1 & 1 & \frac{y}{1!} & \frac{y^2}{2!} & \frac{y^3}{3!} & \cdots & \frac{y^n}{n!} \\
\frac{x}{1!} & \frac{x}{1!} & \frac{xy}{1!\,1!} & \frac{xy^2}{1!\,2!} & \frac{xy^3}{1!\,3!} & \cdots & \frac{xy^n}{1!\,n!} \\
\frac{x^2}{2!} & \frac{x^2}{2!} & \frac{x^2y}{2!\,1!} & \frac{x^2y^2}{2!\,2!} & \frac{x^2y^3}{2!\,3!} & \cdots & \frac{x^2y^n}{2!\,n!} \\
\frac{x^3}{3!} & \frac{x^3}{3!} & \frac{x^3y}{3!\,1!} & \frac{x^3y^2}{3!\,2!} & \frac{x^3y^3}{3!\,3!} & \cdots & \frac{x^3y^n}{3!\,n!} \\
\vdots & \vdots & \vdots & \vdots & \vdots & & \vdots \\
\frac{x^n}{n!} & \frac{x^n}{n!} & \frac{x^ny}{n!\,1!} & \frac{x^ny^2}{n!\,2!} & \frac{x^ny^3}{n!\,3!} & \cdots & \frac{x^ny^n}{n!\,n!}
\end{array}$$

Calculate $s_n(x)s_n(y)$ by writing the terms of $s_n(x)$ down the left side of the diagram in (8) and those of $s_n(y)$ across the top; the terms in the matrix are obtained by multiplying the terms on the left by those at the top. Then $s_n(x)s_n(y)$ equals the sum of the terms in the matrix. Next, sum the terms on and above the diagonal D from the lower left to the upper right by grouping the terms on the diagonals; the binomial theorem shows that the sums of the terms on these diagonals, beginning at the top left, are

$$(9) \qquad 1, \frac{x+y}{1!}, \frac{(x+y)^2}{2!}, \frac{(x+y)^3}{3!}, \ldots, \frac{(x+y)^n}{n!},$$

respectively. Since the sum of the terms in (9) is $s_n(x+y)$ by (7), the sum of the terms in the matrix on and above the diagonal D is $s_n(x+y)$. Thus, since all terms are positive, $s_n(x+y)<s_n(x)s_n(y)$. Extend the diagram by multiplying $s_{2n}(x)$ and $s_{2n}(y)$ in the same way; inspection of the new diagram shows that $s_n(x)s_n(y)<s_{2n}(x+y)$. Thus

$$(10) \qquad s_n(x+y) < s_n(x)s_n(y) < s_{2n}(x+y),$$

$$\lim_{n\to\infty} s_n(x+y) \leqslant \lim_{n\to\infty} s_n(x) \lim_{n\to\infty} s_n(y) \leqslant \lim_{n\to\infty} s_{2n}(x+y),$$

$$\exp(x+y) \leqslant \exp(x)\exp(y) \leqslant \exp(x+y),$$

and the proof of the functional equation (6) is complete in the special case $x>0$ and $y>0$. To prove (6) in the general case, extend the diagram in (8) indefinitely down and to the right; the result is a double series. The case $x>0$ and $y>0$ already treated has shown that this double series is absolutely convergent. Now it is known that all methods of summing an absolutely convergent double series give the same sum for the series. The partial sums by squares are $s_n(x)s_n(y)$ for all n, and the partial sums by triangles on and above the diagonals are $s_n(x+y)$. Therefore, in all cases,

$$(11) \qquad \lim_{n\to\infty} s_n(x+y) = \lim_{n\to\infty} s_n(x)s_n(y) = \lim_{n\to\infty} s_n(x) \lim_{n\to\infty} s_n(y),$$

or $\exp(x+y)=\exp(x)\exp(y)$. The proof of Theorem 17.3 is complete in all cases. □

17.4 THEOREM For all x in $\mathbb{C}_0$,

$$(12) \qquad \exp(x) = e^x,$$

$$(13) \qquad \exp(x) > 0,$$

$$(14) \qquad \frac{d\exp(x)}{dx} = \exp(x).$$

Proof. Now

(15) $$\exp(x) = \sum_{k=0}^{\infty} \frac{x^k}{k!}, \qquad \exp(0) = 1, \qquad \exp(1) = \sum_{k=0}^{\infty} \frac{1}{k!} = e.$$

Then $\exp(2) = \exp(1)\exp(1) = e^2$, and $\exp(n) = e^n$ by induction. Also,

(16) $$e = \exp\left(\frac{1}{q} + \cdots + \frac{1}{q}\right) = \left[\exp\left(\frac{1}{q}\right)\right]^q, \qquad \exp\left(\frac{1}{q}\right) = e^{1/q},$$

$$\exp\left(\frac{p}{q}\right) = \exp\left(\frac{1}{q} + \cdots + \frac{1}{q}\right) = \left[\exp\left(\frac{1}{q}\right)\right]^p = (e^{1/q})^p = e^{p/q}.$$

Thus $e^x = \exp(x)$ if x is rational; if x is irrational, then e^x is defined to be $\exp(x)$. Thus (12) is true for all x in $\mathbb{C}_0$. The definition of $\exp(x)$ in (5) shows that $\exp(x) > 0$ for $x \geqslant 0$. But $\exp(x)\exp(-x) = \exp(0) = 1$; therefore, $\exp(-x) > 0$ for all $x \geqslant 0$. Thus (13) is true for all x in $\mathbb{C}_0$. The definition of the derivative and (6) show that

(17) $$\frac{d\exp(x)}{dx} = \lim_{h \to 0} \frac{\exp(x+h) - \exp(x)}{h}$$

$$= \lim_{h \to 0} \frac{\exp(x)\exp(h) - \exp(x)}{h} = \exp(x) \lim_{h \to 0} \frac{\exp(h) - 1}{h}.$$

Now

$$\frac{\exp(h) - 1}{h} = 1 + \frac{h}{2!} + \frac{h^2}{3!} + \cdots.$$

This equation and the fact that a power series represents a continuous function show that

(18) $$\lim_{h \to 0} \frac{\exp(h) - 1}{h} = 1.$$

Thus (17) and (18) show that (14) is true for all x in $\mathbb{C}_0$. □

17.5 DEFINITION The functions $\cosh : \mathbb{C}_0 \to \mathbb{C}_0$ and $\sinh : \mathbb{C}_0 \to \mathbb{C}_0$, called the *hyperbolic cosine* and *hyperbolic sine*, are defined by the following statements for all x in $\mathbb{C}_0$:

(19) $$\cosh x = 1 + \frac{x^2}{2!} + \frac{x^4}{4!} + \frac{x^6}{6!} + \cdots,$$

(20) $$\sinh x = \frac{x}{1!} + \frac{x^3}{3!} + \frac{x^5}{5!} + \frac{x^7}{7!} + \cdots.$$

17.6 THEOREM For all x in $\mathbb{C}_0$,

$$\frac{d \cosh x}{dx} = \sinh x, \tag{21}$$

$$\frac{d \sinh x}{dx} = \cosh x, \tag{22}$$

$$\cosh x = \frac{e^x + e^{-x}}{2}, \qquad \sinh x = \frac{e^x - e^{-x}}{2}. \tag{23}$$

Proof. Differentiate the power series in (19) and (20) to obtain (21) and (22). Another proof of (21) and (22) can be obtained from (14) and (23). Equations (5), (12), (19), and (20) show that

$$\begin{aligned} \cosh x + \sinh x &= e^x, \\ \cosh x - \sinh x &= e^{-x}. \end{aligned} \tag{24}$$

Solve these equations for $\cosh x$ and $\sinh x$ to obtain (23). □

The next step is to define the cosine and sine, the hyperbolic cosine and hyperbolic sine, and the exponential function for the complex variable z. Definitions are arbitrary, but there is one guiding principle in the present case: it is desirable to make the new definitions so that, when the complex variable has real values x, then the new functions of z reduce to the earlier functions of the real variable x.

17.7 DEFINITION The cosine and sine, the hyperbolic cosine and the hyperbolic sine, and the exponential function are defined for all z in $\mathbb{C}_1$ by the following statements:

$$\cos z = 1 - \frac{z^2}{2!} + \frac{z^4}{4!} - \frac{z^6}{6!} + \cdots; \tag{25}$$

$$\sin z = \frac{z}{1!} - \frac{z^3}{3!} + \frac{z^5}{5!} - \frac{z^7}{7!} + \cdots; \tag{26}$$

$$\cosh z = 1 + \frac{z^2}{2!} + \frac{z^4}{4!} + \frac{z^6}{6!} + \cdots; \tag{27}$$

$$\sinh z = \frac{z}{1!} + \frac{z^3}{3!} + \frac{z^5}{5!} + \frac{z^7}{7!} + \cdots; \tag{28}$$

$$\exp(z) = e^z = \sum_{k=0}^{\infty} \frac{z^k}{k!}. \tag{29}$$

These definitions clearly satisfy the consistency requirement because they become (3), (4), (19), (20), and (5) when z is replaced by x. The series in (25)–

(29) converge for all values of z. The next step is to investigate the properties of these functions of a complex variable.

17.8 THEOREM For all z, z_1, z_2 in $\mathbb{C}_1$,

(30) $$\frac{d\cos z}{dz} = -\sin z, \qquad \frac{d\sin z}{dz} = \cos z;$$

(31) $$\frac{d\cosh z}{dz} = \sinh z, \qquad \frac{d\sinh z}{dz} = \cosh z;$$

(32) $$\frac{d\exp(z)}{dz} = \exp(z), \qquad \text{or } \frac{de^z}{dz} = e^z;$$

(33) $$\exp(z_1 + z_2) = \exp(z_1)\exp(z_2), \qquad \text{or } e^{z_1+z_2} = e^{z_1}e^{z_2}.$$

Proof. It is shown in the theory of functions of a complex variable that the derivative of a function represented by a power series can be found by differentiating the power series term-by-term. The series obtained by differentiating term-by-term the series in (25) is the negative of the series in (26), and the derivative of (26) is (25); thus (30) is true. The statements in (31) and (32) can be proved in the same way. The proof of Theorem 17.3 can be repeated in the present case to establish (33). After (33) has been established, the proof of Theorem 17.4 (14) can be repeated in the complex case to give a second proof of (32). □

Theorem 17.6 (23) states a relation between the real exponential function and the real hyperbolic functions. The next theorem states some of the properties of the complex exponential function, including some of its connections with the real trigonometric functions.

17.9 THEOREM If x and y are in $\mathbb{C}_0$ and z is in $\mathbb{C}_1$, then

(34) $$e^{i_1 y} = \cos y + i_1 \sin y \qquad \text{(Euler's formula);}$$

(35) $$\cos^2 y + \sin^2 y = 1, \qquad \cosh^2 y - \sinh^2 y = 1;$$

(36) $$|e^{i_1 y}| = 1;$$

(37) $$\cos y = \frac{e^{i_1 y} + e^{-i_1 y}}{2}, \qquad \sin y = \frac{e^{i_1 y} - e^{-i_1 y}}{2i_1};$$

(38) $$e^{2\pi i_1} = 1, \qquad e^{z+2\pi i_1} = e^z \qquad \text{for all } z \text{ in } \mathbb{C}_1;$$

(39) $$e^z = e^{x+i_1 y} = e^x(\cos y + i_1 \sin y);$$

(40) $$|e^z| = e^x > 0 \qquad \text{for all } z \text{ in } \mathbb{C}_1.$$

Proof. The trigonometric, hyperbolic, and exponential functions are defined by (25)–(29) for all z in $\mathbb{C}_1$. Then since $i_1^2 = -1$,

$$e^{i_1 y} = 1 + \frac{i_1 y}{1!} + \frac{(i_1 y)^2}{2!} + \frac{(i_1 y)^3}{3!} + \frac{(i_1 y)^4}{4!} + \frac{(i_1 y)^5}{5!} + \cdots,$$

$$e^{i_1 y} = \left(1 - \frac{y^2}{2!} + \frac{y^4}{4!} - \cdots\right) + i_1\left(\frac{y}{1!} - \frac{y^3}{3!} + \frac{y^5}{5!} - \cdots\right),$$

$$e^{i_1 y} = \cos y + i_1 \sin y.$$

Thus (34) is true. By (33) and (34),

$$e^{i_1 y}e^{-1_1 y} = e^0 = 1,$$

$$(\cos y + i_1 \sin y)(\cos y - i_1 \sin y) = 1,$$

$$\cos^2 y + \sin^2 y = 1.$$

In the same way, $\cosh^2 y - \sinh^2 y = 1$ follows from the equations in (24). Thus (35) is true. Also, (34), (35), and the definition of the absolute value prove (36) as follows:

$$|e^{i_1 y}| = |\cos y + i_1 \sin y| = [\cos^2 y + \sin^2 y]^{1/2} = 1.$$

To prove (37), solve the following equations for $\cos y$ and $\sin y$ [compare (24) and (23)]:

$$\cos y + i_1 \sin y = e^{i_1 y},$$

$$\cos y - i_1 \sin y = e^{-i_1 y}.$$

Consider (38). Now $\cos 0 = 1$ and $\sin 0 = 0$ by the elementary definitions of the cosine and sine or by the definitions in (25) and (26); since these functions are periodic with period 2π by Theorem 17.1 (1), then $\cos 2\pi = 1$ and $\sin 2\pi = 0$. Hence (34) shows that

$$e^{2\pi i_1} = \cos 2\pi + i_1 \sin 2\pi = 1.$$

Theorem 17.8 (33) next shows that

$$e^{z + 2\pi i_1} = e^z e^{2\pi i_1} = e^z,$$

and (38) is true. Also, $e^{x + i_1 y} = e^x e^{i_1 y} = e^x(\cos y + i_1 \sin y)$, and (39) is true. Finally, (33), (36), and Theorem 17.4 (13) show that (40) is true since

$$|e^z| = |e^{x + i_1 y}| = |e^x e^{i_1 y}| = |e^x|\,|e^{i_1 y}| = e^x > 0.$$

The proof of all parts of Theorem 17.9 is complete. □

The next theorem illustrates the usefulness of functions of a complex variable for establishing results about functions of a real variable.

17.10 THEOREM If y_1 and y_2 are two numbers in $\mathbb{C}_0$, then

$$\cos(y_1 + y_2) = \cos y_1 \cos y_2 - \sin y_1 \sin y_2, \tag{41}$$

$$\sin(y_1 + y_2) = \sin y_1 \cos y_2 + \cos y_1 \sin y_2, \tag{42}$$

$$\cosh(y_1 + y_2) = \cosh y_1 \cosh y_2 + \sinh y_1 \sinh y_2, \tag{43}$$

$$\sinh(y_1 + y_2) = \sinh y_1 \cosh y_2 + \cosh y_1 \sinh y_2. \tag{44}$$

Proof. By Theorem 17.8 (33)

$$e^{i_1 y_1} e^{i_1 y_2} = e^{i_1(y_1 + y_2)}. \tag{45}$$

By Euler's formula in (34),

$$\begin{aligned} e^{i_1 y_1} e^{i_1 y_2} &= (\cos y_1 + i_1 \sin y_1)(\cos y_2 + i_1 \sin y_2) \\ &= (\cos y_1 \cos y_2 - \sin y_1 \sin y_2) \\ &\qquad + i_1(\sin y_1 \cos y_2 + \cos y_1 \sin y_2). \end{aligned} \tag{46}$$

$$e^{i_1(y_1 + y_2)} = \cos(y_1 + y_2) + i_1 \sin(y_1 + y_2). \tag{47}$$

Since the expressions in (46) and (47) are equal by (45), their real and imaginary parts are equal, and the formulas in (41) and (42) are true. Similar arguments, using the first formula in (24), show that

$$\begin{aligned} e^{y_1} e^{y_2} &= (\cosh y_1 + \sinh y_1)(\cosh y_2 + \sinh y_2) \\ &= (\cosh y_1 \cosh y_2 + \sinh y_1 \sinh y_2) \\ &\qquad + (\sinh y_1 \cosh y_2 + \cosh y_1 \sinh y_2), \\ e^{y_1 + y_2} &= \cosh(y_1 + y_2) + \sinh(y_1 + y_2). \end{aligned} \tag{48}$$

Since the hyperbolic cosine and sine are even and odd functions, respectively, the formulas in (43) and (44) follow from (48). □

The next theorem emphasizes the properties of the cosine and sine, and the hyperbolic cosine and hyperbolic sine, as functions of a complex variable.

17.11 THEOREM The cosine, sine, hyperbolic cosine, hyperbolic sine, and the exponential function defined in (25)–(29) have the following additional properties.

$$e^{i_1 z} = \cos z + i_1 \sin z. \tag{49}$$

$$e^z = \cosh z + \sinh z. \tag{50}$$

$$\cos(-z) = \cos z, \qquad \sin(-z) = -\sin z. \tag{51}$$

$$\cosh(-z) = \cosh z, \qquad \sinh(-z) = -\sinh z. \tag{52}$$

$$\cos^2 z + \sin^2 z = 1, \qquad \cosh^2 z - \sinh^2 z = 1. \tag{53}$$

(54) $\cos z = \dfrac{e^{i_1 z} + e^{-i_1 z}}{2}, \qquad \sin z = \dfrac{e^{i_1 z} - e^{-i_1 z}}{2i_1}.$

(55) $\cosh z = \dfrac{e^{z} + e^{-z}}{2}, \qquad \sinh z = \dfrac{e^{z} - e^{-z}}{2}.$

(56) $e^{i_1 z}$ is periodic with period 2π: $e^{i_1(z+2\pi)} = e^{i_1 z}$.

(57) $\cos z$ and $\sin z$ are periodic with period 2π: for all z in $\mathbb{C}_1$,

$$\cos(z + 2\pi) = \cos z, \qquad \sin(z + 2\pi) = \sin z.$$

(58) $\cosh z$ and $\sinh z$ are periodic with period $2\pi i_1$: for all z in $\mathbb{C}_1$,

$$\cosh(z + 2\pi i_1) = \cosh z, \qquad \sinh(z + 2\pi i_1) = \sinh z.$$

(59) $\cos(z_1 + z_2) = \cos z_1 \cos z_2 - \sin z_1 \sin z_2.$

(60) $\sin(z_1 + z_2) = \sin z_1 \cos z_2 + \cos z_1 \sin z_2.$

(61) $\cosh(z_1 + z_2) = \cosh z_1 \cosh z_2 + \sinh z_1 \sinh z_2.$

(62) $\sinh(z_1 + z_2) = \sinh z_1 \cosh z_2 + \cosh z_1 \sinh z_2.$

Proof. The formulas in (49)–(52) follow from the definitions in (25)–(29). The formulas in (49)–(52) prove (53) as follows:

$$1 = e^0 = e^{i_1 z} e^{-i_1 z} = (\cos z + i_1 \sin z)(\cos z - i_1 \sin z) = \cos^2 z + \sin^2 z;$$

$$1 = e^0 = e^z e^{-z} = (\cosh z + \sinh z)(\cosh z - \sinh z) = \cosh^2 z - \sinh^2 z.$$

As in (37), the formulas in (54) and (55) follow from (49) and (50). Since e^z is periodic with period $2\pi i_1$ by (38), $e^{i_1 z}$ is periodic with period 2π, and (56) is true. Also, (57) and (58) follow from (54), (55), and (38), (56). Finally, the proofs of (59)–(62) are similar to the proofs of (41)–(44) in Theorem 17.10. □

The next step is to define the cosine and sine, the hyperbolic cosine and hyperbolic sine, and the exponential function for a bicomplex variable $z_1 + i_2 z_2$. The same consistency requirement in the guiding principle which applied in extending these functions from real variables to complex variables applies again and dictates that these functions should be defined by the same power series as before.

17.12 DEFINITION The cosine and sine, the hyperbolic cosine and sine, and the exponential function are defined, for bicomplex values $z_1 + i_2 z_2$ of the independent variable, formally by the infinite series in Definition 17.7. Thus, for all values of $z_1 + i_2 z_2$ in $\mathbb{C}_2$ for which the series converge,

(63) $$\cos(z_1 + i_2 z_2) = 1 - \frac{(z_1 + i_2 z_2)^2}{2!} + \frac{(z_1 + i_2 z_2)^4}{4!} - \frac{(z_1 + i_2 z_2)^6}{6!} + \cdots,$$

$$\text{(64)}\qquad \sin(z_1 + i_2z_2) = \frac{(z_1 + i_2z_2)}{1!} - \frac{(z_1 + i_2z_2)^3}{3!} + \frac{(z_1 + i_2z_2)^5}{5!} - \cdots,$$

$$\text{(65)}\qquad \cosh(z_1 + i_2z_2) = 1 + \frac{(z_1 + i_2z_2)^2}{2!} + \frac{(z_1 + i_2z_2)^4}{4!} + \frac{(z_1 + i_2z_2)^6}{6!} + \cdots,$$

$$\text{(66)}\qquad \sinh(z_1 + i_2z_2) = \frac{(z_1 + i_2z_2)}{1!} + \frac{(z_1 + i_2z_2)^3}{3!} + \frac{(z_1 + i_2z_2)^5}{5!} + \cdots,$$

$$\text{(67)}\qquad \exp(z_1 + i_2z_2) = 1 + \frac{(z_1 + i_2z_2)}{1!} + \frac{(z_1 + i_2z_2)^2}{2!} + \frac{(z_1 + i_2z_2)^3}{3!} + \cdots.$$

The next chapter will show that the derivative of a function of a bicomplex variable which is represented by a power series can be obtained by differentiating the series term-by-term as in the case of functions of a complex variable, but a study of the derivatives of these functions will be postponed until later.

17.13 THEOREM The infinite series in (63)–(67) converge for all $z_1 + i_2z_2$ in $\mathbb{C}_2$.

Proof. Theorem 13.6 shows that the series in (63)–(67) converge at $z_1 + i_2z_2$ if and only if the idempotent component series converge at $z_1 - i_1z_2$ and $z_1 + i_1z_2$. But these component series are power series in the complex variables $z_1 - i_1z_2$ and $z_1 + i_1z_2$ which converge for all values of these variables. Thus the series in (63)–(67) converge for all values of $z_1 + i_2z_2$. □

17.14 THEOREM For every $z_1 + i_2z_2$ and $w_1 + i_2w_2$ in $\mathbb{C}_2$,

$$\text{(68)}\qquad e^{[(z_1 + i_2z_2) + (w_1 + i_2w_2)]} = e^{(z_1 + i_2z_2)}e^{(w_1 + i_2w_2)}.$$

Proof. Property (68) is the same property that we have proved already for the exponential function of a real variable and of a complex variable [see Theorems 17.3 and 17.8 (33)]. In these earlier cases it was necessary to go back to first principles in the definition of the exponential function or to use a powerful tool such as Taylor's theorem. A new situation arises in treating the exponential function of a bicomplex variable: special properties of holomorphic functions of a bicomplex variable enable us to show that (68) is true because the corresponding property is true for the exponential function of a complex variable.

By Theorem 13.5 (19),

$$\sum_{k=0}^{\infty}\frac{(z_1+i_2z_2)^k}{k!}=\sum_{k=0}^{\infty}\frac{(z_1-i_1z_2)^k}{k!}e_1+\sum_{k=0}^{\infty}\frac{(z_1+i_1z_2)^k}{k!}e_2, \tag{69}$$

$$\sum_{k=0}^{\infty}\frac{(w_1+i_2w_2)^k}{k!}=\sum_{k=0}^{\infty}\frac{(w_1-i_1w_2)^k}{k!}e_1+\sum_{k=0}^{\infty}\frac{(w_1+i_1w_2)^k}{k!}e_2. \tag{70}$$

In other notation, these equations are the following:

$$e^{z_1+i_2z_2}=e^{z_1-i_1z_2}e_1+e^{z_1+i_1z_2}e_2, \tag{71}$$

$$e^{w_1+i_2w_2}=e^{w_1-i_1w_2}e_1+e^{w_1+i_1w_2}e_2. \tag{72}$$

These equations and Theorem 6.6 (16) show that

$$e^{z_1+i_2z_2}e^{w_1+i_2w_2}=e^{z_1-i_1z_2}e^{w_1-i_1w_2}e_1+e^{z_1+i_1z_2}e^{w_1+i_1w_2}e_2. \tag{73}$$

Now $z_1-i_1z_2$, $w_1-i_1w_2$, $z_1+i_1z_2$, and $w_1+i_1w_2$ are complex numbers in $\mathbb{C}_1$, and by Theorem 17.8 (33),

$$e^{z_1-i_1z_2}e^{w_1-i_1w_2}=e^{(z_1-i_1z_2)+(w_1-i_1w_2)}, \tag{74}$$

$$e^{z_1+i_1z_2}e^{w_1+i_1w_2}=e^{(z_1+i_1z_2)+(w_1+i_1w_2)}. \tag{75}$$

Thus, by (73), (74), and (75),

$$e^{z_1+i_2z_2}e^{w_1+i_2w_2}=e^{(z_1-i_1z_2)+(w_1-i_1w_2)}e_1+e^{(z_1+i_1z_2)+(w_1+i_1w_2)}e_2. \tag{76}$$

The expression on the right in (76) represents $e^{(z_1+i_2z_2)+(w_1+i_2w_2)}$; to verify this statement, write the exponentials on the right as power series [see Definition 17.7 (29)] and apply Theorem 13.5. Thus (76) is the same as (68), and the proof of Theorem 17.14 is complete. □

17.15 COROLLARY For every $z_1+i_2z_2$ in $\mathbb{C}_2$,

$$e^{z_1+i_2z_2}=\frac{e^{z_1+i_1z_2}+e^{z_1-i_1z_2}}{2}+i_2\frac{e^{z_1+i_1z_2}-e^{z_1-i_1z_2}}{2i_1}, \tag{77}$$

$$e^{z_1+i_2z_2}=e^{z_1}(\cos z_2+i_2\sin z_2). \tag{78}$$

Proof. Formula (77) is (71) with the terms on the right rearranged after replacing e_1 and e_2 by their values. Theorem 17.8 (33) and Theorem 17.11 (54) can be used to convert (77) into (78). □

17.16 THEOREM The elementary functions defined in Definition 17.12 have, for every $z_1+i_2z_2$ in $\mathbb{C}_2$, the properties described by the following formulas.

$$e^{i_1(z_1+i_2z_2)} = \cos(z_1+i_2z_2) + i_1 \sin(z_1+i_2z_2). \tag{79}$$

$$e^{i_2(z_1+i_2z_2)} = \cos(z_1+i_2z_2) + i_2 \sin(z_1+i_2z_2). \tag{80}$$

$$e^{i_1i_2(z_1+i_2z_2)} = \cosh(z_1+i_2z_2) + i_1i_2 \sinh(z_1+i_2z_2). \tag{81}$$

$$\cos^2(z_1+i_2z_2) + \sin^2(z_1+i_2z_2) = 1. \tag{82}$$

$$\cosh^2(z_1+i_2z_2) - \sinh^2(z_1+i_2z_2) = 1. \tag{83}$$

$$\begin{aligned}\cos(z_1+i_2z_2) &= \frac{e^{i_1(z_1+i_2z_2)} + e^{-i_1(z_1+i_2z_2)}}{2}\\ &= \frac{e^{i_2(z_1+i_2z_2)} + e^{-i_2(z_1+i_2z_2)}}{2}\\ &= \frac{i_2e^{i_1(z_1+i_2z_2)} + i_1e^{-i_2(z_1+i_2z_2)}}{i_1+i_2}.\end{aligned} \tag{84}$$

$$\begin{aligned}\sin(z_1+i_2z_2) &= \frac{e^{i_1(z_1+i_2z_2)} - e^{-i_1(z_1+i_2z_2)}}{2i_1}\\ &= \frac{e^{i_2(z_1+i_2z_2)} - e^{-i_2(z_1+i_2z_2)}}{2i_2}\\ &= \frac{e^{i_1(z_1+i_2z_2)} - e^{-i_2(z_1+i_2z_2)}}{i_1+i_2}.\end{aligned} \tag{85}$$

$$\cosh(z_1+i_2z_2) = \frac{e^{i_1i_2(z_1+i_2z_2)} + e^{-i_1i_2(z_1+i_2z_2)}}{2}. \tag{86}$$

$$\sinh(z_1+i_2z_2) = \frac{e^{i_1i_2(z_1+i_2z_2)} - e^{-i_1i_2(z_1+i_2z_2)}}{2i_1i_2}. \tag{87}$$

Proof. To prove (79), find the value of $e^{i_1(z_1+i_2z_2)}$ by using the infinite series in (67); simplify the result, remembering $i_1^2=-1$ and the definitions in (63) and (64). The proofs of the formulas in (80) and (81) are similar. The infinite series in (63) and (64) show that the cosine is an even function and that the sine is an odd function. Thus

$$e^{-i_1(z_1+i_2z_2)} = \cos(z_1+i_2z_2) - i_1 \sin(z_1+i_2z_2). \tag{88}$$

Then (79), (88), and Theorem 17.14 show that

$$\begin{aligned}1 = e^0 &= e^{i_1(z_1+i_2z_2)}e^{-i_1(z_1+i_2z_2)}\\ &= [\cos(z_1+i_2z_2)+i_1 \sin(z_1+i_2z_2)][\cos(z_1+i_2z_2)-i_1 \sin(z_1+i_2z_2)]\\ &= \cos^2(z_1+i_2z_2)+\sin^2(z_1+i_2z_2).\end{aligned} \tag{89}$$

The proof of (83) is similar [compare Theorem 17.11 (53)]. To obtain the first formulas in (84) and (85), solve equations (79) and (88) for $\cos(z_1+i_2z_2)$ and $\sin(z_1+i_2z_2)$. The second formulas in (84) and (85) are obtained by solving a similar pair of equations for $e^{i_2(z_1+i_2z_2)}$ and $e^{-i_2(z_1+i_2z_2)}$. The third formulas in (84) and (85) are derived by solving the following equations [see (79) and (80)] for $\cos(z_1+i_2z_2)$ and $\sin(z_1+i_2z_2)$:

(90) $$e^{i_1(z_1+i_2z_2)} = \cos(z_1 + i_2z_2) + i_1 \sin(z_1 + i_2z_2),$$

(91) $$e^{-i_2(z_1+i_2z_2)} = \cos(z_1 + i_2z_2) - i_2 \sin(z_1 + i_2z_2).$$

Finally, the formulas in (86) and (87) are obtained by solving the following equations [see (81)] for $\cosh(z_1+i_2z_2)$ and $\sinh(z_1+i_2z_2)$:

(92) $$e^{i_1i_2(z_1+i_2z_2)} = \cosh(z_1 + i_2z_2) + i_1i_2 \sinh(z_1 + i_2z_2),$$

(93) $$e^{-i_1i_2(z_1+i_2z_2)} = \cosh(z_1 + i_2z_2) - i_1i_2 \sinh(z_1 + i_2z_2).$$

The proof of all parts of 17.16 is complete. □

17.17 THEOREM If $z_1+i_2z_2$ is denoted by $x_1+i_1x_2+i_2x_3+i_1i_2x_4$, then for every $z_1+i_2z_2$ in $\mathbb{C}_2$,

(94) $$e^{z_1+i_2z_2} \notin \mathcal{O}_2,$$

(95) $$\|e^{z_1+i_2z_2}\| \geqslant e^{x_1} > 0.$$

Proof. By (71),

(96) $$e^{z_1+i_2z_2} = e^{z_1-i_1z_2}e_1 + e^{z_1+i_1z_2}e_2,$$

and by Theorem 17.9 (40),

(97) $$|e^{z_1-i_1z_2}| > 0, \qquad |e^{z_1+i_1z_2}| > 0,$$

for every $z_1+i_2z_2$ in $\mathbb{C}_2$. Then (96), (97), and Definition 7.1 show that $e^{z_1+i_2z_2}$ is not in I_1 and not in I_2; therefore it is not in $\mathcal{O}_2$ by Corollary 7.5 (17), and (94) is true. To prove (95), observe that $e^{z_1+i_2z_2}=e^{z_1}e^{i_2z_2}$ by Theorem 17.14. Then

(98) $$e^{z_1+i_2z_2} = e^{z_1}(\cos z_2 + i_2 \sin z_2)$$

by Theorem 17.16 (80). Since $e^{z_1}\in\mathbb{C}_1$ by Theorem 17.9 (39), then Theorem 4.4 (10) shows that

(99) $$\|e^{z_1+i_2z_2}\| = |e^{z_1}|\,\|\cos z_2 + i_2 \sin z_2\|.$$

By Theorem 17.9 (40), for all z_1 in $\mathbb{C}_1$,

(100) $$|e^{z_1}| = |e^{x_1+i_1x_2}| = |e^{x_1}| = e^{x_1} > 0.$$

Thus by (99) and (100),

(101) $\quad \|e^{z_1+i_2z_2}\| = e^{x_1}\|\cos z_2 + i_2 \sin z_2\|.$

By the definition of the norm in Definition 3.1 (see also Exercise 3.5),

(102) $\quad \|\cos z_2 + i_2 \sin z_2\| = [|\cos z_2|^2 + |\sin z_2|^2]^{1/2}.$

Now by Theorem 17.16 (82) and the triangle inequality in $\mathbb{C}_1$,

(103) $\quad 1 = |\cos^2 z_2 + \sin^2 z_2| \leqslant |\cos z_2|^2 + |\sin z_2|^2;$

thus (102) and (103) show that

(104) $\quad \|\cos z_2 + i_2 \sin z_2\| \geqslant 1.$

Therefore, by (101) and (104),

(105) $\quad \|e^{z_1+i_2z_2}\| \geqslant e^{x_1} > 0,$

and the proof of (95) and of the entire theorem is complete. □

17.18 THEOREM For every $z_1+i_2z_2$ or $x_1+i_1x_2+i_2x_3+i_1i_2x_4$, in $\mathbb{C}_2$,

(106) $\quad e^{z_1+i_2z_2} = e^{x_1+i_1x_2+i_2x_3+i_1i_2x_4}$

(107) $\quad = e^{x_1}e^{i_1x_2}e^{i_2x_3}e^{i_1i_2x_4}$

(108) $\quad = (\cosh x_1 + \sinh x_1)(\cos x_2 + i_1 \sin x_2)(\cos x_3 + i_2 \sin x_3)(\cosh x_4 + i_1i_2 \sinh x_4).$

Proof. The bicomplex number $x_1+i_1x_2+i_2x_3+i_1i_2x_4$ can be interpreted as the sum of four bicomplex numbers; thus (107) follows from (106) by repeated application of Theorem 17.14. Now

(109) $\quad e^{x_1} = \cosh x_1 + \sinh x_1$

by (24). Also, by Theorem 17.16 (79), (80), (81) [and compare Theorem 17.9 (34)],

(110) $\quad e^{i_1x_2} = \cos x_2 + i_1 \sin x_2,$

(111) $\quad e^{i_2x_3} = \cos x_3 + i_2 \sin x_3,$

(112) $\quad e^{i_1i_2x_4} = \cosh x_4 + i_1i_2 \sinh x_4.$

Then (108) follows from (107) and (109)–(112). □

It is known that e^x, $\cosh x$, and $\sinh x$, as functions of the real variable x, are not periodic; however, $\cos x$ and $\sin x$ are periodic with period 2π as stated in Theorem 17.1 (1). The periodic properties of e^z, $\cos z$, $\sin z$, $\cosh z$, and $\sinh z$, for z in $\mathbb{C}_1$, are stated in Theorems 17.9 (38) and 17.11 (57) and (58).

The nonperiodic functions e^x, cosh x, sinh x become the periodic functions e^z, cosh z, sinh z when the space is expanded from $\mathbb{C}_0$ to $\mathbb{C}_1$. This fact suggests, correctly, that further periods may be introduced when the space is expanded from $\mathbb{C}_1$ to $\mathbb{C}_2$. The next theorem shows that some functions with a single period in $\mathbb{C}_1$, or no period in $\mathbb{C}_0$, may have two periods in $\mathbb{C}_2$.

17.19 THEOREM Each of the functions in the following tabulation has two independent periods as shown.

	Functions	Periods
(113)	$e^{z_1+i_2z_2}$	$2\pi i_1 e_1$, $2\pi i_1 e_2$
(114)	$e^{i_1(z_1+i_2z_2)}$	$2\pi e_1$, $2\pi e_2$
(115)	$e^{i_2(z_1+i_2z_2)}$	$2\pi e_1$, $2\pi e_2$
(116)	$e^{i_1i_2(z_1+i_2z_2)}$	$2\pi i_1 e_1$, $2\pi i_1 e_2$
(117)	$\cos(z_1+i_2z_2)$	$2\pi e_1$, $2\pi e_2$
(118)	$\sin(z_1+i_2z_2)$	$2\pi e_1$, $2\pi e_2$
(119)	$\cosh(z_1+i_2z_2)$	$2\pi i_1 e_1$, $2\pi i_1 e_2$
(120)	$\sinh(z_1+i_2z_2)$	$2\pi i_1 e_1$, $2\pi i_1 e_2$

Furthermore, the periods can be described in various other ways since

(121) $$2\pi i_1 e_1 = -2\pi i_2 e_1 = \pi(i_1 - i_2),$$

(122) $$2\pi i_1 e_2 = 2\pi i_2 e_2 = \pi(i_1 + i_2).$$

Proof. Simple calculations verify the statements in (121) and (122). Next, consider (113); by (71) or (96),

(123) $$e^{z_1+i_2z_2} = e^{z_1-i_1z_2}e_1 + e^{z_1+i_1z_2}e_2.$$

Now $p+i_2q$, p and q in $\mathbb{C}_1$, is a period of $e^{z_1+i_2z_2}$ if and only if, for all $z_1+i_2z_2$ in $\mathbb{C}_2$,

(124) $$e^{(z_1+i_2z_2)+(p+i_2q)} = e^{z_1+i_2z_2}, \qquad p+i_2q \neq 0.$$

Again, by the idempotent representation,

(125) $$\begin{aligned} &e^{(z_1+i_2z_2)+(p+i_2q)} \\ &= e^{(z_1-i_1z_2)+(p-i_1q)}e_1 + e^{(z_1+i_1z_2)+(p+i_1q)}e_2 \\ &= e^{(z_1-i_1z_2)}e^{(p-i_1q)}e_1 + e^{(z_1+i_1z_2)}e^{(p+i_1q)}e_2. \end{aligned}$$

Then (125) and (123) show that (124) is satisfied if and only if $e^{p-i_1q}=1$ and $e^{p+i_1q}=1$. Since $e^z=1$ if and only if $z=2n\pi i_1$ [see Theorem 17.9 (38)], equation (124) is satisfied if and only if

(126) $\quad p-i_1q=2n\pi i_1, \qquad p+i_1q=2m\pi i_1, \qquad n,m=0,\pm 1,\pm 2,\ldots,$

(127) $\quad p=\pi i_1(m+n), \qquad q=\pi(m-n),$

(128) $\quad p+i_2q=(m+n)\pi i_1+(m-n)\pi i_2.$

There are many periods, but they are not all independent. If $m=0$ and $n=1$, then $p+i_2q=\pi i_1-\pi i_2$; if $m=1$ and $n=0$, then $p+i_2q=\pi i_1+\pi i_2$. Since $n(\pi i_1-\pi i_2)+m(\pi i_1+\pi i_2)=(m+n)\pi i_1+(m-n)\pi i_2$, all of the periods are linear combinations of the two periods $\pi(i_1-i_2)$ and $\pi(i_1+i_2)$. Thus the fundamental periods are $\pi(i_1-i_2)$ and $\pi(i_1+i_2)$, which are the same as $2\pi i_1e_1$ and $2\pi i_1e_2$ by (121) and (122).

The statements in (114) and (115) can be proved by the same method used to prove (113), but they follow more simply from the periods of $e^{z_1+i_2z_2}$ in (113). Since $2\pi i_1e_1$ and $2\pi i_1e_2$ are periods of $e^{z_1+i_2z_2}$, then $2\pi e_1$ and $2\pi e_2$ are periods of $e^{i_1(z_1+i_2z_2)}$. If $e^{i_1(z_1+i_2z_2)}$ had additional independent periods, then $e^{z_1+i_2z_2}$ would have additional independent periods. Similarly, $2\pi i_2e_1$ and $2\pi i_2e_2$ are periods of $e^{z_1+i_2z_2}$ by (113) and (121), (122); then $2\pi e_1$ and $2\pi e_2$ are periods of $e^{i_2(z_1+i_2z_2)}$ and there are no additional fundamental periods. The proofs of (114) and (115) are complete.

A similar proof establishes (116), or it can be proved from first principles as follows. Let $p+i_2q$ denote a period. Then

(129)
$$e^{i_1i_2(z_1+i_2z_2)}=e^{z_1-i_1z_2}e_1+e^{-(z_1+i_1z_2)}e_2,$$
$$e^{i_1i_2[(z_1+i_2z_2)+(p+i_2q)]}=e^{z_1-i_1z_2}e^{p-i_1q}e_1+e^{-(z_1+i_1z_2)}e^{-(p+i_1q)}e_2.$$

The exponentials on the left in these equations are equal if and only if

(130) $\quad e^{p-i_1q}=1, \qquad e^{-(p+i_1q)}=1.$

These equations are satisfied if and only if

(131) $\quad p-i_1q=2n\pi i_1, \qquad p+i_1q=2m\pi i_1.$

As before, these equations show that $e^{i_1i_2(z_1+i_2z_2)}$ has periods $p+i_2q\neq 0$ such that

(132) $\quad p+i_2q=(m+n)\pi i_1+(m-n)\pi i_2, \qquad m,n=0,\pm 1,\pm 2,\ldots.$

These periods are linear combinations of the two independent periods $\pi(i_1-i_2)$ and $\pi(i_1+i_2)$. The proof of (116) is complete.

The statements about the periods of $\cos(z_1+i_2z_2)$ and $\sin(z_1+i_2z_2)$ in (117) and (118) follow from (114) and (115) and the representations of these functions by exponentials in Theorem 17.16 (84) and (85). Since the sum of two periods is a period, and since

(133) $\quad 2\pi e_1 + 2\pi e_2 = 2\pi,$

then $\cos(z_1 + i_2 z_2)$ and $\sin(z_1 + i_2 z_2)$ still have the period 2π of $\cos x$, $\sin x$ and $\cos z$, $\sin z$ [compare Theorem 17.1 (1) and Theorem 17.11 (57)]. The statements about the periods of $\cosh(z_1 + i_2 z_2)$ and $\sinh(z_1 + i_2 z_2)$ in (119) and (120) follow from (116) and the representation of these functions by exponentials in Theorem 17.16 (86) and (87). Since $2\pi i_1 e_1 + 2\pi i_1 e_2 = 2\pi i_1$, and since the sum of two periods is also a period, then $\cosh(z_1 + i_2 z_2)$ and $\sinh(z_1 + i_2 z_2)$ have the period $2\pi i_1$, just as $\cosh z$ and $\sinh z$ do [compare Theorem 17.11 (58)]. The proof of all parts of Theorem 17.19 is complete. □

17.20 THEOREM Let ζ and η denote $z_1 + i_2 z_2$ and $w_1 + i_2 w_2$, respectively. Then

(134) $\quad \cos(\zeta + \eta) = \cos \zeta \cos \eta - \sin \zeta \sin \eta;$

(135) $\quad \sin(\zeta + \eta) = \sin \zeta \cos \eta + \cos \zeta \sin \eta;$

(136) $\quad \cosh(\zeta + \eta) = \cosh \zeta \cosh \eta + \sinh \zeta \sinh \eta;$

(137) $\quad \sinh(\zeta + \eta) = \sinh \zeta \cosh \eta + \cosh \zeta \sinh \eta.$

Proof. The proofs of these formulas are based on the formulas in Theorem 17.16 (79)–(81), and they are similar to the proofs of Theorem 17.11 (59)–(62). For example

(138) $\quad e^{i_1(\zeta+\eta)} = e^{i_1\zeta}e^{i_1\eta}$

by Theorem 17.14. Represent the exponentials on the two sides of this equation by Theorem 17.16 (79). Multiply the expressions which represent the two exponentials on the right, and then compare the result with the left side of the equation. The formulas in (134) and (135) are the result. The formulas in (136) and (137) can be proved in the same way, using the formula in Theorem 17.16 (81). □

Exercises

17.1 The symbol

$$\sum_{m,n=0}^{\infty} (a_{mn} + i_2 b_{mn}), \qquad a_{mn} + i_2 b_{mn} \in \mathbb{C}_2,$$

is called a double series with bicomplex terms. Set

$$A_N = \sum_{m,n=0}^{N} \|a_{mn} + i_2 b_{mn}\|, \qquad B_N = \sum_{m+n=0}^{N} \|a_{mn} + i_2 b_{mn}\|.$$

(a) Prove that $\lim_{N\to\infty} A_N$ exists and equals A if and only if $\lim_{N\to\infty} B_N$ exists and equals A.

(b) Set

$$S_N = \sum_{m,n=0}^{N} (a_{mn} + i_2 b_{mn}), \qquad T_N = \sum_{m+n=0}^{N} (a_{mn} + i_2 b_{mn}).$$

If $\lim_{N\to\infty} A_N$ exists, prove that $\lim_{N\to\infty} S_N$ and $\lim_{N\to\infty} T_N$ exist, and that

$$\lim_{N\to\infty} S_N = \lim_{N\to\infty} T_N.$$

17.2 Prove that $\exp(\zeta+\eta) = \exp(\zeta)\exp(\eta)$, ζ and η in $\mathbb{C}_2$, by the method used to prove Theorem 17.3.

17.3 Use Theorem 17.18 (108) to show that $\pi(i_1 - i_2)$ and $\pi(i_1 + i_2)$ are periods of $e^{z_1 + i_2 z_2}$, and also that $2\pi i_1 e_1$, $2\pi i_1 e_2$, $2\pi i_2 e_1$, and $2\pi i_2 e_2$ are periods. Use these periods to find several other periods.

17.4 Prove that $\cosh x$ and $\sinh x$, x in $\mathbb{C}_0$, are not periodic functions.

17.5 Verify that the following statements are correct.

(a) $e^{(i_1\pi)/2} = i_1, \qquad e^{(i_2\pi)/2} = i_2, \qquad e^{i_1\pi} = -1, \qquad e^{i_2\pi} = -1.$

(b) $e^{(i_1+i_2)\pi/2} = i_1 i_2, \qquad e^{(i_1+i_2)\pi} = 1.$

(c) $e^{i_1\pi + (i_2\pi/2)} = -i_2, \qquad e^{(i_1\pi/2)+i_2\pi} = -i_1.$

(d) $e^{i_1 i_2} = \cosh 1 + i_1 i_2 \sinh 1.$

(e) $\cosh \dfrac{\pi i_1}{2} = 0, \qquad \sinh \dfrac{\pi i_1}{2} = i_1.$

(f) $\cosh \pi i_1 = -1, \qquad \sinh \pi i_1 = 0.$

(g) $e^{i_1 i_2 (i_1\pi/2)} = -i_2, \qquad e^{i_1 i_2 (i_2\pi/2)} = -i_1.$

(h) $e^{i_1 i_2 (\pi i_2)} = -1, \qquad e^{i_1 i_2 (\pi i_1)} = -1.$

17.6 Prove the following theorem. The function $\exp : \mathbb{C}_2 \to \mathbb{C}_2$ is a holomorphic function in $\mathbb{C}_2$. [*Hints*. To prove this theorem, it is necessary to show that exp can be represented by a power series about each point η in $\mathbb{C}_2$. By Theorem 17.14 (68),

$$\begin{aligned}\exp(\zeta) &= \exp[\eta + (\zeta - \eta)]\\ &= \exp(\eta)\exp(\zeta - \eta).\end{aligned}$$

Then use Definition 17.12 (67) to obtain a power series for $\exp(\zeta - \eta)$.]

17.7 Prove the following theorem. The functions $\cos : \mathbb{C}_2 \to \mathbb{C}_2$, $\sin : \mathbb{C}_2 \to \mathbb{C}_2$, $\cosh : \mathbb{C}_2 \to \mathbb{C}_2$, $\sinh : \mathbb{C}_2 \to \mathbb{C}_2$ are holomorphic functions in $\mathbb{C}_2$. [*Hints*. Observe that

$$\cos \zeta = \cos[\eta + (\zeta - \eta)],$$

and that similar statements hold for each of the other functions. Then use Theorem 17.20 and Definition 17.12.]

18. THE LOGARITHM FUNCTION

This section defines the logarithm in $\mathbb{C}_2$ and establishes its elementary properties. It begins by reviewing the logarithm function of a real variable x in $\mathbb{C}_0$; next, it describes the extension from $\log x$ to $\log z$, the logarithm of the complex variable z in $\mathbb{C}_1$; and finally it defines and investigates the logarithm of the bicomplex variable $z_1+i_2z_2$ in $\mathbb{C}_2$. The logarithm function is the inverse of the exponential function; it is defined for all $z_1+i_2z_2$ in $\mathbb{C}_2$ which are not in $\mathcal{O}_2$. Corresponding to the fact that the exponential function is doubly periodic, the logarithm function in $\mathbb{C}_2$ is doubly infinitely many-valued. As in the case of the elementary functions in Section 17, the idempotent representation of elements and holomorphic functions provides the tools and methods needed for the extension of $\log z$ to $\log(z_1+i_2z_2)$.

18.1 DEFINITION If $e^y=x$, then y is called the *logarithm of* x and we write $y=\log x$. The function $\log:\{x\in\mathbb{C}_0: x>0\}\to\mathbb{C}_0$, $x\mapsto\log x$, is called the *logarithm function.*

Some remarks are required. First, the logarithm is defined only for $x>0$ since, by Theorem 17.4 (13), the exponential function is positive for every y in $\mathbb{C}_0$. Next, if $x>0$, then there exists a unique y such that $e^y=x$. To prove this statement, observe that $de^x/dx=e^x>0$ for all x in $\mathbb{C}_0$; then e^x is a strictly increasing continuous function. Furthermore, $\lim_{x\to\infty}e^x=+\infty$ and $\lim_{x\to-\infty}=0$. The intermediate value theorem can now be used to show that the equation $e^y=x$ has a unique solution y for each positive x in $\mathbb{C}_0$. Thus Definition 18.1 defines $\log x$ for $x>0$.

18.2 THEOREM For every positive x,

(1) $$e^{\log x}=x.$$

For every x in $\mathbb{C}_0$,

(2) $$\log e^x=x.$$

For every positive x_1 and x_2,

(3) $$\log x_1x_2=\log x_1+\log x_2.$$

Proof. The statements in (1) and (2) follow from the fact that the logarithm and the exponential are inverse functions. To prove (3), observe that, by (1),

$$e^{\log x_1x_2}=x_1x_2,\qquad e^{\log x_1}=x_1,\qquad e^{\log x_2}=x_2.$$

Then

(4) $$e^{\log x_1x_2}=x_1x_2=e^{\log x_1}e^{\log x_2}=e^{\log x_1+\log x_2}$$

by Theorem 17.3 (6). Since the exponential function is strictly increasing, it is a one-to-one mapping of $\mathbb{C}_0$ onto the positive part of $\mathbb{C}_0$, and (4) implies the statement in (3). □

Turn now to the definition of $\log z$. The function $\log z$ is defined to be the inverse of e^z. Thus if $z = e^w$, then $w = \log z$. In order to show that the equation $e^w = z$ has a solution for w and to find this solution, it is helpful to investigate the mapping of the w-plane into the z-plane by the equation $z = e^w$. Let w and z be $u + i_1 v$ and $x + i_1 y$, respectively. Since e^w is periodic with period $2\pi i_1$ [see Theorem 17.9 (38)], the nature of the entire mapping $z = e^w$, $w \in \mathbb{C}_1$, can be determined by investigating the mapping of the period strip $\{u + i_1 v: -\infty < u < +\infty, 0 \leqslant v < 2\pi\}$ (see Figure 18.1). Since $e^w = e^u(\cos v + i_1 \sin v)$ by Theorem 17.9 (39), the mapping $z = e^w$ is

$$(5) \qquad x = e^u \cos v, \qquad y = e^u \sin v, \qquad x^2 + y^2 = e^{2u}.$$

Thus the points on the vertical segment at u_0 in the period strip are mapped onto the circle with radius e^{u_0} and center at the origin in the z-plane. Also, the line $v = v_0$ in the period strip is mapped by $z = e^w$ onto a ray from the origin, open at $z = 0$, which makes an angle v_0 with the x-axis in the z-plane. The radius of the circle approaches zero as $u_0 \to -\infty$ and it becomes infinite as $u_0 \to +\infty$. Thus $z = e^w$ maps the period strip in the w-plane onto $\mathbb{C}_1 - \mathcal{O}_1$ ($\mathcal{O}_1$, by the definition at the end of Section 4, is the set whose only member is the complex zero $0 + i_1 0$); this mapping is one-to-one since (a) the line $v = v_0$ is mapped onto the ray which makes the angle v_0 with the positive x-axis in the

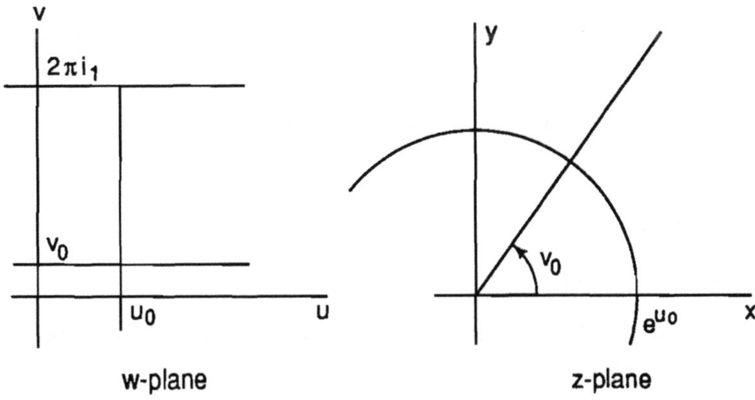

Figure 18.1 The mapping $z = e^w$.

z-plane, and (b) the mapping of the line $v=v_0$ onto the ray with angle v_0 is one-to-one because e^u is a strictly increasing function of u.

To find the inverse of the exponential function it is necessary to solve the equation $e^w=z$ for w in terms of z. Observe first that this equation has no solution if $z=0$. If $z\neq 0$, then it is possible to write z in the polar coordinate form. Thus, if $z\neq 0$,

$$(6)\qquad z = r(\cos\theta + i_1 \sin\theta), \qquad 0 \leqslant \theta < 2\pi,$$
$$e^w = e^u(\cos v + i_1 \sin v).$$

Then $e^w=z$ if $e^u=r$ and $v=\theta$. Thus, if $e^w=z\neq 0$ and $w=u+i_1v$, then $u=\log r=\log|z|$ and $v=\arg z$, and one solution is

$$(7)\qquad w = \log|z| + i_1 \arg z, \qquad 0 \leqslant \arg z < 2\pi.$$

Because e^w is periodic with period $2\pi i_1$ by Theorem 17.9 (38), all solutions of $e^w=z$ are

$$(8)\qquad w = \log|z| + i_1 \arg z + 2n\pi i_1, \qquad n = 0, \pm 1, \pm 2, \ldots.$$

18.3 DEFINITION If $z\neq 0$, then each of the numbers

$$(9)\qquad \log|z| + i_1 \arg z + 2n\pi i_1, \qquad n = 0, \pm 1, \pm 2, \ldots,$$

is called a *logarithm of* z. The *principal determination* of $\log z$ is

$$(10)\qquad \log|z| + i_1 \arg z, \qquad 0 \leqslant \arg z < 2\pi.$$

Also, $\log : \mathbb{C}_1 - \mathcal{O}_1 \to \mathbb{C}_1$ is an infinitely many-valued function called the *logarithm function* $\log z$, and for each n, where $n=0, \pm 1, \pm 2, \ldots$, the function

$$(11)\qquad \log : \mathbb{C}_1 - \mathcal{O}_1 \to \mathbb{C}_1, \qquad z \mapsto \log|z| + i_1 \arg z + 2n\pi i_1,$$

is called a *branch* of $\log z$. The *principal branch* of $\log z$ is the function in (11) with $n = 0$.

18.4 THEOREM For every z in $\mathbb{C}_1 - \mathcal{O}_1$ and every branch of $\log z$,

$$(12)\qquad e^{\log z} = z.$$

For every z in $\mathbb{C}_1$,

$$(13)\qquad \log e^z = z + 2n\pi i_1, \qquad n = 0, \pm 1, \pm 2, \ldots.$$

For every z_1 and z_2 in $\mathbb{C}_1 - \mathcal{O}_1$ and for each value of $\log z_1 z_2$, it is possible to choose values for $\log z_1$ and $\log z_2$ such that

$$(14)\qquad \log z_1 z_2 = \log z_1 + \log z_2.$$

Proof. The statement in (12) is true because the exponential and the logarithm are inverse functions. Also, direct calculation, the definition in (11), and the properties of the exponential function prove (12) as follows:

$$(15)\qquad \begin{aligned} e^{\log z} &= e^{\log|z| + i_1 \arg z + 2n\pi i_1} \\ &= e^{\log|z|} e^{i_1 \arg z} e^{2n\pi i_1} \\ &= |z| [\cos(\arg z) + i_1 \sin(\arg z)] \\ &= z. \end{aligned}$$

To prove (13), solve $e^w = e^z$ for w; one solution is $w = z$ and all solutions are $w = z + 2n\pi i_1$, $n = 0, \pm 1, \pm 2, \ldots$. To prove (14), observe that, by (11),

$$(16)\qquad \log z_1 z_2 = \log|z_1 z_2| + i_1 \arg(z_1 z_2) + 2n\pi i_1.$$

Now a property of the norm in $\mathbb{C}_1$ and (3) show that

$$\log|z_1 z_2| = \log|z_1|\,|z_2| = \log|z_1| + \log|z_2|,$$

and an elementary property of complex multiplication shows that

$$\begin{aligned} \arg(z_1 z_2) &= \arg z_1 + \arg z_2 &\quad& \text{if } 0 \leqslant \arg z_1 + \arg z_2 < 2\pi, \\ &= \arg z_1 + \arg z_2 - 2\pi && \text{if } 2\pi \leqslant \arg z_1 + \arg z_2 < 4\pi. \end{aligned}$$

Then one of the following statements is true:

$$(17)\qquad \log z_1 z_2 = [\log|z_1| + i_1 \arg z_1] + [\log|z_2| + i_1 \arg z_2] + 2n\pi i_1,$$

$$(18)\qquad \log z_1 z_2 = [\log|z_1| + i_1 \arg z_1] + [\log|z_2| + i_1 \arg z_2] + 2(n-1)\pi i_1.$$

Again by (11),

$$(19)\qquad \log z_1 = \log|z_1| + i_1 \arg z_1 + 2n_1 \pi i_1,$$

$$(20)\qquad \log z_2 = \log|z_2| + i_1 \arg z_2 + 2n_2 \pi i_1.$$

Then

$$\begin{aligned} \log z_1 + \log z_2 = {} & [\log|z_1| + i_1 \arg z_1] \\ & + [\log|z_2| + i_1 \arg z_2] + 2(n_1 + n_2)\pi i_1, \end{aligned}$$

and (14) is true if $n_1 + n_2 = n$ in case (17) holds, or if $n_1 + n_2 = n - 1$ in case (18) holds. The proof of all parts of Theorem 18.4 is complete. □

Turn now to the definition and properties of $\log(z_1 + i_2 z_2)$. In $\mathbb{C}_0$ and in $\mathbb{C}_1$, the logarithm is defined to be the inverse of the exponential function; likewise in $\mathbb{C}_2$, $\log(z_1 + i_2 z_2)$ is defined to be the inverse of $e^{z_1 + i_2 z_2}$. Thus if $z_1 = i_2 z_2 = e^{w_1 + i_2 w_2}$, then $w_1 + i_2 w_2 = \log(z_1 + i_2 z_2)$, and $\log(z_1 + i_2 z_2)$ can be found by solving the exponential equation. The following theorem shows, with the help of the idempotent representation in $\mathbb{C}_2$, that $\log(z_1 + i_2 z_2)$ has a representation in terms of the logarithm function in $\mathbb{C}_1$.

18.5 THEOREM If $z_1+i_2z_2\notin\mathcal{O}_2$, that is, if $z_1-i_1z_2\neq 0$ and $z_1+i_1z_2\neq 0$, then $\log(z_1+i_2z_2)$ is defined, and

$$\log(z_1+i_2z_2)=\log(z_1-i_1z_2)e_1+\log(z_1+i_1z_2)e_2. \tag{21}$$

Proof. If $z_1+i_2z_2\in\mathcal{O}_2$, then Theorem 17.17 (94) shows that the equation $e^{w_1+i_2w_2}=z_1+i_2z_2$ has no solution for $w_1+i_2w_2$. Assume that $z_1+i_2z_2\notin\mathcal{O}_2$; then by (6), (7), and (17) in Section 7,

$$z_1-i_1z_2\neq 0,\qquad z_1+i_1z_2\neq 0. \tag{22}$$

If $e^{w_1+i_2w_2}=z_1+i_2z_2$, then by definition,

$$w_1+i_2w_2=\log(z_1+i_2z_2). \tag{23}$$

By the idempotent representation of elements in $\mathbb{C}_2$ [see Theorem 6.4 (11)],

$$w_1+i_2w_2=(w_1-i_1w_2)e_1+(w_1+i_1w_2)e_2, \tag{24}$$

$$z_1+i_2z_2=(z_1-i_1z_2)e_1+(z_1+i_1z_2)e_2. \tag{25}$$

Finally, since $e^{w_1+i_2w_2}$ is a holomorphic function by Definition 15.2 and Exercise 17.6, then Theorem 15.5 [see also (71) in Section 17] shows that

$$e^{w_1+i_2w_2}=e^{w_1-i_1w_2}e_1+e^{w_1+i_1w_2}e_2. \tag{26}$$

Furthermore, Theorem 6.4 shows that $e^{w_1+i_2w_2}$ in (26) equals $z_1+i_2z_2$ in (25) if and only if

$$e^{w_1-i_1w_2}=z_1-i_1z_2,\qquad e^{w_1+i_1w_2}=z_1+i_1z_2. \tag{27}$$

Since the equations in (27) are equations in $\mathbb{C}_1$, they can be solved for $w_1-i_1w_2$ and $w_1+i_1w_2$ in terms of the complex logarithm to obtain

$$w_1-i_1w_2=\log(z_1-i_1z_2),\qquad w_1+i_1w_2=\log(z_1+i_1z_2). \tag{28}$$

These solutions are possible because of the hypotheses in (22) and the definition of $\log:\mathbb{C}_1-\mathcal{O}_1\to\mathbb{C}_1$. Substitute from (23) and (28) in (24) to obtain (21). The proof of Theorem 18.5 is complete. □

18.6 COROLLARY If $z_1+i_2z_2\notin\mathcal{O}_2$, then $\log(z_1+i_2z_2)$ is defined and

$$\begin{aligned}\log(z_1+i_2z_2)=&\{\log|z_1-i_1z_2|+i_1\arg(z_1-i_1z_2)+2n_1\pi i_1\}e_1\\&+\{\log|z_1+i_1z_2|+i_1\arg(z_1+i_1z_2)+2n_2\pi i_1\}e_2,\\&n_1,n_2=0,\pm1,\pm2,\ldots.\end{aligned} \tag{29}$$

Proof. The formula in (29) follows from Theorem 18.5 (21) and the definition of the complex logarithm in Definition 18.3. □

18.7 THEOREM For every $z_1+i_2z_2$ in $\mathbb{C}_2-\mathcal{O}_2$,

$$e^{\log(z_1+i_2z_2)} = z_1 + i_2z_2. \tag{30}$$

For every $z_1+i_2z_2$ in $\mathbb{C}_2$,

$$\log e^{z_1+i_2z_2} = (z_1+i_2z_2)+n_1\pi(i_1-i_2)+n_2\pi(i_1+i_2), \quad n_1, n_2 = 0, \pm 1, \pm 2, \ldots. \tag{31}$$

For every $z_1+i_2z_2$ and $z_1'+i_2z_2'$ in $\mathbb{C}_2-\mathcal{O}_2$, and for each value of $\log(z_1+i_2z_2)(z_1'+i_2z_2')$, it is possible to choose values for $\log(z_1+i_2z_2)$ and $\log(z_1'+i_2z_2')$ such that

$$\log(z_1+i_2z_2)(z_1'+i_2z_2') = \log(z_1+i_2z_2)+\log(z_1'+i_2z_2'). \tag{32}$$

Proof. The logarithm function is defined to be the inverse of the exponential function. Thus $w_1+i_2w_2 = \log(z_1+i_2z_2)$ means $z_1+i_2z_2 = e^{w_1+i_2w_2} = e^{\log(z_1+i_2z_2)}$, and (30) is true. To find $\log e^{z_1+i_2z_2}$, solve $e^{w_1+i_2w_2} = e^{z_1+i_2z_2}$ for $w_1+i_2w_2$. One solution is $w_1+i_1z_2 = z_1+i_2z_2$, and all solutions are

$$w_1+i_2w_2 = (z_1+i_2z_2)+n_1\pi(i_1-i_2)+n_2\pi(i_1+i_2), \quad n_1, n_2 = 0, \pm 1, \pm 2, \ldots, \tag{33}$$

because $e^{w_1+i_2w_2}$ is periodic with periods $\pi(i_1-i_2)$ and $\pi(i_1+i_2)$ by Theorem 17.19 (113), (121), (122); then (31) follows from (33). Consider (32). By (21),

$$\begin{aligned} &\log(z_1 + i_2z_2)(z_1' + i_2z_2') \\ &\quad = \log(z_1 - i_1z_2)(z_1' - i_1z_2')e_1 + \log(z_1 + i_1z_2)(z_1' + i_1z_2')e_2. \end{aligned} \tag{34}$$

The logarithms on the right in this equation have definite, fixed values. Now

$$\log(z_1 + i_2z_2) = \log(z_1 - i_1z_2)e_1 + \log(z_1 + i_1z_2)e_2, \tag{35}$$

$$\log(z_1' + i_2z_2') = \log(z_1' - i_1z_2')e_1 + \log(z_1' + i_1z_2')e_2. \tag{36}$$

Theorem 18.4 shows that it is possible to choose values for the logarithms on the right in (35) and (36) so that

$$\log(z_1-i_1z_2)(z_1'-i_1z_2') = \log(z_1-i_1z_2)+\log(z_1'-i_1z_2'), \tag{37}$$

$$\log(z_1+i_1z_2)(z_1'+i_1z_2') = \log(z_1+i_1z_2)+\log(z_1'+i_1z_2'). \tag{38}$$

Then the formula in (32) follows from (34)–(38), and the proof of Theorem 18.7 is complete. □

18.8 THEOREM If $z_1+i_2z_2 \notin \mathcal{O}_2$, then $\log(z_1+i_2z_2)$ is defined and

$$\text{(39)}\quad \log(z_1+i_2z_2)=\log|z_1^2+z_2^2|^{1/2}+i_1\frac{[\arg(z_1-i_1z_2)+\arg(z_1+i_1z_2)]}{2}$$

$$+i_1i_2\left\{\log\left|\frac{z_1-i_1z_2}{z_1+i_1z_2}\right|^{1/2}+i_1\frac{[\arg(z_1-i_1z_2)-\arg(z_1+i_1z_2)]}{2}\right\}$$

$$+n_1\pi(i_1-i_2)+n_2\pi(i_1+i_2),$$

$$n_1=0,\pm 1,\pm 2,\ldots,n_2=0,\pm 1,\pm 2,\ldots.$$

Proof. The value of $\log(z_1+i_2z_2)$ is given by (29). By rearranging the terms on the right, we find

$$\text{(40)}\quad \log(z_1+i_2z_2)=\frac{1}{2}[\log|z_1-i_1z_2|+i_1\arg(z_1-i_1z_2)+2n_1\pi i_1]$$

$$+\frac{1}{2}[\log|z_1+i_1z_2|+i_1\arg(z_1+i_1z_2)+2n_2\pi i_1]$$

$$+\frac{i_1i_2}{2}[\log|z_1-i_1z_2|+i_1\arg(z_1-i_1z_2)+2n_1\pi i_1]$$

$$-\frac{i_1i_2}{2}[\log|z_1+i_1z_2|+i_1\arg(z_1+i_1z_2)+2n_2\pi i_1].$$

Elementary properties of the real logarithm can be used to simplify (40) to (39). □

18.9 THEOREM The function $\log:\mathbb{C}_2-\mathcal{O}_2\to\mathbb{C}_2$ such that

$$\text{(41)}\quad \log(z_1+i_2z_2)=\log(z_1-i_1z_2)e_1+\log(z_1+i_1z_2)e_2$$

is a holomorphic function of $z_1+i_2z_2$.

Proof. By Definition 15.2, a holomorphic function of $z_1+i_2z_2$ is one which can be represented by a power series in the neighborhood of each point $a_1+i_2a_2$. Now it is known that $\log z$ is a holomorphic function in $\mathbb{C}_1-\mathcal{O}_1$. Let $a_1+i_2a_2$ be a point in $\mathbb{C}_2-\mathcal{O}_2$. Then

$$\text{(42)}\quad a_1+i_2a_2=(a_1-i_1a_2)e_1+(a_1+i_1a_2)e_2,$$

and $a_1-i_1a_2\neq 0$, $a_1+i_1a_2\neq 0$. Then there are power series such that

$$\text{(43)}\quad \sum_{k=0}^{\infty}a_k[(z_1-i_1z_2)-(a_1-i_1a_2)]^k=\log(z_1-i_1z_2),$$

$$a_k\in\mathbb{C}_1,\ k=0,1,\ldots,$$

(44) $$\sum_{k=0}^{\infty} b_k[(z_1 + i_1 z_2) - (a_1 + i_1 a_2)]^k = \log(z_1 + i_1 z_2),$$
$$b_k \in \mathbb{C}_1,\ k = 0, 1, \ldots.$$

These power series converge and represent $\log(z_1 - i_1 z_2)$ and $\log(z_1 + i_1 z_2)$ in sufficiently small circles about $a_1 - i_1 a_2$ and $a_1 + i_1 a_2$. Then (41), (43), and (44) show that

(45) $$\log(z_1 + i_2 z_2) = \sum_{k=0}^{\infty} a_k[(z_1 - i_1 z_2) - (a_1 - i_1 a_2)]^k e_1 + \sum_{k=0}^{\infty} b_k[(z_1 + i_1 z_2) - (a_1 + i_1 a_2)]^k e_2.$$

Now by the results in Theorem 15.3, the right side of this equation is the idempotent representation of the power series

(46) $$\sum_{k=0}^{\infty} (c_k + i_2 d_k)[(z_1 + i_2 z_2) - (a_1 + i_2 a_2)]^k$$

where

(47) $$c_k - i_1 d_k = a_k, \qquad c_k + i_1 d_k = b_k, \qquad k = 0, 1, \ldots.$$

These equations have the unique solution

(48) $$c_k = \frac{a_k + b_k}{2}, \qquad d_k = \frac{i_1(a_k - b_k)}{2}, \qquad k = 0, 1, \ldots.$$

Thus the power series (46) represents $\log(z_1 + i_2 z_2)$ in a discus about $a_1 + i_2 a_2$, and $\log(z_1 + i_2 z_2)$ is a holomorphic function by Definition 15.2. □

Exercises

18.1 Use the formula in (39) to prove that

$$\log e^{z_1 + i_2 z_2} = (z_1 + i_2 z_2) + n_1\pi(i_1 - i_2) + n_2\pi(i_1 + i_2),$$
$$n_1, n_2 = 0, \pm 1, \pm 2, \ldots.$$

18.2 Use the formula in (39) to prove the following [see Theorem 18.7 (32)]: for every $z_1 + i_2 z_2$ and $z_1' + i_2 z_2'$ in $\mathbb{C}_2 - \mathcal{O}_2$, and for each value of $\log(z_1 + i_2 z_2)(z_1' + i_2 z_2')$, it is possible to choose values for $\log(z_1 + i_2 z_2)$ and $\log(z_1' + i_2 z_2')$ such that

$$\log(z_1 + i_2 z_2)(z_1' + i_2 z_2') = \log(z_1 + i_2 z_2) + \log(z_1' + i_2 z_2').$$

18.3 (a) Show that the function $\exp: \mathbb{C}_1 \to \mathbb{C}_1$ maps $\mathbb{C}_1$ onto $\mathbb{C}_1 - \mathcal{O}_1$.
(b) Show that exp maps an infinite number of points in $\mathbb{C}_1$ into each point in $\mathbb{C}_1 - \mathcal{O}_1$.

(c) Find a period strip such that exp, restricted to this period strip, is a one-to-one mapping of the period strip onto $\mathbb{C}_1 - \mathcal{O}_1$.
(d) Explain the significance of this exercise for the function $\log : \mathbb{C}_1 - \mathcal{O}_1 \to \mathbb{C}_1$.

18.4 (a) Show that the function $\exp : \mathbb{C}_2 \to \mathbb{C}_2$ maps $\mathbb{C}_2$ onto $\mathbb{C}_2 - \mathcal{O}_2$.
(b) Show that exp maps a double infinity of points in $\mathbb{C}_2$ into each point in $\mathbb{C}_2 - \mathcal{O}_2$.
(c) Find a period strip in $\mathbb{C}_2$ such that exp, restricted to this period strip, is a one-to-one mapping of the period strip onto $\mathbb{C}_2 - \mathcal{O}_2$.
(d) Explain the significance of this exercise for the function $\log : \mathbb{C}_2 - \mathcal{O}_2$.

18.5 (a) Let X and Y be domains in $\mathbb{C}_2$, and let $f : X \to Y$ be a one-to-one holomorphic mapping

$$w_1 + i_2 w_2 = f(z_1 + i_2 z_2), \qquad z_1 + i_2 z_2 \in X,\ w_1 + i_2 w_2 \in Y,$$

of X onto Y.
(b) Show that there are domains X_1, X_2 and Y_1, Y_2 and also mappings

$$w_1 - i_1 w_2 = f_1(z_1 - i_1 z_2),\quad z_1 - i_1 z_2 \in X_1,\ w_1 - i_1 w_2 \in Y_1,$$
$$w_1 + i_1 w_2 = f_2(z_1 + i_1 z_2),\quad z_1 + i_1 z_2 \in X_2,\ w_1 + i_1 w_2 \in Y_2,$$

such that

$$f(z_1 + i_2 z_2) = f_1(z_1 - i_1 z_2)e_1 + f_2(z_1 + i_1 z_2)e_2,\quad z_1 + i_2 z_2 \in X.$$

(c) If the mappings in (b) have the inverses

$$z_1 - i_1 z_2 = g_1(w_1 - i_1 w_2),\quad w_1 - i_1 w_2 \in Y_1,\ z_1 - i_1 z_2 \in X_1,$$
$$z_1 + i_1 z_2 = g_2(w_1 + i_1 w_2),\quad w_1 + i_1 w_2 \in Y_2,\ z_1 + i_1 z_2 \in X_2,$$

show that the mapping $w_1 + i_2 w_2 = f(z_1 + i_2 z_2)$ in (a) has the inverse

$$z_1 + i_2 z_2 = g(w_1 + i_2 w_2),\quad w_1 + i_2 w_2 \in Y,\ z_1 + i_2 z_2 \in X,$$

where

$$g(w_1 + i_2 w_2) = g_1(w_1 - i_1 w_2)e_1 + g_2(w_1 + i_1 w_2)e_2,$$
$$w_1 - i_1 w_2 \in Y_1,\ w_1 + i_1 w_2 \in Y_2,\ w_1 + i_2 w_2 \in Y.$$

[*Hint.* $z_1 + i_2 z_2 = (z_1 - i_1 z_2)e_1 + (z_1 + i_1 z_2)e_2$.]
(d) If $g_1 : Y_1 \to X_1$ and $g_2 : Y_2 \to X_2$ are holomorphic functions in $\mathbb{C}_1$, show that $g : Y \to X$ is a holomorphic function in $\mathbb{C}_2$. [*Hint.* Theorem 15.3.]

3

Derivatives and Holomorphic Functions

19. INTRODUCTION

The definition of the derivative at ζ_0 of a function $f: X \to \mathbb{C}_2$, $X \subset \mathbb{C}_2$, of a bicomplex variable ζ is formally the same as for a function of a complex variable:

$$(1) \qquad D_\zeta f(\zeta_0) = \lim_{\zeta \to \zeta_0} \frac{f(\zeta) - f(\zeta_0)}{\zeta - \zeta_0}, \qquad \zeta - \zeta_0 \notin \mathcal{O}_2.$$

The theories are remarkably similar in the two cases, but many differences arise in the details because $\mathcal{O}_2$ contains many points rather than a single point as in the complex case. This chapter shows that f is differentiable in X if and only if it satisfies the following (strong) Stolz condition at each ζ_0 in X:

$$(2) \qquad f(\zeta) - f(\zeta_0) = d(\zeta_0)(\zeta - \zeta_0) + r(f; \zeta_0, \zeta)(\zeta - \zeta_0),$$
$$\lim_{\zeta \to \zeta_0} r(f; \zeta_0, \zeta) = 0, \; r(f; \zeta_0, \zeta_0) = 0.$$

The constant $d(\zeta_0)$ equals the derivative $D_\zeta f(\zeta_0)$ of f at ζ_0. As in other studies of differentiation [7], the Stolz condition plays a major role. With the help of the theory of functions of a complex variable, the chapter shows that $f: X \to \mathbb{C}_2$ is a differentiable function in X if and only if there exist differentiable functions f_1, f_2 of a complex variable such that

$$(3) \qquad f(z_1 + i_2 z_2) = f_1(z_1 - i_1 z_2)e_1 + f_2(z_1 + i_1 z_2)e_2.$$

Also, with help from the theory of functions of a complex variable, the chapter establishes (a) necessary and sufficient conditions for differentiability, (b) the calculus of differentiable functions, (c) the representation of f by the usual Taylor series, and (d) the relation of differentiable functions to the holomorphic functions defined in Chapter 2 (they are the same).

A prominent feature of the study of bicomplex systems is the many isomorphic forms of these systems. The bicomplex number ζ is represented by each of the following:

(4) $z_1 + i_2 z_2, \qquad z_1, z_2 \text{ in } \mathbb{C}_1;$

(5) $x_1 + i_1 x_2 + i_2 x_3 + i_1 i_2 x_4, \qquad x_1, \ldots, x_4 \text{ in } \mathbb{C}_0.$

Corresponding to (4) there is a matrix algebra with elements

(6) $$\begin{bmatrix} z_1 & -z_2 \\ z_2 & z_1 \end{bmatrix}$$

which is isomorphic to $\mathbb{C}_2$ under the correspondence

(7) $$z_1 + i_2 z_2 \leftrightarrow \begin{bmatrix} z_1 & -z_2 \\ z_2 & z_1 \end{bmatrix}.$$

Corresponding to (5) there is a matrix algebra with elements

(8) $$\begin{bmatrix} x_1 & -x_2 & -x_3 & x_4 \\ x_2 & x_1 & -x_4 & -x_3 \\ x_3 & -x_4 & x_1 & -x_2 \\ x_4 & x_3 & x_2 & x_1 \end{bmatrix}$$

which is isomorphic to $\mathbb{C}_2$ under the correspondence (5)$\leftrightarrow$(8). The matrices in (6) and (8) are called Cauchy–Riemann matrices in this book; the matrix algebras with these elements are interesting in themselves and they contribute to the development of the theory of functions of a bicomplex variable. The idempotent representation

(9) $z_1 + i_2 z_2 = (z_1 - i_1 z_2)e_1 + (z_1 + i_1 z_2)e_2$

supplies another system which is isomorphic to $\mathbb{C}_2$. Also, the function $f : X \to \mathbb{C}_2$ has the following isomorphic representations:

(10) $f(\zeta) = u(z_1, z_2) + i_2 v(z_1, z_2),$

(11) $f(\zeta) = g_1(x) + i_1 g_2(x) + i_2 g_3(x) + i_1 i_2 g_4(x), \qquad x = (x_1, \ldots, x_4).$

Each of the isomorphic systems contributes something to the total theory. Next, there is a real-valued function $g : X \to \mathbb{C}_0^4$ which corresponds to the differentiable function in (11). If $g: X \to \mathbb{C}_0^4$ is represented as a transformation, then it is

$$(12) \qquad \begin{aligned} y_1 &= g_1(x_1, \ldots, x_4), \\ &\vdots \\ y_4 &= g_4(x_1, \ldots, x_4). \end{aligned}$$

The derivative $D_\zeta f(\zeta)$ corresponds to the jacobian $D_{(1,\ldots,4)}(g_1, \ldots, g_4)(x)$, and the derivative belongs to $\mathcal{O}_2$ if and only if the jacobian belongs to $\mathcal{O}_0$. The chapter uses $g: X \to \mathbb{C}_0^4$ and (12) to give an elementary proof that $f: X \to \mathbb{C}_2$ has an inverse, which is a differentiable function, in the neighborhood of a point ζ_0 at which $D_\zeta f(\zeta_0) \notin \mathcal{O}_2$.

19.1 QUESTION This introduction will be concluded with several questions which the reader may wish to keep in mind while reading the remaining chapters of the book.

(13) Is the theory of functions of a bicomplex variable merely a theory of pairs of holomorphic functions of a complex variable [see (3)], or is there something more to the subject?

(14) Is the theory of functions of a bicomplex variable wholly dependent on the theory of functions of a complex variable for its construction, or can significant parts of the former be constructed without using fundamental parts of the latter?

(15) What is the role of the Stolz condition in the theory of functions of a bicomplex variable?

(16) What is the role of the derivative in the theory of functions of a bicomplex variable? How much of the theory can be constructed without employing the derivative?

(17) Can the theory of holomorphic functions $f: X \to \mathbb{C}_2$, $X \subset \mathbb{C}_2$, be generalized to a theory of holomorphic functions $f: X \to \mathbb{C}_n$, $X \subset \mathbb{C}_n$, for $n = 3, 4, \ldots$ (see the Preface and the Epilogue)? If the generalization is possible, what is its nature and what are the relations among the theories in $\mathbb{C}_1, \mathbb{C}_2, \ldots, \mathbb{C}_n, \ldots$?

Exercises

19.1 Let $f: X \to \mathbb{C}_2$, $X \subset \mathbb{C}_2$, be the function such that $f(\zeta) = \zeta^2$. Show that f has a derivative at ζ_0, and that $D_\zeta f(\zeta_0) = 2\zeta_0$. [*Hint*. For every ζ_0,

$$\begin{aligned} \frac{f(\zeta) - f(\zeta_0)}{\zeta - \zeta_0} = \frac{\zeta^2 - \zeta_0^2}{\zeta - \zeta_0} &= \frac{(\zeta + \zeta_0)(\zeta - \zeta_0)}{\zeta - \zeta_0} \\ &= \zeta + \zeta_0, \qquad \zeta - \zeta_0 \notin \mathcal{O}_2.] \end{aligned}$$

19.2 Let $f: X \to \mathbb{C}_2$, $X \subset \mathbb{C}_2$, be a function which satisfies the Stolz condition (2) at each point ζ_0 in X. Show that f has a derivative at ζ_0, and that $D_\zeta f(\zeta_0) = d(\zeta_0)$ at each ζ_0 in X.

19.3 If $f: X \to \mathbb{C}_2$ is the function in Exercise 19.1, show that f satisfies the Stolz condition (2) with $r(f; \zeta_0, \zeta) = \zeta - \zeta_0$, $r(f; \zeta_0, \zeta_0) = 0$.

19.4 Let $f: X \to \mathbb{C}_2$ be the function in Exercise 19.1. Write ζ in the form $z_1 + i_2 z_2$. Show that $f(\zeta) = u(z_1, z_2) + i_2 v(z_1, z_2)$, where $u(z_1, z_2) = z_1^2 - z_2^2$ and $v(z_1, z_2) = 2z_1 z_2$. Verify that u and v satisfy the following Cauchy–Riemann differential equations: $D_{z_1} u = D_{z_2} v$, $D_{z_2} u = -D_{z_1} v$.

19.5 The space $\mathbb{C}_3$ has elements $\zeta_1 + i_3 \zeta_2$, where ζ_1 and ζ_2 are elements in $\mathbb{C}_2$ and $i_3^2 = -1$. Show that the elements in $\mathbb{C}_3$ can be represented in each of the following forms:

$$\zeta_1 + i_3\zeta_2, \qquad \zeta_1, \zeta_2 \text{ in } \mathbb{C}_2;$$

$$(z_1 + i_2 z_2) + i_3(z_3 + i_2 z_4), \qquad z_1, \ldots, z_4 \text{ in } \mathbb{C}_1;$$

$$[(x_1 + i_1 x_2) + i_2(x_3 + i_1 x_4)] + i_3[(x_5 + i_1 x_6) + i_2(x_7 + i_1 x_8)],$$

$$x_1, \ldots, x_8 \text{ in } \mathbb{C}_0.$$

19.6 Describe all of the representations of the element in $\mathbb{C}_n$. What is the dimension of the real euclidean space in which $\mathbb{C}_n$ is embedded?

20. DERIVATIVES AND THE STOLZ CONDITION

A function is differentiable at a point if it can be approximated, with proper accuracy, by a linear function. The precise statement about the linear approximation of the function is known as the *Stolz condition.* Stolz introduced the condition in 1893 in connection with his study of the differentiation of functions of several real variables [7, pp. 23, 551–552, 646], and it has proved to be important in the theory of differentiation of a variety of types of functions. This chapter will show that the Stolz condition is also important in the study of the differentiation of functions of a bicomplex variable. As preparation for beginning this study, this section contains four examples which review the Stolz condition for functions of real and of complex variables. There is a weak form and a strong form for the Stolz condition; as explained in the four examples, the weak and strong forms are equivalent for many types of functions. After the four examples, this section contains definitions of the derivative and of the (weak and strong) Stolz conditions for functions $f: X \to \mathbb{C}_2$, $X \subset \mathbb{C}_2$, of a bicomplex variable. The weak and strong forms of the Stolz condition are equivalent also for functions of a bicomplex variable, but much work is required in later sections to establish this result.

20.1 EXAMPLE Let X be a domain in $\mathbb{C}_0$, and let $f: X \to \mathbb{C}_0$ be a function defined in X.

20.2 DEFINITION The function $f: X \to \mathbb{C}_0$ satisfies the *strong Stolz condition* at x_0 in X if and only if there exists a constant d in $\mathbb{C}_0$ and a function $r(f; x_0, \cdot)$, defined in a neighborhood of x_0 and with values in $\mathbb{C}_0$, such that

(1) $$f(x) - f(x_0) = d(x - x_0) + r(f; x_0, x)(x - x_0),$$

(2) $$\lim_{x \to x_0} r(f; x_0, x) = 0, \qquad r(f; x_0, x_0) = 0.$$

The function f satisfies the *weak Stolz condition* at x_0 in X if and only if there exists a constant d' in $\mathbb{C}_0$ and a function $r'(f; x_0, \cdot)$, defined in a neighborhood of x_0 and with values in $\mathbb{C}_0$, such that

(3) $$f(x) - f(x_0) = d'(x - x_0) + r'(f; x_0, x)|x - x_0|,$$

(4) $$\lim_{x \to x_0} r'(f; x_0, x) = 0, \qquad r'(f; x_0, x_0) = 0.$$

20.3 THEOREM The following three conditions are equivalent:

(5) f has a derivative $D_x f(x_0)$ at x_0;

(6) f satisfies the strong Stolz condition at x_0;

(7) f satisfies the weak Stolz condition at x_0.

Furthermore,

(8) $$D_x f(x_0) = d = d'.$$

Proof. The theorem will be proved by showing that (5) implies (6), that (6) implies (7), and that (7) implies (5). Assume then that f has a derivative $D_x f(x_0)$ at x_0. Set $d = D_x f(x_0)$ and define $r(f; x_0, \cdot)$ as follows:

(9) $$r(f; x_0, x) = \frac{f(x) - f(x_0)}{x - x_0} - D_x f(x_0), \qquad x - x_0 \notin \mathcal{O}_0,$$

(10) $$r(f; x_0, x_0) = 0.$$

Then since

(11) $$f(x) - f(x_0) = D_x f(x_0)(x - x_0) + \left[\frac{f(x) - f(x_0)}{x - x_0} - D_x f(x_0)\right](x - x_0),$$

it follows from (9), (10), and (11) that (1) and (2) are satisfied with $d = D_x f(x_0)$. Therefore f satisfies the strong Stolz condition at x_0, and (5) implies (6). Assume next that f satisfies (1) and (2). Set $d' = d$, and define $r'(f; x_0, \cdot)$ as follows:

$$(12)\qquad r'(f; x_0, x) = \frac{r(f; x_0, x)(x - x_0)}{|x - x_0|}, \qquad x - x_0 \notin \mathcal{O}_0,$$

$$(13)\qquad r'(f; x_0, x_0) = 0.$$

Then (2) shows that $r'(f; x_0, \cdot)$ satisfies (4). Also, (1) shows that

$$(14)\qquad f(x) - f(x_0) = d'(x - x_0) + \left[\frac{r(f; x_0, x)(x - x_0)}{|x - x_0|}\right]|x - x_0|,$$
$$x - x_0 \notin \mathcal{O}_0.$$

Then (12), (13), (14) show that (3) and (4) are satisfied with $d' = d$, and the proof is complete that (6) implies (7); that is, the strong Stolz condition implies the weak Stolz condition. Next, continue the proof by showing that (7) implies (5) or that the weak Stolz condition implies differentiability. Assume then that f satisfies (3) and (4). Equation (3) shows that

$$(15)\qquad \frac{f(x) - f(x_0)}{x - x_0} = d' + \frac{r'(f; x_0, x)|x - x_0|}{x - x_0}, \qquad x - x_0 \notin \mathcal{O}_0,$$

and (4) and (15) show that

$$(16)\qquad \lim_{x \to x_0} \frac{f(x) - f(x_0)}{x - x_0} = d'.$$

Thus f has a derivative $D_x f(x_0)$, and $D_x f(x_0) = d'$. Therefore (7) implies (5). Finally, the proof has shown that $D_x f(x_0) = d = d'$, and (8) is true. The proof of Theorem 20.3 is complete. □

20.4 EXAMPLE Let X be a domain in $\mathbb{C}_0^n$, and let $f: X \to \mathbb{C}_0$ be a function defined in X.

20.5 DEFINITION The function $f: X \to \mathbb{C}_0$ satisfies the *strong Stolz condition* at $x^0 : (x_1^0, \ldots, x_n^0)$ in X if and only if there exist constants $d_1, \ldots, d_n$ in $\mathbb{C}_0$ and functions $r_k(f; x^0, \cdot)$, $k = 1, \ldots, n$, defined in a neighborhood of x^0 and with values in $\mathbb{C}_0$, such that

$$(17)\qquad f(x) - f(x^0) = \sum_{k=1}^{n} d_k(x_k - x_k^0) + \sum_{k=1}^{n} r_k(f; x^0, x)(x_k - x_k^0),$$

$$(18)\qquad \lim_{x \to x_0} r_k(f; x^0, x) = 0, \qquad r_k(f; x^0, x^0) = 0,\ k = 1, \ldots, n.$$

As a matter of notation, set

$$(19)\qquad |x - x^0| = \left[\sum_{k=1}^{n} |x_k - x_k^0|^2\right]^{1/2}.$$

Then f satisfies the *weak Stolz condition* at x^0 in X if and only if there exist constants $d'_1, \ldots, d'_n$ in $\mathbb{C}_0$ and a function $r(f; x^0, \cdot)$, defined in a neighborhood of x^0 and with values in $\mathbb{C}_0$, such that

$$(20) \qquad f(x) - f(x^0) = \sum_{k=1}^{n} d'_k(x_k - x_k^0) + r(f; x^0, x)|x - x^0|,$$

$$(21) \qquad \lim_{x \to x_0} r(f; x^0, x) = 0, \qquad r(f; x^0, x^0) = 0.$$

20.6 THEOREM The following three conditions are equivalent:

(22) f has derivatives $D_{x_k}f(x^0)$, $k = 1, \ldots, n$, at x^0;

(23) f satisfies the strong Stolz condition at x^0;

(24) f satisfies the weak Stolz condition at x^0.

Furthermore,

$$(25) \qquad D_{x_k}f(x^0) = d_k = d'_k, \qquad k = 1, \ldots, n.$$

Proof. Again the theorem will be proved by showing that (22) implies (23), that (23) implies (24), and that (24) implies (22). Assume then that f has derivatives $D_{x_k}f(x^0)$, $k=1, \ldots, n$, at x^0. These are not partial derivatives; they are rather derivatives of the kind treated in Chapter 1 of [7]. It is shown there [7, pp. 26–27] that there are functions $r_k(f; x^0, \cdot)$, $k=1, \ldots, n$, such that

$$(26) \qquad f(x) - f(x^0) = \sum_{k=1}^{n} D_{x_k}f(x^0)(x_k - x_k^0) + \sum_{k=1}^{n} r_k(f; x^0, x)(x_k - x_k^0),$$

$$(27) \qquad \lim_{x \to x_0} r_k(f; x^0, x) = 0, \qquad r_k(f; x^0, x^0) = 0, \ k = 1, \ldots, n.$$

If $d_k = D_{x_k}f(x^0)$, $k=1, \ldots, n$, then (26) and (27) show that f satisfies (17) and (18); therefore, (22) implies (23) and f satisfies the strong Stolz condition at x^0 if it is differentiable at x^0. Assume next that f satisfies the strong Stolz condition in (17) and (18). Set $d'_k = d_k$, $k=1, \ldots, n$, and define $r(f; x^0, \cdot)$ as follows:

$$(28) \qquad r(f; x^0, x) = \frac{1}{|x - x^0|} \sum_{k=1}^{n} r_k(f; x^0, x)(x_k - x_k^0), \qquad |x - x^0| \notin \mathcal{O}_0,$$

$$(29) \qquad r(f; x^0, x^0) = 0.$$

Then (28) and Schwarz's inequality show that

$$(30) \qquad |r(f; x^0, x)| \leqslant \left[\sum_{k=1}^{n} |r_k(f; x^0, x)|^2 \right]^{1/2},$$

and (18) shows that

$$(31) \qquad \lim_{x \to x_0} r(f; x^0, x) = 0.$$

Thus if f satisfies (17) and (18), then there exist constants $d'_1, \ldots, d'_n$ and a function $r(f; x^0, \cdot)$ which satisfy (20) and (21). Therefore, (23) implies (24), or the strong Stolz condition implies the weak Stolz condition. Next, continue the proof by showing that (24) implies (22), or that the weak Stolz condition implies differentiability. If f satisfies (20) and (21), then f has derivatives $D_{x_k}f(x^0)$, $k=1,\ldots,n$, and

$$(32) \qquad D_{x_k}f(x^0) = d'_k, \qquad k = 1, \ldots, n.$$

The proof can be found in [7, pp. 24–25], and it will not be repeated here. Thus (24) implies (22), and the proof that the conditions in (22)–(24) are equivalent is now complete. The proof has shown that the statements in (25) are true. The proof of all parts of Theorem 20.6 is complete. □

20.7 EXAMPLE Let X be a domain in $\mathbb{C}_1$, and let $f: X \to \mathbb{C}_1$ be a function defined in X.

20.8 DEFINITION The function $f: X \to \mathbb{C}_1$ satisfies the *strong Stolz condition* at z_0 in X if and only if there exists a constant d in $\mathbb{C}_1$ and a function $r(f; z_0, \cdot)$, defined in a neighborhood of z_0 and with values in $\mathbb{C}_1$, such that

$$(33) \qquad f(z) - f(z_0) = d(z - z_0) + r(f; z_0, z)(z - z_0),$$

$$(34) \qquad \lim_{z \to z_0} r(f; z_0, z) = 0, \qquad r(f; z_0, z_0) = 0.$$

The function satisfies the *weak Stolz condition* at z_0 in X if and only if there exists a constant d' in $\mathbb{C}_1$ and a function $r'(f; z_0, \cdot)$, defined in a neighborhood of z_0 and with values in $\mathbb{C}_1$, such that

$$(35) \qquad f(z) - f(z_0) = d'(z - z_0) + r'(f; z_0, z)|z - z_0|,$$

$$(36) \qquad \lim_{z \to z_0} r'(f; z_0, z) = 0, \qquad r'(f; z_0, z_0) = 0.$$

20.9 THEOREM The following three conditions are equivalent:

(37) f has a derivative $D_z f(z_0)$ at z_0;

(38) f satisfies the strong Stolz condition at z_0;

(39) f satisfies the weak Stolz condition at z_0.

Furthermore,

$$(40) \qquad D_z f(z_0) = d = d'.$$

The proof of this theorem is similar to the proof of Theorem 20.3; it is left to the reader as an exercise. For the differentiation of functions of a single complex variable see [7, pp. 496–501].

20.10 EXAMPLE Let X be a domain in $\mathbb{C}_1^n$, and let $f: X \to \mathbb{C}_1$ be a function defined in X.

20.11 DEFINITION The function $f: X \to \mathbb{C}_1$ satisfies the *strong Stolz condition* at $z^0: (z_1^0, \ldots, z_n^0)$ in X if and only if there exist constants $d_1, \ldots, d_n$ in $\mathbb{C}_1$ and functions $r_k(f; z^0, \cdot)$, $k = 1, \ldots, n$, defined in a neighborhood of z^0 and with values in $\mathbb{C}_1$, such that

$$f(z) - f(z^0) = \sum_{k=1}^{n} d_k(z_k - z_k^0) + \sum_{k=1}^{n} r_k(f; z^0, z)(z_k - z_k^0), \tag{41}$$

$$\lim_{z \to z^0} r_k(f; z^0, z) = 0, \qquad r_k(f; z^0, z^0) = 0,\ k = 1, \ldots, n. \tag{42}$$

As a matter of notation, set

$$|z - z^0| = \left[\sum_{k=1}^{n} |z_k - z_k^0|^2\right]^{1/2}. \tag{43}$$

Then f satisfies the *weak Stolz condition* at z^0 in X if and only if there exist constants $d_1', \ldots, d_n'$ in $\mathbb{C}_1$ and a function $r(f; z^0, \cdot)$, defined in a neighborhood of z^0 and with values in $\mathbb{C}_1$, such that

$$f(z) - f(z^0) = \sum_{k=1}^{n} d_k'(z_k - z_k^0) + r(f; z^0, z)|z - z^0|, \tag{44}$$

$$\lim_{z \to z^0} r(f; z^0, z) = 0, \qquad r(f; z^0, z^0) = 0. \tag{45}$$

20.12 THEOREM The following three conditions are equivalent:

(46) f has derivatives $D_{z_k} f(z^0)$, $k = 1, \ldots, n$, at z^0;

(47) f satisfies the strong Stolz condition at z^0;

(48) f satisfies the weak Stolz condition at z^0.

Furthermore,

$$D_{z_k} f(z^0) = d_k = d_k', \qquad k = 1, \ldots, n. \tag{49}$$

The proof of this theorem is similar to the proof of Theorem 20.6; it is left to the reader as an exercise. For the differentiation of functions of several complex variables, see [7, pp. 548–553].

We turn now to a consideration of functions of a bicomplex variable. It would be nice if it were possible to add a fifth example which carried out a similar program for functions $f: X \to \mathbb{C}_2$, $X \subset \mathbb{C}_2$, but the theory of functions of a bicomplex variable encounters many new situations because $\mathbb{C}_2$ contains a large class $\mathcal{O}_2$ of singular elements. The definitions of the derivative and of the strong and weak Stolz conditions follow the pattern suggested by the examples above, but there are differences and much work is required to establish the equivalence of the three conditions.

20.13 DEFINITION Let X be a domain in $\mathbb{C}_2$. The function $f: X \to \mathbb{C}_2$, $\zeta \mapsto f(\zeta)$, has a *derivative* $D_\zeta f(\zeta_0)$ at ζ_0 in X if and only if

$$(50)\qquad \lim_{\zeta\to\zeta_0} \frac{f(\zeta)-f(\zeta_0)}{\zeta-\zeta_0} = D_\zeta f(\zeta_0), \qquad \zeta-\zeta_0 \notin \mathcal{O}_2.$$

Also, f has a *derivative* in X if and only if it has a derivative at each point ζ_0 in X. Finally, f is *differentiable at* ζ_0 *(in* X*)* if and only if it has a derivative at ζ_0 (in X).

20.14 DEFINITION The function $f: X \to \mathbb{C}_2$ satisfies the *strong Stolz condition* at ζ_0 in X if and only if there exists a constant d in $\mathbb{C}_2$ and a function $r(f; \zeta_0, \cdot)$, defined in a neighborhood of ζ_0 and with values in $\mathbb{C}_2$, such that

$$(51)\qquad f(\zeta)-f(\zeta_0) = d(\zeta-\zeta_0) + r(f;\zeta_0,\zeta)(\zeta-\zeta_0),$$

$$(52)\qquad \lim_{\zeta\to\zeta_0} r(f;\zeta_0,\zeta) = 0, \qquad r(f;\zeta_0,\zeta_0) = 0.$$

The function f satisfies the *weak Stolz condition* at ζ_0 in X if and only if there exists a constant d' in $\mathbb{C}_2$ and a function $r'(f; \zeta_0, \cdot)$, defined in a neighborhood of ζ_0 and with values in $\mathbb{C}_2$, such that

$$(53)\qquad f(\zeta)-f(\zeta_0) = d'(\zeta-\zeta_0) + r'(f;\zeta_0,\zeta)\|\zeta-\zeta_0\|,$$

$$(54)\qquad \lim_{\zeta\to\zeta_0} r'(f;\zeta_0,\zeta) = 0, \qquad r'(f;\zeta_0,\zeta_0) = 0.$$

If f satisfies the strong (weak) Stolz condition at each point ζ_0 in X, then it satisfies the *strong (weak) Stolz condition in* X.

20.15 THEOREM If $f: X \to \mathbb{C}_2$, $X \subset \mathbb{C}_2$, satisfies the strong Stolz condition at ζ_0 in X, then f is differentiable at ζ_0, and it satisfies the weak Stolz condition at ζ_0.

Proof. Since f satisfies the strong Stolz condition at ζ_0, then (51) and (52) are satisfied. Thus if $\zeta - \zeta_0 \notin \mathcal{O}_2$, the equation (51) can be divided by $\zeta - \zeta_0$ to obtain

$$(55)\qquad \frac{f(\zeta)-f(\zeta_0)}{\zeta-\zeta_0} = d + r(f;\zeta_0,\zeta), \qquad \zeta-\zeta_0 \notin \mathcal{O}_2.$$

Then (52) shows that

$$(56)\qquad \lim_{\zeta\to\zeta_0} \frac{f(\zeta)-f(\zeta_0)}{\zeta-\zeta_0} = d,$$

and f is differentiable at ζ_0 by Definition 20.13. The proof that the strong Stolz condition implies the weak Stolz condition for $f: X \to \mathbb{C}_2$ is similar to the proof that (6) implies (7) in Theorem 20.3. Since f satisfies the strong Stolz condition at ζ_0, then f satisfies (51) and (52). Then for every $\zeta \neq \zeta_0$ the following equation is valid:

$$(57)\qquad f(\zeta) - f(\zeta_0) = d(\zeta - \zeta_0) + \left[\frac{r(f;\zeta_0,\zeta)(\zeta - \zeta_0)}{\|\zeta - \zeta_0\|}\right]\|\zeta - \zeta_0\|.$$

Set $d' = d$ and define $r'(f;\zeta_0,\cdot)$ as follows:

$$(58)\qquad r'(f;\zeta_0,\zeta) = \frac{r(f;\zeta_0,\zeta)(\zeta - \zeta_0)}{\|\zeta - \zeta_0\|}, \qquad \zeta \neq \zeta_0,$$

$$(59)\qquad r'(f;\zeta_0,\zeta_0) = 0.$$

Then (58) and (59) define $r'(f;\zeta_0,\cdot)$ at all points in a neighborhood of ζ_0. Since

$$(60)\qquad \|r'(f;\zeta_0,\zeta)\| \leqslant \frac{\sqrt{2}\,\|r(f;\zeta_0,\zeta)\|\ \|\zeta - \zeta_0\|}{\|\zeta - \zeta_0\|} = \sqrt{2}\,\|r(f;\zeta_0,\zeta)\|,$$

then (59) and the hypothesis in (52) show that

$$(61)\qquad \lim_{\zeta \to \zeta_0} r'(f;\zeta_0,\zeta) = 0, \qquad r'(f;\zeta_0,\zeta_0) = 0.$$

Since (57) is

$$(62)\qquad f(\zeta) - f(\zeta_0) = d'(\zeta - \zeta_0) + r'(f;\zeta_0,\zeta)\|\zeta - \zeta_0\|,$$

then (62) and (61) show that f satisfies the weak Stolz condition at ζ_0. The proofs of the two conclusions in Theorem 20.15 are complete. □

20.16 EXAMPLE This example explains why a proof similar to the proof that (5) implies (6) in Theorem 20.3 cannot be used to show that differentiability of $f: X \to \mathbb{C}_2$ at ζ_0 implies that f satisfies the strong Stolz condition at ζ_0. The following equation [compare (11)] is valid for all ζ such that $\zeta - \zeta_0$ is not in $\mathcal{O}_2$:

$$(63)\qquad f(\zeta) - f(\zeta_0) = D_\zeta f(\zeta_0)(\zeta - \zeta_0) + \left[\frac{f(\zeta) - f(\zeta_0)}{\zeta - \zeta_0} - D_\zeta f(\zeta_0)\right](\zeta - \zeta_0).$$

The pattern in Theorem 20.3 [see (9) and (10)] suggests that $r'(f;\zeta_0,\cdot)$ be defined as follows:

$$(64)\qquad r'(f;\zeta_0,\zeta) = \frac{f(\zeta) - f(\zeta_0)}{\zeta - \zeta_0} - D_\zeta f(\zeta_0),$$

$$(65)\qquad r'(f;\zeta_0,\zeta_0) = 0.$$

Since the right side of (64) is defined only for $\zeta-\zeta_0\notin\mathcal{O}_2$, these equations do not define the function $r'(f;\zeta_0,\cdot)$ at all points in a neighborhood of ζ_0. Therefore, this effort fails to prove that differentiability implies the strong Stolz condition for functions $f:X\to\mathbb{C}_2$, $X\subset\mathbb{C}_2$.

Exercises

20.1 Write out the proofs of Theorems 20.9 and 20.12.

20.2 Let X be a domain in $\mathbb{C}_1$, and let $f:X\to\mathbb{C}_1$ be a function of the complex variable z. Assume that f satisfies the strong Stolz condition at z_0 in X as stated in (33) and (34) of Definition 20.8.

(a) Prove that f is continuous at z_0.

(b) Let $z=x+i_1y$. Show that there exist functions u and v of x and y such that $f(z)=u(x,y)+i_1v(x,y)$ for z in X.

(c) Let $d=d_1+i_1d_2$ and $r(f;z_0,z)=r_1(f;z_0,z)+i_1r_2(f;z_0,z)$. Show that

$$\lim_{z\to z_0} r_1(f;z_0,z)=0,\qquad r_1(f;z_0,z_0)=0,$$

$$\lim_{z\to z_0} r_2(f;z_0,z)=0,\qquad r_2(f;z_0,z_0)=0.$$

(d) Use (c) and the assumption that f satisfies the strong Stolz condition at z_0 to show that

$$\begin{aligned}u(x,y)-u(x_0,y_0)&=d_1(x-x_0)-d_2(y-y_0)\\&\quad+r_1(f;z_0,z)(x-x_0)-r_2(f;z_0,z)(y-y_0),\\v(x,y)-v(x_0,y_0)&=d_2(x-x_0)+d_1(y-y_0)\\&\quad+r_2(f;z_0,z)(x-x_0)+r_1(f;z_0,z)(y-y_0).\end{aligned}$$

(e) Show that u and v are continuous at z_0: $x_0+i_1y_0$.

(f) Use (d) to show that u and v satisfy the strong Stolz condition (see Definition 20.5) at (x_0,y_0) and therefore by Theorem 20.6 have derivatives

$$D_xu(x_0,y_0)=d_1,\qquad D_yu(x_0,y_0)=-d_2,\qquad D_xv(x_0,y_0)=d_2,$$
$$D_yv(x_0,y_0)=d_1.$$

Observe that these derivatives are not partial derivatives but rather the derivatives defined in [7, pp. 12–23].

(g) Prove the following theorem. If $f:X\to\mathbb{C}_1$ satisfies the strong (or weak) Stolz condition at z_0 in X, then u and v satisfy the following Cauchy–Riemann differential equations:

$$D_xu(x_0,y_0)=D_yv(x_0,y_0),\qquad D_yu(x_0,y_0)=-D_xv(x_0,y_0).$$

(h) Prove the following theorem. If $f: X \to \mathbb{C}_1$ has a derivative $D_z f(z_0)$ at z_0: $x_0 + i_1 y_0$ in X, then u and v have derivatives $D_x u(x_0, y_0)$, $D_y u(x_0, y_0)$, $D_x v(x_0, y_0)$, $D_y v(x_0, y_0)$ which satisfy the Cauchy–Riemann differential equations in (g).

20.3 Let $f: \mathbb{C}_2 \to \mathbb{C}_2$ be the function such that $f(\zeta) = \zeta^2$.

(a) Show that f has a derivative at each point ζ_0 in $\mathbb{C}_2$ and that $D_\zeta f(\zeta_0) = 2\zeta_0$.

(b) Show that f satisfies (51) with $d = 2\zeta_0$ and $r(f; \zeta_0, \zeta) = \zeta - \zeta_0$. Complete the proof that f satisfies the strong Stolz condition [see (52)] at each ζ_0 in $\mathbb{C}_2$.

20.4 Let $f: \mathbb{C}_2 \to \mathbb{C}_2$ be the function such that $f(\zeta) = \zeta^3$.

(a) Show that f has a derivative at each point ζ_0 in $\mathbb{C}_2$ and find its value.

(b) Show that f satisfies the strong Stolz condition in $\mathbb{C}_2$ by finding a d in $\mathbb{C}_2$ and a function $r(f; \zeta_0, \cdot)$, defined in a neighborhood of ζ_0 and with values in $\mathbb{C}_2$, which satisfy (51) and (52).

21. DIFFERENTIABILITY IMPLIES THE STRONG STOLZ CONDITION

This section considers functions $f: X \to \mathbb{C}_2$ defined on a domain X in $\mathbb{C}_2$. Definition 20.13 defines differentiability for these functions, and Definition 20.14 defines the strong and weak Stolz conditions. The goal is to prove that the three conditions are equivalent just as Theorems 20.3, 20.6, 20.9, and 20.12 have shown that they are equivalent for several other classes of functions. Steps to reach this goal have been taken already because Theorem 20.15 has shown that the strong Stolz condition implies differentiability and the weak Stolz condition. The purpose of the present section is to prove that differentiability implies the strong Stolz condition, that is, that $f: X \to \mathbb{C}_2$ satisfies the strong Stolz condition at each point in X at which it has a derivative (compare Example 20.16). The goal is reached in the next section (Section 22), which proves that the weak Stolz condition implies differentiability.

The proof in this section that differentiability implies the strong Stolz condition is based on one of the fundamental theorems of the theory of functions of a bicomplex variable: if $f: X \to \mathbb{C}_2$ is differentiable in X, then f has an idempotent representation by means of two holomorphic functions of a complex variable.

The following additional notation will be useful. Let $u: X \to \mathbb{C}_1$ and $v: X \to \mathbb{C}_1$ be functions such that

$$f(\zeta) = u(\zeta) + i_2 v(\zeta), \qquad \zeta \in X; \tag{1}$$

and let $d_1: X \to \mathbb{C}_1$ and $d_2: X \to \mathbb{C}_1$ be functions such that

$$(2) \qquad D_\zeta f(\zeta) = d_1(\zeta) + i_2 d(\zeta), \qquad \zeta \in X.$$

As usual, ζ is denoted also by $z_1 + i_2 z_2$, and X_1 and X_2 are the sets in $\mathbb{C}_1$ generated by X in $\mathbb{C}_2$.

21.1 THEOREM Let $f: X \to \mathbb{C}_2$, $X \subset \mathbb{C}_2$, be a function which has a derivative $D_\zeta f(\zeta_0)$ at each point ζ_0 in X. Then:

(3) $u - i_1 v: X \to \mathbb{C}_1$ is a differentiable function $f_1: X_1 \to \mathbb{C}_1$ of $z_1 - i_1 z_2$ in X_1;

(4) $u + i_1 v: X \to \mathbb{C}_1$ is a differentiable function $f_2: X_2 \to \mathbb{C}_1$ of $z_1 + i_1 z_2$ in X_2;

(5) $f(z_1 + i_2 z_2) = f_1(z_1 - i_1 z_2)e_1 + f_2(z_1 + i_1 z_2)e_2, \quad z_1 + i_2 z_2$ in X;

(6) $D_{z_1 - i_1 z_2} f_1(z_1 - i_1 z_2) = d_1(\zeta) - i_1 d_2(\zeta), \quad \zeta$ in X;

(7) $D_{z_1 + i_1 z_2} f_2(z_1 + i_1 z_2) = d_1(\zeta) + i_1 d_2(\zeta), \quad \zeta$ in X;

(8) $D_{z_1 + i_2 z_2} f(z_1 + i_2 z_2) = D_{z_1 - i_1 z_2} f_1(z_1 - i_1 z_2)e_1 + D_{z_1 + i_1 z_2} f_2(z_1 + i_1 z_2)e_2, \quad \zeta$ in X.

Proof. Since f has a derivative at ζ_0,

$$(9) \qquad \lim_{\zeta \to \zeta_0} \left\{ \frac{f(\zeta) - f(\zeta_0)}{\zeta - \zeta_0} - D_\zeta f(\zeta_0) \right\} = 0, \qquad \zeta - \zeta_0 \notin \mathcal{O}_2.$$

Now

$$(10) \qquad \zeta - \zeta_0 = [(z_1 - i_1 z_2) - (z_1^0 - i_1 z_2^0)]e_1 + [(z_1 + i_1 z_2) - (z_1^0 + i_1 z_2^0)]e_2,$$

and the restriction $\zeta - \zeta_0 \notin \mathcal{O}_2$ in (9) is equivalent to

$$(11) \qquad |(z_1 - i_1 z_2) - (z_1^0 - i_1 z_2^0)| > 0, \qquad |(z_1 + i_1 z_2) - (z_1^0 + i_1 z_2^0)| > 0.$$

The idempotent representation of elements in $\mathbb{C}_2$ shows that the expression in the braces in (9) is equal to

$$(12) \qquad \left\{ \frac{[u(\zeta) - i_1 v(\zeta)] - [u(\zeta_0) - i_1 v(\zeta_0)]}{(z_1 - i_1 z_2) - (z_1^0 - i_1 z_2^0)} \right\} e_1$$

$$(13) \qquad + \left\{ \frac{[u(\zeta) + i_1 v(\zeta)] - [u(\zeta_0) + i_1 v(\zeta_0)]}{(z_1 + i_1 z_2) - (z_1^0 + i_1 z_2^0)} \right\} e_2.$$

Let A, B, C denote the expressions in the braces in (9), (12), (13), respectively. All of the terms in A, B, C are defined because of the restriction in (11). Then $A = Be_1 + Ce_2$, and by Theorem 6.8,

$$(14) \qquad \|A\| = \left[\frac{|B|^2 + |C|^2}{2} \right]^{1/2},$$

$$(15)\quad \|(z_1+i_2z_2)-(z_1^0+i_2z_2^0)\| = \left[\frac{|(z_1-i_1z_2)-(z_1^0-i_1z_2^0)|^2+|(z_1+i_1z_2)-(z_1^0+i_1z_2^0)|^2}{2}\right]^{1/2}.$$

As stated above,

$$(16)\quad \lim_{\zeta\to\zeta_0} A = 0, \qquad \zeta-\zeta_0\notin \mathcal{O}_2.$$

Thus if $\varepsilon>0$ is given, there exists a $\delta>0$ such that

$$(17)\quad \|A\|<\varepsilon$$

for all $z_1+i_2z_2$ which satisfy (11) and

$$(18)\quad 0<\|(z_1+i_2z_2)-(z_1^0+i_2z_2^0)\|<\delta.$$

Let $z_1-i_1z_2$ and $z_1+i_1z_2$ be arbitrary points in X_1 and X_2 such that

$$(19)\quad \begin{aligned} &0<|(z_1-i_1z_2)-(z_1^0-i_1z_2^0)|<\delta,\\ &0<|(z_1+i_1z_2)-(z_1^0+i_1z_2^0)|<\delta. \end{aligned}$$

Corresponding to each such pair of points $z_1-i_1z_2$ and $z_1+i_1z_2$ there is a point $z_1+i_2z_2$ in X; if $z_1-i_1z_2=w_1$ and $z_1+i_1z_2=w_2$, then

$$(20)\quad z_1+i_2z_2=w_1e_1+w_2e_2=\frac{w_1+w_2}{2}+i_2\frac{w_1-w_2}{2}.$$

Furthermore, (15) and (19) show that (18) holds for this point; also, (19) and (11) show that the restriction $z_1+i_2z_2\notin\mathcal{O}_2$ in (9) is satisfied. Then by (17), $\|A\|<\varepsilon$ for each point (20), and (14) shows that

$$(21)\quad |B|<\sqrt{2}\varepsilon, \qquad |C|<\sqrt{2}\varepsilon.$$

Therefore,

$$(22)\quad \lim_{z_1-i_1z_2\to z_1^0-i_1z_2^0} B=0, \qquad z_1-i_1z_2\neq z_1^0-i_1z_2^0,$$

$$(23)\quad \lim_{z_1+i_1z_2\to z_1^0+i_1z_2^0} C=0, \qquad z_1+i_1z_2\neq z_1^0+i_1z_2^0.$$

These statements are equivalent to the following:

$$(24)\quad \lim_{z_1-i_1z_2\to z_1^0-i_1z_2^0} \frac{[u(\zeta)-i_1v(\zeta)]-[u(\zeta_0)-i_1v(\zeta_0)]}{(z_1-i_1z_2)-(z_1^0-i_1z_2^0)} = d_1(\zeta_0)-i_1d_2(\zeta_0);$$

$$(25)\quad \lim_{z_1+i_1z_2\to z_1^0+i_1z_2^0} \frac{[u(\zeta)+i_1v(\zeta)]-[u(\zeta_0)+i_1v(\zeta_0)]}{(z_1+i_1z_2)-(z_1^0+i_1z_2^0)} = d_1(\zeta_0)+i_1d_2(\zeta_0).$$

Since $z_1^0+i_2z_2^0$ is an arbitrary point in X, equation (24) shows that $u-i_1v$ is a function which has a derivative with respect to $z_1-i_1z_2$ at each point $z_1^0-i_1z_2^0$ in X_1; therefore, $u-i_1v$ is a differentiable function $f_1:X_1\to\mathbb{C}_1$ of the complex variable $z_1-i_1z_2$ in X_1. Similarly, (25) shows that $u+i_1v$ has a derivative with respect to $z_1+i_1z_2$ at each point in X_2; therefore, $u+i_1v$ is a differentiable function $f_2:X\to\mathbb{C}_1$ of the complex variable $z_1+i_1z_2$ in X_2. These statements establish the conclusions in (3) and (4). Since $f_1=u-i_1v$ and $f_2=u+i_1v$, and since $u+i_2v=(u-i_1v)e_1+(u+i_1v)e_2$, then $f=f_1e_1+f_2e_2$, and the proof of (5) is complete. Equations (24) and (25) establish the formulas for the derivatives of f_1 and f_2 in (6) and (7). Finally, since $D_{z_1+i_2z_2}f(z_1+i_2z_2)=D_\zeta f(\zeta)$, equations (2), (6), and (7) show that

$$
\begin{aligned}
(26)\quad D_{z_1+i_2z_2}f(z_1+i_2z_2)&=d_1(\zeta)+i_2d_2(\zeta)\\
&=[d_1(\zeta)-i_1d_2(\zeta)]e_1+[d_1(\zeta)+i_1d_2(\zeta)]e_2\\
&=D_{z_1-i_1z_2}f_1(z_1-i_1z_2)e_1+D_{z_1+i_1z_2}f_2(z_1+i_1z_2)e_2.
\end{aligned}
$$

These equations establish (8), and they complete the proof of all parts of Theorem 21.1. □

The statement in (5) may be considered the fundamental theorem of the theory of functions of a bicomplex variable; (5) follows from (3) and (4). A proof of (3) and (4), and thus (5), more complete than the one explained above, is contained in Theorem 38.5. Also, there are corresponding results for functions $f:X\to\mathbb{C}_n$, $X\subset\mathbb{C}_n$, $n\geqslant 2$; their proofs can be found in Theorem 49.3, Theorem 49.12 and Corollary 49.13, and the remarks which follow the corollary.

21.2 THEOREM Let $f:X\to\mathbb{C}_2$ be a function which has a derivative at each point ζ_0: $z_1^0+i_2z_2^0$ in a domain X in $\mathbb{C}_2$. Then f satisfies the strong Stolz condition (see Definition 20.14) at each point ζ_0 in X.

Proof. By Theorem 21.1 there are differentiable functions $f_1:X_1\to\mathbb{C}_1$ and $f_2:X_2\to\mathbb{C}_1$ such that

$$(27)\qquad f(z_1+i_2z_2)=f_1(z_1-i_1z_2)e_1+f_2(z_1+i_1z_2)e_2,$$

$$
\begin{aligned}
(28)\qquad & f(z_1+i_2z_2)-f(z_1^0+i_2z_2^0)\\
&=[f_1(z_1-i_1z_2)-f_1(z_1^0-i_1z_2^0)]e_1+[f_2(z_1+i_1z_2)-f_2(z_1^0+i_1z_2^0)]e_2.
\end{aligned}
$$

In order to simplify the notation in this proof, set

$$
\begin{aligned}
(29)\qquad & w_1=z_1-i_1z_2, && w_2=z_1+i_1z_2,\\
& w_1^0=z_1^0-i_1z_2^0, && w_2^0=z_1^0+i_1z_2^0.
\end{aligned}
$$

Since a differentiable function of a complex variable satisfies the strong Stolz condition at each point (see Theorem 20.9 in Example 20.7), there are functions $r(f_1; w_1^0, \cdot)$ and $r(f_2; w_2^0, \cdot)$ such that

$$(30) \qquad f_1(w_1) - f_1(w_1^0) = D_{w_1} f_1(w_1^0)(w_1 - w_1^0) + r(f_1; w_1^0, w_1)(w_1 - w_1^0),$$

$$(31) \qquad \lim_{w_1 \to w_1^0} r(f_1; w_1^0, w_1) = 0, \qquad r(f_1; w_1^0, w_1^0) = 0;$$

and

$$(32) \qquad f_2(w_2) - f_2(w_2^0) = D_{w_2} f_2(w_2^0)(w_2 - w_2^0) + r(f_2; w_2^0, w_2)(w_2 - w_2^0),$$

$$(33) \qquad \lim_{w_2 \to w_2^0} r(f_2; w_2^0, w_2) = 0, \qquad r(f_2; w_2^0, w_2^0) = 0.$$

Substitute from (30) and (32) in (28); the result can be simplified to the following:

$$(34) \qquad f(\zeta) - f(\zeta_0) = [D_{z_1 - i_1 z_2} f_1(w_1^0) e_1 + D_{z_1 + i_1 z_2} f_2(w_2^0) e_2](\zeta - \zeta_0) + [r(f_1; w_1^0, w_1) e_1 + r(f_2; w_2^0, w_2) e_2](\zeta - \zeta_0).$$

Now Theorem 21.1 (8) shows that

$$(35) \qquad D_{z_1 - i_1 z_2} f_1(w_1^0) e_1 + D_{z_1 + i_1 z_2} f_2(w_2^0) e_2 = D_{z_1 + i_2 z_2} f(\zeta_0).$$

Then (34) can be written as

$$(36) \qquad f(\zeta) - f(\zeta_0) = D_\zeta f(\zeta_0)(\zeta - \zeta_0) + [r(f_1; w_1^0, w_1) e_1 + r(f_2; w_2^0, w_2) e_2](\zeta - \zeta_0).$$

Define $r(f; \zeta_0, \zeta)$ by the following statement:

$$(37) \qquad r(f; \zeta_0, \zeta) = r(f_1; w_1^0, w_1) e_1 + r(f_2; w_2^0, w_2) e_2.$$

If $\zeta - \zeta_0$ is in $\mathcal{O}_2$, then either $w_1 - w_1^0 = 0$ or $w_2 - w_2^0 = 0$, or both, and (37) defines $r(f; \zeta_0, \zeta)$ even in this case. Thus $r(f; \zeta_0, \zeta)$ is defined for all ζ in a neighborhood of ζ_0, and

$$(38) \qquad r(f; \zeta_0, \zeta_0) = 0.$$

Finally, (37) and Theorem 6.8 show that

$$(39) \qquad \|r(f; \zeta_0, \zeta)\| = [(1/2)|r(f_1; w_1^0, w_1)|^2 + (1/2)|r(f_2; w_2^0, w_2)|^2]^{1/2}.$$

If ζ approaches ζ_0, then w_1 approaches w_1^0 and w_2 approaches w_2^0 [see (15)], and (31), (33), and (39) show that

$$(40) \qquad \lim_{\zeta \to \zeta_0} r(f; \zeta_0, \zeta) = 0.$$

To summarize, if f has a derivative in X, then for each ζ_0 in X there is a function $r(f;\zeta_0,\cdot)$ such that, for every ζ in a neighborhood of ζ_0,

$$(41)\qquad f(\zeta)-f(\zeta_0)=D_\zeta f(\zeta_0)(\zeta-\zeta_0)+r(f;\zeta_0,\zeta)(\zeta-\zeta_0),$$
$$\lim_{\zeta\to\zeta_0} r(f;\zeta_0,\zeta)=0,\qquad r(f;\zeta_0,\zeta_0)=0.$$

Then by Definition 20.14, f satisfies the strong Stolz condition in X, and the proof of Theorem 21.2 is complete. □

Exercises

21.1 Let $f:\mathbb{C}_2\to\mathbb{C}_2$ be the function such that $f(\zeta)=\zeta^2$.
- (a) If $f(\zeta)=u(\zeta)+i_2v(\zeta)$, show that $u(\zeta)=z_1^2-z_2^2$ and $v(\zeta)=2z_1z_2$.
- (b) Show that $u(\zeta)-i_1v(\zeta)=(z_1-i_1z_2)^2$ and thus show that $u-i_1v$ is a differentiable function of the complex variable $z_1-i_1z_2$.
- (c) Show that $u(\zeta)+i_1v(\zeta)=(z_1+i_1z_2)^2$ and thus show that $u+i_1v$ is a differentiable function of the complex variable $z_1+i_1z_2$.

21.2 Let $f:\mathbb{C}_2\to\mathbb{C}_2$ be the function such that $f(\zeta)=\zeta^3$.
- (a) Find the functions $u:\mathbb{C}_2\to\mathbb{C}_1$ and $v:\mathbb{C}_2\to\mathbb{C}_1$.
- (b) Find analytic expressions for $u-i_1v$ and $u+i_1v$ and thus verify that $u-i_1v:X_1\to\mathbb{C}_1$ is a differentiable function of $z_1-i_1z_2$, and that $u+i_1v$ is a differentiable function of $z_1+i_1z_2$.

21.3 Show that the class of functions $f:X\to\mathbb{C}_2$ which satisfy the strong Stolz condition in X is exactly the class of functions which are differentiable in X.

21.4 Let $f:X\to\mathbb{C}_2$ and $g:X\to\mathbb{C}_2$ be two functions which are differentiable in X. Prove that the following functions are differentiable in X and find their derivatives:
- (a) $f+g$;
- (b) cf, where c is a constant in $\mathbb{C}_2$;
- (c) fg.

22. THE WEAK STOLZ CONDITION IMPLIES DIFFERENTIABILITY

The purpose of this section is to prove that a function $f:X\to\mathbb{C}_2$ is differentiable in X if it satisfies the weak Stolz condition in X (see Definition 20.14). This result completes the proof that differentiability and the strong and weak Stolz conditions are equivalent conditions (see Theorems 20.15 and 21.2).

For functions of real variables and complex variables, the proof that the weak Stolz conditions implies differentiability has been simple and direct (see Theorems 20.3, 20.6, 20.9, and 20.12). The complications which arise from the presence of the large set $\mathcal{O}_2$ of singular elements in $\mathbb{C}_2$ explain as follows why a similar proof is not possible in the present case.

Let $f: X \to \mathbb{C}_2$ be a function which satisfies the weak Stolz condition at each point ζ_0 in the domain X in $\mathbb{C}_2$. Then there exists a constant $d(\zeta_0)$ and a function $r(f; \zeta_0, \cdot)$ such that

$$(1)\qquad f(\zeta) - f(\zeta_0) = d(\zeta_0)(\zeta - \zeta_0) + r(f; \zeta_0, \zeta)\|\zeta - \zeta_0\|,$$

$$(2)\qquad \lim_{\zeta \to \zeta_0} r(f; \zeta_0, \zeta) = 0, \qquad r(f; \zeta_0, \zeta_0) = 0.$$

Then

$$(3)\qquad \frac{f(\zeta) - f(\zeta_0)}{\zeta - \zeta_0} = d(\zeta_0) + r(f; \zeta_0, \zeta)\frac{\|\zeta - \zeta_0\|}{\zeta - \zeta_0}, \qquad \zeta - \zeta_0 \notin \mathcal{O}_2.$$

Now

$$(4)\qquad \lim_{\zeta \to \zeta_0} r(f; \zeta_0, \zeta)\frac{\|\zeta - \zeta_0\|}{\zeta - \zeta_0}$$

cannot be shown to be zero if the only restriction on ζ as it approaches ζ_0 is

$$(5)\qquad \zeta - \zeta_0 \notin \mathcal{O}_2.$$

Although (2) is true, the restriction (5) is not enough to guarantee that

$$(6)\qquad \frac{\|\zeta - \zeta_0\|}{\zeta - \zeta_0}$$

remains bounded. It is necessary to examine the bounds of this fraction.

22.1 LEMMA If $\zeta - \zeta_0$ is in $\mathbb{C}_2$ but $\zeta - \zeta_0 \notin \mathcal{O}_2$, then (see Definition 4.12)

$$(7)\qquad \frac{\sqrt{2}}{2} \leqslant \left\|\frac{\|\zeta - \zeta_0\|}{\zeta - \zeta_0}\right\| = \frac{\|\zeta - \zeta_0\|^2}{[V(\zeta - \zeta_0)]^{1/2}}.$$

If $\zeta - \zeta_0$ is a complex number in $\mathbb{C}_2$, then

$$(8)\qquad \left\|\frac{\|\zeta - \zeta_0\|}{\zeta - \zeta_0}\right\| = 1.$$

Proof. If $\zeta - \zeta_0$ is a complex number in $\mathbb{C}_2$, then the norm of $\zeta - \zeta_0$ is the same as its absolute value [see Definition 3.1 and Theorem 4.4 (10)], and

$$\left\|\frac{\|\zeta - \zeta_0\|}{\zeta - \zeta_0}\right\| = |\zeta - \zeta_0|\left|\frac{1}{\zeta - \zeta_0}\right| = \frac{|\zeta - \zeta_0|}{|\zeta - \zeta_0|} = 1.$$

Thus (8) is true. To prove the inequality in (7), observe first that

$$\left\|\frac{\|\zeta-\zeta_0\|}{\zeta-\zeta_0}\right\| = \|\zeta-\zeta_0\|\left\|\frac{1}{\zeta-\zeta_0}\right\|. \tag{9}$$

Since $\zeta-\zeta_0 \notin \mathcal{O}_2$, then

$$1 = \frac{\zeta-\zeta_0}{\zeta-\zeta_0} = (\zeta-\zeta_0)\left(\frac{1}{\zeta-\zeta_0}\right).$$

Therefore

$$1 = \left\|(\zeta-\zeta_0)\left(\frac{1}{\zeta-\zeta_0}\right)\right\| \leqslant \sqrt{2}\|\zeta-\zeta_0\|\left\|\frac{1}{\zeta-\zeta_0}\right\| = \sqrt{2}\left\|\frac{\|\zeta-\zeta_0\|}{\zeta-\zeta_0}\right\|, \tag{10}$$

and the inequality in (7) follows. To prove the equality in (7), introduce the following notation:

$$\zeta = z_1 + i_2 z_2, \qquad \zeta_0 = z_1^0 + i_2 z_2^0. \tag{11}$$

By the idempotent representation of elements in $\mathbb{C}_2$,

$$\zeta-\zeta_0 = [(z_1-z_1^0) - i_1(z_2-z_2^0)]e_1 + [(z_1-z_1^0) + i_1(z_2-z_2^0)]e_2. \tag{12}$$

Since $1 = 1e_1 + 1e_2$, then

$$\frac{1}{\zeta-\zeta_0} = \frac{1}{(z_1-z_1^0)-i_1(z_2-z_2^0)}e_1 + \frac{1}{(z_1-z_1^0)+i_1(z_2-z_2^0)}e_2,$$

$$\begin{aligned}\left\|\frac{1}{\zeta-\zeta_0}\right\| &= \frac{1}{\sqrt{2}}\left[\frac{1}{|(z_1-z_1^0)-i_1(z_2-z_2^0)|^2} + \frac{1}{|(z_1-z_1^0)+i_1(z_2-z_2^0)|^2}\right]^{1/2} \\ &= \frac{\|\zeta-\zeta_0\|}{[V(\zeta-\zeta_0)]^{1/2}},\end{aligned} \tag{13}$$

because

$$\|\zeta-\zeta_0\| = [(1/2)|(z_1-z_1^0) - i_1(z_2-z_2^0)|^2 + (1/2)|(z_1-z_1^0) + i_1(z_2-z_2^0)|^2]^{1/2},$$

$$V(\zeta-\zeta_0) = |(z_1-z_1^0) - i_1(z_2-z_2^0)|^2\,|(z_1-z_1^0) + i_1(z_2-z_2^0)|^2.$$

Then

$$\left\|\frac{\|\zeta-\zeta_0\|}{\zeta-\zeta_0}\right\| = \|\zeta-\zeta_0\|\left\|\frac{1}{\zeta-\zeta_0}\right\| = \frac{\|\zeta-\zeta_0\|^2}{[V(\zeta-\zeta_0)]^{1/2}}, \tag{14}$$

and the proof of (7) and of all parts of Lemma 22.1 is complete. □

If $\zeta-\zeta_0\notin\mathcal{O}_2$, then Corollary 4.14 shows that $V(\zeta-\zeta_0)>0$; but in order to prove that the limit in (4) is zero, it is necessary to restrict $\zeta-\zeta_0$ further so that

$$\left\|\frac{\|\zeta-\zeta_0\|}{\zeta-\zeta_0}\right\| \tag{15}$$

is bounded. This result is accomplished by introducing a regularity condition.

22.2 DEFINITION Let ζ be a number in $\mathbb{C}_2$ such that $\zeta-\zeta_0\notin\mathcal{O}_2$, and let ρ be a number in $\mathbb{R}$ such that $\rho>1$. Then the increment $\zeta-\zeta_0$ is said to satisfy the *regularity condition* with *constant of regularity* ρ if and only if

$$\left\|\frac{\|\zeta-\zeta_0\|}{\zeta-\zeta_0}\right\|<\rho. \tag{16}$$

If (16) is satisfied, then (14) shows that

$$0<\|\zeta-\zeta_0\|^2<\rho[V(\zeta-\zeta_0)]^{1/2}. \tag{17}$$

If $\zeta-\zeta_0$ is any increment which satisfies this inequality, then $\zeta-\zeta_0\notin\mathcal{O}_2$ by Corollary 4.14.

The purpose of the next theorem is to prove that a function $f:X\to\mathbb{C}_2$, $f(\zeta)=u(\zeta)+i_2v(\zeta)$, which satisfies the weak Stolz condition also satisfies the strong Stolz condition; the proof is a refinement of the proof of Theorem 21.2. Let $d(\zeta_0)$ in equations (1) and (3) above be the bicomplex number $d_1(\zeta_0)+i_2d_2(\zeta_0)$. If $f:X\to\mathbb{C}_2$ satisfies the weak Stolz condition in X, then (1)–(4) show that

$$\lim_{\zeta\to\zeta_0}\frac{f(\zeta)-f(\zeta_0)}{\zeta-\zeta_0}=d_1(\zeta_0)+i_2d_2(\zeta_0) \tag{18}$$

provided $\zeta-\zeta_0$ satisfies the regularity condition in (16).

22.3 THEOREM Let $f:X\to\mathbb{C}_2$, $f(\zeta)=u(\zeta)+i_2v(\zeta)$, be a function which satisfies the weak Stolz condition in X. Then:

(19) $u-i_1v$ is a differentiable function $f_1:X_1\to\mathbb{C}_1$ of $z_1-i_1z_2$ in X_1;

(20) $u+i_1v$ is a differentiable function $f_2:X_2\to\mathbb{C}_2$ of $z_1+i_1z_2$ in X_2;

(21) $f(z_1+i_2z_2)=f_1(z_1-i_1z_2)e_1+f_2(z_1+i_1z_2)e_2,\qquad z_1+i_2z_2$ in X;

(22) $D_{z_1-i_1z_2}f_1(z_1-i_1z_2)=d_1(z_1+i_2z_2)-i_1d_2(z_1+i_2z_2)$;

(23) $D_{z_1+i_1z_2}f_2(z_1+i_1z_2)=d_1(z_1+i_2z_2)+i_1d_2(z_1+i_2z_2)$;

(24) $D_{z_1+i_2z_2}f(z_1+i_2z_2)$
$$=D_{z_1-i_1z_2}f_1(z_1-i_1z_2)e_1+D_{z_1+i_1z_2}f_2(z_1+i_1z_2)e_2;$$

(25) f satisfies the strong Stolz condition in X.

Proof. Since f satisfies the weak Stolz condition at each point ζ_0 in X,

$$\lim_{\zeta\to\zeta_0}\left\{\frac{f(\zeta)-f(\zeta_0)}{\zeta-\zeta_0}-[d_1(\zeta_0)+i_2d_2(\zeta_0)]\right\}=0 \tag{26}$$

provided ζ approaches ζ_0 so that

$$0<\left\|\frac{\|\zeta-\zeta_0\|}{\zeta-\zeta_0}\right\|<\rho. \tag{27}$$

Now

$$\zeta-\zeta_0=[(z_1-z_1^0)-i_1(z_2-z_2^0)]e_1+[(z_1-z_1^0)+i_1(z_2-z_2^0)]e_2, \tag{28}$$

and the restriction in (27) is equivalent to

$$0<\|\zeta-\zeta_0\|^2<\rho[V(\zeta-\zeta_0)]^{1/2}. \tag{29}$$

The idempotent representation of elements in $\mathbb{C}_2$ shows that the expression in the braces in (26) is equal to

$$\left\{\frac{[u(\zeta)-i_1v(\zeta)]-[u(\zeta_0)-i_1v(\zeta_0)]}{(z_1-i_1z_2)-(z_1^0-i_1z_2^0)}-[d_1(\zeta_0)-i_1d_2(\zeta_0)]\right\}e_1 \tag{30}$$

$$+\left\{\frac{[u(\zeta)+i_1v(\zeta)]-[u(\zeta_0)+i_1v(\zeta_0)]}{(z_1+i_1z_2)-(z_1^0+i_1z_2^0)}-[d_1(\zeta_0)+i_1d_2(\zeta_0)]\right\}e_2. \tag{31}$$

Let A, B, C denote the expressions in the braces in (26), (30), and (31). Then (26) and (27) show that

$$\lim_{\zeta\to\zeta_0}A=0 \text{ provided } \left\|\frac{\|\zeta-\zeta_0\|}{\zeta-\zeta_0}\right\|<\rho. \tag{32}$$

The strategy of the proof is to use (32) to show that

$$\lim_{z_1-i_1z_2\to z_1^0-i_1z_2^0}B=0, \qquad \lim_{z_1+i_1z_2\to z_1^0+i_1z_2^0}C=0. \tag{33}$$

It is important to observe that, although there is a regularity restriction in taking the limit in (32), there are no restrictions on $z_1-i_1z_2$ and $z_1+i_2z_2$ in (32) except

$$z_1-i_1z_2\neq z_1^0-i_1z_2^0, \qquad z_1+i_1z_2\neq z_1^0+i_1z_2^0. \tag{34}$$

By Theorem 6.8 (25),

$$\|A\|=\left[\frac{|B|^2+|C|^2}{2}\right]^{1/2}. \tag{35}$$

Let $\varepsilon>0$ be given. Then (32) shows that there exists a $\delta>0$ such that, if $\|\zeta-\zeta_0\|<\delta$ and if $\zeta-\zeta_0$ satisfies the regularity condition in (32), then

$$\|A\|<\varepsilon. \tag{36}$$

For this increment, (35) shows that

$$(37)\qquad |B| < \sqrt{2}\varepsilon, \qquad |C| < \sqrt{2}\varepsilon.$$

Now (37) is a beginning toward establishing the limits in (33), but more is required. It is necessary to show that the inequalities in (37) hold for all $z_1 - i_1 z_2$ and $z_1 + i_1 z_2$ such that

$$(38)\qquad 0 < |(z_1 - i_1 z_2) - (z_1^0 - i_1 z_2^0)| < \delta, \qquad 0 < |(z_1 + i_1 z_2) - (z_1^0 + i_1 z_2^0)| < \delta.$$

Let $z_1 - i_1 z_2$ be an arbitrary point in X_1 which satisfies the first inequality in (38). Choose a point $z_1 + i_1 z_2$ in X_2 which satisfies the second inequality in (38) and also

$$(39)\qquad |(z_1 - i_1 z_2) - (z_1^0 - i_1 z_2^0)| = |(z_1 + i_1 z_2) - (z_1^0 + i_1 z_2^0)|.$$

Then these points determine a point $\zeta = z_1 + i_2 z_2$ such that

$$(40)\qquad z_1 + i_2 z_2 = (z_1 - i_1 z_2) e_1 + (z_1 + i_1 z_2) e_2.$$

They determine also an increment $\zeta - \zeta_0$ such that

$$(41)\qquad \|\zeta - \zeta_0\| = [(1/2)|(z_1 - i_1 z_2) - (z_1^0 - i_1 z_2^0)|^2 + (1/2)|(z_1 + i_1 z_2) - (z_1^0 + i_1 z_2^0)|^2]^{1/2} < \delta,$$

$$(42)\qquad \left\| \frac{\|\zeta - \zeta_0\|}{\zeta - \zeta_0} \right\| = \frac{\|\zeta - \zeta_0\|^2}{[V(\zeta - \zeta_0)]^{1/2}} = \frac{|(z_1 - i_1 z_2) - (z_1^0 - i_1 z_2^0)|^2}{|(z_1 - i_1 z_2) - (z_1^0 - i_1 z_2^0)|^2} = 1.$$

Thus (41) shows that $\|\zeta - \zeta_0\| < \delta$, and (42) shows that $\zeta - \zeta_0$ satisfies the regularity condition in (32). Thus ζ is a point for which (36) holds, and therefore the argument already given shows that $|B| < \sqrt{2}\varepsilon$ for the point $z_1 - i_1 z_2$, and that $|C| < \sqrt{2}\varepsilon$ for the point $z_1 + i_1 z_2$. In this discussion, $z_1 - i_1 z_2$ was an arbitrary point which satisfies the first inequality in (38), and $z_1 + i_1 z_2$ was a conveniently chosen auxiliary point. The argument has established the first limit in (33). To establish the second limit in (33), begin by choosing $z_1 + i_1 z_2$ in X_2 as an arbitrary point in X_2 which satisfies the second inequality in (38). Then choose $z_1 - i_1 z_2$ as a point in X_1 which satisfies (39), and proceed as before. Thus the proof is complete that the limits exist as stated in (33). These limits exist at each point $z_1^0 - i_1 z_2^0$ in X_1 and $z_1^0 + i_1 z_2^0$ in X_2, respectively. Thus $u - i_1 v$ is a complex valued function which has a derivative with respect to the complex variable $z_1 - i_1 z_2$ in X_1; it is therefore a differentiable function $f_1 : X_1 \to \mathbb{C}_1$. Similarly, $u + i_1 v$ is a differentiable function $f_2 : X_2 \to \mathbb{C}_1$ of the complex variable $z_1 + i_1 z_2$ in X_2. Thus the statements in (19) and (20) have been established. Also,

$$(43)\qquad f_1 = u - i_1 v, \qquad f_2 = u + i_1 v,$$

$$f = u + i_2 v = (u - i_1 v) e_1 + (u + i_1 v) e_2 = f_1 e_1 + f_2 e_2,$$

and the statement in (21) holds. The statements in (22) and (23) follow from (30), (31), and (33). Now (19), (20), and (21) show that there are differentiable functions $f_1: X_1 \to \mathbb{C}_1$ and $f_2: X_2 \to \mathbb{C}_1$ such that

$$(44)\qquad f(z_1 + i_2 z_2) = f_1(z_1 - i_1 z_2)e_1 + f_2(z_1 + i_1 z_2)e_2,\quad z_1 + i_2 z_2 \text{ in } X.$$

Then the proof of Theorem 21.2 has shown that a function which satisfies (44) also satisfies the strong Stolz condition in X, and (25) has been established. Next, since f satisfies the strong Stolz condition in X, it has a derivative $D_{z_1+i_2z_2}f$ in X by Theorem 20.15, and (26) shows that

$$(45)\qquad \begin{aligned} D_{z_1+i_2z_2}f(z_1 + i_2 z_2) &= d_1(z_1 + i_2 z_2) + i_2 d_2(z_1 + i_2 z_2) \\ &= [d_1(z_1 + i_2 z_2) - i_1 d_2(z_1 + i_2 z_2)]e_1 \\ &\quad + [d_1(z_1 + i_2 z_2) + i_1 d_2(z_1 + i_2 z_2)]e_2. \end{aligned}$$

Then (24) follows from (45), (22), and (23). Thus the proof of all parts of Theorem 22.3 is complete. □

The conclusion in the next theorem is contained in Theorem 22.3, but it is somewhat hidden there. The result is made explicit in the next theorem for ease of use later.

22.4 THEOREM Let $f: X \to \mathbb{C}_2$ satisfy the following weak Stolz condition at each ζ_0 in X:

$$(46)\qquad f(\zeta) - f(\zeta_0) = d(\zeta_0)(\zeta - \zeta_0) + r(f; \zeta_0, \zeta)\|\zeta - \zeta_0\|,$$

$$(47)\qquad \lim_{\zeta \to \zeta_0} r(f; \zeta_0, \zeta) = 0, \qquad r(f; \zeta_0, \zeta_0) = 0.$$

Then f has a derivative $D_\zeta f(\zeta_0)$ at each ζ_0 in X, and

$$(48)\qquad D_\zeta f(\zeta_0) = d(\zeta_0).$$

Proof. Since f satisfies (46) and (47), then, for a fixed $\rho > 1$,

$$(49)\qquad \lim_{\zeta \to \zeta_0} \frac{f(\zeta) - f(\zeta_0)}{\zeta - \zeta_0} = d(\zeta_0) + \lim_{\zeta \to \zeta_0} r(f; \zeta_0, \zeta)\frac{\|\zeta - \zeta_0\|}{\zeta - \zeta_0},$$

$$(50)\qquad \lim_{\zeta \to \zeta_0} \frac{f(\zeta) - f(\zeta_0)}{\zeta - \zeta_0} = d(\zeta_0) \text{ provided } \left\|\frac{\|\zeta - \zeta_0\|}{\zeta - \zeta_0}\right\| < \rho.$$

The proof of Theorem 22.3 has shown that

$$(51)\qquad \lim_{\zeta \to \zeta_0} \frac{f(\zeta) - f(\zeta_0)}{\zeta - \zeta_0}, \qquad \left\|\frac{\|\zeta - \zeta_0\|}{\zeta - \zeta_0}\right\| < \rho$$

$$= \left\{\lim_{z_1 - i_1 z_2 \to z_1^0 - i_1 z_2^0} A\right\} e_1 + \left\{\lim_{z_1 + i_1 z_2 \to z_1^0 + i_1 z_2^0} B\right\} e_2$$

$$(52) \qquad = \lim_{\zeta \to \zeta_0} \frac{f(\zeta) - f(\zeta_0)}{\zeta - \zeta_0}, \qquad (\zeta - \zeta_0) \notin \mathcal{O}_2.$$

Now the limit in (51) is $d(\zeta_0)$ by (50), and the limit in (52) is $D_\zeta f(\zeta_0)$ by Definition 20.13. Then (51) and (52) show that $D_\zeta f(\zeta_0)$ exists, and that (48) is true. □

Exercises

22.1 Let $f: X \to \mathbb{C}_2$, $f(z_1 + i_2 z_2) = u(z_1 + i_2 z_2) + i_2 v(z_1 + i_2 z_2)$, be a function which is differentiable in the domain X in $\mathbb{C}_2$.

(a) Find functions $f_1: X_1 \to \mathbb{C}_1$ and $f_2: X_2 \to \mathbb{C}_1$, differentiable in X_1 and X_2, respectively, such that for all $z_1 + i_2 z_2$ in X,

$$f(z_1 + i_2 z_2) = f_1(z_1 - i_1 z_2)e_1 + f_2(z_1 + i_1 z_2)e_2.$$

(b) If h_1 and h_2 are the functions defined in (2) in Section 8, show that

$$h_1[f(z_1 + i_2 z_2)] = f_1[h_1(z_1 + i_2 z_2)],$$
$$h_2[f(z_1 + i_2 z_2)] = f_2[h_2(z_1 + i_2 z_2)].$$

22.2 (a) Construct a counter example to show that the following statement is false: Every function $g: X \to \mathbb{C}_2$, $X \subset \mathbb{C}_2$, satisfies an equation of the form

$$g(z_1 + i_2 z_2) = g_1(z_1 - i_1 z_2)e_1 + g_2(z_1 + i_1 z_2)e_2.$$

(b) If X is a domain in $\mathbb{C}_2$, prove that each of the following is a sufficient condition that $g: X \to \mathbb{C}_2$ satisfy an equation of the form stated in (a): (i) g is differentiable in X; (ii) g satisfies the strong Stolz condition in X; and (iii) g satisfies the weak Stolz condition in X.

22.3 Let $g_1: X_1 \to \mathbb{C}_1$ and $g_2: X_2 \to \mathbb{C}_1$ be functions defined on domains X_1 and X_2 in $\mathbb{C}_1$.

(a) Show that the following equation defines a function $g: X \to \mathbb{C}_2$:

$$g(z_1 + i_2 z_2) = g_1(z_1 - i_1 z_2)e_1 + g_2(z_1 + i_1 z_2)e_2,$$
$$z_1 + i_2 z_2 \text{ in } X.$$

(b) Show that $g: X \to \mathbb{C}_2$ is differentiable if and only if each of the functions g_1 and g_2 satisfies one of the following conditions: differentiable, strong Stolz condition, weak Stolz condition.

23. DIFFERENTIABILITY: NECESSARY CONDITIONS

In this section we shall assume that $f: X \to \mathbb{C}_2$ is differentiable in X and then derive several properties of f. Since differentiability is equivalent to the strong Stolz condition (see Theorems 20.15 and 21.2), the working hypothesis is that

$f:X\to\mathbb{C}_2$, $X\subset\mathbb{C}_2$, is a function which satisfies the following strong Stolz condition: for each ζ_0 in X, there is a derivative $D_\zeta f(\zeta_0)$ and a function $r(f;\zeta_0,\cdot)$ such that

$$(1)\qquad f(\zeta)-f(\zeta_0)=D_\zeta f(\zeta_0)(\zeta-\zeta_0)+r(f;\zeta_0,\zeta)(\zeta-\zeta_0),$$
$$\lim_{\zeta\to\zeta_0} r(f;\zeta_0,\zeta)=0,\qquad r(f;\zeta_0,\zeta_0)=0.$$

Let x denote the point $(x_1,\ldots,x_4)$ in $\mathbb{C}_0^4$, and let ζ denote the corresponding point $x_1+i_1x_2+i_2x_3+i_1i_2x_4$ in $\mathbb{C}_2$. Let X denote a domain in $\mathbb{C}_2$ and also the corresponding domain in $\mathbb{C}_0^4$. Since $f(\zeta)\in\mathbb{C}_2$, there are functions $g_k:X\to\mathbb{C}_0$, $k=1,\ldots,4$, such that

$$(2)\qquad f(\zeta)=g_1(x)+i_1g_2(x)+i_2g_3(x)+i_1i_2g_4(x),\qquad \zeta\in X.$$

Also, since $D_\zeta f(\zeta)\in\mathbb{C}_2$, there are functions $d_k:X\to\mathbb{C}_0$, $k=1,\ldots,4$, such that

$$(3)\qquad D_\zeta f(\zeta)=d_1(x)+i_1d_2(x)+i_2d_3(x)+i_1i_2d_4(x),\qquad \zeta\in X.$$

Finally, since $r(f;\zeta_0,\zeta)\in\mathbb{C}_2$, there are real-valued functions $r_k(f;x^0,\cdot)$, $k=1,\ldots,4$, defined in a neighborhood of x^0, such that

$$(4)\qquad r(f;\zeta_0,\zeta)=r_1(f;x^0,x)+i_1r_2(f;x^0,x)+i_2r_3(f;x^0,x)$$
$$+\,i_1i_2r_4(f;x^0,x),$$
$$\lim_{x\to x^0} r_k(f;x^0,x)=0,\qquad r_k(f;x^0,x^0)=0,\qquad k=1,\ldots,4.$$

23.1 THEOREM If $f:X\to\mathbb{C}_2$ satisfies the strong Stolz condition in (1), then the functions $g_1,\ldots,g_4$ satisfy the following strong Stolz conditions:

$$(5)\qquad g_1(x)-g_1(x^0)=d_1(x^0)(x_1-x_1^0)-d_2(x^0)(x_2-x_2^0)$$
$$-\,d_3(x^0)(x_3-x_3^0)+d_4(x^0)(x_4-x_4^0)+R_1,$$
$$g_2(x)-g_2(x^0)=d_2(x^0)(x_1-x_1^0)+d_1(x^0)(x_2-x_2^0)$$
$$-\,d_4(x^0)(x_3-x_3^0)-d_3(x^0)(x_4-x_4^0)+R_2,$$
$$g_3(x)-g_3(x^0)=d_3(x^0)(x_1-x_1^0)-d_4(x^0)(x_2-x_2^0)$$
$$+\,d_1(x^0)(x_3-x_3^0)-d_2(x^0)(x_4-x_4^0)+R_3,$$
$$g_4(x)-g_4(x^0)=d_4(x^0)(x_1-x_1^0)+d_3(x^0)(x_2-x_2^0)$$
$$+\,d_2(x^0)(x_3-x_3^0)+d_1(x^0)(x_4-x_4^0)+R_4,$$

where

$$(6)\qquad R_1=r_1(x_1-x_1^0)-r_2(x_2-x_2^0)-r_3(x_3-x_3^0)+r_4(x_4-x_4^0),$$
$$R_2=r_2(x_1-x_1^0)+r_1(x_2-x_2^0)-r_4(x_3-x_3^0)-r_3(x_4-x_4^0),$$
$$R_3=r_3(x_1-x_1^0)-r_4(x_2-x_2^0)+r_1(x_3-x_3^0)-r_2(x_4-x_4^0),$$
$$R_4=r_4(x_1-x_1^0)+r_3(x_2-x_2^0)+r_2(x_3-x_3^0)+r_1(x_4-x_4^0).$$

The functions $g_1,\ldots,g_4$ satisfy the strong Stolz condition in X, they are differentiable in X, and their derivatives satisfy the following Cauchy–Riemann differential equations at each point x in X:

(7) $D_{x_1}g_1(x) = D_{x_2}g_2(x) = D_{x_3}g_3(x) = D_{x_4}g_4(x),$

(8) $D_{x_1}g_2(x) = -D_{x_2}g_1(x) = D_{x_3}g_4(x) = -D_{x_4}g_3(x),$

(9) $D_{x_1}g_3(x) = D_{x_2}g_4(x) = -D_{x_3}g_1(x) = -D_{x_4}g_2(x),$

(10) $D_{x_1}g_4(x) = -D_{x_2}g_3(x) = -D_{x_3}g_2(x) = D_{x_4}g_1(x).$

Proof. The method to be used in proving this theorem is the same as that suggested in Exercise 20.2 for a similar theorem for functions of a complex variable. To prove (5) and (6), proceed as follows: substitute from (2), (3), and (4) in the first equation in (1), carry out the indicated multiplications and additions, and then equate the coefficients of 1, i_1, i_2, i_1i_2 on the two sides of the resulting equation. Equation (5) shows that $g_1,\ldots,g_4$ satisfy the strong Stolz condition for functions of several real variables as stated in Definition 20.5, and Theorem 20.6 shows that these functions are differentiable in X. The derivatives $D_{x_1}g_k(x^0),\ldots,$ $D_{x_4}g_k(x^0)$ are the coefficients of $(x_1-x_1^0),\ldots,$ $(x_4-x_4^0)$, respectively in the Stolz condition which g_k satisfies. Thus (5) shows that the matrix of derivatives of $g_1,\ldots,g_4$ is the following:

(11)
$$\begin{bmatrix} D_{x_1}g_1(x) & D_{x_2}g_1(x) & D_{x_3}g_1(x) & D_{x_4}g_1(x) \\ D_{x_1}g_2(x) & D_{x_2}g_2(x) & D_{x_3}g_2(x) & D_{x_4}g_2(x) \\ D_{x_1}g_3(x) & D_{x_2}g_3(x) & D_{x_3}g_3(x) & D_{x_4}g_3(x) \\ D_{x_1}g_4(x) & D_{x_2}g_4(x) & D_{x_3}g_4(x) & D_{x_4}g_4(x) \end{bmatrix}$$
$$= \begin{bmatrix} d_1(x) & -d_2(x) & -d_3(x) & d_4(x) \\ d_2(x) & d_1(x) & -d_4(x) & -d_3(x) \\ d_3(x) & -d_4(x) & d_1(x) & -d_2(x) \\ d_4(x) & d_3(x) & d_2(x) & d_1(x) \end{bmatrix}.$$

This matrix equation shows that each of the derivatives in (7) is equal to $d_1(x)$; the four derivatives are therefore equal. The equations in (8), (9), and (10) can be verified from (11) in the same way. Observe that $g_1,\ldots,g_4$ have derivatives [7, pp. 19–20] and not merely the weaker partial derivatives. The matrix in (11) may be called a Cauchy–Riemann matrix. The proof of Theorem 23.1 is complete. □

23.2 COROLLARY If $f: X\to\mathbb{C}_2$ is a differentiable function of ζ in X, then $g_1,\ldots,g_4$ in (2) are differentiable functions of $x_1,\ldots,x_4$ in X, and

(12) $D_\zeta f(\zeta) = D_{x_1}g_1(x) + i_1D_{x_1}g_2(x) + i_2D_{x_1}g_3(x) + i_1i_2D_{x_1}g_4(x).$

Other expressions for $D_\zeta f(\zeta)$ can be obtained from (12) and the Cauchy–Riemann equations in (7)–(10).

Proof. Equation (12) follows from (3) and (11). Each derivative of $g_1,\dots,g_4$ in (12) is equal to three other derivatives as stated in (7), …, (10). □

The functions $g_k: X\to\mathbb{C}_0$, $k=1,\dots,4$, can be used to construct the complex-valued functions $g_1+i_1g_2$ and $g_3+i_1g_4$. The next theorem establishes properties of these functions.

23.3 THEOREM If $f: X\to\mathbb{C}_2$ is differentiable in X, then the functions $g_1+i_1g_2$ and $g_3+i_1g_4$ satisfy the following strong Stolz conditions at each x^0 in X:

$$
\begin{aligned}
(13)\quad & [g_1(x)+i_1g_2(x)]-[g_1(x^0)+i_1g_2(x^0)]\\
&=[d_1(x^0)+i_1d_2(x^0)][(x_1+i_1x_2)-(x_1^0+i_1x_2^0)]\\
&\quad-[d_3(x^0)+i_1d_4(x^0)][(x_3+i_1x_4)-(x_3^0+i_1x_4^0)]\\
&\quad+[r_1(f;x^0,x)+i_1r_2(f;x^0,x)][(x_1+i_1x_2)-(x_1^0+i_1x_2^0)]\\
&\quad-[r_3(f;x^0,x)+i_1r_4(f;x^0,x)][(x_3+i_1x_4)-(x_3^0+i_1x_4^0)];
\end{aligned}
$$

$$
\begin{aligned}
(14)\quad & [g_3(x)+i_1g_4(x)]-[g_3(x^0)+i_1g_4(x^0)]\\
&=[d_3(x^0)+i_1d_4(x^0)][(x_1+i_1x_2)-(x_1^0+i_1x_2^0)]\\
&\quad+[d_1(x^0)+i_1d_2(x^0)][(x_3+i_1x_4)-(x_3^0+i_1x_4^0)]\\
&\quad+[r_3(f;x^0,x)+i_1r_4(f;x^0,x)][(x_1+i_1x_2)-(x_1^0+i_1x_2^0)]\\
&\quad+[r_1(f;x^0,x)+i_1r_2(f;x^0,x)][(x_3+i_1x_4)-(x_3^0+i_1x_4^0)].
\end{aligned}
$$

Furthermore, the functions $g_1+i_1g_2$ and $g_3+i_1g_4$ are differentiable functions of the complex variables $x_1+i_1x_2$ and $x_3+i_1x_4$ in X, and

$$(15)\quad D_{x_1+i_1x_2}(g_1+i_1g_2)(x)=D_{x_1}g_1(x)+i_1D_{x_1}g_2(x),$$

$$(16)\quad D_{x_3+i_1x_4}(g_1+i_1g_2)(x)=D_{x_3}g_1(x)+i_1D_{x_3}g_2(x),$$

$$(17)\quad D_{x_1+i_1x_2}(g_3+i_1g_4)(x)=D_{x_1}g_3(x)+i_1D_{x_1}g_4(x),$$

$$(18)\quad D_{x_3+i_1x_4}(g_3+i_1g_4)(x)=D_{x_3}g_3(x)+i_1D_{x_3}g_4(x).$$

Proof. Expressions for $g_k(x)-g_k(x^0)$, $k=1,\dots,4$, can be obtained from (5) and (6); these equations are the strong Stolz conditions for g_k at x^0. These four equations can be used to find expressions for $[g_1(x)+i_1g_2(x)]-[g_1(x^0)+i_1g_2(x^0)]$ and $[g_3(x)+i_1g_4(x)]-[g_3(x^0)+i_1g_4(x^0)]$; the results are shown in (13) and (14). These equations show that $g_1+i_1g_2$ and $g_3+i_1g_4$ satisfy the strong Stolz condition in the variables $x_1+i_1x_2$ and $x_3+i_1x_4$ (see Definition 20.11); therefore, by Theorem 20.12, $g_1+i_1g_2$ and $g_3+i_1g_4$ are differentiable functions of these variables. The Stolz conditions in (13) and (14) show that, at each x^0 in X,

$$D_{x_1+i_1x_2}(g_1 + i_1g_2)(x^0) = d_1(x^0) + i_1d_2(x^0), \tag{19}$$

$$D_{x_3+i_1x_4}(g_1 + i_1g_2)(x^0) = -[d_3(x^0) + i_1d_4(x^0)], \tag{20}$$

$$D_{x_1+i_1x_2}(g_3 + i_1g_4)(x^0) = d_3(x^0) + i_1d_4(x^0), \tag{21}$$

$$D_{x_3+i_1x_4}(g_3 + i_1g_4)(x^0) = d_1(x^0) + i_1d_2(x^0). \tag{22}$$

The values for the derivatives in (15)–(18) follow from these equations and (11). As shown by (11), there are many ways to express the right sides of (19)–(22) as derivatives; the choices made in (15)–(18) express the derivatives of each of the functions $g_1+i_1g_2$ and $g_3+i_1g_4$ in terms of derivatives of the function itself. The proof of Theorem 23.3 is complete. □

As usual, denote $x_1+i_1x_2$ by z_1 and $x_3+i_1x_4$ by z_2; then $\zeta=z_1+i_2z_2$. Then Theorem 23.3 shows that $g_1+i_1g_2$ and $g_3+i_1g_4$ are differentiable functions of z_1 and z_2. Set

$$g_1(x) + i_1g_2(x) = u(z_1,z_2), \qquad g_3(x) + i_1g_4(x) = v(z_1,z_2). \tag{23}$$

23.4 COROLLARY If $f:X\to\mathbb{C}_2$, $X\subset\mathbb{C}_2$, is the differentiable function of ζ in (2), then there exist functions $u:X\to\mathbb{C}_1$ and $v:X\to\mathbb{C}_1$ such that

$$f(\zeta) = u(z_1, z_2) + i_2v(z_1, z_2), \qquad \zeta \text{ in } X. \tag{24}$$

Also, u and v are differentiable functions with derivatives

$$D_{z_1}u(z_1^0, z_2^0) = D_{x_1}g_1(x^0) + i_1D_{x_1}g_2(x^0), \tag{25}$$

$$D_{z_2}u(z_1^0, z_2^0) = D_{x_3}g_1(x^0) + i_1D_{x_3}g_2(x^0), \tag{26}$$

$$D_{z_1}v(z_1^0, z_2^0) = D_{x_1}g_3(x^0) + i_1D_{x_1}g_4(x^0), \tag{27}$$

$$D_{z_2}v(z_1^0, z_2^0) = D_{x_3}g_3(x^0) + i_1D_{x_3}g_4(x^0). \tag{28}$$

Finally, the derivatives of u and v satisfy the following Cauchy–Riemann differential equations:

$$D_{z_1}u(z_1, z_2) = D_{z_2}v(z_1, z_2), \qquad D_{z_2}u(z_1, z_2) = -D_{z_1}v(z_1, z_2). \tag{29}$$

Proof. Equation (24) follows from (2) and (23). Equations (25)–(28) follow from (15)–(18) and (23). Since $D_{x_1}g_1(x)=D_{x_3}g_3(x)$ and $D_{x_1}g_2(x)=D_{x_3}g_4(x)$ by (7) and (8), the first equation in (29) follows from (15)–(18); in the same way, the second equation in (29) follows from (9), (10), and (16), (17). □

If $f:X\to\mathbb{C}_2$ is differentiable in X, then $g_1,\ldots,g_4$ satisfy the Cauchy–Riemann equations in (7)–(10), and u, v satisfy the Cauchy–Riemann equations in (29). If $g_1,\ldots,g_4$ have two continuous derivatives, then these equations imply many other relations, some of which are stated in the next two theorems. These theorems are meaningful because later developments show that $g_1, \ldots, g_4$ and also u and v have derivatives of all orders if f is differentiable in X.

23.5 THEOREM Let $f: X \to \mathbb{C}_2$ be a function which is differentiable in X. If the functions $g_k: X \to \mathbb{C}_0$, $k = 1, \ldots, 4$, have two continuous derivatives, then the following equations are satisfied at each x in X and for $k = 1, \ldots, 4$:

$$(30) \qquad D^2_{x_1} g_k + D^2_{x_2} g_k = 0,$$

$$(31) \qquad D^2_{x_3} g_k + D^2_{x_4} g_k = 0,$$

$$(32) \qquad D^2_{x_1} g_k + D^2_{x_3} g_k = 0,$$

$$(33) \qquad D^2_{x_2} g_k + D^2_{x_4} g_k = 0,$$

$$(34) \qquad D^2_{x_1} g_k - D^2_{x_4} g_k = 0,$$

$$(35) \qquad D^2_{x_2} g_k - D^2_{x_3} g_k = 0,$$

$$(36) \qquad D_{x_1} D_{x_4} g_k - D_{x_2} D_{x_3} g_k = 0.$$

Proof. If g_k has two continuous derivatives, then it is well known that

$$(37) \qquad D_{x_r} D_{x_s} g_k = D_{x_s} D_{x_r} g_k, \qquad r, s = 1, \ldots, 4.$$

All of the equations in (30)–(36) follow from the relations in (7)–(10) and (37). For example, to prove (30) for $k = 1$, differentiate (7) and (8) with respect to x_1 and x_2, respectively to obtain

$$(38) \qquad D^2_{x_1} g_1 = D_{x_1} D_{x_2} g_2 = D_{x_1} D_{x_3} g_3 = D_{x_1} D_{x_4} g_4,$$

$$(39) \qquad D_{x_2} D_{x_1} g_2 = -D^2_{x_2} g_1 = D_{x_2} D_{x_3} g_4 = -D_{x_2} D_{x_4} g_3.$$

Since $D_{x_2} D_{x_1} g_2 = D_{x_1} D_{x_2} g_2$ by (37), these equations show that $D^2_{x_1} g_1 = -D^2_{x_2} g_1$, and the first equation in (30) follows. Differentiate (9) and (10) with respect to x_3 and x_4, respectively, to obtain

$$(40) \qquad D_{x_3} D_{x_1} g_3 = D_{x_3} D_{x_2} g_4 = -D^2_{x_3} g_1 = -D_{x_3} D_{x_4} g_2,$$

$$(41) \qquad D_{x_4} D_{x_1} g_4 = -D_{x_4} D_{x_2} g_3 = -D_{x_4} D_{x_3} g_2 = D^2_{x_4} g_1.$$

These equations show that $D^2_{x_3} g_1 + D^2_{x_4} g_1 = 0$ and (31) is true for $k = 1$. All of the other equations (30)–(35) can be proved in the same way. Equations (38)–(41) show that

$$(42) \qquad D^2_{x_1} g_1 + D^2_{x_2} g_1 + D^2_{x_3} g_1 + D^2_{x_4} g_1$$
$$= D_{x_1} D_{x_4} g_4 - D_{x_2} D_{x_3} g_4 - D_{x_3} D_{x_2} g_4 + D_{x_4} D_{x_1} g_4.$$

Then (30), (31), and (37) show that (36) is true for $k = 4$. Similar proofs can be used to establish (36) for $k = 1, 2, 3$. The proof of Theorem 23.5 is complete.

□

23.6 THEOREM Let $f: X \to \mathbb{C}_2$ be a function which is differentiable in X. If the functions $g_k: X \to \mathbb{C}_0$, $k=1,\ldots,4$, have two continuous derivatives, then the following equations are satisfied at each (z_1, z_2) in X:

$$D_{z_1}^2 u + D_{z_2}^2 u = 0, \tag{43}$$

$$D_{z_1}^2 v + D_{z_2}^2 v = 0, \tag{44}$$

$$D_{z_2} D_{z_1} u = D_{z_1} D_{z_2} u, \tag{45}$$

$$D_{z_2} D_{z_1} v = D_{z_1} D_{z_2} v. \tag{46}$$

Proof. There are two steps in the proof of (43)–(46); first, show that $D_{z_1}u$, $D_{z_2}u$, $D_{z_1}v$, $D_{z_2}v$ have derivatives with respect to z_1 and z_2; second, show that these derivatives satisfy (43)–(46). Now the four first derivatives of u and v are given by the formulas in (25)–(28). To show that these derivatives are differentiable, we shall show that they satisfy the weak Stolz condition. Thus

$$D_{z_1} u(z_1, z_2) = D_{x_1} g_1(x) + i_1 D_{x_1} g_2(x), \tag{47}$$

$$\begin{aligned} &D_{z_1} u(z_1, z_2) - D_{z_1} u(z_1^0, z_2^0) \\ &\quad = [D_{x_1} g_1(x) - D_{x_1} g_1(x^0)] + i_1 [D_{x_1} g_2(x) - D_{x_1} g_2(x^0)]. \end{aligned} \tag{48}$$

Apply the mean-value theorem [7, pp. 70–71] to obtain

$$\begin{aligned} &D_{x_1} g_1(x) - D_{x_1} g_1(x^0) \\ &\quad = D_{x_1} D_{x_1} g_1(\xi_1)(x_1 - x_1^0) + D_{x_2} D_{x_1} g_1(\xi_1)(x_2 - x_2^0) \\ &\qquad + D_{x_3} D_{x_1} g_1(\xi_1)(x_3 - x_3^0) + D_{x_4} D_{x_1} g_1(\xi_1)(x_4 - x_4^0), \end{aligned} \tag{49}$$

$$\begin{aligned} &D_{x_1} g_2(x) - D_{x_1} g_2(x^0) \\ &\quad = D_{x_1} D_{x_1} g_2(\xi_2)(x_1 - x_1^0) + D_{x_2} D_{x_1} g_2(\xi_2)(x_2 - x_2^0) \\ &\qquad + D_{x_3} D_{x_1} g_2(\xi_2)(x_3 - x_3^0) + D_{x_4} D_{x_1} g_2(\xi_2)(x_4 - x_4^0). \end{aligned} \tag{50}$$

Since g_1 and g_2 have two continuous derivatives, it is easy to deduce from (49) and (50) that $D_{x_1} g_1$ and $D_{x_1} g_2$ satisfy the weak Stolz condition. Then (47), (49), (50), the Cauchy–Riemann equations in (7)–(10), and the formula in (37) show that $D_{z_1} u$ satisfies the following weak Stolz condition:

$$\begin{aligned} &D_{z_1} u(z_1, z_2) - D_{z_1} u(z_1^0, z_2^0) \\ &\quad = [D_{x_1}^2 g_1(x^0) + i_1 D_{x_1}^2 g_2(x^0)][(x_1 - x_1^0) + i_1(x_2 - x_2^0)] \\ &\qquad + [D_{x_1} D_{x_3} g_1(x^0) + i_1 D_{x_1} D_{x_3} g_2(x^0)][(x_3 - x_3^0) + i_1(x_4 - x_4^0)] \\ &\qquad + \text{remainder terms.} \end{aligned} \tag{51}$$

Next, by (26)

$$D_{z_2} u(z_1, z_2) = D_{x_3} g_1(x) + i_1 D_{x_3} g_2(x), \tag{52}$$

and similar calculations show that

$$(53)\quad \begin{aligned} &D_{z_2}u(z_1, z_2) - D_{z_2}u(z_1^0, z_2^0) \\ &\quad = [D_{x_1}D_{x_3}g_1(x^0) + i_1 D_{x_1}D_{x_3}g_2(x^0)][(x_1 - x_1^0) + i_1(x_2 - x_2^0)] \\ &\quad + [D_{x_3}^2 g_1(x^0) + i_1 D_{x_3}^2 g_2(x^0)][(x_3 - x_3^0) + i_1(x_4 - x_4^0)] \\ &\quad + \text{remainder terms.} \end{aligned}$$

Next, by (27),

$$(54)\quad D_{z_1}v(z_1, z_2) = D_{x_1}g_3(x) + i_1 D_{x_1}g_4(x),$$

and once more similar calculations show that

$$(55)\quad \begin{aligned} &D_{z_1}v(z_1, z_2) - D_{z_1}v(z_1^0, z_2^0) \\ &\quad = [D_{x_1}^2 g_3(x^0) + i_1 D_{x_1}^2 g_4(x^0)][(x_1 - x_1^0) + i_1(x_2 - x_2^0)] \\ &\quad + [D_{x_1}D_{x_3}g_3(x^0) + i_1 D_{x_1}D_{x_3}g_4(x^0)][(x_3 - x_3^0) + i_1(x_4 - x_4^0)] \\ &\quad + \text{remainder terms.} \end{aligned}$$

Finally, by (28),

$$(56)\quad D_{z_2}v(z_1, z_2) = D_{x_3}g_3(x) + i_1 D_{x_3}g_4(x),$$

and the same analysis shows that

$$(57)\quad \begin{aligned} &D_{z_2}v(z_1, z_2) - D_{z_2}v(z_1^0, z_2^0) \\ &\quad = [D_{x_1}D_{x_3}g_3(x^0) + i_1 D_{x_1}D_{x_3}g_4(x^0)][(x_1 - x_1^0) + i_1(x_2 - x_2^0)] \\ &\quad + [D_{x_3}^2 g_3(x^0) + i_1 D_{x_3}^2 g_4(x^0)][(x_3 - x_3^0) + i_1(x_4 - x_4^0)] \\ &\quad + \text{remainder terms.} \end{aligned}$$

Equations (51), (53), (55), (57) show that the derivatives $D_{z_1}u$, $D_{z_2}u$, $D_{z_1}v$, $D_{z_2}v$ satisfy the weak Stolz condition. Then Theorem 20.12 states that these derivatives are differentiable with respect to z_1 and z_2 and that

$$(58)\quad D_{z_1}D_{z_1}u(z_1^0, z_2^0) = D_{x_1}^2 g_1(x^0) + i_1 D_{x_1}^2 g_2(x^0),$$

$$(59)\quad D_{z_2}D_{z_1}u(z_1^0, z_2^0) = D_{x_1}D_{x_3}g_1(x^0) + i_1 D_{x_1}D_{x_3}g_2(x^0),$$

$$(60)\quad D_{z_1}D_{z_2}u(z_1^0, z_2^0) = D_{x_1}D_{x_3}g_1(x^0) + i_1 D_{x_1}D_{x_3}g_2(x^0),$$

$$(61)\quad D_{z_2}D_{z_2}u(z_1^0, z_2^0) = D_{x_3}^2 g_1(x^0) + i_1 D_{x_3}^2 g_2(x^0),$$

$$(62)\quad D_{z_1}D_{z_1}v(z_1^0, z_2^0) = D_{x_1}^2 g_3(x^0) + i_1 D_{x_1}^2 g_4(x^0),$$

$$(63)\quad D_{z_2}D_{z_1}v(z_1^0, z_2^0) = D_{x_1}D_{x_3}g_3(x^0) + i_1 D_{x_1}D_{x_3}g_4(x^0),$$

$$(64)\quad D_{z_1}D_{z_2}v(z_1^0, z_2^0) = D_{x_1}D_{x_3}g_3(x^0) + i_1 D_{x_1}D_{x_3}g_4(x^0),$$

$$(65)\quad D_{z_2}D_{z_2}v(z_1^0, z_2^0) = D_{x_3}^2 g_3(x^0) + i_1 D_{x_3}^2 g_4(x^0).$$

Then (58), (61), and (32) show that

(66) $$D^2_{z_1}u + D^2_{z_2}u = (D^2_{x_1}g_1 + D^2_{x_3}g_1) + i_1(D^2_{x_1}g_2 + D^2_{x_3}g_2) = 0,$$

and (43) is true. Similarly, (62), (65), and (32) show that

(67) $$D^2_{z_1}v + D^2_{z_2}v = (D^2_{x_1}g_3 + D^2_{x_3}g_3) + i_1(D^2_{x_1}g_4 + D^2_{x_3}g_4) = 0,$$

and (44) is true. Finally, (45) follows from (59), (60), and (46) follows from (63), (64). The proof of all parts of Theorem 23.6 is complete. □

Exercises

23.1 The function $f:\mathbb{C}_2\to\mathbb{C}_2$ such that $f(\zeta)=\zeta^2$ has a derivative at each point ζ_0 in $\mathbb{C}_2$.

(a) If $f(\zeta)=g_1(x)+i_1g_2(x)+i_2g_3(x)+i_1i_2g_4(x)$, show that

$$g_1(x)=x_1^2-x_2^2-x_3^2+x_4^2, \qquad g_2(x)=2x_1x_2-2x_3x_4,$$
$$g_3(x)=2x_1x_3-2x_2x_4, \qquad g_4(x)=2x_1x_4+2x_2x_3.$$

(b) Show that $g_1,\dots,g_4$ satisfy the Cauchy–Riemann differential equations in (7)–(10).

(c) Show that each of the functions $g_1,\dots,g_4$ satisfies the equations in (30)–(36).

23.2 If ζ is written as $z_1+i_2z_2$, show that the function f in Exercise 23.1 is $f(z_1+i_2z_2)=(z_1+i_2z_2)^2$.

(a) Show that $u(z_1,z_2)=z_1^2-z_2^2$ and $v(z_1,z_2)=2z_1z_2$.

(b) Show that u and v satisfy the Cauchy–Riemann equations in (29).

(c) Show that u and v satisfy Laplace's equation in (43) and (44) and the equations in (45) and (46).

23.3 If $f:\mathbb{C}_2\to\mathbb{C}_2$ is the function such that $f(\zeta)=e^{z_1+i_2z_2}=e^{z_1}(\cos z_2 + i_2\sin z_2)$, show that $u(z_1,z_2)=e^{z_1}\cos z_2$ and $v(z_1,z_2)=e^{z_1}\sin z_2$. Verify that u and v satisfy the Cauchy–Riemann equations (29), Laplace's equation in (43) and (44), and the equality of the cross derivatives in (45) and (46).

23.4 Let $f:X\to\mathbb{C}_2$ be a function such that $g_k:X\to\mathbb{C}_0$, $k=1,\dots,4$, have continuous first and second derivatives with the values shown in (58)–(65). Use the Cauchy–Riemann equations in (29) and the equality of the cross derivatives in (45), (46) to give a short proof that u and v satisfy Laplace's equation as stated in (43) and (44).

23.5 Let $f:X\to\mathbb{C}_2$ and $h:X\to\mathbb{C}_2$ be two functions which satisfy the necessary conditions in (7)–(10) for the existence of the derivative. Prove that the following functions also satisfy these necessary conditions for the existence of the derivative:

(a) $f+h$;

(b) cf, where c is a constant in $\mathbb{C}_2$;

(c) fh.

23.6 Complete the proof in Theorem 23.6 that $D_{z_1}u$, $D_{z_2}u$, $D_{z_1}v$, and $D_{z_2}v$ satisfy the weak Stolz condition by showing that the remainder terms in (51), (53), (55), (57) can be given the form stated in (44) and (45) in Definition 20.11.

24. DIFFERENTIABILITY: SUFFICIENT CONDITIONS

This section establishes several theorems which state sufficient conditions for the differentiability of a function $f: X \to \mathbb{C}_2$. The chapter has shown already that differentiability and the strong and weak Stolz conditions are equivalent conditions for these functions (see Theorems 20.15, 21.2, and 22.4). In many cases the easiest way to show that a function is differentiable is to show that it satisfies the weak Stolz condition. Accordingly, this section establishes the differentiability of several classes of functions, usually by showing that they satisfy the weak Stolz condition.

Consider again the function

$$(1) \qquad f(\zeta)=g_1(x)+i_1g_2(x)+i_2g_3(x)+i_1i_2g_4(x), \qquad \zeta\in X\subset\mathbb{C}_2,$$

in (2) in Section 23 and the notation employed there.

24.1 THEOREM Let $f: X\to\mathbb{C}_2$ be the function in (1). Assume that $g_k: X\to\mathbb{C}_0$, $k=1,\ldots,4$, are differentiable and that their derivatives satisfy the following Cauchy–Riemann differential equations in X:

$$(2) \qquad D_{x_1}g_1(x) = D_{x_2}g_2(x) = D_{x_3}g_3(x) = D_{x_4}g_4(x),$$

$$(3) \qquad D_{x_1}g_2(x) = -D_{x_2}g_1(x) = D_{x_3}g_4(x) = -D_{x_4}g_3(x),$$

$$(4) \qquad D_{x_1}g_3(x) = D_{x_2}g_4(x) = -D_{x_3}g_1(x) = -D_{x_4}g_2(x),$$

$$(5) \qquad D_{x_1}g_4(x) = -D_{x_2}g_3(x) = -D_{x_3}g_2(x) = D_{x_4}g_1(x).$$

Then f satisfies the following weak Stolz condition in X:

$$(6) \qquad f(\zeta)-f(\zeta_0) = D_\zeta f(\zeta_0)(\zeta-\zeta_0) + r(f;\zeta_0,\zeta)\|\zeta-\zeta_0\|.$$

Furthermore, f is differentiable in X and

$$(7) \qquad D_\zeta f(\zeta) = D_{x_1}g_1(x) + i_1D_{x_1}g_2(x) + i_2D_{x_1}g_3(x) + i_1i_2D_{x_1}g_4(x).$$

Proof. Since $g_1, \ldots, g_4$ are differentiable by hypothesis, they satisfy the following weak Stolz condition [7, pp. 27–28]:

$$(8) \qquad g_k(x) - g_k(x^0) = \sum_{t=1}^{4} D_{x_t}g_k(x^0)(x_t - x_t^0) + r(g_k; x^0, x)\|x - x^0\|,$$

$$(9) \qquad \lim_{x \to x^0} r(g_k; x^0, x) = 0, \qquad r(g_k; x^0, x^0) = 0,\ k = 1, \ldots, 4.$$

Now by (1),

$$\begin{aligned}(10) \qquad & f(\zeta) - f(\zeta_0) \\ &= [g_1(x) - g_1(x^0)] + i_1[g_2(x) - g_2(x^0)] \\ &\quad + i_2[g_3(x) - g_3(x^0)] + i_1 i_2[g_4(x) - g_4(x^0)].\end{aligned}$$

Replace $[g_k(x) - g_k(x^0)]$, $k = 1, \ldots, 4$, by its value in (8). Then

$$\begin{aligned}(11) \quad f(\zeta) - f(\zeta_0) = {} & \sum_{t=1}^{4} D_{x_t}g_1(x^0)(x_t - x_t^0) + i_1 \sum_{t=1}^{4} D_{x_t}g_2(x^0)(x_t - x_t^0) \\ & + i_2 \sum_{t=1}^{4} D_{x_t}g_3(x^0)(x_t - x_t^0) \\ & + i_1 i_2 \sum_{t=1}^{4} D_{x_t}g_4(x^0)(x_t - x_t^0) \\ & + [r(g_1; x^0, x) + i_1 r(g_2; x^0, x) + i_2 r(g_3; x^0, x) \\ & \qquad + i_1 i_2 r(g_4; x^0, x)]\|\zeta - \zeta^0\|.\end{aligned}$$

Use the Cauchy–Riemann equations in (2)–(5) to express each derivative in (11) as a derivative with respect to x_1. Then (11) can be written in the following form:

$$\begin{aligned}(12) \quad & f(\zeta) - f(\zeta_0) \\ &= [D_{x_1}g_1(x^0) + i_1 D_{x_1}g_2(x^0) + i_2 D_{x_1}g_3(x^0) + i_1 i_2 D_{x_1}g_4(x^0)] \\ &\quad [(x_1 - x_1^0) + i_1(x_2 - x_2^0) + i_2(x_3 - x_3^0) + i_1 i_2(x_4 - x_4^0)] \\ &\quad + r(f; \zeta_0, \zeta)\|\zeta - \zeta_0\|,\end{aligned}$$

where

$$\begin{aligned}(13) \qquad r(f; \zeta_0, \zeta) = {} & r(g_1; x^0, x) + i_1 r(g_2; x^0, x) + i_2 r(g_3; x^0, x) \\ & + i_1 i_2 r(g_4; x^0, x)\end{aligned}$$

and $\|\zeta - \zeta_0\|$ is $\|x - x^0\|$. Then (9) and (13) show that

$$(14) \qquad \lim_{\zeta \to \zeta_0} r(f; \zeta_0, \zeta) = 0, \qquad r(f; \zeta_0, \zeta_0) = 0.$$

Then f satisfies the weak Stolz condition (6), f is differentiable in X by Theorem 22.4, and $D_\zeta f(\zeta)$ has the value shown in (7). The proof is complete. □

The next theorem treats functions $f: X \to \mathbb{C}_2$ such that

$$(15)\quad f(\zeta) = u(z_1, z_2) + i_2 v(z_1, z_2), \qquad \zeta = z_1 + i_2 z_2,\ \zeta \in X \subset \mathbb{C}_2.$$

Now (z_1, z_2) is a point in $\mathbb{C}_1^2$, and $\mathbb{C}_1^2$ is imbedded in $\mathbb{C}_0^4$. Also, $\mathbb{C}_2$ is imbedded in $\mathbb{C}_0^4$, and it is convenient to consider that $\mathbb{C}_1^2$ is imbedded in $\mathbb{C}_2$. Then (z_1, z_2) and (z_1^0, z_2^0) correspond to $z_1 + i_2 z_2$ and $z_1^0 + i_2 z_2^0$, which are ζ and ζ_0. The distance

$$(16)\qquad (|z_1 - z_1^0|^2 + |z_2 - z_2^0|^2)^{1/2}$$

between (z_1, z_2) and (z_1^0, z_2^0) in $\mathbb{C}_1^2$ is the same as the distance between ζ and ζ_0 in $\mathbb{C}_2$, and thus it can be denoted by $\|\zeta - \zeta_0\|$.

24.2 THEOREM Let $f: X \to \mathbb{C}_2$ be the function in (15). Assume that $u: X \to \mathbb{C}_1$ and $v: X \to \mathbb{C}_1$ are differentiable and that their derivatives satisfy the following Cauchy–Riemann differential equations in X:

$$(17)\qquad D_{z_1} u(z_1, z_2) = D_{z_2} v(z_1, z_2), \qquad D_{z_2} u(z_1, z_2) = -D_{z_1} v(z_1, z_2).$$

Then f satisfies the following weak Stolz condition in X:

$$(18)\qquad f(\zeta) - f(\zeta_0) = D_\zeta f(\zeta_0)(\zeta - \zeta_0) + r(f; \zeta_0, \zeta)\|\zeta - \zeta_0\|.$$

Furthermore, f is differentiable in X and

$$(19)\qquad D_\zeta f(\zeta) = D_{z_1} u(\zeta) + i_2 D_{z_1} v(\zeta).$$

Proof. Since u and v are differentiable by hypothesis, they satisfy the following weak Stolz conditions [7, pp. 550–552]:

$$(20)\qquad u(\zeta) - u(\zeta_0) = \sum_{t=1}^{2} D_{z_t} u(\zeta_0)(z_t - z_t^0) + r(u; \zeta_0, \zeta)\|\zeta - \zeta_0\|,$$

$$\lim_{\zeta \to \zeta_0} r(u; \zeta_0, \zeta) = 0, \qquad r(u; \zeta_0, \zeta_0) = 0;$$

$$(21)\qquad v(\zeta) - v(\zeta_0) = \sum_{t=1}^{2} D_{z_t} v(\zeta_0)(z_t - z_t^0) + r(v; \zeta_0, \zeta)\|\zeta - \zeta_0\|,$$

$$\lim_{\zeta \to \zeta_0} r(v; \zeta_0, \zeta) = 0, \qquad r(v; \zeta_0, \zeta_0) = 0.$$

Now by (15),

$$(22)\qquad f(\zeta) - f(\zeta_0) = [u(\zeta) - u(\zeta_0)] + i_2[v(\zeta) - v(\zeta_0)].$$

Substitute from (20) and (21) in (22). Then

$$(23)\quad f(\zeta) - f(\zeta_0) = \sum_{t=1}^{2} D_{z_t}u(\zeta_0)(z_t - z_t^0) + i_2 \sum_{t=1}^{2} D_{z_t}v(\zeta_0)(z_t - z_t^0) + [r(u; \zeta_0, \zeta) + i_2 r(v; \zeta_0, \zeta)]\|\zeta - \zeta_0\|.$$

Use the Cauchy–Riemann equations in (17) to express all derivatives in (23) as derivatives with respect to z_1. Then (23) can be simplified to the following:

$$(24)\quad f(\zeta) - f(\zeta_0) = [D_{z_1}u(\zeta_0) + i_2 D_{z_1}v(\zeta_0)](\zeta - \zeta_0) + [r(u; \zeta_0, \zeta) + i_2 r(v; \zeta_0, \zeta)]\|\zeta - \zeta_0\|.$$

Then f satisfies the weak Stolz condition (18) with

$$(25)\quad D_\zeta f(\zeta) = D_{z_1}u(\zeta_0) + i_2 D_{z_1}v(\zeta_0),$$

$$(26)\quad r(f; \zeta_0, \zeta) = r(u; \zeta_0, \zeta) + i_2 r(v; \zeta_0, \zeta).$$

Since f satisfies the weak Stolz condition in X, it is differentiable by Theorem 22.4, and (24) shows that $D_\zeta f(\zeta)$ has the value shown in (19). The proof is complete. □

Although numerous theorems established above contain sufficient conditions for the differentiability of functions $f: X \to \mathbb{C}_2$ (see Theorems 24.1 and 24.2 but also Theorems 20.15, 22.3, and 22.4), the great abundance of differentiable functions of a bicomplex variable is not readily apparent from them. There is another sufficient condition, hidden in the proofs of Theorems 21.1 and 22.3, which emphasizes this abundance. This sufficient condition is made explicit and proved in the next theorem.

24.3 THEOREM Let X_1 and X_2 be domains in $\mathbb{C}_1$, and let X be the domain in $\mathbb{C}_2$ generated by X_1 and X_2. Let $f_1: X_1 \to \mathbb{C}_1$, $z_1 - i_1 z_2 \mapsto f_1(z_1 - i_1 z_2)$, and $f_2: X_2 \to \mathbb{C}_1$, $z_1 + i_1 z_2 \mapsto f_2(z_1 + i_1 z_2)$, be differentiable functions of the complex variables $z_1 - i_1 z_2$ and $z_1 + i_1 z_2$. Then the function $f: X \to \mathbb{C}_2$ such that

$$(27)\quad f(z_1 + i_2 z_2) = f_1(z_1 - i_1 z_2)e_1 + f_2(z_1 + i_1 z_2)e_2, \quad z_1 + i_2 z_2 \text{ in } X,$$

is a differentiable function of the bicomplex variable $z_1 + i_2 z_2$, and

$$(28)\quad D_{z_1 + i_2 z_2} f(z_1 + i_2 z_2) = D_{z_1 - i_1 z_2} f_1(z_1 - i_1 z_2)e_1 + D_{z_1 + i_1 z_2} f_2(z_1 + i_1 z_2)e_2.$$

Proof. The theorem will be proved by showing that f satisfies the strong Stolz condition. By hypothesis, f_1 and f_2 are differentiable. Then Theorem

20.9 shows that f_1 and f_2 satisfy the following strong Stolz conditions:

$$(29)\quad f_1(z_1 - i_1 z_2) - f_1(z_1^0 - i_1 z_2^0) = D_{z_1 - i_1 z_2} f_1(z_1^0 - i_1 z_2^0)[(z_1 - i_1 z_2) - (z_1^0 - i_1 z_2^0)] + r(f_1; z_1^0 - i_1 z_2^0, z_1 - i_1 z_2)[(z_1 - i_1 z_2) - (z_1^0 - i_1 z_2^0)],$$

$$(30)\quad \lim_{z_1 - i_1 z_2 \to z_1^0 - i_1 z_2^0} r(f_1; z_1^0 - i_1 z_2^0, z_1 - i_1 z_2) = 0,$$
$$r(f_1; z_1^0 - i_1 z_2^0, z_1^0 - i_1 z_2^0) = 0.$$

$$(31)\quad f_2(z_1 + i_1 z_2) - f_2(z_1^0 + i_1 z_2^0) = D_{z_1 + i_1 z_2} f_2(z_1^0 + i_1 z_2^0)[(z_1 + i_1 z_2) - (z_1^0 + i_1 z_2^0)] + r(f_2; z_1^0 + i_1 z_2^0, z_1 + i_1 z_2)[(z_1 + i_1 z_2) - (z_1^0 + i_1 z_2^0)],$$

$$(32)\quad \lim_{z_1 + i_1 z_2 \to z_1^0 + i_1 z_2^0} r(f_2; z_1^0 + i_1 z_2^0, z_1 + i_1 z_2) = 0,$$
$$r(f_2; z_1^0 + i_1 z_2^0, z_1^0 + i_1 z_2^0) = 0.$$

Now (27) shows that

$$(33)\quad f(z + i_2 z_2) - f(z_1^0 + i_2 z_2^0) = [f_1(z_1 - i_1 z_2) - f_1(z_1^0 - i_1 z_2^0)]e_1 + [f_2(z_1 + i_1 z_2) - f_2(z_1^0 + i_1 z_2^0)]e_2.$$

Substitute from (29) and (31) in (33) and simplify as follows:

$$(34)\quad f(z_1 + i_2 z_2) - f(z_1^0 + i_2 z_2^0) = [D_{z_1 - i_1 z_2} f_1(z_1^0 - i_1 z_2^0)e_1 + D_{z_1 + i_1 z_2} f_2(z_1^0 + i_1 z_2^0)e_2][(z_1 + i_2 z_2) - (z_1^0 + i_2 z_2^0)] + [r(f_1; z_1^0 - i_1 z_2^0, z_1 - i_1 z_2)e_1 + r(f_2; z_1^0 + i_1 z_2^0, z_1 + i_1 z_2)e_2][(z_1 + i_2 z_2) - (z_1^0 + i_2 z_2^0)].$$

Let $\zeta = z_1 + i_2 z_2$ and $\zeta_0 = z_1^0 + i_2 z_2^0$, and set

$$(35)\quad D(\zeta_0) = D_{z_1 - i_1 z_2} f_1(z_1^0 - i_1 z_2^0)e_1 + D_{z_1 + i_1 z_2} f_2(z_1^0 + i_1 z_2^0)e_2,$$

$$(36)\quad r(f; \zeta_0, \zeta) = r(f_1; z_1^0 - i_1 z_2^0, z_1 - i_1 z_2)e_1 + r(f_2; z_1^0 + i_1 z_2^0, z_1 + i_1 z_2)e_2.$$

Substitute from (35) and (36) in (34) to obtain

$$(37)\quad f(\zeta) - f(\zeta_0) = D(\zeta_0)(\zeta - \zeta_0) + r(f; \zeta_0, \zeta)(\zeta - \zeta_0).$$

Now (36) shows that

$$(38)\quad \|r(f; \zeta_0, \zeta)\| = [(1/2)|r(f_1; z_1^0 - i_1 z_2^0, z_1 - i_1 z_2)|^2 + (1/2)|r(f_2; z_1^0 + i_1 z_2^0, z_1 + i_1 z_2)|^2]^{1/2}.$$

Then (30), (32), and (38) show that

(39) $$\lim_{\zeta\to\zeta_0} r(f;\zeta_0,\zeta)=0, \qquad r(f;\zeta_0,\zeta_0)=0.$$

Definition 20.14 and (37), (39) now show that f satisfies the strong Stolz condition; therefore, f is differentiable in X by Theorem 20.15, its derivative $D_\zeta f(\zeta_0)$ is $D(\zeta_0)$, which by (35) is the value stated in (28). The proof of Theorem 24.3 is complete. □

The remainder of this section treats a problem concerning sufficient conditions of a different type. The problem is the following. If a function $g_1: X\to\mathbb{C}_0$, $x\mapsto g_1(x)$, is given, under what conditions do there exist functions $g_k: X\to\mathbb{C}_0$, $x\mapsto g_k(x)$, $k=2,3,4$, such that the function $f: X\to\mathbb{C}_2$,

(40) $$f(\zeta)=g_1(x)+i_1g_2(x)+i_2g_3(x)+i_1i_2g_4(x), \qquad \zeta\in X,$$

is differentiable in X? If a solution exists, is it unique? In preparation for answering these questions, the following theorem summarizes some necessary conditions, already established, which a differentiable function $f: X\to\mathbb{C}_2$ satisfies.

24.4 THEOREM Assume that each of the functions $g_1,\ldots,g_4$ has two continuous derivatives with respect to $x_1,\ldots,x_4$, and that $f: X\to\mathbb{C}_2$ in (40) is differentiable in X. Then

(41) $$D_{x_r}D_{x_s}g_k(x)=D_{x_s}D_{x_r}g_k(x), \qquad r,s=1,\ldots,4,\ k=1,\ldots,4,$$

for all x in X. Also, $g_1,\ldots,g_4$ satisfy the Cauchy–Riemann differential equations:

(42) $$\begin{aligned} D_{x_1}g_1 &= D_{x_2}g_2 = D_{x_3}g_3 = D_{x_4}g_4,\\ D_{x_2}g_1 &= -D_{x_1}g_2 = D_{x_4}g_3 = -D_{x_3}g_4,\\ D_{x_3}g_1 &= D_{x_4}g_2 = -D_{x_1}g_3 = -D_{x_2}g_4,\\ D_{x_4}g_1 &= -D_{x_3}g_2 = -D_{x_2}g_3 = D_{x_1}g_4. \end{aligned}$$

Furthermore, $g_1,\ldots,g_4$ satisfy the equations in (30)–(36) in Theorem 23.5, and also the additional identities implied by (41) and (42).

Proof. Equation (41) is (37) in Section 23. The identities in (42) are the Cauchy–Riemann differential equations in (7)–(10) in Section 23, but arranged so that $g_1,\ldots,g_4$ appear in order in the four columns in (42). The identities in (30)–(36) in Section 23 are proved in Theorem 23.5. Further identities can be derived by using (41) in connection with the identities in (42).

□

24.5 THEOREM Let X be a star-shaped domain in $\mathbb{C}_0^4$, and let $g_1: X \to \mathbb{C}_0$ be a function which has two continuous derivatives with respect to $x_1, \ldots, x_4$, and which satisfies all of the identities referred to in Theorem 24.4. Then there exist functions $g_k: X \to \mathbb{C}_0$, $k = 2, 3, 4$, which have the following properties:

(43) They are unique up to an additive constant;

(44) They have two continuous derivatives and therefore satisfy the strong Stolz condition in X.

Furthermore, $g_1, \ldots, g_4$ satisfy the Cauchy–Riemann equations in (42), and the function $f: X \to \mathbb{C}_2$ such that

$$f(\zeta) = g_1(x) + i_1 g_2(x) + i_2 g_3(x) + i_1 i_2 g_4(x) \tag{45}$$

is differentiable in X.

Proof. The first step in the proof is to show that there exists a function $h_2: X \to \mathbb{C}_0$, $x \mapsto h_2(x)$, which has two continuous derivatives and satisfies the following equations:

$$\begin{aligned} D_{x_1} g_1 &= D_{x_2} h_2, & D_{x_2} g_1 &= -D_{x_1} h_2, \\ D_{x_3} g_1 &= D_{x_4} h_2, & D_{x_4} g_1 &= -D_{x_3} h_2. \end{aligned} \tag{46}$$

Observe that the left-hand sides form the first column in (42). Also, if h_2 exists and is taken as g_2, then (46) shows that the equations in the first two columns of (42) are satisfied. Write the equations in (46) as follows:

$$\begin{aligned} D_{x_1} h_2 &= -D_{x_2} g_1, & D_{x_2} h_2 &= D_{x_1} g_1, \\ D_{x_3} h_2 &= -D_{x_4} g_1, & D_{x_4} h_2 &= D_{x_3} g_1. \end{aligned} \tag{47}$$

The functions on the right in these equations are given, and the problem is to find a function h_2 which satisfies the equations. Now sufficient conditions are known [7, pp. 422–427] for the existence of a function $h_2: X \to \mathbb{C}_0$ with two continuous derivatives which satisfy (47). If

$$\begin{aligned} -D_{x_2} D_{x_2} g_1 &= D_{x_1} D_{x_1} g_1, & -D_{x_3} D_{x_2} g_1 &= -D_{x_1} D_{x_4} g_1, \\ -D_{x_4} D_{x_2} g_1 &= D_{x_1} D_{x_3} g_1, & D_{x_3} D_{x_1} g_1 &= -D_{x_2} D_{x_4} g_1, \\ D_{x_4} D_{x_1} g_1 &= D_{x_2} D_{x_3} g_1, & -D_{x_4} D_{x_4} g_1 &= D_{x_3} D_{x_3} g_1, \end{aligned} \tag{48}$$

then there is a functions h_2 with two continuous derivatives which satisfies (47) and therefore (46). But the relations in (48) are among the necessary conditions which g_1 is assumed to satisfy. Therefore h_2 with the desired properties exists; this function is unique up to an additive constant [7, pp. 426–427]. Henceforth let h_2 be called g_2. Thus there are functions g_1 and g_2 which satisfy the equations in the first two columns of (42).

The next step is to find a function $h_3: X \to \mathbb{C}_0$ such that

$$(49)\qquad \begin{aligned} &D_{x_1}g_1 = D_{x_2}g_2 = D_{x_3}h_3, \qquad D_{x_2}g_1 = -D_{x_1}g_2 = D_{x_4}h_3,\\ &D_{x_3}g_1 = D_{x_4}g_2 = -D_{x_1}h_3, \qquad D_{x_4}g_1 = -D_{x_3}g_2 = -D_{x_2}h_3. \end{aligned}$$

Here g_1 and g_2 are known functions, and the problem is to find a function h_3 with two continuous derivatives which satisfies these equations. The sufficient conditions for the existence of a solution are the same as before. Write the equations in the following normal order to simplify the verification of the sufficient conditions.

$$(50)\qquad \begin{aligned} &D_{x_1}h_3 = -D_{x_3}g_1 = -D_{x_4}g_2, \qquad D_{x_2}h_3 = -D_{x_4}g_1 = D_{x_3}g_2,\\ &D_{x_3}h_3 = D_{x_1}g_1 = D_{x_2}g_2, \qquad D_{x_4}h_3 = D_{x_2}g_1 = -D_{x_1}g_2. \end{aligned}$$

If

$$(51)\qquad \begin{aligned} &-D_{x_2}D_{x_3}g_1 = -D_{x_1}D_{x_4}g_1, \qquad -D_{x_3}D_{x_3}g_1 = D_{x_1}D_{x_1}g_1,\\ &-D_{x_4}D_{x_3}g_1 = D_{x_1}D_{x_2}g_1, \qquad -D_{x_3}D_{x_4}g_1 = D_{x_2}D_{x_1}g_1,\\ &-D_{x_4}D_{x_4}g_1 = D_{x_2}D_{x_2}g_1, \qquad D_{x_4}D_{x_1}g_1 = D_{x_3}D_{x_2}g_1, \end{aligned}$$

then there exists a function h_3 with two continuous derivatives which satisfies (50) and therefore (49). But the relations in (51) are among the necessary conditions which g_1 is assumed to satisfy. Therefore, h_3 with the desired properties exists; as before, this function also is unique up to an additive constant. Henceforth, let h_3 be called g_3. Thus there are functions g_1, g_2, g_3 which satisfy the equations in the first three columns of (42).

The final step is to find a function $h_4: X \to \mathbb{C}_0$ such that

$$(52)\qquad \begin{aligned} &D_{x_1}g_1 = D_{x_2}g_2 = D_{x_3}g_3 = D_{x_4}h_4,\\ &D_{x_2}g_1 = -D_{x_1}g_2 = D_{x_4}g_3 = -D_{x_3}h_4,\\ &D_{x_3}g_1 = D_{x_4}g_2 = -D_{x_1}g_3 = -D_{x_2}h_4,\\ &D_{x_4}g_1 = -D_{x_3}g_2 = -D_{x_2}g_3 = D_{x_1}h_4. \end{aligned}$$

Here g_1, g_2, and g_3 are known functions, and the problem is to find a function h_4 with two continuous derivatives which satisfies these equations. The sufficient conditions for the existence of a solution are the same as before. In order to simplify the verification that the sufficient conditions are satisfied, write the equations in normal order as follows:

$$(53)\qquad \begin{aligned} &D_{x_1}h_4 = D_{x_4}g_1 = -D_{x_3}g_2 = -D_{x_2}g_3,\\ &D_{x_2}h_4 = -D_{x_3}g_1 = -D_{x_4}g_2 = D_{x_1}g_3,\\ &D_{x_3}h_4 = -D_{x_2}g_1 = D_{x_1}g_2 = -D_{x_4}g_3,\\ &D_{x_4}h_4 = D_{x_1}g_1 = D_{x_2}g_2 = D_{x_3}g_3. \end{aligned}$$

If

$$(54)\qquad \begin{aligned} D_{x_2}D_{x_4}g_1 &= -D_{x_1}D_{x_3}g_1, & D_{x_3}D_{x_4}g_1 &= -D_{x_1}D_{x_2}g_1,\\ D_{x_4}D_{x_4}g_1 &= D_{x_1}D_{x_1}g_1, & -D_{x_3}D_{x_3}g_1 &= -D_{x_2}D_{x_2}g_1,\\ -D_{x_4}D_{x_3}g_1 &= D_{x_2}D_{x_1}g_1, & -D_{x_4}D_{x_2}g_1 &= D_{x_3}D_{x_1}g_1, \end{aligned}$$

then there exists a function h_4 with two continuous derivatives which satisfies (53) and (52). But the relations in (54) are among the necessary conditions which g_1 is assumed to satisfy. Therefore h_4 with the desired properties exists; this function is unique up to an additive constant. Let h_4 be the required function g_4. Thus there are functions $g_1, \dots, g_4$ which have two continuous derivatives and satisfy the Cauchy–Riemann equations in (42).

Finally, since $g_1, \dots, g_4$ have continuous derivatives in X, they satisfy the weak and strong Stolz condition (see Theorem 20.6), [7, pp. 34–35]. Then the function $f: X \to \mathbb{C}_2$ in (45) satisfies all of the hypotheses of Theorem 24.1, and that theorem shows that f is differentiable in X. The proof of Theorem 24.5 is complete. □

Exercises

24.1 Prove the following theorem. Let X be a domain in $\mathbb{C}_2$, and let $w: X \to \mathbb{C}_1, (z_1, z_2) \mapsto w(z_1, z_2)$, be a function which has continuous partial derivatives w_{z_1} and w_{z_2} in X. Then w is a differentiable function of z_1, z_2 with derivatives $D_{z_1}w$ and $D_{z_2}w$, and it satisfies the following weak Stolz condition at each point $\zeta_0: (z_1^0, z_2^0)$ in X:

$$w(\zeta) - w(\zeta_0) = \sum_{t=1}^{2} D_{z_t}w(\zeta_0)(z_t - z_t^0) + r(w; \zeta_0, \zeta)\|\zeta - \zeta_0\|,$$

$$\lim_{\zeta \to \zeta_0} r(w; \zeta_0, \zeta) = 0, \qquad r(w; \zeta_0, \zeta_0) = 0.$$

[*Hints.* The corresponding theorem for functions of real variables can be proved easily with the help of the mean-value theorem; see [7, pp. 34–35]. Since there seems to be no corresponding mean-value theorem for functions of complex variables, another method is required. The following statement is an obvious identity:

$$w(z_1, z_2) - w(z_1^0, z_2^0) = [w(z_1, z_2) - w(z_1^0, z_2)] + [w(z_1^0, z_2) - w(z_1^0, z_2^0)].$$

By the fundamental theorem of the integral calculus,

$$\begin{aligned} &w(z_1, z_2) - w(z_1^0, z_2)\\ &\quad = w_{z_1}(z_1^0, z_2^0)(z_1 - z_1^0) + \int_{z_1^0}^{z_1} [w_{z_1}(\xi_1, z_2) - w_{z_1}(z_1^0, z_2^0)]\, d\xi_1, \end{aligned}$$

$$w(z_1^0, z_2) - w(z_1^0, z_2^0)$$
$$= w_{z_2}(z_1^0, z_2^0)(z_2 - z_2^0) + \int_{z_2^0}^{z_2} [w_{z_2}(z_1^0, \xi_2) - w_{z_2}(z_1^0, z_2^0)]\, d\xi_2.$$

To complete the proof, use the standard inequality for the absolute value of the integral of a complex function and also Schwarz's inequality.]

24.2 Let $f: \mathbb{C}_2 \to \mathbb{C}_2$ be the function such that

$$f(z_1 + i_2 z_2) = e^{z_1 + i_2 z_2} = e^{z_1}(\cos z_2 + i_2 \sin z_2).$$

(a) Use Exercise 24.1 to show that f is differentiable in $\mathbb{C}_2$.
(b) Use Theorem 24.2 to show that

$$D_{z_1 + i_2 z_2} e^{z_1 + i_2 z_2} = e^{z_1}(\cos z_2 + i_2 \sin z_2) = e^{z_1 + i_2 z_2}.$$

24.3 Let $f_1: X_1 \to \mathbb{C}_1$ and $f_2: X_2 \to \mathbb{C}_1$ be differentiable functions of the complex variables $z_1 - i_1 z_2$ and $z_1 + i_1 z_2$, respectively. Let $f: X \to \mathbb{C}_2$ be the function such that

$$f(z_1 + i_2 z_2) = f_1(z_1 - i_1 z_2)e_1 + f_2(z_1 + i_1 z_2)e_2.$$

(a) Show that

$$u(z_1, z_2) = \frac{f_1(z_1 - i_1 z_2) + f_2(z_1 + i_1 z_2)}{2},$$

$$v(z_1, z_2) = \frac{i_1[f_1(z_1 - i_1 z_2) - f_2(z_1 + i_1 z_2)]}{2}.$$

(b) Show that u and v satisfy the Cauchy–Riemann equations in Theorem 24.2 (17).
(c) Use Theorem 24.2 to show that f is differentiable in X.

24.4 Let $f: \mathbb{C}_2 \to \mathbb{C}_2$ be the function such that $f(\zeta) = \zeta^n$.
(a) Show that

$$f(\zeta) = (z_1 - i_1 z_2)^n e_1 + (z_1 + i_1 z_2)^n e_2.$$

(b) Use Theorem 24.3 to show that

$$D_\zeta f(\zeta) = n(z_1 - i_1 z_2)^{n-1} e_1 + n(z_1 + i_1 z_2)^{n-1} e_2 = n\zeta^{n-1}.$$

24.5 Let $g_k: \mathbb{C}_0^4 \to \mathbb{C}_0$, $k = 1, \ldots, 4$, be functions such that

$$g_1(x) = x_1^2 - x_2^2 - x_3^2 + x_4^2, \qquad g_2(x) = 2x_1x_2 - 2x_3x_4,$$
$$g_3(x) = 2x_1x_3 - 2x_2x_4, \qquad g_4(x) = 2x_2x_3 + 2x_1x_4.$$

(a) Show that each of the functions $g_1, \ldots, g_4$ has continuous partial derivatives and therefore [7, pp. 34–35] satisfies the weak Stolz condition.
(b) Show that $g_1, \ldots, g_4$ satisfy the Cauchy–Riemann equations in Theorem 24.1 (2)–(5).
(c) Show that the function $f: \mathbb{C}_2 \to \mathbb{C}_2$ such that

$$f(\zeta) = g_1(x) + i_1 g_2(x) + i_2 g_3(x) + i_1 i_2 g_4(x)$$

is differentiable in $\mathbb{C}_2$.
(d) Use Theorem 24.1 (7) to show that $D_\zeta f(\zeta) = 2(x_1 + i_1 x_2 + i_2 x_3 + i_1 i_2 x_4)$. Use this fact to show that $f(\zeta) = \zeta^2$.
(e) Show that g_1 satisfies all of the relations in Theorem 24.4 (41) and Theorem 23.5 (30)–(36) (compare the hypotheses in Theorem 24.5).

24.6 The function $f: \mathbb{C}_2 \to \mathbb{C}_2$, $\zeta \mapsto \zeta^2$, is differentiable in $\mathbb{C}_2$; it therefore satisfies an equation of the form

$$f(\zeta) - f(\zeta_0) = D_\zeta f(\zeta_0)(\zeta - \zeta_0) + r(f; \zeta_0, \zeta)(\zeta - \zeta_0)$$

at each point ζ_0 in $\mathbb{C}_2$.
(a) Show that $r(f; \zeta_0, \zeta) = \zeta - \zeta_0$.
(b) Show that $r(f; \zeta_0, \zeta_0) = 0$ and $\lim_{\zeta \to \zeta_0} r(f; \zeta_0, \zeta) = 0$.

24.7 The function $f: \mathbb{C}_2 \to \mathbb{C}_2$, $\zeta \mapsto \zeta^3$, is differentiable in $\mathbb{C}_2$. Show that

$$r(f; \zeta_0, \zeta) = \zeta^2 + \zeta\zeta_0 - 2\zeta_0^2 = (\zeta - \zeta_0)(\zeta + 2\zeta_0),$$

and therefore that $\lim_{\zeta \to \zeta_0} r(f; \zeta_0, \zeta) = 0$.

24.8 If $f: X \to \mathbb{C}_2$, $\zeta \mapsto f(\zeta)$, has a derivative in X, prove that for each ζ_0 in X there exists a function $r(f; \zeta_0, \cdot)$ such that

$$f(\zeta) - f(\zeta_0) = D_\zeta f(\zeta_0)(\zeta - \zeta_0) + r(f; \zeta_0, \zeta)(\zeta - \zeta_0),$$
$$\lim_{\zeta \to \zeta_0} r(f; \zeta_0, \zeta) = 0, \qquad r(f; \zeta_0, \zeta_0) = 0.$$

24.9 Let $f_1: \mathbb{C}_1 \to \mathbb{C}_1$ and $f_2: \mathbb{C}_1 \to \mathbb{C}_1$ be the functions such that $f_1(z) = z^m$ and $f_2(z) = z^n$. Define $f: \mathbb{C}_2 \to \mathbb{C}_2$ to be the function such that

$$f(z_1 + i_2 z_2) = f_1(z_1 - i_1 z_2)e_1 + f_2(z_1 + i_1 z_2)e_2,$$
$$z_1 + i_2 z_2 \in \mathbb{C}_2.$$

(a) If $f(z_1 + i_2 z_2) = u(z_1, z_2) + i_2 v(z_1, z_2)$, show that

$$u(z_1, z_2) = (1/2)[(z_1 - i_1 z_2)^m + (z_1 + i_1 z_2)^n],$$
$$v(z_1, z_2) = (i_1/2)[(z_1 - i_1 z_2)^m - (z_1 + i_1 z_2)^n].$$

(b) Verify that $u:\mathbb{C}_2\to\mathbb{C}_1$ and $v:\mathbb{C}_2\to\mathbb{C}_1$ satisfy Laplace's equation and the Cauchy–Riemann equations.
(c) Show that f is a differentiable function of $z_1+i_2z_2$ and find its derivative.
(d) Prove that f has an infinite number of derivatives and find these derivatives.

25. HOLOMORPHIC AND DIFFERENTIABLE FUNCTIONS

Four significant classes of functions $f:X\to\mathbb{C}_2$ have been identified and investigated thus far; they are the functions which satisfy (a) the strong Stolz condition, or (b) the weak Stolz condition (see Definition 20.14); (c) functions which have a derivative (see Definition 20.13); and (d) functions which are holomorphic (see Definition 15.2). The purpose of this section is to prove that these four classes of functions are identical.

The functions to be investigated in this section are defined on a domain X in $\mathbb{C}_2$. As in (2) in Section 8, set $z_1-i_1z_2=h_1(z_1+i_2z_2)$ and $z_1+i_1z_2=h_2(z_1+i_2z_2)$. Define sets X_1 and X_2 as follows:

$$X_1=\{z_1-i_1z_2 \text{ in } \mathbb{C}_1\colon z_1-i_1z_2=h_1(z_1+i_2z_2),\ z_1+i_2z_2\in X\},$$
$$X_2=\{z_1+i_1z_2 \text{ in } \mathbb{C}_1\colon z_1+i_1z_2=h_2(z_1+i_2z_2),\ z_1+i_2z_2\in X\}.$$

The sets X_1 and X_2 are said to be generated by X; since X is a domain, then X_1 and X_2 are domains (see Theorems 8.7 and 8.11). Observe, however, that X is not required to be a cartesian set.

25.1 THEOREM Let X be a domain in $\mathbb{C}_2$. The following four classes of functions $f:X\to\mathbb{C}_2$ are identical:

(1) The class of functions which satisfy the strong Stolz condition;

(2) The class of functions which satisfy the weak Stolz condition;

(3) The class of differentiable functions;

(4) The class of holomorphic functions.

Proof. The proof will be given by showing that

(5) $(1)\subset(2)\subset(3)\subset(4)\subset(1)$.

If $f\in(1)$, then f satisfies the weak Stolz condition by Theorem 20.15; therefore, $f\in(2)$, and

(6) $(1)\subset(2)$.

Next, let $f \in (2)$. Then f is a differentiable function by Theorem 22.3, and $f \in (3)$; therefore,

(7) $\quad (2) \subset (3).$

If f is a differentiable function in (3), then Theorem 21.1 shows that there are holomorphic functions $f_1: X_1 \to \mathbb{C}_1$ and $f_2: X_2 \to \mathbb{C}_1$ such that

(8)
$$f(z_1 + i_2 z_2) = f_1(z_1 - i_1 z_2)e_1 + f_2(z_1 + i_1 z_2)e_2,$$
$$z_1 + i_2 z_2 \in X.$$

Then Theorem 15.3 shows that f is a holomorphic function of a bicomplex variable. Thus, if $f \in (3)$, then $f \in (4)$ and

(9) $\quad (3) \subset (4).$

Finally, let f be a holomorphic function of a bicomplex variable in (4). Then by Theorem 15.5, there exist holomorphic functions f_1 and f_2 of a complex variable which satisfy (8); therefore, by the proof of Theorem 24.3, f satisfies the strong Stolz condition and belongs to (1). Thus

(10) $\quad (4) \subset (1).$

The statements in (6), (7), (9), and (10) show that

(11) $\quad (1) = (2) = (3) = (4),$

and the proof of Theorem 25.1 is complete. □

Holomorphic functions of a bicomplex variable have been defined in Definition 15.2, and Theorem 25.1 shows that each of the classes (1), (2), (3) is the class of holomorphic functions.

Exercises

25.1 Prove the following theorem. The function $f: X \to \mathbb{C}_2$ is a holomorphic function if and only if there exist holomorphic functions $f_1: X_1 \to \mathbb{C}_1$ and $f_2: X_2 \to \mathbb{C}_1$ such that

$$f(z_1 + i_2 z_2) = f_1(z_1 - i_1 z_2)e_1 + f_2(z_1 + i_1 z_2)e_2, \qquad z_1 + i_2 z_2 \text{ in } X.$$

25.2 Let $g: Y \to \mathbb{C}_1$, $Y \subset \mathbb{C}_1$, be a holomorphic function of a complex variable. Let X be the set of points $z_1 + i_2 z_2$ in $\mathbb{C}_2$ such that

$$z_1 + i_2 z_2 = (z_1 - i_1 z_2)e_1 + (z_1 + i_1 z_2)e_2,$$
$$z_1 - i_1 z_2 \in Y,\ z_1 + i_1 z_2 \in Y.$$

Define a function $f: X \to \mathbb{C}_2$ as follows:

$$f(z_1 + i_2 z_2) = g(z_1 - i_1 z_2)e_1 + g(z_1 + i_1 z_2)e_2,$$
$$z_1 + i_2 z_2 \text{ in } X.$$

(a) Prove that f is a holomorphic function of a bicomplex variable.
(b) Show that Y can be considered the subset of X obtained by setting $z_2 = 0$.
(c) Show that f is the holomorphic extension of g from Y in $\mathbb{C}_1$ into X in $\mathbb{C}_2$, and that $f(z_1 + i_2 z_2) = g(z_1)$ for all $z_1 + i_2 z_2$ in Y.

25.3 Show that the function $f: \mathbb{C}_2 \to \mathbb{C}_2$, $f(z_1 + i_2 z_2) = (z_1 + i_2 z_2)^n$, has a derivative and that $D_{z_1 + i_2 z_2} f(z_1 + i_2 z_2) = n(z_1 + i_2 z_2)^{n-1}$.

25.4 Prove the following theorem. If $D(\zeta_0; r_1, r_2)$ is a discus with $r_1 > 0$ and $r_2 > 0$, and if

$$\sum_{k=0}^{\infty} (a_k + i_2 b_k)(\zeta - \zeta_0)^k$$

converges in $D(\zeta_0; r_1, r_2)$ and represents a function f there, then f has a derivative and

$$D_\zeta f(\zeta) = \sum_{k=0}^{\infty} k(a_k + i_2 b_k)(\zeta - \zeta_0)^{k-1}$$

for each ζ in $D(\zeta_0; r_1, r_2)$.

25.5 Prove that the functions $f: \mathbb{C}_2 \to \mathbb{C}_2$ whose values are given by the following formulas are holomorphic in $\mathbb{C}_2$ and find their derivatives: (a) $e^{z_1 + i_2 z_2}$; (b) $\cos(z_1 + i_2 z_2)$; (c) $\sin(z_1 + i_2 z_2)$; (d) $\cosh(z_1 + i_2 z_2)$; (e) $\sinh(z_1 + i_2 z_2)$.

25.6 Let $f: \mathbb{C}_2 \to \mathbb{C}_2$ be a polynomial such that

$$f(z_1 + i_2 z_2) = \sum_{k=0}^{n} (a_k + i_2 b_k)(z_1 + i_2 z_2)^k,$$
$$a_n + i_2 b_n \notin \mathcal{O}_2.$$

Show that f_1 and f_2 (see Exercise 25.1) are polynomials of degree n in $z_1 - i_1 z_2$ and $z_1 + i_1 z_2$, respectively, and find these polynomials.

25.7 Let

$$f_1(z_1 - i_1 z_2) = \frac{1}{1 - (z_1 - i_1 z_2)}, \qquad z_1 - i_1 z_2 \neq 1,$$

$$f_2(z_1 + i_1 z_2) = \frac{1}{1 - (z_1 + i_1 z_2)}, \qquad z_1 + i_1 z_2 \neq 1.$$

(a) If $f(z_1+i_2z_2)=f_1(z_1-i_1z_2)e_1+f_2(z_1+i_1z_2)e_2$, show that

$$f(z_1+i_2z_2)=\frac{1}{1-(z_1+i_2z_2)}, \qquad [1-(z_1+i_2z_2)]\notin\mathcal{O}_2.$$

(b) Show that f can be represented by a power series in a neighborhood of $0+i_20$. Describe the regions of convergence and divergence of this power series.
(c) Find the first three terms in the power series in (b).
(d) Find the region in which f is a holomorphic function of $z_1+i_2z_2$.

25.8 Let

$$f(z_1+i_2z_2)=\frac{1}{1-(z_1+i_2z_2)}+\frac{1}{1+(z_1+i_2z_2)},$$
$$[1-(z_1+i_2z_2)]\notin\mathcal{O}_2,\ [1+(z_1+i_2z_2)]\notin\mathcal{O}_2.$$

(a) Show that

$$f_1(z_1-i_1z_2)=\frac{1}{1-(z_1-i_1z_2)}+\frac{1}{1+(z_1-i_1z_2)},$$
$$(z_1-i_1z_2)\neq 1,\ -1,$$

$$f_2(z_1+i_1z_2)=\frac{1}{1-(z_1+i_1z_2)}+\frac{1}{1+(z_1+i_1z_2)},$$
$$(z_1+i_1z_2)\neq 1,\ -1.$$

(b) Show that f_1 and f_2 can be expanded in power series about the origin; find the radii of convergence of these power series.
(c) Show that f can be represented by a power series in a neighborhood of $0+i_20$. Find the first three terms of this power series. Describe the regions of convergence and divergence of the power series.

25.9 Let

$$f_1(z_1-i_1z_2)=\frac{1}{1-(z_1-i_1z_2)}, \qquad z_1-i_1z_2\neq 1,$$

$$f_2(z_1+i_1z_2)=\frac{1}{4+(z_1+i_1z_2)}, \qquad z_1+i_1z_2\neq -4.$$

(a) Find the function f.
(b) Show that f can be represented by a power series in a neighborhood of $0+i_20$. Describe the regions of convergence and divergence of this power series.
(c) Find the first four terms of the power series in (b).

26. THE CALCULUS OF DERIVATIVES

The purpose of this section is to prove that a differentiable (or holomorphic) function of a bicomplex variable is continuous, and to prove that the sums, products, quotients, and compositions of differentiable functions are differentiable. The formulas for the derivatives of sums, products, quotients, and compositions are corollaries of the differentiability statements. The exercises contain indications of alternate methods of proof of some of the results in the section. The section emphasizes the importance of the strong Stolz condition.

26.1 THEOREM If $f: X \to \mathbb{C}_2$ is differentiable at ζ_0 in X, then f is continuous at ζ_0.

Proof. Since f is differentiable at ζ_0, then f satisfies the strong Stolz condition at ζ_0 and there is a function $r(f;\zeta_0,\cdot)$, defined for all ζ in a neighborhood of ζ_0, such that

$$(1) \qquad f(\zeta) - f(\zeta_0) = D_\zeta f(\zeta_0)(\zeta - \zeta_0) + r(f;\zeta_0,\zeta)(\zeta - \zeta_0),$$
$$\lim_{\zeta\to\zeta_0} r(f;\zeta_0,\zeta) = 0, \qquad r(f;\zeta_0,\zeta_0) = 0.$$

If $D_\zeta f(\zeta_0)=0$, then (1) shows that the theorem is true. If $D_\zeta f(\zeta_0)\neq 0$, choose $\delta_1>0$ so that

$$(2) \qquad \|r(f;\zeta_0,\zeta)\| < \|D_\zeta f(\zeta_0)\|, \qquad \|\zeta - \zeta_0\| < \delta_1.$$

Then (1) and (2) show that

$$(3) \qquad \|f(\zeta) - f(\zeta_0)\| < 2\sqrt{2}\,\|D_\zeta f(\zeta_0)\|\,\|\zeta - \zeta_0\|, \qquad \|\zeta - \zeta_0\| < \delta_1.$$

Let $\varepsilon>0$ be given. Choose δ_2 so that $\delta_2 \leqslant \delta_1$ and so that

$$(4) \qquad 2\sqrt{2}\,\|D_\zeta f(\zeta_0)\|\,\delta_2 < \varepsilon.$$

Then (3) and (4) show that $\|f(\zeta)-f(\zeta_0)\|<\varepsilon$ for $\|\zeta-\zeta_0\|<\delta_2$, and f is continuous at ζ_0. □

26.2 THEOREM Let $f: X\to\mathbb{C}_2$ and $g: X\to\mathbb{C}_2$ be functions which are differentiable at ζ_0. Then $f+g$ and fg are differentiable at ζ_0. Furthermore, if $g(\zeta_0)\notin\mathcal{O}_2$, then f/g is differentiable at ζ_0.

Proof. Since f and g are differentiable at ζ_0, they satisfy the strong Stolz condition at ζ_0, and

$$(5) \qquad f(\zeta) - f(\zeta_0) = D_\zeta f(\zeta_0)(\zeta - \zeta_0) + r(f;\zeta_0,\zeta)(\zeta - \zeta_0),$$
$$(6) \qquad g(\zeta) - g(\zeta_0) = D_\zeta g(\zeta_0)(\zeta - \zeta_0) + r(g;\zeta_0,\zeta)(\zeta - \zeta_0).$$

Equations (5) and (6) show that

$$(7)\quad [f(\zeta)+g(\zeta)]-[f(\zeta_0)+g(\zeta_0)] = [D_\zeta f(\zeta_0)+D_\zeta g(\zeta_0)](\zeta-\zeta_0)+[r(f;\,\zeta_0,\zeta)+r(g;\,\zeta_0,\zeta)](\zeta-\zeta_0),$$

and this equation implies that $f+g$ is differentiable at ζ_0.

Consider fg. Since

$$(8)\quad f(\zeta)=[f(\zeta)-f(\zeta_0)]+f(\zeta_0), \qquad g(\zeta)=[g(\zeta)-g(\zeta_0)]+g(\zeta_0),$$

then

$$(9)\quad f(\zeta)g(\zeta)-f(\zeta_0)g(\zeta_0) = f(\zeta_0)[g(\zeta)-g(\zeta_0)]+g(\zeta_0)[f(\zeta)-f(\zeta_0)] + [f(\zeta)-f(\zeta_0)][g(\zeta)-g(\zeta_0)].$$

On the right side of this equation, replace $[f(\zeta)-f(\zeta_0)]$ and $[g(\zeta)-g(\zeta_0)]$ by their values in (5) and (6). The resulting equation can be simplified to an equation of the following form:

$$(10)\quad f(\zeta)g(\zeta)-f(\zeta_0)g(\zeta_0) = [f(\zeta_0)D_\zeta g(\zeta_0)+g(\zeta_0)D_\zeta f(\zeta_0)](\zeta-\zeta_0)+r(fg;\,\zeta_0,\zeta)(\zeta-\zeta_0),$$

$$(11)\quad \lim_{\zeta\to\zeta_0} r(fg;\,\zeta_0,\zeta)=0, \qquad r(fg;\,\zeta_0,\zeta_0)=0.$$

Equations (10) and (11) show that fg satisfies the strong Stolz condition at ζ_0 and is therefore differentiable at ζ_0.

Consider f/g. Since $g(\zeta_0)\notin \mathcal{O}_2$ by hypothesis, there is a neighborhood of ζ_0 in which $g(\zeta)$ is not in $\mathcal{O}_2$; restrict ζ to be in this neighborhood. Then

$$(12)\quad \frac{f(\zeta)}{g(\zeta)}-\frac{f(\zeta_0)}{g(\zeta_0)}=\frac{f(\zeta)g(\zeta_0)-f(\zeta_0)g(\zeta)}{g(\zeta_0)g(\zeta)} = \frac{g(\zeta_0)[f(\zeta)-f(\zeta_0)]-f(\zeta_0)[g(\zeta)-g(\zeta_0)]}{g(\zeta_0)g(\zeta)}.$$

To the right side of the equation, add and subtract the term

$$(13)\quad \frac{g(\zeta_0)[f(\zeta)-f(\zeta_0)]-f(\zeta_0)[g(\zeta)-g(\zeta_0)]}{[g(\zeta_0)]^2}.$$

The resulting equation can be simplified to an equation of the following form:

$$(14)\quad \frac{f(\zeta)}{g(\zeta)}-\frac{f(\zeta_0)}{g(\zeta_0)} = \left[\frac{g(\zeta_0)D_\zeta f(\zeta_0)-f(\zeta_0)D_\zeta g(\zeta_0)}{[g(\zeta_0)]^2}\right](\zeta-\zeta_0)+r\left(\frac{f}{g};\,\zeta_0,\zeta\right)(\zeta-\zeta_0).$$

Thus f/g satisfies the strong Stolz condition at ζ_0, and f/g is differentiable at ζ_0. The proof of all parts of Theorem 26.2 is complete. □

26.3 COROLLARY Let $f:X\to\mathbb{C}_2$ and $g:X\to\mathbb{C}_2$ be functions which are differentiable at ζ_0. Then $f+g$ and fg are differentiable at ζ_0, and

$$D_\zeta(f+g)(\zeta_0)=D_\zeta f(\zeta_0)+D_\zeta g(\zeta_0), \tag{15}$$

$$D_\zeta(fg)(\zeta_0)=f(\zeta_0)D_\zeta g(\zeta_0)+g(\zeta_0)D_\zeta f(\zeta_0). \tag{16}$$

Furthermore, if $g(\zeta_0)\notin\mathcal{O}_2$, then f/g has a derivative at ζ_0, and

$$D_\zeta(f/g)(\zeta_0)=\frac{g(\zeta_0)D_\zeta f(\zeta_0)-f(\zeta_0)D_\zeta g(\zeta_0)}{[g(\zeta_0)]^2}. \tag{17}$$

Proof. The formulas in (15), (16), (17) follow from the proof of Theorem 20.15 and the strong Stolz conditions in (7), (10), and (14), respectively. □

The next theorem concerns the differentiability of a composite function. Let X and Y be domains in $\mathbb{C}_2$, and let $g:X\to Y$, $\zeta\mapsto g(\zeta)$, and $f:Y\to\mathbb{C}_2$, $\eta\mapsto f(\eta)$, be functions such that

$$g(\zeta)\in Y,\qquad \forall\zeta \text{ in } X. \tag{18}$$

Then $f[g(\zeta)]$, *or* $f\circ g(\zeta)$, is defined for all ζ in X. Let ζ_0 be a point in X, and let $\eta_0=g(\zeta_0)$.

26.4 THEOREM Let $g:X\to Y$ and $f:Y\to\mathbb{C}_2$ be the functions just described. If g is differentiable at ζ_0 and f is differentiable at η_0, then $f\circ g$ is differentiable at ζ_0.

Proof. Since g and f are differentiable at ζ_0 and η_0 by hypothesis, there are functions $r(g;\zeta_0,\cdot)$ and $r(f;\eta_0,\cdot)$ such that

$$\begin{aligned} g(\zeta)-g(\zeta_0)&=D_\zeta g(\zeta_0)(\zeta-\zeta_0)+r(g;\zeta_0,\zeta)(\zeta-\zeta_0),\\ \lim_{\zeta\to\zeta_0} r(g;\zeta_0,\zeta)&=0,\qquad r(g;\zeta_0,\zeta_0)=0. \end{aligned} \tag{19}$$

$$\begin{aligned} f(\eta)-f(\eta_0)&=D_\eta f(\eta_0)(\eta-\eta_0)+r(f;\eta_0,\eta)(\eta-\eta_0),\\ \lim_{\eta\to\eta_0} r(f;\eta_0,\eta)&=0,\qquad r(f;\eta_0,\eta_0)=0. \end{aligned} \tag{20}$$

Now g maps X into Y by hypothesis; then $\eta=g(\zeta)$ and $\eta_0=g(\zeta_0)$, and η is in Y for all ζ in X. Then (20) shows that

$$\begin{aligned} f[g(\zeta)]-f[g(\zeta_0)]&=D_\eta f[g(\zeta_0)][g(\zeta)-g(\zeta_0)]\\ &\quad+r(f;g(\zeta_0),g(\zeta))[g(\zeta)-g(\zeta_0)]. \end{aligned} \tag{21}$$

Replace $g(\zeta)-g(\zeta_0)$ by its value from (19); then

$$(22)\qquad f[g(\zeta)]-f[g(\zeta_0)]=D_\eta f[g(\zeta_0)]D_\zeta g(\zeta_0)(\zeta-\zeta_0) \\ +r(f\circ g;\zeta_0,\zeta)(\zeta-\zeta_0),$$

$$(23)\qquad r(f\circ g;\zeta_0,\zeta)=D_\eta f[g(\zeta_0)]r(g;\zeta_0,\zeta)(\zeta-\zeta_0) \\ +r(f;g(\zeta_0),g(\zeta))[D_\zeta g(\zeta_0)+r(g;\zeta_0,\zeta)].$$

Now g is differentiable at ζ_0 by hypothesis; then Theorem 26.1 shows that g is continuous at ζ_0 and that $\lim_{\zeta\to\zeta_0} g(\zeta)=g(\zeta_0)$. Therefore,

$$(24)\qquad \lim_{\zeta\to\zeta_0} r(f\circ g;\zeta_0,\zeta)=0,\qquad r(f\circ g;\zeta_0,\zeta_0)=0.$$

Finally, (22) and (24) show that $f\circ g$ satisfies the strong Stolz condition at ζ_0 and is therefore differentiable at ζ_0. The proof of Theorem 26.4 is complete. □

26.5 COROLLARY Let $g:X\to Y$ and $f:Y\to\mathbb{C}_2$ be the functions in Theorem 26.4. Then $f\circ g$ has a derivative at ζ_0, and

$$(25)\qquad D_\zeta(f\circ g)(\zeta_0)=D_\eta f[g(\zeta_0)]D_\zeta g(\zeta_0).$$

Proof. Since f and g are differentiable by hypothesis, the derivatives $D_\eta f(\eta_0)$ and $D_\zeta g(\zeta_0)$ exist. Also, $D_\zeta(f\circ g)(\zeta_0)$ exists and has the value shown in (25) as a result of (22) and (24). □

26.6 THEOREM Let $f:X\to\mathbb{C}_2$ be differentiable in X, and let $f_1:X_1\to\mathbb{C}_1$ and $f_2:X_2\to\mathbb{C}_1$ be the holomorphic functions of a complex variable such that

$$(26)\qquad f(z_1+i_2z_2)=f_1(z_1-i_1z_2)e_1+f_2(z_1+i_1z_2)e_2, \\ \forall z_1+i_2z_2 \text{ in } X.$$

Then f, f_1, and f_2 have an infinite number of derivatives, and

$$(27)\qquad D^n_{z_1+i_2z_2}f(z_1+i_2z_2)=D^n_{z_1-i_1z_2}f_1(z_1-i_1z_2)e_1 \\ +D^n_{z_1+i_1z_2}f_2(z_1+i_1z_2)e_2$$

for $n=0,1,2,\ldots$.

Proof. This formula has been proved already for $n=1$ in Theorem 21.1 (8). Since f_1 and f_2 are holomorphic functions of a complex variable, they have an infinite number of derivatives and the proof of (27) can be completed by induction. □

Exercises

26.1 If $f(z_1+i_2z_2)=c+i_2d$, a constant in $\mathbb{C}_2$, for every $z_1+i_2z_2$ in X, prove that f is differentiable in X and that $D_{z_1+i_2z_2}f(z_1+i_2z_2)=0$ in X.

26.2 Let $f:\mathbb{C}_2\to\mathbb{C}_2$ be the function such that $f(z_1+i_2z_2)=z_1+i_2z_2$. It is known that $D_{z_1+i_2z_2}f(z_1+i_2z_2)=1$. Use induction, Theorem 26.2, and Corollary 26.3 to show that

$$D_{z_1+i_2z_2}(z_1+i_2z_2)^n = n(z_1+i_2z_2)^{n-1}, \qquad n=0,\pm1,\pm2,\ldots.$$

For what values of $z_1+i_2z_2$ is this formula valid?

26.3 If $f:\mathbb{C}_2\to\mathbb{C}_2$ is a function such that

$$f(z_1+i_2z_2)=\sum_{k=0}^{n}(a_k+i_2b_k)(z_1+i_2z_2)^k$$

for all $z_1+i_2z_2$ in $\mathbb{C}_2$, show that f is differentiable in $\mathbb{C}_2$ and find its derivative.

26.4 Let $g:X\to\mathbb{C}_2$ be a function which is differentiable in X, and let $f:X\to\mathbb{C}_2$ be the function such that $f(z_1+i_2z_2)=e^{g(z_1+i_2z_2)}$. Show that f is differentiable in X and that

$$D_{z_1+i_2z_2}f(z_1+i_2z_2)=e^{g(z_1+i_2z_2)}D_{z_1+i_2z_2}g(z_1+i_2z_2).$$

26.5 Define a function f of a bicomplex variable as follows:

$$f(z_1+i_2z_2)=\frac{1}{1-(z_1+i_2z_2)},$$

$$z_1+i_2z_2\in\mathbb{C}_2,\ [1-(z_1+i_2z_2)]\notin\mathcal{O}_2.$$

(a) Prove that f has a derivative at every point in $\mathbb{C}_2$ except those for which $[1-(z_1+i_2z_2)]\in\mathcal{O}_2$.

(b) Find the first three derivatives of f at $0+i_20$.

(c) Compare the derivatives in (b) with the coefficients of the first three terms of the power series found in Exercise 25.7 (c).

26.6 Use (26) to give an ε,δ proof of the following theorem. If $f:X\to\mathbb{C}_2$ is differentiable in X, then f is continuous at each point ζ_0 in X. Assume that a holomorphic function of a complex variable is continuous.

26.7 Use the formulas in (26) and (27) to establish the formulas in (15), (16), and (17). Assume that the formulas for derivatives of functions of a complex variable are known.

26.8 Use the formulas in (26) and (27) to prove Corollary 26.5. Assume that the formula is known for the derivative of composite holomorphic functions of a complex variable.

26.9 Establish the following formula:

$$D_\zeta^n\left(\frac{2}{1-\zeta^2}\right)=\frac{n!(1+\zeta)^{n+1}+(-1)^n n!(1-\zeta)^{n+1}}{(1-\zeta^2)^{n+1}},$$

$n=0,1,2,\ldots.$

For what values of ζ in $\mathbb{C}_2$ is this formula valid?

27. THE TAYLOR SERIES OF A HOLOMORPHIC FUNCTION

Let X be a domain in $\mathbb{C}_2$, and let $f:X\to\mathbb{C}_2$ be a function which is differentiable in X (see Definition 20.13). Then Theorem 25.1 proves that f is a holomorphic function (see Definition 15.2); therefore f can be represented by a power series in a neighborhood of each point in X. The purpose of this section is to show that this power series is the usual Taylor series.

27.1 THEOREM Let $f:X\to\mathbb{C}_2$, $\zeta\mapsto f(\zeta)$, be a differentiable function in X. Then for each ζ_0 in X there is a discus $D(\zeta_0; r_1, r_2)$, with $r_1>0$ and $r_2>0$, such that

$$(1)\qquad f(\zeta)=\sum_{k=0}^{\infty}\frac{D_\zeta^k f(\zeta_0)}{k!}(\zeta-\zeta_0)^k$$

for all ζ in $D(\zeta_0; r_1, r_2)$.

Proof. Since f is differentiable in X, Theorem 21.1 shows that there are holomorphic functions $f_1:X_1\to\mathbb{C}_1$ and $f_2:X_2\to\mathbb{C}_1$ such that

$$(2)\qquad f(z_1+i_2z_2)=f_1(z_1-i_1z_2)e_1+f_2(z_1+i_1z_2)e_2,\qquad z_1+i_2z_2 \text{ in } X.$$

Let ζ_0 be a point $z_1^0+i_2z_2^0$ in X. Then

$$(3)\qquad z_1^0+i_2z_2^0=(z_1^0-i_1z_2^0)e_1+(z_1^0+i_1z_2^0)e_2.$$

Since f_1 and f_2 are holomorphic functions, they can be represented by power series in neighborhoods of $z_1^0-i_1z_2^0$ and $z_1^0+i_1z_2^0$. Thus there exist constants $r_1>0$ and $r_2>0$ such that

$$(4)\qquad f_1(z_1-i_1z_2)=\sum_{k=0}^{\infty}\frac{D_{z_1-i_1z_2}^k f_1(z_1^0-i_1z_2^0)}{k!}[(z_1-i_1z_2)-(z_1^0-i_1z_2^0)]^k,$$

$$|(z_1-i_1z_2)-(z_1^0-i_1z_2^0)|<r_1;$$

$$(5)\qquad f_2(z_1+i_1z_2)=\sum_{k=0}^{\infty}\frac{D_{z_1+i_1z_2}^k f_2(z_1^0+i_1z_2^0)}{k!}[(z_1+i_1z_2)-(z_1^0+i_1z_2^0)]^k,$$

$$|(z_1+i_1z_2)-(z_1^0+i_1z_2^0)|<r_2.$$

Substitute from (4) and (5) in (2); the result is

$$(6)\quad f(\zeta)=\sum_{k=0}^{\infty}\frac{D^k_{z_1-i_1z_2}f_1(z_1^0-i_1z_2^0)}{k!}[(z_1-i_1z_2)-(z_1^0-i_1z_2^0)]e_1$$

$$+\sum_{k=0}^{\infty}\frac{D^k_{z_1+i_1z_2}f_2(z_1^0+i_1z_2^0)}{k!}[(z_1+i_1z_2)-(z_1^0+i_1z_2^0)]e_2.$$

The proof can be completed by showing that the expression on the right is the Taylor series in (1). By (27) in Section 26,

$$(7)\quad D^k_{z_1-i_1z_2}f_1(z_1^0-i_1z_2^0)e_1+D^k_{z_1+i_1z_2}f_2(z_1^0+i_1z_2^0)e_2=D^k_\zeta f(\zeta_0).$$

Also,

$$(8)\quad [(z_1-i_1z_2)-(z_1^0-i_1z_2^0)]^ke_1+[(z_1+i_1z_2)-(z_1^0+i_1z_2^0)]^ke_2=(\zeta-\zeta_0)^k.$$

Then

$$(9)\quad \frac{1}{k!}D^k_{z_1-i_1z_2}f_1(z_1^0-i_1z_2^0)[(z_1-i_1z_2)-(z_1^0-i_1z_2^0)]^ke_1$$

$$+\frac{1}{k!}D^k_{z_1+i_1z_2}f_2(z_1^0+i_1z_2^0)[(z_1+i_1z_2)-(z_1^0+i_1z_2^0)]^ke_2$$

$$=\frac{1}{k!}D^k_\zeta f(\zeta_0)(\zeta-\zeta_0)^k.$$

This equation shows that (6) simplifies to

$$(10)\quad f(\zeta)=\sum_{k=0}^{\infty}\frac{D^k_\zeta f(\zeta_0)}{k!}(\zeta-\zeta_0)^k.$$

By (4) and (5) and Theorems 13.5–13.7, the Taylor series in (10) converges for all $\zeta: z_1+i_2z_2$ such that

$$(11)\quad |(z_1-i_1z_2)-(z_1^0-i_1z_2^0)|<r_1,\qquad |(z_1+i_1z_2)-(z_1^0+i_1z_2^0)|<r_2;$$

that is, the Taylor series in (1) and (10) converges in the discus $D(\zeta_0; r_1, r_2)$. The proof of Theorem 27.1 is complete. □

Exercises

27.1 Use Exercise 24.2 and Theorem 27.1 to expand $e^{z_1+i_2z_2}$ in a Taylor series about $0+i_20$. Find the discus in which this series converges. Compare the series you obtain with the series used to define $e^{z_1+i_2z_2}$ in Definition 17.12 (67).

27.2 Expand $e^{(z_1+i_2z_2)^2}$ in a Taylor series about $0+i_20$. Find the discus in which this series converges.

27.3 If f is a function such that

$$f(\zeta) = \frac{1}{1-\zeta},$$

what is the region in which f is defined? Show that

$$f(\zeta) = \sum_{k=0}^{\infty} \zeta^k$$

for all ζ in $D(0+i_2 0; 1, 1)$.

27.4 Show that

$$\frac{1}{1+\zeta} = \sum_{k=0}^{\infty} (-1)^k \zeta^k, \qquad \frac{1}{1-\zeta} + \frac{1}{1+\zeta} = 2\sum_{k=0}^{\infty} \zeta^{2k},$$

for all ζ in $D(0+i_2 0; 1, 1)$.

27.5 Let $f: X \to \mathbf{C}_2$ be a holomorphic function such that $f(z_1+i_2 z_2) = u(z_1, z_2) + i_2 v(z_1, z_2)$. Prove that

$$D^k_{z_1+i_2 z_2} f(z_1+i_2 z_2) = D^k_{z_1} u(z_1, z_2) + i_2 D^k_{z_2} v(z_1, z_2),$$
$$k = 0, 1, 2, \ldots.$$

[*Hint.* Theorem 24.2 (19.]

27.6 Let

$$g(z_1) = \sum_{k=0}^{\infty} \frac{D^k_{z_1} g(z_1^0)}{k!} (z_1 - z_1^0),$$

and assume that this series converges for $|z_1 - z_1^0| < r$.

(a) Define a function f of the bicomplex variable $z_1 + i_2 z_2$ as follows

$$f(z_1+i_2 z_2) = g(z_1 - i_1 z_2)e_1 + g(z_1 + i_1 z_2)e_2.$$

Show that this equation defines f in the discus $D(z_1^0 + i_2 0; r, r)$.

(b) Show that the Taylor series expansion of f about $z_1^0 + i_2 0$ is the following:

$$f(z_1+i_2 z_2) = \sum_{k=0}^{\infty} \frac{D^k_{z_1} g(z_1^0)}{k!} [(z_1+i_2 z_2) - (z_1^0 + i_2 0)]^k.$$

Prove that this series converges in the discus $D(z_1^0 + i_2 0; r, r)$.

(c) Explain why f is described as the extension of the complex function g in the disk $\{z_1 \text{ in } \mathbf{C}_1: |z_1 - z_1^0| < r\}$, considered as a subset of $\mathbf{C}_2$ in $z_2 = 0$, into the holomorphic function f of the bicomplex variable in the discus $D(z_1^0 + i_2 0; r, r)$.

28. ISOMORPHIC BICOMPLEX ALGEBRAS AND CAUCHY–RIEMANN MATRICES

The bicomplex algebra is a special Banach algebra, and ideally there would be an axiom system which would distinguish the bicomplex algebra from other Banach algebras and provide a complete abstract description of it. Perhaps such an axiom system could be based on special assumptions about the singular elements and the idempotent elements. Lacking an axiomatic description, the bicomplex algebra is not a unique algebra but rather any Banach algebra which is isomorphic to $\mathbb{C}_2$. According to (2) in Section 1, the elements in $\mathbb{C}_2$ are the bicomplex numbers $x: x_1 + i_1x_2 + i_2x_3 + i_1i_2x_4$, but throughout the book we have employed also an isomorphic representation of $\mathbb{C}_2$ in which the element is $z: z_1 + i_2z_2$. Furthermore, the idempotent representation of elements in $\mathbb{C}_2$ leads to another Banach algebra which is isomorphic to $\mathbb{C}_2$ (see Exercise 6.15). A standard theorem on linear algebras states that every linear algebra can be represented by a matrix algebra [1, pp. 240–241]; thus each of the two linear algebra representations x and z of the bicomplex algebra leads to a matrix algebra which is isomorphic to $\mathbb{C}_2$. The matrices in these matrix algebras are called Cauchy–Riemann matrices. The purpose of this section is to establish these matrix representations of $\mathbb{C}_2$ and to investigate the properties of the Cauchy–Riemann matrices.

28.1 DEFINITION Two algebras A and B are said to *isomorphic* if and only if there exists a one-to-one mapping $f: A \to B$, $a \mapsto f(a)$, of A onto B such that

(1) $f(a_1 + a_2) = f(a_1) + f(a_2)$, $\quad a_1, a_2$ in A; $f(a_1)$, $f(a_2)$ in B;

(2) $f(a_1a_2) = f(a_1)f(a_2)$.

If A and B are normed algebras and the norm of $f(a)$ is equal to the norm of a, then A and B are said to be *isometric*.

Multiplication of all of the elements in $\mathbb{C}_2$ by an element $z_1 + i_2z_2$ performs a linear transformation on $\mathbb{C}_2$; this property of $\mathbb{C}_2$ results from the properties of multiplication described in Theorem 4.3. If

(3) $(z_1 + i_2z_2)(w_1 + i_2w_2) = s_1 + i_2s_2$,

then

(4) $z_1w_1 - z_2w_2 = s_1$, $\quad z_2w_1 + z_1w_2 = s_2$.

Thus multiplication by $z_1 + i_2z_2$ corresponds to the linear transformation (4) with matrix

(5) $\begin{bmatrix} z_1 & -z_2 \\ z_2 & z_1 \end{bmatrix}$.

28.2 THEOREM Let $z:z_1+i_2z_2$ be an element in $\mathbb{C}_2$, and let M be a function defined on $\mathbb{C}_2$ as follows:

$$(6)\qquad M(z)=[z],\qquad [z]=\begin{bmatrix} z_1 & -z_2 \\ z_2 & z_1 \end{bmatrix}.$$

Then the set of 2×2 complex Cauchy–Riemann matrices $[z]$ with the operations of matrix addition and multiplication, and with norm

$$(7)\qquad \|[z]\|=(|z_1|^2+|z_2|^2)^{1/2},$$

is a Banach algebra which is isomorphic and isometric to the bicomplex algebra $\mathbb{C}_2$.

Proof. First, the mapping (6) is one-to-one: given an element $z:z_1+i_2z_2$ in $\mathbb{C}_2$, there is a unique corresponding matrix $[z]$ by (b); also, given $[z]$ in (5), there is a unique corresponding element $z:z_1+i_2z_2$ in $\mathbb{C}_2$. Next, M in (6) has the property (1) since

$$(8)\qquad M(z+z')=\begin{bmatrix} z_1+z_1' & -z_2-z_2' \\ z_2+z_2' & z_1+z_1' \end{bmatrix}$$
$$=\begin{bmatrix} z_1 & -z_2 \\ z_2 & z_1 \end{bmatrix}+\begin{bmatrix} z_1' & -z_2' \\ z_2' & z_1' \end{bmatrix}=M(z)+M(z').$$

Furthermore, M in (6) has property (2) for the following reasons: Since $(zz')w=z(z'w)$, the transformation with matrix $M(zz')$ is the same as the transformation with matrix $M(z')$ followed by the transformation with matrix $M(z)$. Known properties of linear transformations show that

$$(9)\qquad M(zz')=M(z)M(z').$$

Of course, in the present case it is a trivial matter to verify (9) by direct calculation. Since $zz'=(z_1z_1'-z_2z_2')+i_2(z_1z_2'+z_2z_1')$, then

$$(10)\qquad M(zz')=\begin{bmatrix} z_1z_1'-z_2z_2' & -(z_1z_2'+z_1'z_2) \\ z_1z_2'+z_1'z_2 & z_1z_1'-z_2z_2' \end{bmatrix}$$
$$=\begin{bmatrix} z_1 & -z_2 \\ z_2 & z_1 \end{bmatrix}\begin{bmatrix} z_1' & -z_2' \\ z_2' & z_1' \end{bmatrix}=M(z)M(z').$$

Thus $\mathbb{C}_2$ and the matrix algebra of the matrices $[z]$ are isomorphic by Definition 28.1. Also, since

$$\|z\|=\|z_1+i_2z_2\|=[|z_1|^2+|z_2|^2]^{1/2}=\|[z]\|$$

by (7), the algebras are isometric. The proof of Theorem 28.2 is complete.

□

28.3 COROLLARY The following table displays corresponding operations and properties of $\mathbb{C}_2$ and two of its isomorphic representations by linear algebras:

(11)	Element	ζ	$z_1 + i_2 z_2$	$[z]$
(12)	Addition	$\zeta + \zeta'$	$(z_1 + i_2 z_2) + (z_1' + i_2 z_2')$	$[z] + [z']$
(13)	Multiplication	$\zeta\zeta'$	$(z_1 + i_2 z_2)(z_1' + i_2 z_2')$	$[z]\,[z']$
(14)	Norm	$\lVert\zeta\rVert$	$[\lvert z_1\rvert^2 + \lvert z_2\rvert^2]^{1/2}$	$\lVert[z]\rVert$
(15)	Singularity	$\zeta \in \mathcal{O}_2$	$(z_1^2 + z_2^2) \in \mathcal{O}_1$	$\det[z] \in \mathcal{O}_1$

Proof. In (11)–(13), the entries in the columns correspond. In (14), the norms are equal nonnegative numbers in $\mathbb{C}_0$. Theorem 4.8 shows that $z_1 + i_2 z_2 \in \mathcal{O}_2$ if and only if $z_1^2 + z_2^2 \in \mathcal{O}_1$. Then since

$$\det\begin{bmatrix} z_1 & -z_2 \\ z_2 & z_1 \end{bmatrix} = z_1^2 + z_2^2, \tag{16}$$

$z_1 + i_2 z_2$ is a singular element if and only if $\det[z] \in \mathcal{O}_1$. □

28.4 THEOREM Let $x : x_1 + i_1 x_2 + i_2 x_3 + i_1 i_2 x_4$ be an element in $\mathbb{C}_2$, and let N be a function defined on $\mathbb{C}_2$ as follows:

$$N(x) = [x], \qquad [x] = \begin{bmatrix} x_1 & -x_2 & -x_3 & x_4 \\ x_2 & x_1 & -x_4 & -x_3 \\ x_3 & -x_4 & x_1 & -x_2 \\ x_4 & x_3 & x_2 & x_1 \end{bmatrix}. \tag{17}$$

Then the set of 4×4 real Cauchy–Riemann matrices $[x]$ with the operations of matrix addition and multiplication, and with norm

$$\lVert[x]\rVert = (x_1^2 + \cdots + x_4^2)^{1/2}, \tag{18}$$

is a Banach algebra which is isomorphic and isometric to the bicomplex algebra $\mathbb{C}_2$.

Proof. If

$$(x_1 + i_1 x_2 + i_2 x_3 + i_1 i_2 x_4)(y_1 + i_1 y_2 + i_2 y_3 + i_1 i_2 y_4) = r_1 + i_1 r_2 + i_2 r_3 + i_1 i_2 r_4, \tag{19}$$

then Definition 4.1 shows that multiplication of all of the elements y in $\mathbb{C}_2$ by x corresponds to the following linear transformation on $\mathbb{C}_2$ (compare

Theorem 4.3):

$$(20)\qquad \begin{aligned} x_1y_1 - x_2y_2 - x_3y_3 + x_4y_4 &= r_1,\\ x_2y_1 + x_1y_2 - x_4y_3 - x_3y_4 &= r_2,\\ x_3y_1 - x_4y_2 + x_1y_3 - x_2y_4 &= r_3,\\ x_4y_1 + x_3y_2 + x_2y_3 + x_1y_4 &= r_4. \end{aligned}$$

The matrix of this transformation is the matrix $[x]$ in (17). The mapping (17) of x into $N(x)$ is one-to-one: given an element $x: x_1 + i_1x_2 + i_2x_3 + i_1i_2x_4$ in $\mathbb{C}_2$, there is a unique corresponding matrix $[x]$; also, given $[x]$ in (17), there is a unique corresponding element x in $\mathbb{C}_2$. Next, N in (17) has the property (1) since

$$(21)\qquad N(x + x') = N(x) + N(x').$$

Furthermore, N in (17) has property (2) for the same reason as before; since $(xx')y = x(x'y)$, the transformation with matrix $N(xx')$ is the same as the transformation with matrix $N(x')$ followed by the transformation with matrix $N(x)$. Known properties of linear transformations show that

$$(22)\qquad N(xx') = N(x)N(x').$$

Of course, (22) can be verified by direct calculation, but the details are long and tedious. Thus (21) and (22) show that $\mathbb{C}_2$ and the matrix algebra of matrices $[x]$ are isomorphic by Definition 28.1. Also, since

$$(23)\qquad \|x\| = (x_1^2 + \cdots + x_4^2)^{1/2} = \|[x]\|$$

by (18), the algebras are isometric. The proof of Theorem 28.4 is complete.

□

28.5 COROLLARY The following table displays corresponding operations and properties of $\mathbb{C}_2$ and two of its isomorphic representations by linear algebra:

(24)	Element	ζ	x	$[x]$
(25)	Addition	$\zeta + \zeta'$	$x + x'$	$[x] + [x']$
(26)	Multiplication	$\zeta\zeta'$	xx'	$[x][x']$
(27)	Norm	$\|\zeta\|$	$(x_1^2 + \cdots + x_4^2)^{1/2}$	$\|[x]\|$
(28)	Singularity	$\zeta \in \mathcal{O}_2$	$V(x) \in \mathcal{O}_0$	$\det[x] \in \mathcal{O}_0$

Proof. In (24)–(26) the entries in the columns correspond. In (27), the norms are equal nonnegative numbers in $\mathbb{C}_0$. Corollary 4.14 shows that x is in $\mathcal{O}_2$ if and only if $V(x)$ is in $\mathcal{O}_0$. Since $\det[x]=V(x)$ by Theorem 28.6 which follows, then $x\in\mathcal{O}_2$ if and only if $\det[x]\in\mathcal{O}_0$. □

28.6 THEOREM The set of matrices $[x]$ has the following properties:

(29) It is closed under scalar multiplication by scalars in $\mathbb{C}_0$;

(30) It is closed under matrix addition and multiplication; matrix multiplication is commutative;

$$(31)\quad \det[x]=\det\begin{bmatrix} x_1 & -x_2 & -x_3 & x_4 \\ x_2 & x_1 & -x_4 & -x_3 \\ x_3 & -x_4 & x_1 & -x_2 \\ x_4 & x_3 & x_2 & x_1 \end{bmatrix}$$

$$=\det\begin{bmatrix} x_1+x_4 & -(x_2-x_3) \\ x_2-x_3 & x_1+x_4 \end{bmatrix}\det\begin{bmatrix} x_1-x_4 & -(x_2+x_3) \\ x_2+x_3 & x_1-x_4 \end{bmatrix};$$

(32) $\det[x] = V(x)$;

(33) $\det[x] \geqslant 0$;

(34) $\det[x]=0$ if and only if x is a singular element in $\mathbb{C}_2$ and $[x]$ is a singular matrix.

Proof. The statements in (29) and (30) are true because the set of matrices $[x]$ is isomorphic to $\mathbb{C}_2$ and is therefore an algebra. To prove the statements in (31), use row and column operations on the matrix in (17) as follows: first, subtract column 1 from column 4 and add column 2 to column 3 to show that $\det[x]$ equals

$$(35)\qquad \det\begin{bmatrix} x_1 & -x_2 & -(x_2+x_3) & -(x_1-x_4) \\ x_2 & x_1 & x_1-x_4 & -(x_2+x_3) \\ x_3 & -x_4 & x_1-x_4 & -(x_2+x_3) \\ x_4 & x_3 & x_2+x_3 & x_1-x_4 \end{bmatrix}.$$

Next, add row 4 to row 1 and subtract row 3 from row 2 to show that (35) equals

$$(36)\qquad \det\begin{bmatrix} x_1+x_4 & -(x_2-x_3) & 0 & 0 \\ x_2-x_3 & x_1+x_4 & 0 & 0 \\ x_3 & -x_4 & x_1-x_4 & -(x_2+x_3) \\ x_4 & x_3 & x_2+x_3 & x_1-x_4 \end{bmatrix}.$$

Expand this determinant by Laplace's expansion [7, pp. 586–588] to complete the proof of (31). By Definition 4.12, properties of complex numbers, and (31),

$$(37)\quad V(x)=|(x_1+i_1x_2)^2+(x_3+i_1x_4)^2|^2$$
$$=|(x_1+i_1x_2)-i_1(x_3+i_1x_4)|^2|(x_1+i_1x_2)+i_1(x_3+i_1x_4)|^2$$
$$=\det\begin{bmatrix} x_1+x_4 & -(x_2-x_3) \\ x_2-x_3 & x_1+x_4 \end{bmatrix}\det\begin{bmatrix} x_1-x_4 & -(x_2+x_3) \\ x_2+x_3 & x_1-x_4 \end{bmatrix}=\det[x].$$

Then (32) is true, and (33) follows from (32) since $V(x)\geqslant 0$. Also, each of the determinants in (37) is nonnegative. Finally, Corollary 4.14 states that x is singular if and only if $V(x)=0$; then (32) shows that $[x]$ is a singular matrix if and only if $\det[x]=0$. The proof of Theorem 28.6 is complete.

□

To each theory developed in one representation of $\mathbb{C}_2$ there is a corresponding theory in each of the other representations. It frequently happens, however, that the development is much easier, or more intuitive, in one of the representations than in the others. As a result, the development is given its principal setting in one representation and assisted there by special results derived in other representations. For example, the theory of holomorphic functions of a bicomplex variable has been developed in $\mathbb{C}_2$ with elements usually represented as $z_1+i_2z_2$ or $x_1+i_1x_2+i_2x_3+i_1i_2x_4$. Nevertheless, important parts of this theory were obtained from the idempotent representation of $\mathbb{C}_2$. The theory of holomorphic functions could be developed in the matrix algebra representation of $\mathbb{C}_2$, but there the results would look very strange. If

$$f(\zeta) = g_1(x) + i_1g_2(x) + i_2g_3(x) + i_1i_2g_4(x),$$

then in the matrix algebra representation, $f(\zeta)=[g(x)]$. If $\det[x-x^0]\notin\mathcal{O}_0$ and $[x-x^0]^{-1}$ denotes the inverse of $[x-x^0]$, then

$$D_\zeta f(\zeta) = \lim_{\zeta\to\zeta_0}\frac{f(\zeta)-f(\zeta_0)}{\zeta-\zeta_0} = \lim_{x\to x_0}\,[g(x)-g(x^0)][x-x^0]^{-1}.$$

Even this definition of the derivative shows that the theory of holomorphic functions seems less intuitive and more difficult in the matrix algebra representation. It is possible, however, that the development might provide some significant results and insights, but it will not be pursued further.

Exercises

28.1 If 0 and 1 are the zero and unit elements in $\mathbb{C}_2$, show that $N(0)$ is the 4×4 real zero matrix, and that $N(1)$ is the 4×4 real identity matrix.

28.2 Show that

$$N(e_1)=\begin{bmatrix} \frac{1}{2} & 0 & 0 & \frac{1}{2} \\ 0 & \frac{1}{2} & -\frac{1}{2} & 0 \\ 0 & -\frac{1}{2} & \frac{1}{2} & 0 \\ \frac{1}{2} & 0 & 0 & \frac{1}{2} \end{bmatrix}, \qquad N(e_2)=\begin{bmatrix} \frac{1}{2} & 0 & 0 & -\frac{1}{2} \\ 0 & \frac{1}{2} & \frac{1}{2} & 0 \\ 0 & \frac{1}{2} & \frac{1}{2} & 0 \\ -\frac{1}{2} & 0 & 0 & \frac{1}{2} \end{bmatrix}.$$

Verify that

$$N(e_1)N(e_1) = N(e_1), \qquad N(e_2)N(e_2) = N(e_2),$$
$$N(e_1)N(e_2) = N(0).$$

28.3 (a) If $x=x_1+i_1x_2+i_2x_3+i_1i_2x_4$, prove the following identity:

$$N(x)=N[(x_1+x_4)+i_1(x_2-x_3)]N(e_1)+N[(x_1-x_4)+i_1(x_2+x_3)]N(e_2).$$

(b) Show that

$$N[(x_1+x_4)+i_1(x_2-x_3)] = \begin{bmatrix} (x_1+x_4) & -(x_2-x_3) & 0 & 0 \\ (x_2-x_3) & (x_1+x_4) & 0 & 0 \\ 0 & 0 & (x_1+x_4) & -(x_2-x_3) \\ 0 & 0 & (x_2-x_3) & (x_1+x_4) \end{bmatrix}.$$

$$N[(x_1-x_4)+i_1(x_2+x_3)] = \begin{bmatrix} (x_1-x_4) & -(x_2+x_3) & 0 & 0 \\ (x_2+x_3) & (x_1-x_4) & 0 & 0 \\ 0 & 0 & (x_1-x_4) & -(x_2+x_3) \\ 0 & 0 & (x_2+x_3) & (x_1-x_4) \end{bmatrix}.$$

(c) Verify the identity in (a) by replacing the indicated matrices by their values and evaluating the expressions on the two sides.

28.4 Construct a matrix algebra of 2×2 real matrices which is isomorphic to the complex algebra $\mathbb{C}_1$.

28.5 Prove the following theorems:

(a) If $x=x_1+i_1x_2+i_2x_3+i_1i_2x_4$, then $\det N(x)=0$ if and only if either (i) $x_1=x_4$ and $x_2=-x_3$, or (ii) $x_1=-x_4$ and $x_2=x_3$.

(b) $\det N(x)=0$ if and only if $x\in\mathcal{O}_2$. [*Hint.* Corollary 7.5 and Theorem 7.6.]

28.6 (a) Show that x in $\mathbb{C}_2$ has an inverse if and only if $x\notin\mathcal{O}_2$. [*Hints.* Now x has an inverse if and only if there is a y such that $xy=1$. If $x\notin\mathcal{O}_2$, then this equation has the solution $y=1/x$ (Theorem 5.2). If $x\in\mathcal{O}_2$, then xy is in $\mathcal{O}_2$ (and not equal to 1) for every y in $\mathbb{C}_2$ by Definition 7.1 and Corollary 7.5.]

(b) Show that $[x]$ in the matrix algebra isomorphic to $\mathbb{C}_2$ has an inverse if and only if $\det[x]\notin\mathcal{O}_0$, and that the inverse of $[x]$, when it exists, is in the algebra. [*Hint.* Use the isomorphism with (a).]

28.7 For each element $x: x_1+i_1x_2+i_2x_3+i_1i_2x_4$ define matrices A_1 and A_2 as follows:

$$A_1=\begin{bmatrix} x_1+x_4 & -(x_2-x_3) \\ x_2-x_3 & x_1+x_4 \end{bmatrix}, \qquad A_2=\begin{bmatrix} x_1-x_4 & -(x_2+x_3) \\ x_2+x_3 & x_1-x_4 \end{bmatrix}.$$

(a) Verify the following formulas:

$$\det[x] = V(x) = \det A_1 \det A_2,$$
$$\|x\|^2 = \frac{\det A_1 + \det A_2}{2}.$$

(b) Show that

$$\det A_1 = |(x_1 + i_1x_2) - i_1(x_3 + i_1x_4)|^2 \geqslant 0,$$
$$\det A_2 = |(x_1 + i_1x_2) + i_1(x_3 + i_1x_4)|^2 \geqslant 0,$$
$$\det[x] \geqslant 0.$$

(c) Use (b) to show that the second formula in (a) is the same as the formula for $\|x\|$ in terms of the idempotent components of x.

29. HOLOMORPHIC FUNCTIONS AND THEIR INVERSES

Let X be a domain in $\mathbb{C}_2$, and let $f: X\to\mathbb{C}_2$ be a holomorphic function in X. If

$$(1)\qquad \begin{aligned} &\zeta = x_1 + i_1x_2 + i_2x_3 + i_1i_2x_4, \qquad x = (x_1, \dots, x_4), \\ &f(\zeta) = g_1(x) + i_1g_2(x) + i_2g_3(x) + i_1i_2g_4(x), \end{aligned}$$

then f corresponds to the following mapping $y=g(x)$ of X into $\mathbb{C}_0^4$:

$$(2)\qquad \begin{aligned} &y_1 = g_1(x), \\ &\vdots \qquad\qquad\qquad x\in X,\ y = (y_1, \dots, y_4). \\ &y_4 = g_4(x), \end{aligned}$$

Since f is holomorphic, then $g_1,\ldots,g_4$ have derivatives $D_{x_i}g_j$ by Theorem 23.1, and the transformation (2) has the jacobian

$$(3)\qquad D_{(x_1,\ldots,x_4)}(g_1,\ldots,g_4)(x)=\det\begin{bmatrix} D_{x_1}g_1(x) & \cdots & D_{x_4}g_1(x) \\ \vdots & & \vdots \\ D_{x_1}g_4(x) & \cdots & D_{x_4}g_4(x)\end{bmatrix}.$$

The purpose of this section is to study the relations among the derivative $D_\zeta f$ of f, the jacobian (3) of the mapping $y=g(x)$, and the mapping $y=g(x)$ itself. In particular, the section establishes the following results: (a) $D_{(x_1,\ldots,x_4)}(g_1,\ldots,g_4)(x)\geqslant 0$ for every x in X, and $D_{(x_1,\ldots,x_4)}(g_1,\ldots,g_4)(x)=0$ if and only if $D_\zeta f(\zeta)\in\mathcal{O}_2$, $\zeta\leftrightarrow x$; (b) the mapping $y=g(x)$ is one-to-one in a neighborhood of x^0 if $D_\zeta f(\zeta_0)\notin\mathcal{O}_2$; (c) if $D_\zeta f(\zeta_0)\notin\mathcal{O}_2$, then f has an inverse in a neighborhood of ζ_0 and this inverse is a holomorphic function; and (d) the points x at which $D_{(x_1,\ldots,x_4)}(g_1,\ldots,g_4)(x)=0$ are never isolated.

29.1 THEOREM If $f:X\to\mathbb{C}_2$ is a holomorphic function in X, then:

(4) $g_1,\ldots,g_4$ are differentiable in X;

(5) the jacobian matrix in (3) is a Cauchy–Riemann matrix;

(6) $D_{(x_1,\ldots,x_4)}(g_1,\ldots,g_4)(x)\geqslant 0,\qquad \forall x$ in X;

(7) $D_{(x_1,\ldots,x_4)}(g_1,\ldots,g_4)(x)=0\qquad$ if and only if $D_\zeta f(\zeta)\in\mathcal{O}_2$, $\zeta\leftrightarrow x$.

Proof. The functions $g_1,\ldots,g_4$ are differentiable by Theorem 23.1, and (4) is true. By Theorem 23.1, the derivatives of $g_1,\ldots,g_4$ satisfy the Cauchy–Riemann differential equations, and the jacobian matrix in (3) is a Cauchy–Riemann matrix [see (11) in Section 23]; thus (5) is true. Then $D_{(x_1,\ldots,x_4)}(g_1,\ldots,g_4)(x)\geqslant 0$ for every x in X by Theorem 28.6 (33). Now $D_\zeta f(\zeta)=D_{x_1}g_1(x)+i_1D_{x_1}g_2(x)+i_2D_{x_1}g_3(x)+i_1i_2D_{x_1}g_4(x)$, and the Cauchy–Riemann differential equations [see (11) in Section 23] show that $D_\zeta f(\zeta)$ and the jacobian matrix in (3) are corresponding elements in the isomorphism between $\mathbb{C}_2$ and the matrix algebra. Then (6) and (7) are true by (33) and (34) in Theorem 28.6. The proof of Theorem 29.1 is complete. □

If $f:X\to\mathbb{C}_2$ is differentiable in X, then Theorem 23.1 has shown that $g_1,\ldots,g_4$ have derivatives $D_{x_i}g_j$, $i,j=1,\ldots,4$. The next theorem assumes that these derivatives are continuous. Chapter 4 will show that $g_1,\ldots,g_4$ have an infinite number of continuous derivatives.

29.2 THEOREM Make the following hypotheses:

(8) $f:X\to\mathbb{C}_2$ is differentiable in X;

(9) $D_\zeta f(\zeta_0)\notin\mathcal{O}_2,\qquad \zeta_0=x_1^0+i_1x_2^0+i_2x_3^0+i_1i_2x_4^0;$

(10) The derivatives $D_{x_i}g_j$, $i, j = 1, \ldots, 4$, are continuous in X.

The conclusions are the following:

(11) There is an $r_2 > 0$ such that g maps the neighborhood $N(x^0, r_2)$ in a one-to-one manner onto a neighborhood U of y^0, where $y^0 = g(x^0)$, and $D_{(x_1,\ldots,x_4)}(g_1, \ldots, g_4)(x) > 0$ in $N(x^0, r_2)$.

(12) The function $g: N(x^0, r_2) \to U$ has an inverse $h: U \to N(x^0, r_2)$ such that

$$\begin{aligned} x_1 &= h_1(y_1, \ldots, y_4), \\ &\vdots \qquad\qquad\qquad\qquad y \in U,\ x \in N(x^0, r_2). \\ x_4 &= h_4(y_1, \ldots, y_4), \end{aligned}$$

(13) The functions $h_1, \ldots, h_4$ are differentiable, and their jacobian matrix

$$\begin{bmatrix} D_{y_1}h_1 & \cdots & D_{y_4}h_1 \\ \vdots & & \vdots \\ D_{y_1}h_4 & \cdots & D_{y_4}h_4 \end{bmatrix}$$

is a Cauchy–Riemann matrix.

(14) The function $F: U \to N(x^0, r_2)$ such that

$$F(\eta) = h_1(y) + i_1 h_2(y) + i_2 h_3(y) + i_1 i_2 h_4(y),$$

$$\eta = y_1 + i_1 y_2 + i_2 y_3 + i_1 i_2 y_4,$$

is a holomorphic function.

(15) $F[f(\zeta)] = \zeta, \qquad f[F(\eta)] = \eta, \qquad D_\eta F(\eta) D_\zeta F(\zeta) = 1.$

Proof. Hypothesis (9) and Theorem 29.1 (7) show that

(16) $D_{(x_1,\ldots,x_4)}(g_1, \ldots, g_4)(x) > 0.$

Then a standard theorem [19, pp. 35–39] states that the image under the transformation $y = g(x)$ of every neighborhood $N(x^0, r)$ in X contains a neighborhood of y^0. Let $N(x^0, r_1)$ be a neighborhood for which this statement is true. Next, we shall show that $N(x^0, r_1)$ contains a neighborhood $N(x^0, r_2)$ so small that g in (2) maps $N(x^0, r_2)$ onto a neighborhood U of y^0 in a one-to-one manner. Assume that g maps x and x' in $N(x^0, r_1)$ into the same point. Then

(17)
$$\begin{aligned} g_1(x) - g_1(x') &= 0, \\ &\vdots \\ g_4(x) - g_4(x') &= 0. \end{aligned}$$

The mean-value theorem and (17) show that there are points $\xi_1, \dots, \xi_4$ on the segment with ends x and x' such that

$$\text{(18)} \qquad \begin{aligned} &\sum_{k=1}^{4} D_{x_k} g_1(\xi_1)(x_k - x_k') = 0, \\ &\qquad\vdots \\ &\sum_{k=1}^{4} D_{x_k} g_4(\xi_4)(x_k - x_k') = 0. \end{aligned}$$

This is a system of homogeneous linear equations in the unknowns $(x_1 - x_1'), \dots, (x_4 - x_4')$ whose determinant is

$$\text{(19)} \qquad \det \begin{bmatrix} D_{x_1} g_1(\xi_1) & \cdots & D_{x_4} g_1(\xi_1) \\ \vdots & & \vdots \\ D_{x_1} g_4(\xi_4) & \cdots & D_{x_4} g_4(\xi_4) \end{bmatrix}.$$

Choose r_2 so that $r_2 \leqslant r_1$ and so that the determinant in (19) is positive for all $\xi_1, \dots, \xi_4$ in $N(x^0, r_2)$. This choice is possible for the following reasons: (a) by (16) the determinant (19) is positive if $\xi_1 = \cdots = \xi_4 = x^0$; (b) the derivatives $D_{x_i} g_j$ are continuous in X by hypothesis (10); and (c) a determinant is a continuous function of the elements in its matrix. Observe that $D_{(x_1, \dots, x_4)}(g_1, \dots, g_4)(x) > 0$ for all x in $N(x^0, r_2)$ as a result of the choice of r_2. Thus the choice of r_2 shows that if x and x' are in $N(x^0, r_2)$, then the determinant in (19) is positive, and (18) has only the trivial solution $x_k - x_k' = 0$, $k = 1, \dots, 4$, and $x = x'$. Then distinct points in $N(x^0, r_2)$ are mapped by g into distinct points, and thus g maps $N(x^0, r_2)$ in a one-to-one manner onto a neighborhood U of y^0, and the function $g: N(x^0, r_2) \to U$ has an inverse $h: U \to N(x^0, r_2)$ such that

$$\text{(20)} \qquad \begin{aligned} &x_1 = h_1(y_1, \dots, y_4), \\ &\quad\vdots \\ &x_4 = h_4(y_1, \dots, y_4), \qquad (y_1, \dots, y_4) \in U,\ (x_1, \dots, x_4) \in N(x^0, r_2). \end{aligned}$$

Conclusion (12) is true.

Consider (13). Since $g_1, \dots, g_4$ are differentiable by (8), and since

$$\text{(21)} \qquad D_{(x_1, \dots, x_4)}(g_1, \dots, g_4)(x) > 0, \qquad \forall x \text{ in } N(x^0, r_2),$$

the functions $h_1, \dots, h_4$ are differentiable [7, pp. 250–256] and their derivatives have the following values for $i, j = 1, \dots, 4$.

(22) $D_{y_i}h_j(y)$

$$= \frac{1}{D_{(x_1,\ldots,x_4)}(g_1,\ldots,g_4)(x)} \det \begin{bmatrix} D_{x_1}g_1(x) & \cdots & 0 & \cdots & D_{x_4}g_1(x) \\ \vdots & & & & \vdots \\ D_{x_1}g_i(x) & \cdots & 1 & \cdots & D_{x_4}g_i(x) \\ \vdots & & & & \vdots \\ D_{x_1}g_4(x) & \cdots & 0 & \cdots & D_{x_4}g_4(x) \end{bmatrix}.$$

The 1 in the matrix is in the ith row and jth column. Now $g_1,\ldots,g_4$ satisfy the Cauchy–Riemann differential equations; this fact can be used with the formula in (22) to show that $h_1,\ldots,h_4$ also satisfy the Cauchy–Riemann differential equations. Therefore the jacobian matrix of $h_1,\ldots,h_4$ is a Cauchy–Riemann matrix, and the proof of (13) is complete.

Consider (14). Since $h_1,\ldots,h_4$ have derivatives (not *partial* derivatives), they satisfy the strong Stolz condition by Theorem 20.6. Since $h_1,\ldots,h_4$ also satisfy the Cauchy–Riemann differential equations, Theorem 24.1 shows that the function $F: U \to N(x^0, r_2)$ is a differentiable function of the bicomplex variable $\eta = y_1 + i_1y_2 + i_2y_3 + i_1i_2y_4$. Thus F is a holomorphic function, and the proof of (14) is complete.

Finally, consider (15). Since

$$\begin{aligned} (23)\quad & y_1 = g_1(x_1,\ldots,x_4), && x_1 = h_1(y_1,\ldots,y_4), \\ & \vdots && \vdots \\ & y_4 = g_4(x_1,\ldots,x_4), && x_4 = h_4(y_1,\ldots,y_4), \\ & \zeta = x_1 + i_1x_2 + i_2x_3 + i_1i_2x_4, && \eta = y_1 + i_1y_2 + i_2y_3 + i_1i_2y_4, \\ & f(\zeta) = \eta, \qquad F(\eta) = \zeta, \end{aligned}$$

then f and F are inverse functions and each is holomorphic; therefore

$$(24) \qquad f[F(\eta)] = \eta, \qquad F[f(\zeta)] = \zeta.$$

The functions on the left in these equations have derivatives with respect to η and ζ, respectively by Theorem 26.4, and the formula in Corollary 26.5 (25) gives the following results:

$$(25) \qquad D_\zeta f(\zeta) D_\eta F(\eta) = 1, \qquad D_\eta F(\eta) D_\zeta f(\zeta) = 1.$$

The proof of (15) and all parts of Theorem 29.2 is complete. □

29.3 EXAMPLE Let f be the function such that $f(\zeta)=\zeta^2$. Then the values given for $g_1,\dots,g_4$ in Exercise 23.1 show that

$$(26)\quad D_{(x_1,\dots,x_4)}(g_1,\dots,g_4)(x)=\det\begin{bmatrix} 2x_1 & -2x_2 & -2x_3 & 2x_4 \\ 2x_2 & 2x_1 & -2x_4 & -2x_3 \\ 2x_3 & -2x_4 & 2x_1 & -2x_2 \\ 2x_4 & 2x_3 & 2x_2 & 2x_1 \end{bmatrix}.$$

The jacobian matrix in (26) is a Cauchy–Riemann matrix [see (11) in Section 23] as required by Theorem 29.1 (5). Now $D_\zeta f(\zeta)=2\zeta$, and the jacobian matrix in (26) is the Cauchy–Riemann matrix which corresponds to 2ζ in the matrix algebra which is isomorphic to $\mathbb{C}_2$. By (26) and Theorem 28.6 (31),

$$(27)\quad D_{(x_1,\dots,x_4)}(g_1,\dots,g_4)(x)$$
$$=2^4\det\begin{bmatrix} x_1+x_4 & -(x_2-x_3) \\ x_2-x_3 & x_1+x_4 \end{bmatrix}\det\begin{bmatrix} x_1-x_4 & -(x_2+x_3) \\ x_2+x_3 & x_1-x_4 \end{bmatrix}.$$

This equation shows that $D_{(x_1,\dots,x_4)}(g_1,\dots,g_4)(x)$ equals zero at every point $x:(x_1,\dots,x_4)$ such that $x_1+x_4=0$, $x_2-x_3=0$, and also at every point x such that $x_1-x_4=0$, $x_2+x_3=0$. At these points $D_\zeta f(\zeta)$ belongs to $\mathcal{O}_2$ by Theorem 29.1 (7); this fact is easily verified by direct calculation. This example illustrates the fact that the points x at which $D_{(x_1,\dots,x_4)}(g_1,\dots,g_4)(x)=0$, and also the points ζ at which $D_\zeta f(\zeta)$ is in $\mathcal{O}_2$, are never isolated.

29.4 THEOREM If $f:X\to\mathbb{C}_2$ is a holomorphic function, then the points ζ and x in the sets

$$(28)\qquad \{\zeta \text{ in } \mathbb{C}_2\colon D_\zeta f(\zeta)\in\mathcal{O}_2\},$$

$$(29)\qquad \{x \text{ in } \mathbb{C}_0^4\colon D_{(x_1,\dots,x_4)}(g_1,\dots,g_4)(x)=0\},$$

respectively, are never isolated points in these sets.

Proof. Since f is a holomorphic function of a bicomplex variable, Theorem 21.1 shows that there exist holomorphic functions $f_1:X_1\to\mathbb{C}_1$ and $f_2:X_2\to\mathbb{C}_1$ of a complex variable such that

$$(30)\quad f(z_1+i_2z_2)=f_1(z_1-i_1z_2)e_1+f_2(z_1+i_1z_2)e_2,$$
$$D_{z_1+i_2z_2}f(z_1+i_2z_2)$$
$$=D_{z_1-i_1z_2}f_1(z_1-i_1z_2)e_1+D_{z_1+i_1z_2}f_2(z_1+i_1z_2)e_2,$$
$$z_1+i_2z_2\in X,\ z_1-i_1z_2\in X_1,\ z_1+i_1z_2\in X_2.$$

Now $D_{z_1+i_2z_2}f(z_1+i_2z_2)\in\mathcal{O}_2$ if and only if $D_{z_1-i_1z_2}f_1(z_1-i_1z_2)=0$ or $D_{z_1+i_1z_2}f_2(z_1+i_1z_2)=0$. If

(31) $\quad D_{z_1-i_1z_2}f_1(z_1^0-i_1z_2^0)=0,$

then $D_{z_1+i_2z_2}f(z_1+i_2z_2)\in\mathcal{O}_2$ at all points $z_1+i_2z_2$ in the set

(32) $\quad \{z_1+i_2z_2 \text{ in } X\colon z_1-i_1z_2=z_1^0-i_1z_2^0\},$

and this set is contained in the set (28). Then Theorem 29.1 (7) shows that

(33) $\quad D_{(x_1,\dots,x_4)}(g_1,\dots,g_4)(x)=0$

at the points in X which correspond to the points $z_1+i_2z_2$ in (32); this set is contained in the set (29). Since X is a domain, the points in the set (32), and the points in the set of corresponding points x, are clearly not isolated points; thus the theorem is true in this case. Similarly, if

(34) $\quad D_{z_1+i_1z_2}f_2(z_1^0+i_1z_2^0)=0,$

then $D_{z_1+i_2z_2}f(z_1+i_2z_2)\in\mathcal{O}_2$ at all points $z_1+i_2z_2$ in the set

(35) $\quad \{z_1+i_2z_2 \text{ in } X\colon z_1+i_1z_2=z_1^0+i_1z_2^0\}.$

If x is a point which corresponds to the point $z_1+i_2z_2$ in (35), then Theorem 29.1 (7) shows again that (33) is true. The points in the set (35), and also in the set of corresponding points x, are not isolated points, and the theorem is true in this case also. □

Exercises

29.1 Let $f\colon\mathbb{C}_2\to\mathbb{C}_2$ be a function such that $f(\zeta)=\zeta^m$, $m\geqslant 1$.

(a) Show that f is a holomorphic function and find its derivative.

(b) Show that the mapping $w_1+i_2w_2=f(z_1+i_2z_2)$ is equivalent to the two mappings

$$w_1-i_1w_2=(z_1-i_1z_2)^m,\qquad w_1+i_1w_2=(z_1+i_1z_2)^m.$$

(c) Let $w_1^0+i_2w_2^0$ be a point near 0 in $\mathbb{C}_2$. By examining the two mappings in (b) show that there are m^2 points $z_1+i_2z_2$ in $\mathbb{C}_2$ which are mapped by f into $w_1^0+i_2w_2^0$.

(d) Establish the results in (c) by solving the algebraic equation $\zeta^m=w_1^0+i_2w_2^0$ for ζ. [*Hint.* Exercise 6.4.]

(e) Part (c) shows that f maps m^2 points near 0 in $\mathbb{C}_2$ into each point $w_1^0+i_2w_2^0$ near 0 in $\mathbb{C}_2$. Does this conclusion contradict Theorem 29.2 (11)? Why?

29.2 Let f_1, f_2, and f be functions such that

$$f_1(z_1-i_1z_2)=z_1-i_1z_2, \qquad f_2(z_1+i_1z_2)=(z_1+i_1z_2)^m, \qquad m\geqslant 1,$$
$$f(z_1+i_2z_2)=f_1(z_1-i_1z_2)e_1+f_2(z_1+i_1z_2)e_2.$$

(a) Determine the nature of the mapping $f:\mathbb{C}_2\to\mathbb{C}_2$ in the neighborhood of $\zeta=0$.

(b) Find $D_\zeta f(0)$, and show that the mapping by f is one-to-one near $\zeta=0$ if and only if $D_\zeta f(0)\notin\mathcal{O}_2$.

29.3 Let $f:\mathbb{C}_2\to\mathbb{C}_2$ be a function such that $f(\zeta)=a\zeta+b$, where a and b are constants in $\mathbb{C}_2$.

(a) Show that f is a holomorphic function and that $D_\zeta f(\zeta)=a$.

(b) If $D_\zeta f(\zeta)\notin\mathcal{O}_2$, show that the inverse of f can be found by solving the equation $a\zeta+b=\eta$ for ζ; find this inverse.

(c) Let

$$a=a_1+i_1a_2+i_2a_3+i_1i_2a_4, \qquad b=b_1+i_1b_2+i_2b_3+i_1i_2b_4,$$
$$\zeta=x_1+i_1x_2+i_2x_3+i_1i_2x_4, \qquad \eta=y_1+i_1y_2+i_2y_3+i_1i_2y_4.$$

Show that the equation $\eta=a\zeta+b$ is equivalent to the following system of equations:

$$\begin{aligned}
y_1 &= a_1x_1 - a_2x_2 - a_3x_3 + a_4x_4 + b_1,\\
y_2 &= a_2x_1 + a_1x_2 - a_4x_3 - a_3x_4 + b_2,\\
y_3 &= a_3x_1 - a_4x_2 + a_1x_3 - a_2x_4 + b_3,\\
y_4 &= a_4x_1 + a_3x_2 + a_2x_3 + a_1x_4 + b_4.
\end{aligned}$$

(d) If $D_\zeta f(\zeta)\notin\mathcal{O}_2$, find the inverse of f by solving the system of equations in (c); show that this inverse is a holomorphic function. Compare this inverse with the one found in (b).

29.4 Verify that the functions $h_1,\ldots,h_4$ in (20), whose derivatives have the values shown in (22), satisfy the Cauchy–Riemann differential equations. In other words, using (22), carry out the details of proving that the matrix in Theorem 29.2 (13) is a Cauchy–Riemann matrix.

4

Integrals and Holomorphic Functions

30. INTRODUCTION

The subject matter of this chapter is similar to that in the theory of functions of a complex variable which centers around integration theory. Broadly stated, the topics are integrals, Cauchy's integral theorem, Cauchy's integral formula, and Taylor's series, but the details contain much more than this brief outline suggests.

A significant feature of the theory continues to be the multiple representations of bicomplex numbers and of holomorphic functions with values in $\mathbb{C}_2$. A bicomplex number ζ is represented by the following:

(1) $\quad x_1 + i_1 x_2 + i_2 x_3 + i_1 i_2 x_4,$

(2) $\quad z_1 + i_2 z_2,$

(3) $\quad (z_1 - i_1 z_2)e_1 + (z_1 + i_1 z_2)e_2.$

The values $f(\zeta)$ of a holomorphic function $f: X \to \mathbb{C}_2$ are represented also as follows:

(4) $\quad g_1(x) + i_1 g_2(x) + i_2 g_3(x) + i_1 i_2 g_4(x), \qquad x = (x_1, \ldots, x_4),$

(5) $\quad u(z_1, z_2) + i_2 v(z_1, z_2),$

(6) $\quad f_1(z_1 - i_1 z_2)e_1 + f_2(z_1 + i_1 z_2)e_2.$

In order to obtain the maximum generality and applicability of the results, an effort has been made to develop the theory in terms of the representations ζ and $f(\zeta)$, but, since some parts of the theory are not true in every Banach algebra, it is necessary occasionally to employ one of the representations which identifies the result with the theory of functions of a bicomplex variable. For example, (4) is used in Section 33 to show that a function $f: X \to \mathbb{C}_2$, $X \subset \mathbb{C}_2$, which has a continuous derivative $D_\zeta f$ satisfies the fundamental theorem of the integral calculus. A related property is the uniform Stolz condition (see Definition 33.2). A function which has a continuous derivative satisfies the fundamental theorem of the integral calculus and also the uniform Stolz condition; conversely, a function which satisfies the uniform Stolz condition has a continuous derivative and satisfies the fundamental theorem of the integral calculus. The two properties, fundamental theorem and uniform Stolz condition, are used to establish some basic properties of functions $f: X \to \mathbb{C}_2$. For example, if $F: X \to \mathbb{C}_2$ and $G: X \to \mathbb{C}_2$ have continuous derivatives and $D_\zeta F(\zeta) = D_\zeta G(\zeta)$ for all ζ in X, then $F(\zeta) = G(\zeta) + \text{constant}$ on X.

Integrals which are independent of the path play an important role, and there are two approaches to the subject. A standard theorem of complex variable theory [5, I, pp. 163–169] states that the integral around the boundary of a triangle is zero if the function has a derivative in the triangle. Since a holomorphic function of a bicomplex variable satisfies the strong Stolz condition, the proof holds without change, for functions $f: X \to \mathbb{C}_2$, to establish Cauchy's integral theorem (see Section 34). The fundamental theorem states that the integral of a continuous derivative is independent of the path; thus the integral of a continuous function which has a primitive is independent of the path. Section 35 uses Cauchy's integral theorem in Section 34 to show that every function which is holomorphic in a star-shaped region has a primitive; Section 36 uses this result to establish very general forms of Cauchy's integral theorem. All of these results are established without any use of the representation in (6).

The theory of functions of a bicomplex variable can be used sometimes to establish theorems for functions of several real variables. For example, if the integral of the holomorphic function $f: X \to \mathbb{C}_2$ is independent of the path in a certain set of curves C (and equal to zero around a closed curve), then automatically the integrals of $g_1, \ldots, g_4$ are independent of the path with respect to the same set of curves (and equal to zero around closed curves). Since $g_1, \ldots, g_4$ satisfy the Cauchy–Riemann differential equations, at least some part of this result can be established by Stokes' theorem. Thus Section 37 shows that bicomplex variable theory contributes some simple proofs and general results in real-variable theory.

Section 38 for the first time treats the integrals of functions in the idempotent representation (6). Section 39 establishes the relation between Cauchy's integral theorem for $f:X\to\mathbb{C}_2$ and the same theorem for $f_1:X_1\to\mathbb{C}_1$ and $f_2:X_2\to\mathbb{C}_1$. It would be possible to derive the entire theory for f from that for f_1 and f_2, but the exposition presented here provides an independent proof derived from first principles which makes minimum use of the special representations of ζ and $f(\zeta)$. Some use of the representations in (2), (3), and (6) seems to be necessary, however, especially in dealing with the geometry of curves and regions in the four-dimensional space $\mathbb{C}_2$.

Sections 40, 41, and 42 treat Cauchy's integral formula and its applications. These include the following: every holomorphic function has an infinite number of derivatives; corresponding to each point $a: a_1+i_2a_2$ in X, there is a discus $D(a; r_1, r_2)$ in which f is represented by its Taylor series; the power series which represents a holomorphic function is unique, and it is the Taylor series.

Exercise

30.1 Let $f:\mathbb{C}_2\to\mathbb{C}_2$ be the function such that $f(\zeta)=\zeta^2$. Find the representations of f corresponding to (4), (5), and (6).

31. CURVES IN $\mathbb{C}_2$

This section establishes some basic properties of curves in $\mathbb{C}_2$ in preparation for treating the integrals of functions on such curves in later sections.

31.1 DEFINITION Let $[a,b]$ be an interval in $\mathbb{C}_0$. A *curve* C in $\mathbb{C}_2$ is a mapping $\zeta:[a,b]\to\mathbb{C}_2$, $t\mapsto\zeta(t)$. The *trace* of C is the set $\{\zeta(t) \text{ in } \mathbb{C}_2: t\in[a,b]\}$.

If C is the mapping $\zeta:[a,b]\to\mathbb{C}_2$, there are functions $x_k:[a,b]\to\mathbb{C}_0$, $t\mapsto x_k(t)$, for $k=1,\dots,4$ such that

(1) $$\zeta(t)=x_1(t)+i_1x_2(t)+i_2x_3(t)+i_1i_2x_4(t).$$

Now

(2) $$\begin{aligned}\frac{\zeta(t)-\zeta(t_0)}{t-t_0} &= \frac{x_1(t)-x_1(t_0)}{t-t_0}+i_1\frac{x_2(t)-x_2(t_0)}{t-t_0}+i_2\frac{x_3(t)-x_3(t_0)}{t-t_0}\\ &\quad+i_1i_2\frac{x_4(t)-x_4(t_0)}{t-t_0}.\end{aligned}$$

Let primes denote derivatives; then

(3) $$\zeta'(t_0)=x_1'(t_0)+i_1x_2'(t_0)+i_2x_3'(t_0)+i_1i_2x_4'(t_0).$$

These derivatives are assumed to be continuous in this chapter, but a somewhat weaker hypothesis would suffice for many of the results.

31.2 THEOREM If $\zeta:[a,b]\to\mathbb{C}_2$ has a continuous derivative, then

$$(4)\qquad \int_{t_1}^{t_2}\zeta'(t)\,dt=\zeta(t_2)-\zeta(t_1).$$

Proof. The integral $\int_{t_1}^{t_2}\zeta'(t)\,dt$ is defined in the obvious way as the limit of the sum

$$(5)\qquad \sum_{i=1}^{n}\zeta'(a_{i-1})(a_i-a_{i-1}),\qquad t_1=a_0<a_1<\cdots<a_{i-1}<a_i<\cdots<a_n=t_2.$$

The properties of the Riemann integral show that

$$\begin{aligned}(6)\qquad \int_{t_1}^{t_2}\zeta'(t)\,dt&=\int_{t_1}^{t_2}x_1'(t)\,dt+i_1\int_{t_1}^{t_2}x_2'(t)\,dt\\&\quad+i_2\int_{t_1}^{t_2}x_3'(t)\,dt+i_1i_2\int_{t_1}^{t_2}x_4'(t)\,dt,\\&=[x_1(t_2)-x_1(t_1)]+i_1[x_2(t_2)-x_2(t_1)]\\&\quad+i_2[x_3(t_2)-x_3(t_1)]+i_1i_2[x_4(t_2)-x_4(t_1)],\\&=\zeta(t_2)-\zeta(t_1).\end{aligned}$$

The proof of (4) is complete. □

The mean-value theorem $f(b)-f(a)=f'(\xi)(b-a)$, $a<\xi<b$, is a powerful tool for functions $f:[a,b]\to\mathbb{C}_0$, but there seems to be no precise equivalent for functions with values in $\mathbb{C}_2$. For this reason the proof of Theorem 31.2 resorts to the representation of $\zeta(t)$ in terms of the real-valued components $x_1(t),\ldots,x_4(t)$. Then application of the mean-value theorem to the components does not yield a useful result since

$$(7)\quad \zeta(t_2)-\zeta(t_1)=[x_1'(\xi_1)+i_1x_2'(\xi_2)+i_2x_3'(\xi_3)+i_1i_2x_4'(\xi_4)](t_2-t_1).$$

Now $t_1<\xi_k<t_2$, $k=1,\ldots,4$, and the fact that $\xi_1,\ldots,\xi_4$ in (7) may be four different points largely nullifies the usefulness of (7) in many situations. In many cases the Stolz condition can be used as a replacement for the missing mean-value theorem. The next theorem establishes this condition in one of its most useful forms.

31.3 THEOREM Let $\zeta:[a,b]\to\mathbb{C}_2$ be a function which has a continuous derivative. Then for each t_1 in $[a, b]$ there exists a function $r(\zeta; t_1, \cdot)$ such that

(8) $\quad \zeta(t_2) - \zeta(t_1) = \zeta'(t_1)(t_2 - t_1) + r(\zeta; t_1, t_2)(t_2 - t_1),$

(9) $\quad \lim_{t_2 \to t_1} r(\zeta; t_1, t_2) = 0, \qquad r(\zeta; t_1, t_1) = 0.$

Furthermore, given $\varepsilon > 0$, there exists a $\delta > 0$ such that

(10) $\quad \|r(\zeta; t_1, t_2)\| < \varepsilon$

for every pair of points t_1, t_2 in $[a, b]$ such that $|t_2 - t_1| < \delta$.

Proof. By Theorem 31.2 (4),

(11) $$\zeta(t_2) - \zeta(t_1) = \int_{t_1}^{t_2} \zeta'(t)\, dt.$$

Add and subtract the term $\zeta'(t_1)(t_2 - t_1)$ as follows:

(12) $$\zeta(t_2) - \zeta(t_1) = \zeta'(t_1)(t_2 - t_1) + \int_{t_1}^{t_2} [\zeta'(t) - \zeta'(t_1)]\, dt.$$

Make the following change of variable in the integral:

(13) $\quad t = t_1 + (t_2 - t_1)s, \qquad dt = (t_2 - t_1)\, ds.$

Then (12) becomes

(14) $$\zeta(t_2) - \zeta(t_1) = \zeta'(t_1)(t_2 - t_1) + \int_0^1 \{\zeta'[t_1 + s(t_2 - t_1)] - \zeta'(t_1)\}\, ds (t_2 - t_1).$$

Define $r(\zeta; t_1, t_2)$ as follows:

(15) $$r(\zeta; t_1, t_2) = \int_0^1 \{\zeta'[t_1 + s(t_2 - t_1)] - \zeta'(t_1)\}\, ds.$$

Now ζ' is continuous on $[a, b]$; therefore, it is uniformly continuous. To each $\varepsilon > 0$ there corresponds a $\delta > 0$ such that, for every t_1, t_2 in $[a, b]$ for which $|t_2 - t_1| < \delta$,

(16) $\quad \|\zeta'[t_1 + s(t_2 - t_1)] - \zeta'(t_1)\| < \varepsilon, \qquad 0 \leqslant s \leqslant 1.$

Then (14) and (15) show that

(17) $$\|r(\zeta; t_1, t_2)\| \leqslant \int_0^1 \|\zeta'[t_1 + s(t_2 - t_1)] - \zeta'(t_1)\|\, ds < \int_0^1 \varepsilon\, ds = \varepsilon, \qquad |t_2 - t_1| < \delta.$$

Now (14) and (15) show that (8) is true, and (9) follows from (15) and (17). Finally, (17) shows that, given $\varepsilon > 0$, there is a $\delta > 0$ such that (10) is true for every pair of points t_1, t_2 in $[a,b]$ for which $|t_2 - t_1| < \delta$. The proof of all parts of Theorem 31.3 is complete. □

The definition of the length of a curve employs sequences of subdivisions of $[a,b]$. A subdivision P of $[a,b]$ is a set of points

$$a = t_0 < t_1 < \cdots < t_{i-1} < t_i < \cdots < t_n = b, \tag{18}$$

and the norm of P is

$$\max\{t_i - t_{i-1} : i = 1, \ldots, n\}. \tag{19}$$

31.4 DEFINITION Let the curve C be the mapping $\zeta : [a,b] \to \mathbb{C}_2$, and let P be the subdivision (18) of $[a,b]$. Define the sum $S(C,P)$ as follows:

$$S(C,P) = \sum_{i=1}^{n} \|\zeta(t_i) - \zeta(t_{i-1})\|. \tag{20}$$

If $S(C,P)$ approaches a finite limit for a sequence of subdivisions P whose norms approach zero, and if this limit is the same for every such sequence of subdivisions, then this limit is called the *length* $L(C)$ *of* C. Thus

$$L(C) = \lim_{k\to\infty} S(C, P_k). \tag{21}$$

31.5 THEOREM If C is the mapping $\zeta : [a,b] \to \mathbb{C}_2$, and if ζ has a continuous derivative $\zeta' : [a,b] \to \mathbb{C}_2$, then C has length $L(C)$ and

$$L(C) = \int_a^b \|\zeta'(t)\|\, dt. \tag{22}$$

Proof. Let P be the subdivision in (18). Since $\zeta : [a,b] \to \mathbb{C}_2$ has a continuous derivative by hypothesis, it satisfies the Stolz condition in (8) and (9). Then (20) shows that

$$S(C,P) = \sum_{i=1}^{n} \|\zeta'(t_{i-1})(t_i - t_{i-1}) + r(\zeta; t_{i-1}, t_i)(t_i - t_{i-1})\|. \tag{23}$$

Next, by the inequalities satisfied by the norm in $\mathbb{C}_2$,

$$\sum_{i=1}^{n} \|\zeta'(t_{i-1})(t_i - t_{i-1})\| - \sum_{i=1}^{n} \|r(\zeta; t_{i-1}, t_i)(t_i - t_{i-1})\| \tag{24}$$
$$\leqslant S(C,P) \leqslant \sum_{i=1}^{n} \|\zeta'(t_{i-1})(t_i - t_{i-1})\| + \sum_{i=1}^{n} \|r(\zeta; t_{i-1}, t_i)(t_i - t_{i-1})\|.$$

Now

$$(25)\qquad \sum_{i=1}^{n} \|\zeta'(t_{i-1})(t_i - t_{i-1})\| = \sum_{i=1}^{n} \|\zeta'(t_{i-1})\|(t_i - t_{i-1}).$$

Since ζ' is a continuous function by hypothesis, the function on $[a,b]$ whose value at t is $\|\zeta'(t)\|$ is continuous. Then (25) and familiar properties of the Riemann integral show that, for a sequence of subdivisions P whose norms tend to zero,

$$(26)\qquad \lim \sum_{i=1}^{n} \|\zeta'(t_{i-1})(t_i - t_{i-1})\| = \int_a^b \|\zeta'(t)\|\,dt.$$

Furthermore, this limit is the same for every sequence of subdivisions whose norms tend to zero. Next, let $\varepsilon>0$ be given; choose $\delta>0$ so that (10) is satisfied. This choice is possible because ζ' is continuous on the closed interval $[a,b]$. Then

$$(27)\qquad \begin{aligned}\sum_{i=1}^{n} \|r(\zeta;\, t_{i-1},\, t_i)(t_i - t_{i-1})\| &= \sum_{i=1}^{n} \|r(\zeta;\, t_{i-1},\, t_i)\|(t_i - t_{i-1})\\ &< \sum_{i=1}^{n} \varepsilon(t_i - t_{i-1}) = \varepsilon(b-a).\end{aligned}$$

Therefore, for every sequence of subdivisions whose norms approach zero,

$$(28)\qquad \lim \sum_{i=1}^{n} \|r(\zeta;\, t_{i-1},\, t_i)(t_i - t_{i-1})\| = 0.$$

Then (24), (26), and (28) show that

$$(29)\qquad \int_a^b \|\zeta'(t)\|\,dt - 0 \leqslant \lim_{k\to\infty} S(C, P_k) \leqslant \int_a^b \|\zeta'(t)\|\,dt + 0,$$

for every sequence of subdivisions P_k, $k = 1, 2, \ldots$, whose norms tend to zero. Thus (29) and (21) show that

$$(30)\qquad \lim_{k\to\infty} S(C, P_k) = \int_a^b \|\zeta'(t)\|\,dt,$$

$$(31)\qquad L(C) = \int_a^b \|\zeta'(t)\|\,dt,$$

and the proof of Theorem 31.5 is complete. □

Exercises

31.1 Let C be the mapping $\zeta:[a,b]\to\mathbb{C}_2$, and assume that ζ has a continuous derivative ζ'. Show that

$$L(C)=\int_a^b\left\{\sum_{k=1}^{4}[x_k'(t)]^2\right\}^{1/2}dt.$$

[*Hint.* Equations (1) and (3).]

31.2 Let the curves C_1 and C_2 be the mappings $\zeta_1:[a,b]\to\mathbb{C}_2$ and $\zeta_2:[b,c]\to\mathbb{C}_2$, respectively. Assume that ζ_1 and ζ_2 have continuous derivatives, and that $\zeta_1(b)=\zeta_2(b)$. Define a new mapping $\zeta:[a,c]\to\mathbb{C}_2$ as follows:

$$\zeta(t)=\begin{cases}\zeta_1(t), & a\leqslant t\leqslant b,\\ \zeta_2(t), & b\leqslant t\leqslant c.\end{cases}$$

Then $\zeta:[a,c]\to\mathbb{C}_2$ is a continuous curve C which has a continuous derivative except perhaps at $t=b$. Show that the length $L(C)$ of C is defined and that

$$L(C)=L(C_1)+L(C_2)=\int_a^b\|\zeta_1'(t)\|\,dt+\int_b^c\|\zeta_2'(t)\|\,dt=\int_a^c\|\zeta'(t)\|\,dt.$$

31.3 Let ζ_0 and ζ_1 be two points in $\mathbb{C}_2$, and let C be the mapping $\zeta:[0,1]\to\mathbb{C}_2$, $\zeta(t)=(1-t)\zeta_0+t\zeta_1$.
(a) Show that the trace of C is the segment connecting ζ_0 and ζ_1.
(b) Show that $\zeta:[0,1]\to\mathbb{C}_2$ has a continuous derivative and find it.
(c) Show that $L(C)=\|\zeta_1-\zeta_0\|$.

31.4 A function $\zeta:[a,b]\to\mathbb{C}_2$ is differentiable at t_1 and its derivative there is $\zeta'(t_1)$ if and only if for each $\varepsilon>0$ there exists a $\delta(\varepsilon,t_1)$ such that

$$\left\|\frac{\zeta(t)-\zeta(t_1)}{t-t_1}-\zeta'(t_1)\right\|<\varepsilon$$

for every t in $[a,b]$ for which $0<|t-t_1|<\delta(\varepsilon,t_1)$. Furthermore, ζ is said to be uniformly differentiable on $[a,b]$ if and only if δ depends on ε but not on t_1.
(a) Prove the following theorem: If $\zeta:[a,b]\to\mathbb{C}_2$ is uniformly differentiable on $[a,b]$, then ζ has a continuous derivative on $[a,b]$.
(b) Give two proofs of the following theorem. If $\zeta:[a,b]\to\mathbb{C}_2$ has a continuous derivative, then ζ is uniformly differentiable. [*Hints.* For one proof, use (1), (2), and the mean-value theorem; for the other, use Theorem 31.3.]

32. INTEGRALS OF FUNCTIONS WITH VALUES IN $\mathbb{C}_2$

Let X be a domain in $\mathbb{C}_2$, and let $f: X \to \mathbb{C}_2$, $\zeta \mapsto f(\zeta)$, be a function which is continuous. Let the mapping $\zeta: [a,b] \to X$ be a curve C whose trace is in X, and assume that C has a continuous derivative $\zeta': [a,b] \to \mathbb{C}_2$. This section defines the integral of f on C, denoted by $\int_C f(\zeta)\,d\zeta$, proves its existence, and establishes its fundamental properties. Later sections treat the special properties of $\int_C f(\zeta)\,d\zeta$ which follow from the assumption that f is a holomorphic function.

32.1 DEFINITION Let $f: X \to \mathbb{C}_2$ be a continuous function, and let $\zeta: [a,b] \to X$ be a curve C with a continuous derivative $\zeta': [a,b] \to \mathbb{C}_2$. Let P denote a subdivision

(1) $$a = t_0 < t_1 < \cdots < t_{i-1} < t_i < \cdots < t_n = b$$

of $[a,b]$, and let t_i^* be a point such that $t_{i-1} \leqslant t_i^* \leqslant t_i$. Form the sum $S(f,P)$, where

(2) $$S(f,P) = \sum_{i=1}^{n} f[\zeta(t_i^*)][\zeta(t_i) - \zeta(t_{i-1})].$$

If

(3) $$\lim_{k \to \infty} S(f, P_k)$$

exists and has the same value for every choice of the points t_i^* and for every sequence $P_1, P_2, \ldots$ of subdivisions of $[a,b]$ whose norms have the limit zero, then f has an *integral on C*, denoted by $\int_C f(\zeta)\,d\zeta$, and

(4) $$\int_C f(\zeta)\,d\zeta = \lim_{k \to \infty} S(f, P_k).$$

32.2 THEOREM If $f: X \to \mathbb{C}_2$ is continuous and the curve C, defined by $\zeta: [a,b] \to X$, has a continuous derivative, then f has an integral on C and $\int_C f(\zeta)\,d\zeta$ exists.

Proof. The proof will be given first for the case in which t_i^* is chosen to be t_{i-1}; then it will be shown that every other choice for t_i^* leads to the same result. There are two steps in the proof. First, let Q_k, $k = 1, 2, \ldots$, be a specific sequence of subdivisions such that (a) Q_k is a refinement of Q_{k-1}, and (b) the norm of Q_k tends to zero as $k \to \infty$. The proof will show that

(5) $$S(f, Q_k), \qquad k = 1, 2, \ldots,$$

is a Cauchy sequence in $\mathbb{C}_2$. Then since $\mathbb{C}_2$ is a Banach space, $\mathbb{C}_2$ is complete, and the sequence (5) has a limit in $\mathbb{C}_2$. Second, the proof will show that, for every sequence of subdivisions P_k of $[a,b]$ whose norms tend to zero,

(6) $$\lim_{k\to\infty} S(f,P_k) = \lim_{k\to\infty} S(f,Q_k).$$

Since $\zeta:[a,b]\to X$ has a continuous derivative, Theorem 31.5 states that the length of C is defined; denote it by L. Let $\varepsilon>0$ be given. Since $f:X\to\mathbb{C}_2$ and $\zeta:[a,b]\to X$ are continuous, then $f\circ\zeta:[a,b]\to\mathbb{C}_2$ is uniformly continuous; thus there exists a $\delta(\varepsilon)>0$ such that

(7) $$\|f[\zeta(t_2)] - f[\zeta(t_1)]\| < \frac{\varepsilon}{L}$$

for every t_1, t_2 in $[a,b]$ for which $|t_2-t_1|<\delta(\varepsilon)$.

32.3 LEMMA If Q_1 is a subdivision of $[a,b]$ whose norm is less than $\delta(\varepsilon)$, and if Q_2 is a refinement of Q_1, then

(8) $$\|S(f,Q_1) - S(f,Q_2)\| < \sqrt{2}\varepsilon.$$

Proof of Lemma 32.3. Let $[t_{i-1},t_i]$ be a subinterval in Q_1, and recall that $t_i^*=t_{i-1}$ in this first case in the proof. Now Q_2 is a refinement of Q_1; then there are points of subdivision

(9) $$t_{i-1} = s_0 < s_1 < \cdots < s_r = t_i$$

of Q_2 on the interval $[t_{i-1},t_i]$ of Q_1. The sum $S(f,Q_1)$ contains the term

(10) $$f[\zeta(t_{i-1})][\zeta(t_i) - \zeta(t_{i-1})],$$

and (9) shows that $S(f,Q_2)$ contains the corresponding sum

(11) $$\sum_{j=1}^{r} f[\zeta(s_{j-1})][\zeta(s_j) - \zeta(s_{j-1})].$$

Then the norm of the difference of $S(f,Q_1)$ and $S(f,Q_2)$ on the interval $[t_{i-1},t_i]$ is

(12) $$\|f[\zeta(t_{i-1})][\zeta(t_i) - \zeta(t_{i-1})] - \sum_{j=1}^{r} f[\zeta(s_{j-1})][\zeta(s_j) - \zeta(s_{j-1})]\|.$$

Since

(13) $$\zeta(t_i) - \zeta(t_{i-1}) = \sum_{j=1}^{r} [\zeta(s_j) - \zeta(s_{j-1})],$$

then (12) can be written as follows:

(14) $$\left\|\sum_{j=1}^{r} f[\zeta(t_{i-1})][\zeta(s_j) - \zeta(s_{j-1})] - \sum_{j=1}^{r} f[\zeta(s_{j-1})][\zeta(s_j) - \zeta(s_{j-1})]\right\|.$$

Since $|s_{j-1} - t_{i-1}| < \delta(\varepsilon)$ for $j = 1, \ldots, r$, then

$$(15) \qquad \|f[\zeta(t_{i-1})] - f[\zeta(s_{j-1})]\| < \frac{\varepsilon}{L}$$

by (7) and the choice of $\delta(\varepsilon)$. Then (14), (15), and the triangle inequality show that (12) is less than

$$(16) \qquad \sum_{j=1}^{r} \|[\zeta(s_j) - \zeta(s_{j-1})]\| \sqrt{2}\frac{\varepsilon}{L}.$$

Now

$$(17) \qquad \sum_{j=1}^{r} \|[\zeta(s_j) - \zeta(s_{j-1})]\| \leqslant L[\zeta(t_{i-1}), \zeta(t_i)],$$

where $L[\zeta(t_{i-1}), \zeta(t_i)]$ is the length of C between the points $\zeta(t_{i-1})$ and $\zeta(t_i)$. Then (16) and (17) show that (12) is less than

$$(18) \qquad \frac{\sqrt{2}\varepsilon L[\zeta(t_{i-1}), \zeta(t_i)]}{L}.$$

This analysis holds for each interval $[t_{i-1}, t_i]$ in Q_1; therefore,

$$(19) \qquad \|S(f, Q_1) - S(f, Q_2)\| < \sum_{i=1}^{n} \frac{\sqrt{2}\varepsilon L[\zeta(t_{i-1}), \zeta(t_i)]}{L}.$$

Since

$$(20) \qquad \sum_{i=1}^{n} L[\zeta(t_{i-1}), \zeta(t_i)] = L$$

by Exercise 31.2, then (19) and (20) show that

$$(21) \qquad \|S(f, Q_1) - S(f, Q_2)\| < \sqrt{2}\varepsilon,$$

and the proof of (8) and of Lemma 32.3 is complete. □

32.4 LEMMA Let Q_k, $k=1,2,\ldots$, be a sequence of subdivisions of $[a,b]$ with these properties: (a) the norm of Q_k tends to zero as $k\to\infty$; and (b) Q_k is a refinement of Q_{k-1}. Then $S(f, Q_k)$, $k=1,2,\ldots$, is a Cauchy sequence in $\mathbb{C}_2$ which has a limit $S(f)$:

$$(22) \qquad \lim_{k\to\infty} S(f, Q_k) = S(f).$$

Proof. Let $\varepsilon>0$ be given. Choose K so large that the norm of Q_K is less than $\delta(\varepsilon)$; this choice is possible since the norm of Q_k tends to 0 as $k\to\infty$. Since Q_k is a refinement of Q_{k-1}, $k=2,3,\ldots$, by hypothesis, then the norm of Q_k is less than $\delta(\varepsilon)$ for all $k \geqslant K$. Therefore, by Lemma 32.3,

$$\|S(f, Q_m) - S(f, Q_n)\| < \sqrt{2}\varepsilon \tag{23}$$

for all m, n such that $m \geqslant K$ and $n \geqslant K$, and the sequence $S(f, Q_k)$, $k = 1, 2, \ldots$, is a Cauchy sequence. Since $\mathbb{C}_2$ is a Banach space and therefore complete, the sequence has a limit in $\mathbb{C}_2$. Call the limit $S(f)$; thus

$$\lim_{k\to\infty} S(f, Q_k) = S(f), \tag{24}$$

and the proof of Lemma 32.4 is complete. □

32.5 LEMMA If P_k, $k = 1, 2, \ldots$, is a sequence of subdivisions of $[a, b]$ whose norms approach zero, then

$$\lim_{k\to\infty} S(f, P_k) = S(f). \tag{25}$$

Proof. Let P_kQ_k denote the product subdivision of P_k and Q_k; that is, P_kQ_k is the subdivision of $[a, b]$ which contains every point of subdivision in P_k or in Q_k. Then P_kQ_k is a refinement of P_k and also a refinement of Q_k. Now

$$\begin{aligned} &\|S(f, P_k) - S(f)\| \\ &\leqslant \|S(f, P_k) - S(f, P_kQ_k)\| + \|S(f, P_kQ_k) - S(f, Q_k)\| + \|S(f, Q_k) - S(f)\|. \end{aligned} \tag{26}$$

Let $\varepsilon > 0$ be given. By (22) in Lemma 32.4 there is a k_1 such that

$$\|S(f, Q_k) - S(f)\| < \sqrt{2}\frac{\varepsilon}{3}, \qquad \forall k \geqslant k_1. \tag{27}$$

Choose k_2 so that the norm of Q_k is less than $\delta(\varepsilon/3)$ for $k \geqslant k_2$; this choice is possible since the norm of Q_k tends to zero as $k \to \infty$. Then since P_kQ_k is a refinement of Q_k, Lemma 32.3 shows that

$$\|S(f, P_kQ_k) - S(f, Q_k)\| < \sqrt{2}\frac{\varepsilon}{3}, \qquad \forall k \geqslant k_2. \tag{28}$$

Next choose k_3 so large that the norm of P_k is less than $\delta(\varepsilon/3)$ for $k \geqslant k_3$; this choice is possible since the norm of P_k tends to zero as $k \to \infty$. Then since P_kQ_k is a refinement of P_k, Lemma 32.3 shows that

$$\|S(f, P_k) - S(f, P_kQ_k)\| < \sqrt{2}\frac{\varepsilon}{3}, \qquad \forall k \geqslant k_3. \tag{29}$$

Set $K = \max(k_1, k_2, k_3)$. Then each of the inequalities in (27), (28), (29) holds for $k \geqslant K$, and (26) shows that

$$\|S(f, P_k) - S(f)\| < \sqrt{2}\varepsilon, \qquad \forall k \geqslant K. \tag{30}$$

This statement establishes (25) and completes the proof of Lemma 32.5. □

Lemma 32.5 completes the proof of Theorem 32.2 in the first case because it shows that the limit of the sum

$$\sum_{i=1}^{n} f[\zeta(t_{i-1})][\zeta(t_i) - \zeta(t_{i-1})] \tag{31}$$

exists for every sequence P_k, $k=1,2,\ldots$, of subdivisions of $[a,b]$ whose norms tend to zero as $k\to\infty$; furthermore, Lemma 32.5 shows that the limit of (31) is the same for every such sequence.

To complete the proof of Theorem 32.2, it is necessary to show that every other choice for t_i^* in (2) leads to the same result. Let P_k, $k=1,2,\ldots$, be a sequence of subdivisions whose norms approach zero. For the subdivisions P_k set

$$S(f, P_k) = \sum_{i=1}^{n} f[\zeta(t_{i-1})][\zeta(t_i) - \zeta(t_{i-1})], \tag{32}$$

$$S^*(f, P_k) = \sum_{i=1}^{n} f[\zeta(t_i^*)][\zeta(t_i) - \zeta(t_{i-1})], \qquad t_{i-1} \leqslant t_i^* \leqslant t_i. \tag{33}$$

Let $\varepsilon>0$ be given. Then there is a $\delta(\varepsilon)$ such that

$$\|f[\zeta(t_2)] - f[\zeta(t_1)]\| < \frac{\varepsilon}{L} \tag{34}$$

for every pair of points t_1, t_2 in $[a,b]$ such that $|t_2-t_1|<\delta(\varepsilon)$. Choose k_1 so large that the norm of P_k is less than $\delta(\varepsilon)$ for $k\geqslant k_1$; this choice is possible since the norm of P_k tends to zero as $k\to\infty$. Now

$$\|S^*(f, P_k) - S(f)\| \leqslant \|S^*(f, P_k) - S(f, P_k)\| + \|S(f, P_k) - S(f)\|. \tag{35}$$

Now (32), (33), (34) show that, for $k\geqslant k_1$,

$$\begin{aligned} &\|S^*(f, P_k) - S(f, P_k)\| \\ &= \left\| \sum_{i=1}^{n} \{f[\zeta(t_i^*)] - f[\zeta(t_{i-1})]\}[\zeta(t_i) - \zeta(t_{i-1})] \right\| \\ &\leqslant \frac{\sqrt{2}\varepsilon}{L} \sum_{i=1}^{n} \|\zeta(t_i) - \zeta(t_{i-1})\| \\ &\leqslant \sqrt{2}\varepsilon. \end{aligned} \tag{36}$$

Choose k_2 so that

$$\|S(f, P_k) - S(f)\| < \sqrt{2}\varepsilon \tag{37}$$

for $k \geqslant k_2$. Set $K = \max(k_1, k_2)$. Then (35), (36), and (37) show that

$$\|S^*(f, P_k) - S(f)\| < 2\sqrt{2}\varepsilon \tag{38}$$

for $k \geqslant K$; therefore

$$\lim_{k \to \infty} S(f, P_k) = S(f). \tag{39}$$

Thus the limit of the sums in (2) is the same for every sequence of subdivisions and for every choice of t_i^* in the interval $[t_{i-1}, t_i]$. Then the integral $\int_C f(\zeta)\,d\zeta$ exists by Definition 32.1, and the proof of Theorem 32.2 is complete. □

Now C is the mapping $\zeta : [a, b] \to \mathbb{C}_2$; therefore, C is oriented because it is a mapping of the oriented interval $[a, b]$ in $\mathbb{C}_0$. Also, $\int_C f(\zeta)\,d\zeta$ is an oriented integral. It receives its orientation through the term $\zeta(t_i) - \zeta(t_{i-1})$ in the sum in (2). The curve with the opposite orientation is denoted by $-C$, and the sum for the subdivision $a = t_0 < t_1 < \cdots < t_{i-1} < t_i < \cdots < t_n = b$ which leads to the integral of f on $-C$ is

$$\sum_{i=1}^{n} f[\zeta(t_i^*)][\zeta(t_{i-1}) - \zeta(t_i)]. \tag{40}$$

Clearly this sum is the negative of the sum in (2) for the curve C. Denote the integral of f on $-C$ by $\int_{-C} f(\zeta)\,d\zeta$.

32.6 THEOREM If $f : X \to \mathbb{C}_2$ is continuous and the curve C, defined by the mapping $\zeta : [a, b] \to X$, has a continuous derivative, then the integral $\int_{-C} f(\zeta)\,d\zeta$ of f on $-C$ exists, and

$$\int_{-C} f(\zeta)\,d\zeta = -\int_C f(\zeta)\,d\zeta. \tag{41}$$

Proof. The proof follows from the definition of the integral and the fact that the sum in (40) is the negative of the sum in (2). □

If the terminal point of C_1 is the initial point of C_2, then these curves can be combined to form a new curve C, and the next theorem treats the integrals on C_1, C_2, and C. Let X be a domain in $\mathbb{C}_2$, and let $f : X \to \mathbb{C}_2$ be a continuous function. Let the curves C_1 and C_2 be the mappings $\zeta_1 : [a, b] \to X$ and $\zeta_2 : [b, c] \to X$, respectively. Assume that C_1 and C_2 have continuous derivatives, and that $\zeta_1(b) = \zeta_2(b)$. Define a new mapping $\zeta : [a, c] \to X$ as follows:

$$\zeta(t) = \begin{cases} \zeta_1(t), & a \leqslant t \leqslant b, \\ \zeta_2(t), & b \leqslant t \leqslant c. \end{cases} \tag{42}$$

Then $\zeta : [a, c] \to X$ is a continuous curve C which has a continuous derivative except perhaps at $t = b$. From Definition 31.4 it is easy to see that C has a length $L(C)$ and that $L(C) = L(C_1) + L(C_2)$ (compare Exercise 31.2).

32.7 THEOREM Let $f: X \to \mathbb{C}_2$ be a continuous function, and let C_1, C_2, and C be the curves just described. Then the integrals of f on C_1, C_2, and C exist, and

$$\int_C f(\zeta)\,d\zeta = \int_{C_1} f(\zeta)\,d\zeta + \int_{C_2} f(\zeta)\,d\zeta. \tag{43}$$

Proof. The integrals on C_1 and C_2 exist by Theorem 32.2. To show that f has an integral on C and that (43) holds, use subdivisions of $[a, c]$ formed by subdivisions of $[a, b]$ and $[b, c]$ as follows:

$$\begin{aligned} a = t_0 < t_1 < \cdots < t_{i-1} < t_i < \cdots < t_n = b \\ = s_0 < s_1 < \cdots < s_{j-1} < s_j < \cdots < s_m = c. \end{aligned} \tag{44}$$

Then

$$\sum_{i=1}^{n} f[\zeta(t_i^*)][\zeta(t_i) - \zeta(t_{i-1})] + \sum_{j=1}^{m} f[\zeta(s_j^*)][\zeta(s_j) - \zeta(s_{j-1})] \tag{45}$$

is a sum used in the definition of the integral of f on C. Take the limit of (45) for a sequence of subdivisions (44); this limit exists because the integrals $\int_{C_1} f(\zeta)\,d\zeta$ and $\int_{C_2} f(\zeta)\,d\zeta$ exist. Furthermore, the limit of (45), corresponding to a sequence of subdivisions (44), is both $\int_C f(\zeta)\,d\zeta$ and $\int_{C_1} f(\zeta)\,d\zeta + \int_{C_2} f(\zeta)\,d\zeta$; therefore, the integrals in (43) exist and the equality holds. The proof of Theorem 32.7 is complete. □

32.8 THEOREM Let $f_1: X \to \mathbb{C}_2$ and $f_2: X \to \mathbb{C}_2$ be continuous functions, and let C be a curve $\zeta: [a, b] \to X$ which has a continuous derivative. Let a_1 and a_2 be constants in $\mathbb{C}_2$. Then the function $a_1 f_1 + a_2 f_2$ has an integral on C, and

$$\int_C (a_1 f_1 + a_2 f_2)(\zeta)\,d\zeta = a_1 \int_C f_1(\zeta)\,d\zeta + a_2 \int_C f_2(\zeta)\,d\zeta. \tag{45a}$$

Proof. Since f_1, f_2, and $a_1 f_1 + a_2 f_2$ are continuous functions, and since C has a continuous derivative by hypothesis, the three integrals in (45a) exist by Theorem 32.2. To establish the equality, let $a = t_0 < t_1 < \cdots < t_n = b$ be a subdivision of $[a, b]$ and observe that

$$\begin{aligned} &\sum_{i=1}^{n} (a_1 f_1 + a_2 f_2)[\zeta(t_i^*)][\zeta(t_i) - \zeta(t_{i-1})] \\ &= a_1 \sum_{i=1}^{n} f_1[\zeta(t_i^*)][\zeta(t_i) - \zeta(t_{i-1})] + a_2 \sum_{i=1}^{n} f_2[\zeta(t_i^*)][\zeta(t_i) - \zeta(t_{i-1})]. \end{aligned} \tag{45b}$$

A similar relation holds for each subdivision in a sequence of subdivisions whose norms approach zero; each sum has a limit which is the corresponding integral in (45a). Thus (45a) is the limit of (45b) for a sequence of subdivisions of $[a,b]$ whose norms tend to zero. The proof of Theorem 32.8 is complete. □

The next theorem shows that it is possible to introduce the parameter t as the variable of integration in the integral $\int_C f(\zeta)\,d\zeta$.

32.9 THEOREM Let $f: X \to \mathbb{C}_2$ be a continuous function, and let C be a curve defined by the mapping $\zeta: [a,b] \to X$, $t \mapsto \zeta(t)$. If C has a continuous derivative $\zeta': [a,b] \to \mathbb{C}_2$, then the integral $\int_a^b f[\zeta(t)]\zeta'(t)\,dt$ exists, and

$$(46) \qquad \int_C f(\zeta)\,d\zeta = \int_a^b f[\zeta(t))]\zeta'(t)\,dt.$$

Proof. Let $a = t_0 < t_1 < \cdots < t_n = b$ be a subdivision of $[a,b]$. Then $\int_C f(\zeta)\,d\zeta$ is the limit of sums of the following type:

$$(47) \qquad \sum_{i=1}^{n} f[\zeta(t_{i-1})][\zeta(t_i) - \zeta(t_{i-1})].$$

By Theorem 31.3,

$$(48) \qquad \zeta(t_i) - \zeta(t_{i-1}) = \zeta'(t_{i-1})(t_i - t_{i-1}) + r(\zeta;\, t_{i-1},\, t_i)(t_i - t_{i-1}).$$

Substitute in (47) to obtain

$$(49) \qquad \sum_{i=1}^{n} f[\zeta(t_{i-1})][\zeta(t_i) - \zeta(t_{i-1})]$$

$$= \sum_{i=1}^{n} f[\zeta(t_{i-1})]\zeta'(t_{i-1})(t_i - t_{i-1}) + \sum_{i=1}^{n} f[\zeta(t_{i-1})]r(\zeta;\, t_{i-1},\, t_i)(t_i - t_{i-1}).$$

Set

$$(50) \qquad M = \max\{\|f[\zeta(t)]\| : a \leqslant t \leqslant b\}.$$

Now M exists since $\|f[\zeta(t)]\|$ is a continuous function of t for t in the compact set $[a,b]$. Next, let $\varepsilon > 0$ be given. Then Theorem 31.3 states that there is a $\delta(\varepsilon) > 0$ such that

$$(51) \qquad \|r(\zeta;\, t_{i-1},\, t_i)\| < \varepsilon,$$

for every interval $[t_{i-1}, t_i]$ in $[a,b]$ for which $|t_i - t_{i-1}| < \delta(\varepsilon)$. Let P_k, $k = 1, 2, \ldots$, be a sequence of subdivisions of $[a,b]$ whose norms tend to zero as $k \to \infty$. Choose k_1 as an integer so large that the norm of P_k is less than $\delta(\varepsilon)$

for $k \geqslant k_1$. Then for a P_k with $k \geqslant k_1$,

$$\left\| \sum_{i=1}^{n} f[\zeta(t_{i-1})] r(\zeta; t_{i-1}, t_i)(t_i - t_{i-1}) \right\| \leqslant \sqrt{2} M\varepsilon \sum_{i=1}^{n} |t_i - t_{i-1}| = \sqrt{2} M\varepsilon |b - a|. \tag{52}$$

This inequality shows that, for a sequence of subdivisions P_k whose norms tend to zero as $k \to \infty$, the limit of the sums

$$\sum_{i=1}^{n} f[\zeta(t_{i-1})] r(\zeta; t_{i-1}, t_i)(t_i - t_{i-1}) \tag{53}$$

is zero. The limit of the sum on the left in (49) is $\int_C f(\zeta)\, d\zeta$ by Theorem 32.2; therefore the limit of the first sum on the right in (49) exists and equals $\int_C f(\zeta)\, d\zeta$. The limit on the right is the following integral of the real variable t:

$$\int_a^b f[\zeta(t)]\zeta'(t)\, dt. \tag{54}$$

These statements prove (46) and complete the proof of Theorem 32.9. □

Finally, there is a useful inequality for the norm of the integral $\int_C f(\zeta)\, d\zeta$.

32.10 THEOREM Let $f: X \to \mathbb{C}_2$ be a continuous function, and let C be a curve defined by the mapping $\zeta: [a, b] \to X$, $t \mapsto \zeta(t)$. Let C have a continuous derivative $\zeta': [a, b] \to \mathbb{C}_2$, and let $M = \max\{\|f[\zeta(t)]\|: t \in [a, b]\}$. Then $\int_a^b \|f[\zeta(t)]\|\; \|\zeta'(t)\|\, dt$ exists, and

$$\left\| \int_C f(\zeta)\, d\zeta \right\| \leqslant \sqrt{2} \int_a^b \|f[\zeta(t)]\|\; \|\zeta'(t)\|\, dt \leqslant \sqrt{2} M L(C). \tag{55}$$

Proof. The integral $\int_a^b \|f[\zeta(t)]\|\; \|\zeta'(t)\|\, dt$ exists because the integrand is a continuous function. Equation (49) and the triangle inequality show that

$$\begin{aligned} \left\| \sum_{i=1}^{n} f[\zeta(t_{i-1})][\zeta(t_i) - \zeta(t_{i-1})] \right\| &\leqslant \sum_{i=1}^{n} \sqrt{2}\, \|f[\zeta(t_{i-1})]\|\; \|\zeta'(t_{i-1})\| (t_i - t_{i-1}) \\ &\quad + \left\| \sum_{i=1}^{n} f[\zeta(t_{i-1})] r(\zeta; t_{i-1}, t_i)(t_i - t_{i-1}) \right\|. \end{aligned} \tag{56}$$

Consider this inequality for a sequence of subdivisions P_k whose norms tend to zero as $k \to \infty$. The expression on the left has the limit $\|\int_C f(\zeta)\, d\zeta\|$; the first sum on the right has the limit $\sqrt{2} \int_a^b \|f[\zeta(t)]\|\; \|\zeta'(t)\|\, dt$ by the theory of integrals of real-valued functions; and the limit of the remaining term on the right in (56) is zero by (52). Therefore

$$(57)\qquad \left\|\int_C f(\zeta)\,d\zeta\right\| \leqslant \sqrt{2}\int_a^b \|f[\zeta(t)]\|\,\|\zeta'(t)\|\,dt.$$

Since

$$(58)\qquad \|f[\zeta(t)]\| \leqslant M$$

for t in $[a,b]$, then

$$(59)\qquad \sqrt{2}\int_a^b \|f[\zeta(t)]\|\,\|\zeta'(t)\|\,dt \leqslant \sqrt{2}\int_a^b M\|\zeta'(t)\|\,dt = \sqrt{2}M\int_a^b \|\zeta'(t)\|\,dt.$$

Because $\int_a^b \|\zeta'(t)\|\,dt = L(C)$ by Theorem 31.5, then

$$(60)\qquad \sqrt{2}\int_a^b \|f[\zeta(t)]\|\,\|\zeta'(t)\|\,dt \leqslant \sqrt{2}ML(C).$$

Finally, (55) follows from (57) and (60), and the proof of Theorem 32.10 is complete. □

Exercises

32.1 Let ζ_0 and ζ_1 be two points in $\mathbb{C}_2$, and let C be the mapping $\zeta:[0,1]\to\mathbb{C}_2$, $\zeta(t)=(1-t)\zeta_0+t\zeta_1$ (compare Exercise 31.3). Let $f: X\to\mathbb{C}_2$ be a continuous function, and assume that the trace of C lies in X.
(a) Show that the integral $\int_C f(\zeta)\,d\zeta$ exists.
(b) Show that

$$\int_C f(\zeta)\,d\zeta = (\zeta_1-\zeta_0)\int_0^1 f[(1-t)\zeta_0+t\zeta_1]\,dt.$$

(c) Show that

$$\left\|\int_C f(\zeta)\,d\zeta\right\| \leqslant \sqrt{2}\,\|\zeta_1-\zeta_0\|\int_0^1 \|f[(1-t)\zeta_0+t\zeta_1]\|\,dt$$
$$\leqslant \sqrt{2}M\|\zeta_1-\zeta_0\|.$$

(d) Explain why the inequalities in (c) are special cases of the inequalities in Theorem 32.10 (55).

32.2 Let $f: X\to\mathbb{C}_2$ be a continuous function, and let $\zeta:[a,b]\to X$ be a curve C whose trace is in X and which has a continuous derivative. Then

$$\zeta = x_1+i_1x_2+i_2x_3+i_1i_2x_4,$$
$$f(\zeta)=g_1(x)+i_1g_2(x)+i_2g_3(x)+i_1i_2g_4(x),\qquad x=(x_1,\dots,x_4).$$

Also, there are functions $x_k:[a,b]\to\mathbb{C}_0$, $k=1,\ldots,4$, such that the curve C is the mapping $\zeta:[a,b]\to X$ defined by [see (1) in Section 31]

$$\zeta(t)=x_1(t)+i_1x_2(t)+i_2x_3(t)+i_1i_2x_4(t),\qquad a\leqslant t\leqslant b.$$

(a) Show that $g_1,\ldots,g_4$ are continuous functions of $x:(x_1,\ldots,x_4)$.
(b) Show that the functions $x_k:[a,b]\to\mathbb{C}_0$ have continuous derivatives.
(c) Let $a=t_0<t_1<\cdots<t_{i-1}<t_i<\cdots<t_n=b$ be a subdivision of $[a,b]$. Then the sum used in defining the integral $\int_C f(\zeta)\,d\zeta$ is

$$\sum_{i=1}^{n}\{g_1[x(t_{i-1})]+\cdots+i_1i_2g_4[x(t_{i-1})]\}$$
$$\{[x_1(t_i)-x_1(t_{i-1})]+\cdots+[x_4(t_i)-x_4(t_{i-1})]\}.$$

(d) The integral $\int_C f(\zeta)\,d\zeta$ exists by Theorem 32.2. Prove that the line integrals in the following equation exist and establish the equality.

$$\begin{aligned}\int_C f(\zeta)\,d\zeta=&\int_C g_1(x)\,dx_1-g_2(x)\,dx_2-g_3(x)\,dx_3+g_4(x)\,dx_4\\&+i_1\int_C g_2(x)\,dx_1+g_1(x)\,dx_2-g_4(x)\,dx_3-g_3(x)\,dx_4\\&+i_2\int_C g_3(x)\,dx_1-g_4(x)\,dx_2+g_1(x)\,dx_3-g_2(x)\,dx_4\\&+i_1i_2\int_C g_4(x)\,dx_1+g_3(x)\,dx_2+g_2(x)\,dx_3+g_1(x)\,dx_4.\end{aligned}$$

(e) Show that the line integrals in (d) can be evaluated as integrals in the variable t by replacing x by $x(t)$ and dx_k by $x_k'(t)\,dt$, $k=1,\ldots,4$.

32.3 The integral $\int_C f(\zeta)\,d\zeta$ in Exercise 32.2 has still another representation. Set

$$\zeta=z_1+i_2z_2,\qquad z_1,\,z_2\text{ in }\mathbb{C}_1,$$
$$f(\zeta)=u(z_1,z_2)+i_2v(z_1,z_2),\qquad z=(z_1,z_2).$$

Also, there are functions $z_k:[a,b]\to\mathbb{C}_1$, $k=1,2$, such that the curve C is the mapping $\zeta:[a,b]\to X$ defined by $\zeta(t)=z_1(t)+i_2(t)$, $a\leqslant t\leqslant b$.
(a) Show that

$$\int_C f(\zeta)\,d\zeta=\int_C u(z)\,dz_1-v(z)\,dz_2+i_2\int_C v(z)\,dz_1+u(z)\,dz_2.$$

(b) Explain how to evaluate the integrals in (a) as functions of t.

32.4 Let $f:\mathbb{C}_2\to\mathbb{C}_2$ be the function such that $f(\zeta)=\zeta$, and let C be the curve $\zeta:[a,b]\to\mathbb{C}_2$, $t\mapsto\zeta(t)$, such that $\zeta(a)=\zeta_0$ and $\zeta(b)=\zeta_1$. Show that $\int_C f(\zeta)\,d\zeta=(\zeta_1^2-\zeta_0^2)/2$. [*Hint*. Use Exercise 32.2 to show that

$$\begin{aligned}\int_C f(\zeta)\,d\zeta &= \int_a^b [x_1(t)x_1'(t)-x_2(t)x_2'(t)-x_3(t)x_3'(t)+x_4(t)x_4'(t)]\,dt\\ &\quad + i_1\int_a^b [x_2(t)x_1'(t)+x_1(t)x_2'(t)-x_4(t)x_3'(t)-x_3(t)x_4'(t)]\,dt\\ &\quad + i_2\int_a^b [x_3(t)x_1'(t)-x_4(t)x_2'(t)+x_1(t)x_3'(t)-x_2(t)x_4'(t)]\,dt\\ &\quad + i_1i_2\int_a^b [x_4(t)x_1'(t)+x_3(t)x_2'(t)+x_2(t)x_3'(t)+x_1(t)x_4'(t)]\,dt\\ &=(\zeta_1^2-\zeta_0^2)/2.]\end{aligned}$$

32.5 Give a second proof that $\int_C f(\zeta)\,d\zeta=(\zeta_1^2-\zeta_0^2)/2$. [*Hints*. Show that

$$\sum_{i=1}^n \zeta(t_{i-1})[\zeta(t_i)-\zeta(t_{i-1})]\to\int_C f(\zeta)\,d\zeta,$$

$$\sum_{i=1}^n \zeta(t_i)[\zeta(t_i)-\zeta(t_{i-1})]\to\int_C f(\zeta)\,d\zeta,$$

$$\sum_{i=1}^n [\zeta(t_i)+\zeta(t_{i-1})][\zeta(t_i)-\zeta(t_{i-1})]\to 2\int_C f(\zeta)\,d\zeta,$$

$$\sum_{i=1}^n [\zeta(t_i)+\zeta(t_{i-1})][\zeta(t_i)-\zeta(t_{i-1})]=[\zeta(t_n)]^2-[\zeta(t_0)]^2=\zeta_1^2-\zeta_0^2.]$$

32.6 Prove the following theorem. Let $f^n:X\to\mathbb{C}_2$, $n=1,2,\ldots$, be a sequence of bounded continuous functions which converges uniformly to the function $f:X\to\mathbb{C}_2$. Let C be a curve $\zeta:[a,b]\to X$ which has a continuous derivative. Then f is continuous, and

$$\lim_{n\to\infty}\int_C f^n(\zeta)\,d\zeta=\int_C f(\zeta)\,d\zeta.$$

[*Hint*. Show that

$$\left\|\int_C f^n(\zeta)\,d\zeta-\int_C f(\zeta)\,d\zeta\right\|=\left\|\int_C [f^n(\zeta)-f(\zeta)]\,d\zeta\right\|$$

and use Theorem 32.10.]

33. THE FUNDAMENTAL THEOREM OF THE INTEGRAL CALCULUS

The preceding section has treated the properties of integrals of functions $f: X \to \mathbb{C}_2$ which are only continuous rather than holomorphic. As expected, the integral $\int_C f(\zeta)\,d\zeta$ has important additional properties if f is holomorphic. One of these is contained in the fundamental theorem of the integral calculus, which is the subject of this section. Now the fundamental theorem concerns the integral $\int_C D_\zeta f(\zeta)\,d\zeta$, and the proof of the theorem requires that the derivative $D_\zeta f$ be continuous. Later it will be shown that every holomorphic function has a continuous derivative, but this section contains not only the hypothesis that f is holomorphic but also the additional hypothesis that f has a continuous derivative. The additional hypothesis of the continuity of the derivative will be removed later.

Three special representations of a holomorphic function $f: X \to \mathbb{C}_2$, $X \subset \mathbb{C}_2$, have been employed in earlier chapters; they are the following:

(1) $$f(\zeta) = g_1(x) + i_1 g_2(x) + i_2 g_3(x) + i_1 i_2 g_4(x), \qquad x = (x_1, \ldots, x_4),$$
$$\zeta = x_1 + i_1 x_2 + i_2 x_3 + i_1 i_2 x_4;$$

(2) $$f(\zeta) = u(z_1, z_2) + i_2 v(z_1, z_2), \qquad \zeta = z_1 + i_2 z_2;$$

(3) $$f(z_1 + i_2 z_2) = f_1(z_1 - i_1 z_2)e_1 + f_2(z_1 + i_1 z_2)e_2.$$

For aesthetic reasons it is desirable to develop the theory without resort to special representations, but it is not always possible to do so. Furthermore, a topic is sometimes treated in more than one representation in order to provide additional insight into the theorem. In particular, this section contains two treatments of the fundamental theorem of the integral calculus for functions $f: X \to \mathbb{C}_2$.

33.1 THEOREM (Fundamental Theorem of the Integral Calculus) Let $F: X \to \mathbb{C}_2$ be a holomorphic function whose derivative $D_\zeta F$ is continuous in X, and let C be a curve $\zeta: [a, b] \to X$ which has a continuous derivative and whose trace is in X. Then

(4) $$\int_C D_\zeta F(\zeta)\,d\zeta = F[\zeta(b)] - F[\zeta(a)].$$

Proof. The first proof of this theorem employs the special representation in (1); thus

(5) $$F(\zeta) = g_1(x) + i_1 g_2(x) + i_2 g_3(x) + i_1 i_2 g_4(x).$$

Then Corollary 23.2 (12) shows that

$$D_\zeta F(\zeta)=D_{x_1}g_1(x)+i_1D_{x_1}g_2(x)+i_2D_{x_1}g_3(x)+i_1i_2D_{x_1}g_4(x). \tag{6}$$

Also, by (1) and (3) in Section 31,

$$\zeta(t) = x_1(t) + i_1x_2(t) + i_2x_3(t) + i_1i_2x_4(t), \tag{7}$$

$$\zeta'(t) = x_1'(t) + i_1x_2'(t) + i_2x_3'(t) + i_1i_2x_4'(t). \tag{8}$$

Since $D_\zeta F$ and ζ' are continuous, then $D_{x_1}g_k$, $k=1,\dots,4$, are continuous functions of $x_1,\dots,x_4$, and x_k', $k=1,\dots,4$, are continuous functions of t. Then $\int_C D_\zeta F(\zeta)\,d\zeta$ exists by Theorem 32.2, and Theorem 32.9 shows that

$$\int_C D_\zeta F(\zeta)\,d\zeta = \int_a^b D_\zeta F[\zeta(t)]\zeta'(t)\,dt. \tag{9}$$

Replace $D_\zeta F[\zeta(t)]$ and $\zeta'(t)$ by their values in (6) and (8); then (9) shows that

$$\begin{aligned}
&\int_C D_\zeta F(\zeta)\,d\zeta \\
&\quad=\int_a^b [D_{x_1}g_1(x)x_1'(t)-D_{x_1}g_2(x)x_2'(t)-D_{x_1}g_3(x)x_3'(t) \\
&\quad+D_{x_1}g_4(x)x_4'(t)]\,dt \\
&\quad+i_1\int_a^b [D_{x_1}g_2(x)x_1'(t)+D_{x_1}g_1(x)x_2'(t)-D_{x_1}g_4(x)x_3'(t) \\
&\quad-D_{x_1}g_3(x)x_4'(t)]\,dt \\
&\quad+i_2\int_a^b [D_{x_1}g_3(x)x_1'(t)-D_{x_1}g_4(x)x_2'(t)+D_{x_1}g_1(x)x_3'(t) \\
&\quad-D_{x_1}g_2(x)x_4'(t)]\,dt \\
&\quad+i_1i_2\int_a^b [D_{x_1}g_4(x)x_1'(t)+D_{x_1}g_3(x)x_2'(t)+D_{x_1}g_2(x)x_3'(t) \\
&\quad+D_{x_1}g_1(x)x_4'(t)]\,dt.
\end{aligned} \tag{10}$$

Now F is a holomorphic function by hypothesis, and Theorem 23.1 (7)–(10) shows that $g_1,\dots,g_4$ satisfy the Cauchy–Riemann differential equations. Use these equations to write each integrand as a derivative as follows:

$$(11)\quad \int_C D_\zeta F(\zeta)\,d\zeta$$

$$= \int_a^b [D_{x_1}g_1(x)x_1'(t) + D_{x_2}g_1(x)x_2'(t) + D_{x_3}g_1(x)x_3'(t)$$

$$+ D_{x_4}g_1(x)x_4'(t)]\,dt$$

$$+ i_1 \int_a^b [D_{x_1}g_2(x)x_1'(t) + D_{x_2}g_2(x)x_2'(t) + D_{x_3}g_2(x)x_3'(t)$$

$$+ D_{x_4}g_2(x)x_4'(t)]\,dt$$

$$+ i_2 \int_a^b [D_{x_1}g_3(x)x_1'(t) + D_{x_2}g_3(x)x_2'(t) + D_{x_3}g_3(x)x_3'(t)$$

$$+ D_{x_4}g_3(x)x_4'(t)]\,dt$$

$$+ i_1 i_2 \int_a^b [D_{x_1}g_4(x)x_1'(t) + D_{x_2}g_4(x)x_2'(t) + D_{x_3}g_4(x)x_3'(t)$$

$$+ D_{x_4}g_4(x)x_4'(t)]\,dt.$$

Now

$$(12)\qquad \sum_{k=1}^{4} D_{x_k}g[x_1(t), \ldots, x_4(t)]x_k'(t) = D_t g[x_1(t), \ldots, x_4(t)];$$

therefore (11) can be written as follows:

$$(13)\quad \int_C D_\zeta F(\zeta)\,d\zeta = \int_a^b D_t g_1[x_1(t), \ldots, x_4(t)]\,dt + i_1 \int_a^b D_t g_2[x_1(t), \ldots, x_4(t)]\,dt$$

$$+ i_2 \int_a^b D_t g_3[x_1(t), \ldots, x_4(t)]\,dt + i_1 i_2 \int_a^b D_t g_4[x_1(t), \ldots, x_4(t)]\,dt.$$

Then (13) and the fundamental theorem of the integral calculus for functions of a single real variable show that

$$\begin{aligned}(14)\quad \int_C D_\zeta F(\zeta)\,d\zeta &= \{g_1[x(b)] - g_1[x(a)]\} + i_1\{g_2[x(b)] - g_2[x(a)]\} \\ &\quad + i_2\{g_3[x(b)] - g_3[x(a)]\} + i_1 i_2\{g_4[x(b)] - g_4[x(a)]\} \\ &= \{g_1[x(b)] + i_1 g_2[x(b)] + i_2 g_3[x(b)] + i_1 i_2 g_4[x(b)]\} \\ &\quad - \{g_1[x(a)] + i_1 g_2[x(a)] + i_2 g_3[x(a)] + i_1 i_2 g_4[x(a)]\}, \\ &= F[\zeta(b)] - F[\zeta(a)].\end{aligned}$$

The proof of (4) and of Theorem 33.1 is complete. □

The next two theorems contain applications of the fundamental theorem of the integral calculus.

33.2 THEOREM Let $F: X \to \mathbb{C}_2$ and $G: X \to \mathbb{C}_2$ be two holomorphic functions which have continuous derivatives $D_\zeta F$ and $D_\zeta G$ in X such that

$$D_\zeta F(\zeta) = D_\zeta G(\zeta), \qquad \zeta \in X. \tag{15}$$

Then there is a constant c in $\mathbb{C}_2$ such that

$$F(\zeta) = G(\zeta) + c, \qquad \zeta \in X. \tag{16}$$

Proof. Since X is a domain, it is connected. If ζ_1 and ζ_2 are two points in X, then there is a polygonal curve in X which connects ζ_1 and ζ_2. Since F and G have continuous derivatives, then the formula in (4) holds for them (see Exercise 33.1) on this curve; therefore

$$\int_C D_\zeta F(\zeta)\,d\zeta = F(\zeta_2) - F(\zeta_1), \qquad \int_C D_\zeta G(\zeta)\,d\zeta = G(\zeta_2) - G(\zeta_1). \tag{17}$$

Subtract the second equation from the first; thus

$$\int_C D_\zeta F(\zeta)\,d\zeta - \int_C D_\zeta G(\zeta)\,d\zeta = [F(\zeta_2) - F(\zeta_1)] - [G(\zeta_2) - G(\zeta_1)]. \tag{18}$$

Then by Theorem 32.8,

$$\int_C [D_\zeta F(\zeta) - D_\zeta G(\zeta)]\,d\zeta = [F(\zeta_2) - F(\zeta_1)] - [G(\zeta_2) - G(\zeta_1)]. \tag{19}$$

The hypothesis in (15) shows that the integrand of the integral in (19) is identically zero; therefore this integral is zero and

$$[F(\zeta_2) - F(\zeta_1)] - [G(\zeta_2) - G(\zeta_1)] = 0, \tag{20}$$

$$F(\zeta_2) = G(\zeta_2) + [F(\zeta_1) - G(\zeta_1)]. \tag{21}$$

This equation holds for ζ_1 fixed and every ζ_2 in X. If $c = F(\zeta_1) - G(\zeta_1)$, then (21) shows that (16) is true. The proof of Theorem 33.2 is complete. □

Before the theorem which contains the second application of the fundamental theorem of the integral calculus can be stated and proved, it is necessary to define the uniform strong Stolz condition. Let $f: X \to \mathbb{C}_2$ be a holomorphic function. Then f satisfies the following strong Stolz condition at each ζ_0 in X:

$$f(\zeta) - f(\zeta_0) = D_\zeta f(\zeta)(\zeta - \zeta_0) + r(f; \zeta_0, \zeta)(\zeta - \zeta_0), \tag{22}$$

$$\lim_{\zeta \to \zeta_0} r(f; \zeta_0, \zeta) = 0, \qquad r(f; \zeta_0, \zeta_0) = 0.$$

Thus, for each $\varepsilon>0$, there exists a $\delta(\varepsilon, \zeta_0)$, such that

$$(23) \qquad \|r(f; \zeta_0, \zeta)\| < \varepsilon, \qquad \|\zeta - \zeta_0\| < \delta(\varepsilon, \zeta_0).$$

There are situations in which δ does not depend on ζ_0 but only on ε; then f satisfies the uniform Stolz condition as stated in the next definition.

33.3 DEFINITION Let $f: X \to \mathbb{C}_2$ be a holomorphic function defined in the domain X, and let S be a nonempty set such that $S \subset X$. If for each $\varepsilon>0$ there exists a $\delta(\varepsilon)$, which does not depend on ζ_0, such that

$$(24) \qquad \|r(f; \zeta_0, \zeta)\| < \varepsilon$$

for every ζ_0 and ζ in X for which ζ_0 is in S and $\|\zeta - \zeta_0\| < \delta(\varepsilon)$, then f is said to satisfy the *uniform strong Stolz condition* in S.

33.4 THEOREM Let $F: X \to \mathbb{C}_2$ be a holomorphic function which has a continuous derivative, and let S be a compact set in X. Then F satisfies the uniform strong Stolz condition in S.

Proof. Since S is compact, it is closed and bounded. Then the distance from every point of S to the complement of X is equal to or greater than a constant denoted by $2d>0$. Let S_d be the set of points whose distance from S is equal to or less than d. Then $S \subset S_d \subset X$, and S_d is compact. Let ζ_0 be a point in S, and let ζ be a point such that $\|\zeta - \zeta_0\| < d$; then $\zeta \in S_d$. Let C be the curve whose equation is

$$(25) \qquad \zeta(t) = \zeta_0 + t(\zeta - \zeta_0), \qquad 0 \leqslant t \leqslant 1.$$

Then the trace of C is the segment which connects ζ_0 to ζ; this segment is in S_d. Since F has a continuous derivative, the integral $\int_C D_\zeta F(\zeta)\, d\zeta$ is defined, and F satisfies the fundamental theorem of the integral calculus by Theorem 33.1; thus

$$(26) \qquad F(\zeta) - F(\zeta_0) = \int_C D_\zeta F(\zeta)\, d\zeta.$$

Introduce t as the variable of integration by (25) to obtain

$$(27) \qquad F(\zeta) - F(\zeta_0) = \left\{ \int_0^1 D_\zeta F[\zeta_0 + t(\zeta - \zeta_0)]\, dt \right\} (\zeta - \zeta_0).$$

Add and subtract $D_\zeta F(\zeta_0)(\zeta - \zeta_0)$ on the right in this equation as follows:

$$(28) \qquad F(\zeta) - F(\zeta_0) = D_\zeta F(\zeta_0)(\zeta - \zeta_0) + \int_0^1 \{D_\zeta F[\zeta_0 + t(\zeta - \zeta_0)] - D_\zeta F(\zeta_0)\}\, dt (\zeta - \zeta_0).$$

Define $r(F; \zeta_0, \zeta)$ as follows:

$$(29) \qquad r(F; \zeta_0, \zeta) = \int_0^1 \{D_\zeta F[\zeta_0 + t(\zeta - \zeta_0)] - D_\zeta F(\zeta_0)\}\, dt.$$

Then $r(F; \zeta_0, \zeta_0)=0$, and the proof can be completed by showing that $r(F; \zeta_0, \zeta)$ is uniformly small in S. Let $\varepsilon>0$ be given. Since $D_\zeta F$ is continuous in X, it is uniformly continuous in S_d. Then there exists a $\delta_0(\varepsilon)$ such that

$$(30) \qquad \|D_\zeta F(\zeta) - D_\zeta F(\zeta_0)\| < \varepsilon$$

for each two points ζ_0, ζ in S_d for which $\|\zeta-\zeta_0\|<\delta_0(\varepsilon)$. Let $\delta(\varepsilon)=\min[\delta_0(\varepsilon), d]$. Finally, let ζ_0 be a point in S, and let ζ be a point for which $\|\zeta-\zeta_0\|<\delta(\varepsilon)$. Then ζ is in S_d and

$$(31) \qquad \|D_\zeta F[\zeta_0 + t(\zeta - \zeta_0)] - D_\zeta F(\zeta_0)\| < \varepsilon, \qquad 0 \leqslant t \leqslant 1.$$

This inequality and (29) show that

$$(32) \qquad \|r(F; \zeta_0, \zeta)\| < \int_0^1 \varepsilon\, dt = \varepsilon.$$

Therefore, F satisfies the uniform strong Stolz condition in S by Definition 33.3, and the proof of Theorem 33.4 is complete. □

33.5 DEFINITION Let $f: X \to \mathbb{C}_2$ be a holomorphic function which satisfies the weak Stolz condition in X as follows:

$$f(\zeta) - f(\zeta_0) = D_\zeta f(\zeta_0)(\zeta - \zeta_0) + r(f; \zeta_0, \zeta)\|\zeta - \zeta_0\|,$$

$$\lim_{\zeta\to\zeta_0} r(f; \zeta_0, \zeta) = 0, \qquad r(f; \zeta_0, \zeta_0) = 0.$$

If for each $\varepsilon>0$ there exists a $\delta(\varepsilon)$, which does not depend on ζ_0, such that

$$\|r(f; \zeta_0, \zeta)\| < \varepsilon$$

for every ζ_0 and ζ in X for which ζ_0 is in S and $\|\zeta-\zeta_0\|<\delta(\varepsilon)$, then f is said to satisfy the *uniform weak Stolz condition* in S.

The proof of Theorem 33.4 shows that, if F has a continuous derivative in X, then the fundamental theorem of the integral calculus can be used to prove that F satisfies the uniform strong Stolz condition in S. We shall now prove that F satisfies the uniform weak Stolz condition without using the fundamental theorem. The proof employs the special representation of $F: X \to \mathbb{C}_2$ in (5).

33.6 THEOREM Let $F: X \to \mathbb{C}_2$ be a holomorphic function which has a continuous derivative, and let S be a compact set in X. Then F satisfies the uniform weak Stolz condition in S.

Proof. Theorem 33.6 is a trivial consequence of Theorem 33.4; the object here is to exhibit a proof which does not employ the fundamental theorem of the integral calculus.

Equation (6) shows that

$$\|D_\zeta F(\zeta) - D_\zeta F(\zeta_0)\| = \left[\sum_{k=1}^{4} |D_{x_1} g_k(x) - D_{x_1} g_k(x^0)|^2\right]^{1/2}. \tag{33}$$

By hypothesis, $D_\zeta F$ is continuous in X; then (33) shows that $D_{x_1} g_k$, $k=1,\ldots,4$, are continuous. Next, the Cauchy–Riemann equations in (7)–(10) in Theorem 23.1 show that each of the derivatives

$$D_{x_1} g_k, \ldots, D_{x_4} g_k, \qquad k = 1, \ldots, 4, \tag{34}$$

is continuous in X. Then these derivatives are uniformly continuous in S_d. Now by (5),

$$\begin{aligned} F(\zeta) - F(\zeta_0) = {} & [g_1(x) - g_1(x^0)] + i_1[g_2(x) - g_2(x^0)] \\ & + i_2[g_3(x) - g_3(x^0)] + i_1 i_2[g_4(x) - g_4(x^0)]. \end{aligned} \tag{35}$$

Let x^0 be a point in S, and let x be a point such that $\|x - x^0\| < d$. Then x is in S_d, and the segment connecting x^0 and x is in S_d. Each of the functions $g_1, \ldots, g_4$ satisfies the hypothesis of the mean-value theorem [7, p. 74] (compare Theorem 20.6); therefore, for $k = 1, \ldots, 4$,

$$g_k(x) - g_k(x^0) = \sum_{r=1}^{4} D_{x_r} g_k(x^0)(x_r - x_r^0) + r(g_k; x^0, x)\|x - x^0\| \tag{36}$$

$$|r(g_k; x^0, x)| \leqslant \left\{\sum_{r=1}^{4} [D_{x_r} g_k(x_k^*) - D_{x_r} g_k(x^0)]^2\right\}^{1/2}, \; r(g_k; x^0, x^0) = 0 \tag{37}$$

$$x_k^* = x^0 + t_k^*(x - x^0), \qquad 0 \leqslant t_k^* \leqslant 1. \tag{38}$$

Substitute from (36) in (35); the Cauchy–Riemann equations can be used to simplify the resulting equation to the following:

$$F(\zeta) - F(\zeta_0) = D_\zeta F(\zeta_0)(\zeta - \zeta_0) + r(F; \zeta_0, \zeta)\|\zeta - \zeta_0\|, \tag{39}$$

where

$$D_\zeta F(\zeta_0) = D_{x_1} g_1(x^0) + i_1 D_{x_1} g_2(x^0) + i_2 D_{x_1} g_3(x^0) + i_1 i_2 D_{x_1} g_4(x^0), \tag{40}$$

$$\zeta - \zeta_0 = (x_1 - x_1^0) + i_1(x_2 - x_2^0) + i_2(x_3 - x_3^0) + i_1 i_2(x_4 - x_4^0), \tag{41}$$

$$r(F; \zeta_0, \zeta) = r(g_1; x^0, x) + i_1 r(g_2; x^0, x) + i_2 r(g_3; x^0, x) + i_1 i_2 r(g_4; x^0, x), \tag{42}$$

$$\|\zeta - \zeta_0\| = \|x - x^0\|. \tag{43}$$

Equation (39) shows that the proof of the theorem can be completed by showing that $r(F;\zeta_0,\zeta)$ is uniformly small in S. Now (42) shows that

$$(44)\qquad \|r(F;\zeta_0,\zeta)\| = \left\{\sum_{k=1}^{4} |r(g_k;x^0,x)|^2\right\}^{1/2}.$$

The derivatives in (34) are uniformly continuous in S_d. Let $\varepsilon>0$ be given; then there exists a $\delta(\varepsilon)<d$, which depends on ε but not on x^0, such that by (37),

$$(45)\qquad |r(g_k;x^0,x)| < \varepsilon/2, \qquad k=1,\dots,4,$$

for every x^0 in S and x such that $\|x-x^0\|<\delta(\varepsilon)$. Then (44) and (45) show that

$$(46)\qquad \|r(F;\zeta_0,\zeta)\| < \varepsilon$$

for every $\zeta_0\ (=x^0)$ and $\zeta\ (=x)$ such that $\|\zeta-\zeta_0\|<\delta(\varepsilon)$. Now (39) shows that F satisfies the weak Stolz condition in X, and (46) shows that F satisfies the uniform weak Stolz condition in S. The proof of Theorem 33.6 is complete. □

Thus far the results in this section have been derived from the assumption that $f:X\to\mathbb{C}_2$ has a continuous derivative. For example, Theorem 33.1 uses the hypothesis that $D_\zeta F$ is continuous to establish the fundamental theorem of the integral calculus, and Theorems 33.4 and 33.6 use the same hypothesis to show that f satisfies the uniform strong and weak Stolz conditions, respectively, in a compact set S. The proofs of Theorems 33.1 and 33.6 have employed the representation of f in (1); the details are excessive and not very elegant. There is an alternative way to proceed: the results in the section can be derived from the assumption that f satisfies the uniform strong Stolz condition. Thus Theorem 33.7 (below) demonstrates that a function which satisfies this condition has a continuous derivative, and Theorem 33.8 uses the condition to give a more elegant proof of the fundamental theorem of the integral calculus.

33.7 THEOREM If the function $f:X\to\mathbb{C}_2$ satisfies the uniform strong Stolz condition in S, then f has a continuous derivative $D_\zeta f$ in S.

Proof. By hypothesis, f has a derivative $D_\zeta f$ in X. Also,

$$(47)\qquad f(\zeta)-f(\zeta_0) = D_\zeta f(\zeta_0)(\zeta-\zeta_0) + r(f;\zeta_0,\zeta)(\zeta-\zeta_0).$$

Since f satisfies the uniform strong Stolz condition in S, then for each $\varepsilon>0$ there exists a $\delta(\varepsilon)$ such that

$$(48)\qquad \|r(f;\zeta_0,\zeta)\| < \varepsilon$$

for each ζ_0 in S and ζ provided

$$(49)\qquad \|\zeta-\zeta_0\| < \delta(\varepsilon).$$

Now

$$(50)\qquad \zeta = z_1 + i_2 z_2, \qquad \zeta_0 = z_1^0 + i_2 z_2^0,$$
$$\zeta = (z_1 - i_1 z_2)e_1 + (z_1 + i_1 z_2)e_2.$$

For the sake of convenience, employ the following notation:

$$(51)\qquad w_1 = z_1 - i_1 z_2, \qquad w_2 = z_1 + i_1 z_2,$$
$$w_1^0 = z_1^0 - i_1 z_2^0, \qquad w_2^0 = z_1^0 + i_1 z_2^0,$$
$$\zeta = w_1 e_1 + w_2 e_2, \qquad \zeta_0 = w_1^0 e_1 + w_2^0 e_2.$$

The following formulas are derived from the idempotent representation of elements in $\mathbb{C}_2$:

$$(52)\qquad f(\zeta) = f_1(w_1)e_1 + f_2(w_2)e_2,$$

$$(53)\qquad D_\zeta f(\zeta) = D_{w_1} f_1(w_1^0)e_1 + D_{w_2} f_2(w_2^0)e_2,$$

$$(54)\qquad r(f; \zeta_0, \zeta) = r(f_1; w_1^0, w_1)e_1 + r(f_2; w_2^0, w_2)e_2,$$

$$(55)\qquad \|r(f; \zeta_0, \zeta)\| = \left[1/2 \sum_{k=1}^{2} |r(f_k; w_k^0, w_k)|^2\right]^{1/2},$$

$$(56)\qquad \|\zeta - \zeta_0\| = \left[1/2 \sum_{k=1}^{2} |w_k - w_k^0|^2\right]^{1/2}.$$

The idempotent representations of the expressions on the two sides of (47) are equal, and corresponding components in these representations are equal. Thus the strong Stolz condition in (47) is equivalent to the following two conditions:

$$(57)\quad f_1(w_1) - f_1(w_1^0) = D_{w_1} f_1(w_1^0)(w_1 - w_1^0) + r(f_1; w_1^0, w_1)(w_1 - w_1^0),$$

$$(58)\quad f_2(w_2) - f_2(w_2^0) = D_{w_2} f_2(w_2^0)(w_2 - w_2^0) + r(f_2; w_2^0, w_2)(w_2 - w_2^0).$$

Define sets S_1 and S_2 as follows:

$$(59)\qquad S_1 = \{w_1 \text{ in } X_1 \colon w_1 e_1 + w_2 e_2 \in S\},$$
$$S_2 = \{w_2 \text{ in } X_2 \colon w_1 e_1 + w_2 e_2 \in S\}.$$

The first step in the proof is to show that f_1 and f_2 satisfy the uniform strong Stolz condition in S_1 and S_2, respectively. Let $\varepsilon > 0$ be given, and let ζ_0 be a point in S as in (48). If $\zeta_0 = w_1^0 e_1 + w_2^0 e_2$ as in (51), then $w_1^0 \in S_1$ and $w_2^0 \in S_2$. Let w_1 in S_1 and w_2 in S_2 be points such that

$$(60)\qquad |w_1 - w_1^0| < \delta(\varepsilon), \qquad |w_2 - w_2^0| < \delta(\varepsilon).$$

Then (56) shows that $\|\zeta-\zeta_0\|<\delta(\varepsilon)$, and (49) shows that (48) holds. Then (55) shows that

$$(61)\qquad |r(f_1;\, w_1^0,\, w_1)|<\sqrt{2}\varepsilon,\qquad |r(f_2;\, w_2^0,\, w_2)|<\sqrt{2}\varepsilon.$$

Since the inequalities in (60) imply the inequalities in (61), then f_1 and f_2 satisfy the uniform strong Stolz condition in S_1 and S_2, respectively. The second step in the proof is to show that $D_{w_1}f_1$ and $D_{w_2}f_2$ are continuous in S_1 and S_2. Let w_1^0 and w_1 be points in S_1 such that $w_1-w_1^0\notin\mathcal{O}_1$. Then (57) shows that

$$(62)\qquad \begin{aligned} f_1(w_1)-f_1(w_1^0) &= D_{w_1}f_1(w_1^0)(w_1-w_1^0)\\ &\quad + r(f_1;\, w_1^0,\, w_1)(w_1-w_1^0),\\ f_1(w_1^0)-f_1(w_1) &= D_{w_1}f_1(w_1)(w_1^0-w_1)\\ &\quad + r(f_1;\, w_1,\, w_1^0)(w_1^0-w_1),\qquad w_1-w_1^0\notin\mathcal{O}_1. \end{aligned}$$

Add the two equations in (62). Divide the resulting equation by $w_1-w_1^0$; this step is possible since $w_1-w_1^0\in\mathbb{C}_1$ but $w_1-w_1^0\notin\mathcal{O}_1$. The final equation can be given the following form:

$$(63)\qquad D_{w_1}f_1(w_1)-D_{w_1}f_1(w_1^0)=r(f_1;\, w_1^0,\, w_1)-r(f_1;\, w_1,\, w_1^0).$$

Let $\varepsilon>0$ be given. If w_1^0 and w_1 are two points in S_1 such that $0<|w_1-w_1^0|<\delta(\varepsilon)$ as in (60), then because f_1 satisfies the uniform strong Stolz condition in S_1, we have, by (61),

$$(64)\qquad |r(f_1;\, w_1^0,\, w_1)|<\sqrt{2}\varepsilon,\qquad |r(f_1;\, w_1,\, w_1^0)|<\sqrt{2}\varepsilon.$$

Thus (63) and (64) show that

$$(65)\qquad |D_{w_1}f_1(w_1)-D_{w_1}f_1(w_1^0)|\leqslant |r(f_1;\, w_1^0,\, w_1)|+|r(f_1;\, w_1,\, w_1^0)|<2\sqrt{2}\varepsilon$$

provided $|w_1-w_1^0|<\delta(\varepsilon)$. This statement proves that $D_{w_1}f_1$ is continuous at w_1^0. Since w_1^0 is an arbitrary point in S_1, then $D_{w_1}f_1$ is continuous in S_1. A similar argument, based on (58), shows that

$$(66)\qquad \begin{aligned} |D_{w_2}f_2(w_2)-D_{w_2}f_2(w_2^0)| &\leqslant |r(f_2;\, w_2^0,\, w_2)|\\ &\quad + |r(f_2;\, w_2,\, w_2^0)|<2\sqrt{2}\varepsilon \end{aligned}$$

provided $|w_2-w_2^0|<\delta(\varepsilon)$. Thus $D_{w_2}f_2$ is continuous in S_2. The third and final step in the proof is to show that $D_\zeta f$ is continuous. Let ζ_0 be a point in S, and let $\varepsilon>0$ be a given. Equation (53) shows that

$$(67)\qquad \|D_\zeta f(\zeta)-D_\zeta f(\zeta_0)\|=\left[(1/2)\sum_{k=1}^{2}|D_{w_k}f_k(w_k)-D_{w_k}f_k(w_k^0)|^2\right]^{1/2}.$$

If ζ is an arbitrary point such that (49) holds, then (65), (66), and (67) show that

$$\|D_\zeta f(\zeta) - D_\zeta f(\zeta_0)\| < 2\sqrt{2}\varepsilon, \tag{68}$$

and therefore $D_\zeta f$ is continuous in S. The proof of Theorem 33.7 is complete. □

One version of the fundamental theorem of the integral calculus has been proved already in Theorem 33.1; in this theorem the function is represented as in (1), and the proof is based on the theory of functions of a real variable. The next theorem contains another version of the fundamental theorem. The interest is in the proof, which is simple and elegant. It is based on the assumption that the function satisfies the uniform strong Stolz condition, and it uses only properties of holomorphic functions in $\mathbb{C}_2$.

33.8 THEOREM (Fundamental Theorem of the Integral Calculus) Let $F: X \to \mathbb{C}_2$ be a holomorphic function which satisfies the uniform strong Stolz condition in every compact set S in X, and let C be a curve $\zeta: [a, b] \to X$ which has a continuous derivative and whose trace is in X. Then

$$\int_C D_\zeta F(\zeta)\, d\zeta = F[\zeta(b)] - F[\zeta(a)]. \tag{69}$$

Proof. Let S denote the trace of C in X; since C, or $\zeta: [a, b] \to X$, is continuous, then S is a closed and bounded set, that is, a compact set. Since X is open, there is a number $d > 0$ such that the distance of each point of S from the complement of X is equal to or greater than $2d$. Let S_d denote the set of points in X whose distance from S is equal to or less than d. Then $S \subset S_d \subset X$. Now, by hypothesis, F satisfies the uniform strong Stolz condition in S; therefore, by Theorem 33.7, F has a continuous derivative in S and the integral $\int_C D_\zeta F(\zeta)\, d\zeta$ exists by Theorem 32.2. The proof of the theorem will be completed by establishing the formula in (69). Let $a = t_0 < t_1 < \cdots < t_{i-1} < t_i < \cdots < t_n = b$ be a subdivision of $[a, b]$. Then for every such subdivision,

$$\sum_{i=1}^{n} \{F[\zeta(t_i)] - F[\zeta(t_{i-1})]\} = F[\zeta(t_n)] - F[\zeta(t_0)] = F[\zeta(b)] - F[\zeta(a)]. \tag{70}$$

The proof will show that the limit of the sum on the left in (70) is $\int_C D_\zeta F(\zeta)\, d\zeta$. Now F satisfies the uniform strong Stolz condition in S. Thus for $i = 1, \ldots, n$,

$$\begin{aligned} &F[\zeta(t_i)] - F[\zeta(t_{i-1})] \\ &\quad = D_\zeta F[\zeta(t_{i-1})][\zeta(t_i) - \zeta(t_{i-1})] \\ &\qquad + r[F; \zeta(t_{i-1}), \zeta(t_i)][\zeta(t_i) - \zeta(t_{i-1})]. \end{aligned} \tag{71}$$

Add these equations; by (70) the result is

$$(72)\qquad \begin{aligned} &F[\zeta(b)] - F[\zeta(a)] \\ &= \sum_{i=1}^{n} D_\zeta F[\zeta(t_{i-1})][\zeta(t_i) - \zeta(t_{i-1})] \\ &\quad + \sum_{i=1}^{n} r[F;\, \zeta(t_{i-1}),\, \zeta(t_i)][\zeta(t_i) - \zeta(t_{i-1})]. \end{aligned}$$

For a sequence of subdivisions whose norms approach zero, the limit of the first sum on the right in (72) is $\int_C D_\zeta F(\zeta)\, d\zeta$ by Theorem 32.2. Thus the proof of (69) can be completed by showing that the limit of the second sum on the right in (72) is zero. Now by hypothesis, F satisfies the uniform strong Stolz condition in S. Thus for each $\varepsilon > 0$ there exists a $\delta(\varepsilon)$ such that, if

$$(73)\qquad \|\zeta(t_i) - \zeta(t_{i-1})\| < \delta(\varepsilon),$$

then

$$(74)\qquad \|r[F;\, \zeta(t_{i-1}),\, \zeta(t_i)]\| < \varepsilon.$$

Now $\zeta : [a, b] \to X$ is a continuous function on the compact set $[a, b]$; it is therefore uniformly continuous. Thus if the norm of the subdivision of $[a, b]$ is sufficiently small, then (73) is satisfied for $i = 1, \ldots, n$. Let P_k, $k = 1, 2, \ldots$, be a sequence of subdivisions of $[a, b]$ whose norms approach zero. Then for all sufficiently large k, (73) holds for $i = 1, \ldots, n$, and

$$(75)\qquad \begin{aligned} &\left\| \sum_{i=1}^{n} r[F;\, \zeta(t_{i-1}),\, \zeta(t_i)][\zeta(t_i) - \zeta(t_{i-1})] \right\| \\ &\quad < \sqrt{2}\varepsilon \sum_{i=1}^{n} \|\zeta(t_i) - \zeta(t_{i-1})\| < \sqrt{2}\varepsilon L(C). \end{aligned}$$

Here $L(C)$ is the length of the curve C. Thus (75) shows that the limit of the second sum on the right in (72) is zero, and (69) is true. The proof of Theorem 33.8 is complete. □

The following summary of the principal results in this section may be helpful. The topics treated are these:

(a) Continuous derivative;
(b) The fundamental theorem of the integral calculus;
(c) The uniform strong Stolz condition;
(d) The uniform weak Stolz condition.

Starting with functions $F : X \to \mathbb{C}_2$ represented in the real-variable form (1), the following results have been established:

(76) (a) implies (b) (Theorem 33.1); (a) implies (c) (Theorem 33.4); (c) implies (a) (Theorem 33.7);

(77) (a) implies (d) (Theorem 33.6).

Using only the methods of holomorphic functions in $\mathbb{C}_2$, the following results have been established:

(78) (c) implies (a) (Theorem 33.7); (c) implies (b) (Theorem 33.8).

Furthermore, it is clear from the proof of Theorem 33.8 that (d) implies (b).

Exercises

33.1 Let the curves C_1 and C_2 be the mappings $\zeta_1:[a,b]\to X$ and $\zeta_2:[b,c]\to X$, respectively. Assume that C_1 and C_2 have continuous derivatives, and that $\zeta_1(b)=\zeta_2(b)$. Join C_1 and C_2 together to form a curve $\zeta:[a,c]\to X$ which is continuous, but whose derivative may have a discontinuity at $t=b$. Let $F:X\to\mathbb{C}_2$ be a function which has a continuous derivative $D_\zeta F$ in X. Show that $\int_C D_\zeta F(\zeta)\,d\zeta = F[\zeta(c)]-F[\zeta(a)]$. [*Hints.* Theorems 32.7 and 33.1.]

33.2 Prove the following theorem, using the real-variable representation in (1) for the functions F and G. Let $F:X\to\mathbb{C}_2$ and $G:X\to\mathbb{C}_2$ be holomorphic functions which have continuous derivatives $D_\zeta F$ and $D_\zeta G$ in X. If $D_\zeta F(\zeta)=D_\zeta G(\zeta)$ for all ζ in X, then there is a constant c in $\mathbb{C}_2$ such that $F(\zeta)=G(\zeta)+c$ for all ζ in X.

33.3 Prove the following theorem. Let C be a curve $\zeta:[a,b]\to\mathbb{C}_2$, $t\mapsto\zeta(t)$, which has a continuous derivative. Then

$$\int_C \zeta^n\,d\zeta = \frac{[\zeta(b)]^{n+1}-[\zeta(a)]^{n+1}}{n+1}.$$

[*Hints.* Exercise 26.2 and Theorem 33.1; compare Exercise 32.5.]

33.4 Let e^ζ denote the exponential function of ζ defined in Definition 17.7 (29), and let C be a curve with a continuous derivative which connects the origin to the point ζ_0 in $\mathbb{C}_2$.

(a) Show that $\int_C e^\zeta\,d\zeta = e^{\zeta_0}-1$. [*Hint.* $D_\zeta e^\zeta = e^\zeta$.]

(b) Show that the integral in (a) has the same value for every curve C which connects the origin to ζ_0.

34. CAUCHY'S INTEGRAL THEOREM: A SPECIAL CASE

This section continues the proof of properties of the integral $\int_C f(\zeta)\,d\zeta$ which result from the assumption that $f: X \to \mathbb{C}_2$ is a holomorphic function. More specifically, the theorem in this section is a special case of Cauchy's integral theorem; the proof depends critically on the fact that a holomorphic function satisfies the strong Stolz condition. General forms of this fundamental theorem are found in Section 36.

Let $f: X \to \mathbb{C}_2$ be a function which is holomorphic (and therefore continuous by Theorem 26.1) in the domain X, and let T denote the oriented curve whose trace consists of the segments $[\zeta_0, \zeta_1]$, $[\zeta_1, \zeta_2]$, and $[\zeta_2, \zeta_0]$ of the triangle with vertices $\zeta_0, \zeta_1, \zeta_2$. The equations of the three sides S_0, S_1, S_2 of T are (compare Exercise 31.3)

$$(1)\qquad \begin{aligned} &S_0\colon \zeta(t) = (1-t)\zeta_0 + t\zeta_1, && 0 \leqslant t \leqslant 1,\\ &S_1\colon \zeta(t) = (1-t)\zeta_1 + t\zeta_2, && 0 \leqslant t \leqslant 1,\\ &S_2\colon \zeta(t) = (1-t)\zeta_2 + t\zeta_0, && 0 \leqslant t \leqslant 1. \end{aligned}$$

Let $c(T)$ denote the convex extension of $\{\zeta_0, \zeta_1, \zeta_2\}$; it is the solid triangle with vertices $\zeta_0, \zeta_1, \zeta_2$. Assume that $c(T)$ is contained in X. Then Theorem 32.2 shows that f has an integral on each of the curves S_0, S_1, S_2 and Theorem 32.7 shows that

$$(2)\qquad \int_T f(\zeta)\,d\zeta = \int_{S_0} f(\zeta)\,d\zeta + \int_{S_1} f(\zeta)\,d\zeta + \int_{S_2} f(\zeta)\,d\zeta.$$

34.1 THEOREM If $f: X \to \mathbb{C}_2$ is a holomorphic function, and if T is a curve such that $c(T) \subset X$, then

$$(3)\qquad \int_T f(\zeta)\,d\zeta = 0.$$

Proof. The method of proof is similar to that used in the proof of the corresponding theorem in complex-variable theory (see [5, I, pp. 163–166]), but the proof uses no results in the theory of holomorphic functions of a complex variable. The proof is an indirect one; it employs the nesting process. In the last analysis, the proof depends on the fact that a holomorphic function satisfies the strong Stolz condition (see Definitions 15.2 and 20.14 and Theorem 25.1).

Assume then that the theorem is false. Denote the triangle T by T_0; then

$$(4)\qquad \left\| \int_{T_0} f(\zeta)\,d\zeta \right\| > 0.$$

The proof of the theorem consists of showing that this assumption leads to a contradiction.

Set

$$\zeta_{00} = \frac{\zeta_0 + \zeta_1}{2}, \qquad \zeta_{01} = \frac{\zeta_1 + \zeta_2}{2}, \qquad \zeta_{02} = \frac{\zeta_2 + \zeta_0}{2}. \tag{5}$$

The reader will verify easily by drawing a figure that the following triangles have the same orientation as $[\zeta_0, \zeta_1, \zeta_2]$:

$$\begin{aligned} &T_{01}\colon [\zeta_{02}, \zeta_0, \zeta_{00}], \quad T_{02}\colon [\zeta_{00}, \zeta_1, \zeta_{01}], \\ &T_{03}\colon [\zeta_{01}, \zeta_2, \zeta_{02}], \quad T_{04}\colon [\zeta_{00}, \zeta_{01}, \zeta_{02}]. \end{aligned} \tag{6}$$

Then

$$\int_{T_0} f(\zeta)\,d\zeta = \int_{T_{01}} f(\zeta)\,d\zeta + \int_{T_{02}} f(\zeta)\,d\zeta + \int_{T_{03}} f(\zeta)\,d\zeta + \int_{T_{04}} f(\zeta)\,d\zeta \tag{7}$$

for the following reasons. If a segment is a side of two triangles, it has opposite orientations in the two triangles and the integrals on this side in the two triangles cancel by Theorem 32.6. After all such cancellations have been made, the integrals which remain add up, by Theorem 32.7, to the integrals on the three sides of T_0; thus (7) is true. Next, by the triangle inequality,

$$\left\| \int_{T_0} f(\zeta)\,d\zeta \right\| \leqslant \left\| \int_{T_{01}} f(\zeta)\,d\zeta \right\| + \cdots + \left\| \int_{T_{04}} f(\zeta)\,d\zeta \right\|. \tag{8}$$

Since the left side of this inequality is positive by (4), equation (8) shows that at least one of the integrals on the right in (7) is not zero. Let $\|\int_{T_{0k}} f(\zeta)\,d\zeta\|$ be the maximum term, or one of the maximum terms, on the right in (8). Let T_1 be new notation for T_{0k}; then

$$\left\| \int_{T_1} f(\zeta)\,d\zeta \right\| > 0, \tag{9}$$

$$\left\| \int_{T_0} f(\zeta)\,d\zeta \right\| \leqslant 4 \left\| \int_{T_1} f(\zeta)\,d\zeta \right\|. \tag{10}$$

Use the mid-points of the sides of T_1 to divide T_1 into four triangles. A repetition of the analysis used to find T_1 shows that there is a triangle T_2 such that

$$\left\| \int_{T_2} f(\zeta)\,d\zeta \right\| > 0, \tag{11}$$

$$\left\| \int_{T_1} f(\zeta)\,d\zeta \right\| \leqslant 4 \left\| \int_{T_2} f(\zeta)\,d\zeta \right\|. \tag{12}$$

Equations (10) and (12) show that

$$\left\| \int_{T_0} f(\zeta)\,d\zeta \right\| \leqslant 4^2 \left\| \int_{T_2} f(\zeta)\,d\zeta \right\|. \tag{13}$$

A continuation of this process shows that

$$\left\| \int_{T_0} f(\zeta)\,d\zeta \right\| \leqslant 4^n \left\| \int_{T_n} f(\zeta)\,d\zeta \right\|. \tag{14}$$

If $L(T_n)$ is the length of T_n (that is, the sum of the lengths of the sides of T_n; compare Exercise 31.3), then $L(T_1)=\frac{1}{2}L(T_0)$, $L(T_2)=\frac{1}{2}L(T_1)=(\frac{1}{2})^2L(T_0)$, and

$$L(T_n) = (\tfrac{1}{2})^n L(T_0), \qquad n = 1,2,\ldots. \tag{15}$$

Also,

$$c(T_0) \supset c(T_1) \supset \cdots \supset c(T_n) \supset \cdots. \tag{16}$$

Since the diameter of $c(T_n)$ tends to zero as $n \to \infty$, there is a single point ζ^* which belongs to all of the sets $c(T_n)$.

Now ζ^* is in X, and f is holomorphic in X by hypothesis; therefore f satisfies the strong Stolz condition by Theorem 25.1. Let $\varepsilon > 0$ be given; then there is a $\delta(\varepsilon)>0$ such that

$$f(\zeta) - f(\zeta^*) = D_\zeta f(\zeta^*)(\zeta - \zeta^*) + r(f;\zeta^*,\zeta)(\zeta - \zeta^*), \tag{17}$$

$$\|r(f;\zeta^*,\zeta)\| < \varepsilon, \qquad \forall\zeta \text{ such that } \|\zeta - \zeta^*\| < \delta(\varepsilon). \tag{18}$$

Choose n so large that the diameter of $c(T_n)$ is less that $\delta(\varepsilon)$. Now ζ^* is in $c(T_n)$; then $\|\zeta - \zeta^*\| < \delta(\varepsilon)$ for every point ζ in $c(T_n)$. Use (17) to evaluate $\int_{T_n} f(\zeta)\,d\zeta$ in (14) as follows:

$$\begin{aligned} \int_{T_n} f(\zeta)\,d\zeta &= \int_{T_n} f(\zeta^*)\,d\zeta + \int_{T_n} D_\zeta f(\zeta^*)(\zeta - \zeta^*)\,d\zeta + \int_{T_n} r(f;\zeta^*,\zeta)(\zeta-\zeta^*)\,d\zeta, \\ &= [f(\zeta^*) - D_\zeta f(\zeta^*)\zeta^*] \int_{T_n} d\zeta + D_\zeta f(\zeta^*) \int_{T_n} \zeta\,d\zeta + \int_{T_n} r(f;\zeta^*,\zeta)(\zeta-\zeta^*)\,d\zeta. \end{aligned} \tag{19}$$

Now T_n is a closed curve and $\int_{T_n} d\zeta = 0$ by the definition of the integral; furthermore, $\int_{T_n} \zeta\,d\zeta = 0$ by Exercise 32.4 or 32.5. Thus

$$\int_{T_n} f(\zeta)\,d\zeta = \int_{T_n} r(f;\zeta^*,\zeta)(\zeta-\zeta^*)\,d\zeta. \tag{20}$$

Use Theorem 32.10 to estimate the norm of the integral on the right. Now (18) shows that $\|r(f;\zeta^*,\zeta)\|<\varepsilon$ for all ζ on T_n. Also, $\|\zeta-\zeta^*\|$ is equal to or less than the diameter of $c(T_n)$, which is less than $\frac{1}{2}L(T_n)$, for all ζ on T_n; then (15) shows that

(21) $\quad \|\zeta-\zeta^*\| < \frac{1}{2}(\frac{1}{2})^n L(T_0), \qquad \zeta \text{ on } T_n.$

Therefore, by (18) and (21), Theorem 32.10, and (15),

(22) $\quad \|r(f;\zeta^*,\zeta)(\zeta-\zeta^*)\| < \sqrt{2}\varepsilon(\frac{1}{2})^{n+1}L(T_0),$

(23)
$$\left\|\int_{T_n} r(f;\zeta^*,\zeta)(\zeta-\zeta^*)\,d\zeta\right\| < \sqrt{2}[\sqrt{2}\varepsilon(\tfrac{1}{2})^{n+1}L(T_0)]L(T_n),$$
$$< (\tfrac{1}{2})^n\varepsilon L(T_0)(\tfrac{1}{2})^n L(T_0),$$
$$< \frac{1}{4^n}\varepsilon[L(T_0)]^2.$$

Then (20) and (23) show that

(24)
$$\left\|\int_{T_n} f(\zeta)\,d\zeta\right\| < \frac{1}{4^n}\varepsilon[L(T_0)]^2,$$

and (14) shows that

(25)
$$\left\|\int_{T_0} f(\zeta)\,d\zeta\right\| < \varepsilon[L(T_0)]^2.$$

Since this inequality holds for every $\varepsilon>0$, then

(26)
$$\left\|\int_{T_0} f(\zeta)\,d\zeta\right\| = 0.$$

This contradicts the assumption in (4); hence (4) is impossible. Therefore,

(27)
$$\left\|\int_{T_0} f(\zeta)\,d\zeta\right\| = 0, \qquad \int_{T_0} f(\zeta)\,d\zeta = 0,$$

and the proof of (3) and of the entire Theorem 34.1 is complete. □

Exercises

34.1 Let f be a holomorphic function. An oriented, piecewise linear surface is formed by the four oriented triangles $\zeta_0\zeta_1\zeta_4$, $\zeta_1\zeta_2\zeta_4$, $\zeta_2\zeta_3\zeta_4$, $\zeta_3\zeta_0\zeta_4$. Observe that these triangles need not lie in a two-dimensional plane in $\mathbf{C}_2$. Assume that the convex extension of each triangle is contained in X, the domain in which f is holomorphic. Let C denote the polygonal curve with segments $\zeta_0\zeta_1$, $\zeta_1\zeta_2$, $\zeta_2\zeta_3$, $\zeta_3\zeta_0$ which forms the boundary of the surface. Show that $\int_C f(\zeta)\,d\zeta$ exists and that $\int_C f(\zeta)\,d\zeta=0$.

34.2 Exercise 34.1 presents a special case of a more general piecewise linear surface and related conclusion. Describe this surface and state and prove the conclusion.

34.3 Let C be a curve $\zeta:[a,b]\to\mathbb{C}_2$, $t\mapsto\zeta(t)$, which has a continuous derivative. Let $a=t_0<t_1<\cdots<t_{i-1}<t_i<\cdots<t_n=b$ be a subdivision of $[a,b]$.

(a) Establish the following identities:

$$\begin{aligned}[\zeta(b)]^2-[\zeta(a)]^2 &= \sum_{i=1}^{n}\{[\zeta(t_i)]^2-[\zeta(t_{i-1})]^2\}\\ &= \sum_{i=1}^{n}[\zeta(t_i)+\zeta(t_{i-1})][\zeta(t_i)-\zeta(t_{i-1})]\\ &= \sum_{i=1}^{n}\zeta(t_i)[\zeta(t_i)-\zeta(t_{i-1})]\\ &\quad+\sum_{i=1}^{n}\zeta(t_{i-1})[\zeta(t_i)-\zeta(t_{i-1})].\end{aligned}$$

(b) Use the identities in (a) to show that

$$\int_C \zeta\,d\zeta=\tfrac{1}{2}[\zeta(b)]^2-\tfrac{1}{2}[\zeta(a)]^2.$$

(c) If C is a closed curve, show that $\int_C\zeta\,d\zeta=0$.

34.4 Let C be the curve in Exercise 34.3.

(a) Establish the following identities:

$$\begin{aligned}&[\zeta(b)]^3-[\zeta(a)]^3\\ &\quad=\sum_{i=1}^{n}\{[\zeta(t_i)]^3-[\zeta(t_{i-1})]^3\}\\ &\quad=\sum_{i=1}^{n}\{[\zeta(t_i)]^2+\zeta(t_i)\zeta(t_{i-1})+[\zeta(t_{i-1})]^2\}[\zeta(t_1)-\zeta(t_{i-1})].\end{aligned}$$

(b) Use the identities in (a) to show that

$$\int_C \zeta^2\,d\zeta=\tfrac{1}{3}[\zeta(b)]^3-\tfrac{1}{3}[\zeta(a)]^3.$$

(c) If C is a closed curve, show that $\int_C\zeta^2\,d\zeta=0$.

34.5 Use the method suggested by Exercises 34.3 and 34.4 to show that

$$\int_C\zeta^n\,d\zeta=\frac{[\zeta(b)]^{n+1}-[\zeta(a)]^{n+1}}{n+1},$$

and that $\int_C\zeta^n\,d\zeta=0$ if C is a closed curve.

34.6 Let C be a curve $\zeta:[a,b]\to\mathbb{C}_2$, $t\mapsto\zeta(t)$, which has a continuous derivative, and let $f:\mathbb{C}_2\to\mathbb{C}_2$ be a polynomial in ζ of degree n.
(a) Find the value of $\int_C f(\zeta)\,d\zeta$.
(b) If C is a closed curve, show that $\int_C f(\zeta)\,d\zeta=0$.

35. EXISTENCE OF PRIMITIVES

Theorems 33.1 and 33.8 contain the fundamental theorem of the integral calculus, which states that

$$(1)\qquad \int_C D_\zeta F(\zeta)\,d\zeta = F[\zeta(b)] - F[\zeta(a)].$$

This formula, as in elementary calculus, suggests a method for evaluating the integral $\int_C f(\zeta)\,d\zeta$ of an arbitrary holomorphic function $f:X\to\mathbb{C}_2$. If there exists a function $F:X\to\mathbb{C}_2$ such that

$$(2)\qquad D_\zeta F(\zeta) = f(\zeta), \qquad \forall\zeta \text{ in } X,$$

then

$$(3)\qquad \int_C f(\zeta)\,d\zeta = \int_C D_\zeta F(\zeta)\,d\zeta = F[\zeta(b)] - F[\zeta(a)].$$

This section proves that, if $f:X\to\mathbb{C}_2$ is holomorphic, then in some regions X there exists a holomorphic function $F:X\to\mathbb{C}_2$ which satisfies (2) and (3); also, it considers the relation of this proof to some of the results in Section 34. The existence of F for a given function f is used in the next section to prove a very general form of Cauchy's integral theorem. But first a definition is needed.

35.1 DEFINITION If $f:X\to\mathbb{C}_2$ and $F:X\to\mathbb{C}_2$ are holomorphic functions which satisfy (2), then F is called a *primitive of f.*

35.2 THEOREM Let X be a domain in $\mathbb{C}_2$ which is star-shaped with respect to the point ζ^*, and let $f:X\to\mathbb{C}_2$ be a holomorphic function. Then there exists a primitive function F of f.

Proof. Let ζ_1 be a fixed point in X, and let ζ be an arbitrary point. Connect ζ_1 to ζ by a polygonal curve P in X with segments

$$(4)\qquad p_0p_1, p_1p_2, \dots, p_{k-1}p_k, \dots, p_{n-1}p_n, \qquad p_0=\zeta_1,\ p_n=\zeta.$$

Such polygonal curves exist; since X is star-shaped, one such curve consists of the segments $\zeta_1\zeta^*$ and $\zeta^*\zeta$. Then f has an integral on each segment in (4) (see Exercise 32.1) and

$$(5)\qquad \int_P f(\eta)\,d\eta = \sum_{k=1}^{n} \int_{p_{k-1}p_k} f(\eta)\,d\eta.$$

The value of the integral on the left in (5) appears to depend on the polygonal curve P as well as on ζ, but the proof will show that $\int_P f(\eta)\,d\eta$ has the same value for every polygonal curve P which connects ζ_1 to ζ; therefore $\int_P f(\eta)\,d\eta$ is the value of a function $F: X \to \mathbb{C}_2$ at ζ. Let Q, with segments

$$(6)\quad q_0q_1, q_1q_2, \ldots, q_{k-1}q_k, \ldots, q_{m-1}q_m, \qquad q_0 = \zeta_1,\ q_m = \zeta,$$

be a second polygonal curve which connects ζ_1 to ζ. Reverse the orientation of each segment $q_{k-1}q_k$ in Q to form a closed curve P-Q. Connect ζ^* to each point p_k and q_k to form triangles (curves) $\zeta^*p_{k-1}p_k$ and $\zeta^*q_kq_{k-1}$. Since X is star-shaped by hypothesis, then the convex extensions $c(\zeta^*p_{k-1}p_k)$ and $c(\zeta^*q_kq_{k-1})$ are contained in X. Then the special case of Cauchy's integral theorem in Theorem 34.1 shows that

$$(7)\qquad \int_{\zeta^*p_{k-1}p_k} f(\eta)\,d\eta = 0, \qquad \int_{\zeta^*q_kq_{k-1}} f(\eta)\,d\eta = 0,$$

$$(8)\qquad \sum_{k=1}^{n} \int_{\zeta^*p_{k-1}p_k} f(\eta)\,d\eta + \sum_{k=1}^{m} \int_{\zeta^*q_kq_{k-1}} f(\eta)\,d\eta = 0.$$

Now the segment ζ^*p_k belongs to $\zeta^*p_{k-1}p_k$ and also to $\zeta^*p_kp_{k+1}$, and it has opposite orientations in these triangles; then the corresponding integrals on these segments cancel by Theorem 32.6. Also, ζ^*q_k belongs to $\zeta^*q_kq_{k-1}$ and $\zeta^*q_{k+1}q_k$ and it has opposite orientation in these triangles; again the integrals cancel. The result of making all such cancellations is to reduce (8) to the following:

$$(9)\qquad \sum_{k=1}^{n} \int_{p_{k-1}p_k} f(\eta)\,d\eta + \sum_{k=1}^{m} \int_{q_kq_{k-1}} f(\eta)\,d\eta = 0.$$

Reverse the orientation of the segments q_kq_{k-1} in the second sum; then by Theorem 32.6,

$$(10)\qquad \sum_{k=1}^{n} \int_{p_{k-1}p_k} f(\eta)\,d\eta - \sum_{k=1}^{m} \int_{q_{k-1}q_k} f(\eta)\,d\eta = 0.$$

$$(11)\qquad \sum_{k=1}^{n} \int_{p_{k-1}p_k} f(\eta)\,d\eta = \sum_{k=1}^{m} \int_{q_{k-1}q_k} f(\eta)\,d\eta.$$

This equation shows that $\int_P f(\eta)\,d\eta$ has the same value for every polygonal curve P which connects ζ_1 to ζ, and the value of the integral depends on ζ alone. Define the function $F: X \to \mathbf{C}_2$ as follows:

$$\text{(12)} \qquad F(\zeta) = \int_P f(\eta)\, d\eta, \qquad P \text{ connects } \zeta_1 \text{ to } \zeta.$$

To complete the proof of Theorem 35.2, it is necessary and sufficient to prove that $D_\zeta F(\zeta) = f(\zeta)$, that is, to show that

$$\text{(13)} \qquad \lim_{h\to 0} \frac{F(\zeta + h) - F(\zeta)}{h} = f(\zeta), \qquad h \notin \mathcal{O}_2.$$

If $F(\zeta) = \int_P f(\eta)\, d\eta$, then

$$\text{(14)} \qquad F(\zeta + h) = \int_P f(\eta)\, d\eta + \int_{\zeta(\zeta+h)} f(\eta)\, d\eta = F(\zeta) + \int_{\zeta(\zeta+h)} f(\eta)\, d\eta.$$

Use Theorem 32.9 to make the change of variable $\eta = \zeta + th$, $d\eta = h\, dt$, $0 \leqslant t \leqslant 1$. Then

$$\text{(15)} \qquad F(\zeta + h) = F(\zeta) + \int_0^1 f(\zeta + th)h\, dt = F(\zeta) + h\int_0^1 f(\zeta + th)\, dt,$$

$$\text{(16)} \qquad \frac{F(\zeta + h) - F(\zeta)}{h} = \int_0^1 f(\zeta + th)\, dt, \qquad h \notin \mathcal{O}_2,$$

$$\text{(17)} \qquad \frac{F(\zeta + h) - F(\zeta)}{h} - f(\zeta) = \int_0^1 [f(\zeta + th) - f(\zeta)]\, dt.$$

Now f is holomorphic and therefore continuous in X. Let $\varepsilon > 0$ be given. Then there exists a $\delta(\varepsilon, \zeta)$ such that $\|f(\eta) - f(\zeta)\| < \varepsilon$ for every η for which $\|\eta - \zeta\| < \delta(\varepsilon, \zeta)$. Next, Theorem 32.10 and (17) show that

$$\text{(18)} \qquad \left\| \frac{F(\zeta + h) - F(\zeta)}{h} - f(\zeta) \right\| \leqslant \sqrt{2} \int_0^1 \|f(\zeta + th) - f(\zeta)\|\, dt$$

$$< \sqrt{2}\varepsilon, \qquad \|h\| < \delta(\varepsilon, \zeta),\ h \notin \mathcal{O}_2.$$

Therefore, $D_\zeta F(\zeta)$ exists, and $D_\zeta F(\zeta) = f(\zeta)$. The proof of Theorem 35.2 is complete. □

Corresponding to the point ζ_1 in X, a function whose derivative is f has been constructed by integrating f on polygonal curves P which connect ζ_1 to ζ; denote this function by $F_1: X \to \mathbf{C}_2$. Another function whose derivative is f can be constructed by integrating f on polygonal curves which connect ζ_2 in X to ζ; denote this second function $F_2: X \to \mathbf{C}_2$. Then both F_1 and F_2 are primitives of f, and the next corollary establishes the relation between them. Another proof of the corollary can be obtained from Theorem 33.2 as in Corollary 35.4.

35.3 COROLLARY If $F_1 : X \to \mathbb{C}_2$ and $F_2 : X \to \mathbb{C}_2$ are the functions just described, then there is a constant c in $\mathbb{C}_2$ such that

$$(19) \qquad F_2(\zeta) = F_1(\zeta) + c, \qquad \forall \zeta \text{ in } X.$$

Proof. Let P be a polygonal curve which connects ζ_1 to ζ and serves to define F_1. Then since X is star-shaped, one polygonal curve which connects ζ_2 to ζ consists of the segments $\zeta_2\zeta^*$ and $\zeta^*\zeta_1$ followed by the curve P. Then

$$(20) \qquad F_2(\zeta) = \int_{\zeta_2\zeta^*} f(\eta)\, d\eta + \int_{\zeta^*\zeta_1} f(\eta)\, d\eta + \int_P f(\eta)\, d\eta.$$

Now

$$(21) \qquad \int_P f(\eta)\, d\eta = F_1(\zeta),$$

$$(22) \qquad \int_{\zeta_2\zeta^*} f(\eta)\, d\eta + \int_{\zeta^*\zeta_1} f(\eta)\, d\eta = c.$$

Then (20)–(22) show that (19) is true. □

35.4 COROLLARY Let X be a star-shaped domain, and let $f : X \to \mathbb{C}_2$ be a holomorphic function. If $F : X \to \mathbb{C}_2$ is the primitive of f constructed in the proof of Theorem 35.2, and if $G : X \to \mathbb{C}_2$ is any other primitive of f, then there is a constant c in $\mathbb{C}_2$ such that

$$(23) \qquad F(\zeta) = G(\zeta) + c, \qquad \forall \zeta \text{ in } X.$$

Proof. Since f is a holomorphic function, it is continuous. Then since F and G are primitives of f, equation (2) and Definition 35.1 show that F and G have continuous derivatives such that

$$(24) \qquad D_\zeta F(\zeta) = D_\zeta G(\zeta) = f(\zeta), \qquad \forall \zeta \text{ in } X.$$

Then the fundamental theorem of the integral calculus holds for F and G by Theorem 33.1, and (23) and Corollary 35.4 are true by Theorem 33.2. □

Exercises

35.1 Find a primitive of the function $f : \mathbb{C}_2 \to \mathbb{C}_2$, $\zeta \mapsto \zeta^n$, and use this primitive and the fundamental theorem of the integral calculus to show that

$$\int_C \zeta^n\, d\zeta = \frac{[\zeta(b)]^{n+1} - [\zeta(a)]^{n+1}}{n+1}.$$

(Compare Exercise 34.5.)

35.2 Verify the following statement. If F and G are primitives of f and g, and if a and b are constants in $\mathbb{C}_2$, then $aF+bG$ is a primitive of $af+bg$.

35.3 Let F be the primitive of f constructed in the proof of Theorem 35.2. Without using the representations of ζ and f in equations (1), (2), (3) in Section 33, prove that F satisfies the uniform strong Stolz condition and the fundamental theorem of the integral calculus. [*Hint*. In the formula in (15) in this section, make the following change in notation: $\zeta \to \zeta_0$, $\zeta + h \to \zeta$, $h \to \zeta - \zeta_0$.]

36. CAUCHY'S INTEGRAL THEOREM: THE GENERAL CASE

Cauchy's integral theorem has been proved for a triangle in Theorem 34.1, but the theorem is true for much more general curves, including curves whose traces consist of more than a single connected set. The purpose of this section is to use the fundamental theorem of the integral calculus to establish these generalizations.

36.1 THEOREM (Cauchy's Integral Theorem) Let X be a domain in $\mathbb{C}_2$ which is star-shaped with respect to a point ζ^*; let $f: X \to \mathbb{C}_2$ be a holomorphic function; and let C be a curve $\zeta: [a, b] \to X$ which is continuous and closed, and which has a piecewise continuous derivative. Then the integral $\int_C f(\zeta)\,d\zeta$ exists, and

$$(1) \qquad \int_C f(\zeta)\,d\zeta = 0.$$

Proof. The first step is to show that the integral $\int_C f(\zeta)\,d\zeta$ exists. Since f is a holomorphic function, it is continuous. By hypothesis, C has a continuous derivative except at a finite number of points on $[a, b]$; therefore the integral $\int_C f(\zeta)\,d\zeta$ exists by Theorem 32.7. The proof can be completed by showing that the value of this integral is zero.

Since f is a holomorphic function and X is a star-shaped domain, Theorem 35.2 shows that f has a primitive $F: X \to \mathbb{C}_2$ such that

$$(2) \qquad D_\zeta F(\zeta) = f(\zeta), \qquad \forall \zeta \text{ in } X.$$

Then

$$(3) \qquad \int_C f(\zeta)\,d\zeta = \int_C D_\zeta F(\zeta)\,d\zeta.$$

Now $D_\zeta F$ is continuous by (2) since f is continuous. Also, F satisfies the uniform strong Stolz condition by Theorem 33.4, and it satisfies the fundamental theorem of the integral calculus by Theorems 33.1 and 33.8 and Exercise 33.1. Therefore,

$$(4)\qquad \int_C D_\zeta F(\zeta)\,d\zeta = F[\zeta(b)] - F[\zeta(a)].$$

But C is a closed curve by hypothesis; hence $\zeta(b)=\zeta(a)$, and (3) and (4) show that

$$(5)\qquad \int_C f(\zeta)\,d\zeta = \int_C D_\zeta F(\zeta)\,d\zeta = 0.$$

The proof of (1) and of Theorem 36.1 is complete. □

Theorem 36.1 is an exceptionally general theorem. The curve C is not required to be the boundary of an oriented two-dimensional surface. It may be twisted and knotted. The hypotheses are light: f is holomorphic in a star-shaped region X; C is a continuous, closed curve with a piecewise continuous derivative in this region X. This form of Cauchy's integral theorem goes far beyond the original special case in Theorem 34.1, but observe, however, that the special case serves as the foundation for the generalization in Theorem 36.1. Another procedure which can be used to generalize the theorem will now be explained.

36.2 THEOREM Let $f: X \to \mathbb{C}_2$ be a holomorphic function, and let ABC and CBD be two triangles whose convex extensions $c(ABC)$ and $c(CBD)$ are contained in X. If C is the curve which consists of the segments AB, BD, DC, CA, then

$$(6)\qquad \int_C f(\zeta)\,d\zeta = 0.$$

Proof. The integral of f around each of the triangles ABC and CBD equals zero by the special case of Cauchy's integral theorem in Theorem 34.1. Thus

$$(7)\qquad \int_{ABC} f(\zeta)\,d\zeta + \int_{CBD} f(\zeta)\,d\zeta = 0.$$

Now

$$(8)\qquad \int_{ABC} f(\zeta)\,d\zeta = \left\{\int_{AB} + \int_{BC} + \int_{CA}\right\} f(\zeta)\,d\zeta,$$

$$(9)\qquad \int_{CBD} f(\zeta)\,d\zeta = \left\{\int_{CB} + \int_{BD} + \int_{DC}\right\} f(\zeta)\,d\zeta.$$

Theorem 32.6 shows that $\int_{BC} f(\zeta)\,d\zeta + \int_{CB} f(\zeta)\,d\zeta = 0$; then (7), (8), (9) show that

$$\left(\int_{AB} + \int_{BD} + \int_{DC} + \int_{CA}\right) f(\zeta)\,d\zeta = 0. \tag{10}$$

This equation is (6), and the proof of Theorem 36.2 is complete. □

36.3 COROLLARY Let $f: X \to \mathbb{C}_2$ be a holomorphic function, and let K be an oriented simplicial complex such that the convex extension of each simplex in K is contained in X. If ∂K is the boundary of K, then

$$\int_{\partial K} f(\zeta)\,d\zeta = 0. \tag{11}$$

Proof. The proof of Theorem 36.2 illustrates the method to be used in proving the corollary. If a segment belongs to two simplexes, then the integrals on this segment contributed by the two simplexes cancel, and only the sum of the integrals on the boundary segments remains. Since the integral on each simplex is zero by Theorem 34.1, the formula in (11) results. □

The principle involved in the proof of Theorem 36.2 and Corollary 36.3 has wider application. These proofs are based on Theorem 32.6, which states that

$$\int_{-C} f(\zeta)\,d\zeta = -\int_{C} f(\zeta)\,d\zeta. \tag{12}$$

Nothing in the theorem requires that the trace of C be a line segment. This observation suffices to extend the proofs of Theorem 36.2 and Corollary 36.3 to include the following corollary.

36.4 COROLLARY Let $f: X \to \mathbb{C}_2$ be a holomorphic function. Let $C_1 + C$ be a closed curve in a star-shaped sub-domain X_1 of X, and let $C_2 - C$ be a closed curve in a star-shaped sub-domain X_2 of X. Then $C_1 + C_2$ is a closed curve in X, and

$$\int_{C_1 + C_2} f(\zeta)\,d\zeta = 0. \tag{13}$$

Proof. Theorem 35.2 shows that f has primitives $F_1: X_1 \to \mathbb{C}_2$ and $F_2: X_2 \to \mathbb{C}_2$ in the star-shaped regions X_1 and X_2, respectively. Then Theorem 36.1 shows that

$$\int_{C_1 + C} f(\zeta)\,d\zeta = \int_{C_1 + C} D_\zeta F_1(\zeta)\,d\zeta = 0, \tag{14}$$

$$\int_{C_2 - C} f(\zeta)\,d\zeta = \int_{C_2 - C} D_\zeta F_2(\zeta)\,d\zeta = 0. \tag{15}$$

Also,

$$(16)\qquad \int_{C_1+C} f(\zeta)\,d\zeta = \int_{C_1} f(\zeta)\,d\zeta + \int_C f(\zeta)\,d\zeta = 0,$$

$$(17)\qquad \int_{C_2-C} f(\zeta)\,d\zeta = \int_{C_2} f(\zeta)\,d\zeta - \int_C f(\zeta)\,d\zeta = 0.$$

Add these equations; then

$$(18)\qquad \int_{C_1+C_2} f(\zeta)\,d\zeta = \int_{C_1} f(\zeta)\,d\zeta + \int_{C_2} f(\zeta)\,d\zeta = 0.$$

The proof of (13) and of Corollary 36.4 is complete. □

Exercises

36.1 Let X be a domain in $\mathbb{C}_2$, and let $f: X \to \mathbb{C}_2$ be a function which is assumed to be only continuous. Let $ABCD$ be a tetrahedron whose six edges are contained in X. Let $\int_{ABC} f(\zeta)\,d\zeta$ denote the integral of f on the oriented edges AB, BC, and CA of face ABC.

(a) Prove that

$$\left(\int_{ABD} + \int_{BCD} + \int_{CAD} + \int_{ACB}\right) f(\zeta)\,d\zeta = 0.$$

(b) Explain why the equation in (a) is true even though f is not assumed to be holomorphic, and the convex extensions of the triangles ABD, BCD, CAD, and ACB are not assumed to be in X.

36.2 Let X be a domain in $\mathbb{C}_2$, and let $f: X \to \mathbb{C}_2$ be a holomorphic function. Let A, B, C, D be the vertices of a tetrahedron, and assume that the convex extensions of ABD, BCD, CAD are contained in X.

(a) Show that

$$\int_{AB} f(\zeta)\,d\zeta + \int_{BC} f(\zeta)\,d\zeta + \int_{CA} f(\zeta)\,d\zeta = 0.$$

(b) Is the equation in (a) true also for the function f described in Exercise 36.1? Compare and explain the two exercises.

36.3 Let ζ_0 be a point in $\mathbb{C}_2$ which is not in $\mathcal{O}_2$.

(a) Show that there is a ball $B(\zeta_0, r)$, $r > 0$, such that $B(\zeta_0, r) \cap \mathcal{O}_2 = \varnothing$.

(b) Show that the function f such that $f(\zeta) = 1/\zeta^3$ is holomorphic in $B(\zeta_0, r)$.

(c) Find a function $f: B(\zeta_0, r) \to \mathbb{C}_2$ which is a primitive of f.

(d) Let C be a closed curve which has a continuous derivative and whose trace is in $B(\zeta_0, r)$. Use the fundamental theorem of the integral calculus to show that $\int_C f(\zeta)\,d\zeta = 0$.

(e) Show that

$$\int_C f(\zeta)\,d\zeta = \int_C D_\zeta F(\zeta)\,d\zeta = 0,$$

and thus verify Cauchy's integral theorem for the function f.

37. INTEGRALS INDEPENDENT OF THE PATH

Let $F: X \to \mathbb{C}_2$ be a holomorphic function whose derivative $D_\zeta F$ is continuous in X, and let C be a curve $\zeta:[a,b]\to X$ which has a continuous derivative and whose trace is in X. Then the fundamental theorem of the integral calculus in Theorem 33.1 states that

$$(1) \qquad \int_C D_\zeta F(\zeta)\,d\zeta = F[\zeta(b)] - F[\zeta(a)].$$

This formula shows that the value of the integral depends on the values of F at the two end points of C but not on the curve which connects these points. For this reason, the integral $\int_C D_\zeta F(\zeta)\,d\zeta$ is said to be independent of the path. The integral in (1) is a very special integral because the integrand is a derivative $D_\zeta F$, but there is a large class of integrals which are independent of the path. The purpose of this section is to investigate these integrals.

37.1 THEOREM Let X be a domain in $\mathbb{C}_2$ which is star-shaped with respect to a point ζ^*, and let $f: X\to\mathbb{C}_2$ be a holomorphic function. If C is a curve $\zeta:[a,b]\to X$ which has a continuous derivative, then $\int_C f(\zeta)\,d\zeta$ is independent of the path.

Proof. By Theorem 35.2, the function f has a primitive $F: X\to\mathbb{C}_2$ such that $D_\zeta F(\zeta) = f(\zeta)$ in X. Then by (1),

$$(2) \qquad \int_C f(\zeta)\,d\zeta = \int_C D_\zeta F(\zeta)\,d\zeta = F[\zeta(b)] - F[\zeta(a)],$$

and $\int_C f(\zeta)\,d\zeta$ is independent of the path. □

37.2 COROLLARY If $\int_C f(\zeta)\,d\zeta$ is the integral in Theorem 37.1, and if C is a closed curve, then $\int_C f(\zeta)\,d\zeta = 0$.

Proof. If C is a closed curve, then $\zeta(b)=\zeta(a)$, and (2) shows that $\int_C f(\zeta)\,d\zeta = 0$. Compare the results in Section 36. □

Now these results have important meaning for other representations of the integral. If

(3) $\zeta = x_1 + i_1x_2 + i_2x_3 + i_1i_2x_4,$

(4) $\zeta(t) = x_1(t) + i_1x_2(t) + i_2x_3(t) + i_1i_2x_4(t),$

(5) $f(t) = g_1(x) + i_1g_2(x) + i_2g_3(x) + i_1i_2g_4(x), \qquad x = (x_1, \ldots, x_4),$

as in Exercise 32.2, then $\int_C f(\zeta)\,d\zeta$ can be evaluated as integrals of real-valued functions of the variable t (compare the proof of Theorem 33.1). Thus

(6)
$$\begin{aligned}\int_C f(\zeta)\,d\zeta = &\int_C g_1(x)\,dx_1 - g_2(x)\,dx_2 - g_3(x)\,dx_3 + g_4(x)\,dx_4\\ &+ i_1\int_C g_2(x)\,dx_1 + g_1(x)\,dx_2 - g_4(x)\,dx_3 - g_3(x)\,dx_4\\ &+ i_2\int_C g_3(x)\,dx_1 - g_4(x)\,dx_2 + g_1(x)\,dx_3 - g_2(x)\,dx_4\\ &+ i_1i_2\int_C g_4(x)\,dx_1 + g_3(x)\,dx_2 + g_2(x)\,dx_3 + g_1(x)\,dx_4.\end{aligned}$$

37.3 THEOREM Let X be a domain in $\mathbb{C}_2$ which is star-shaped with respect to a point ζ^*, and let $f: X \to \mathbb{C}_2$ be a holomorphic function. If C is a curve $\zeta: [a, b] \to X$ which has a continuous derivative, then each of the following line integrals is independent of the path:

(7) $\int_C g_1(x)\,dx_1 - g_2(x)\,dx_2 - g_3(x)\,dx_3 + g_4(x)\,dx_4,$

(8) $\int_C g_2(x)\,dx_1 + g_1(x)\,dx_2 - g_4(x)\,dx_3 - g_3(x)\,dx_4,$

(9) $\int_C g_3(x)\,dx_1 - g_4(x)\,dx_2 + g_1(x)\,dx_3 - g_2(x)\,dx_4,$

(10) $\int_C g_4(x)\,dx_1 + g_3(x)\,dx_2 + g_2(x)\,dx_3 + g_1(x)\,dx_4.$

Proof. There is more than one approach to the proof of this theorem. A first approach uses a primitive of f to show that each of the integrals in (7)–(10) can be evaluated explicitly by the fundamental theorem of the integral calculus. Since X is a star-shaped domain, then f has a primitive F, and $D_\zeta F(\zeta) = f(\zeta)$. If

(11) $F(\zeta) = F_1(x) + i_1F_2(x) + i_2F_3(x) + i_1i_2F_4(x),$

then

$$(12)\qquad D_\zeta F(\zeta) = D_{x_1}F_1(x) + i_1 D_{x_1}F_2(x) + i_2 D_{x_1}F_3(x) + i_1 i_2 D_{x_1}F_4(x)$$

by Corollary 23.2, and the equation $D_\zeta F(\zeta) = f(\zeta)$ implies that

$$(13)\qquad D_{x_1}F_k(x) = g_k(x), \qquad k = 1,\ldots,4,\ x \in X.$$

Also, $F: X \to \mathbb{C}_2$ satisfies the Cauchy–Riemann differential equations in Theorem 23.1, and

$$
\begin{aligned}
(14)\qquad \int_C & g_1(x)\,dx_1 - g_2(x)\,dx_2 - g_3(x)\,dx_3 + g_4(x)\,dx_4 \\
&= \int_a^b [D_{x_1}F_1(x)x_1'(t) - D_{x_1}F_2(x)x_2'(t) - D_{x_1}F_3(x)x_3'(t) \\
&\quad + D_{x_1}F_4(x)x_4'(t)]\,dt \\
&= \int_a^b [D_{x_1}F_1(x)x_1'(t) + D_{x_2}F_1(x)x_2'(t) + D_{x_3}F_1(x)x_3'(t) \\
&\quad + D_{x_4}F_1(x)x_4'(t)]\,dt \\
&= F_1[x(t)]\big|_a^b \\
&= F_1[x(b)] - F_1[x(a)].
\end{aligned}
$$

These equations show that the integral in (7) is independent of the path, and similar analyses show that those in (8), (9), and (10) are also independent of the path.

But this elaborate proof is unnecessary. The theory of holomorphic functions of a bicomplex variable has contributed a proof of (2), and the equation $\int_C f(\zeta)\,d\zeta = F[\zeta(b)] - F[\zeta(a)]$ shows directly that the integrals in (7)–(10) are independent of the path. By equating corresponding components on the two sides of this equation we find that

$$(15)\qquad \int_C g_1(x)\,dx_1 - g_2(x)\,dx_2 - g_3(x)\,dx_3 + g_4(x)\,dx_4 = F_1[x(b)] - F_1[x(a)],$$

$$(16)\qquad \int_C g_2(x)\,dx_1 + g_1(x)\,dx_2 - g_4(x)\,dx_3 - g_3(x)\,dx_4 = F_2[x(b)] - F_2[x(a)],$$

$$(17)\qquad \int_C g_3(x)\,dx_1 - g_4(x)\,dx_2 + g_1(x)\,dx_3 - g_2(x)\,dx_4 = F_3[x(b)] - F_3[x(a)],$$

$$(18)\qquad \int_C g_4(x)\,dx_1 + g_3(x)\,dx_2 + g_2(x)\,dx_3 + g_1(x)\,dx_4 = F_4[x(b)] - F_4[x(a)],$$

and these equations state that the integrals on the left are independent of the path. The second proof of Theorem 37.3 is complete. □

The integrals in (7)–(10) are line integrals of real valued functions of real variables, and, since they are independent of path, their value around a closed curve C is zero. Stokes' theorem can be used to show that certain line integrals around closed curves equal zero, and the following question requires an answer: does Stokes' theorem show that the integrals in (7)–(10) around a closed curve C have the value zero, or are they members of some new class of such integrals? The answer is contained in the next theorem. To prove this theorem, however, it is necessary to assume not only that $f: X \to \mathbb{C}_2$ in (5) is holomorphic, but also that $g_1, \ldots, g_4$ have continuous derivatives with respect to $x_1, \ldots, x_4$. Of course, later theorems will show that the continuity of the derivatives of $g_1, \ldots, g_4$ follows from the hypothesis that f is holomorphic.

37.4 THEOREM Let $f: X \to \mathbb{C}_2$ be a holomorphic function, and assume that $g_1, \ldots, g_4$ in (5) have continuous derivatives with respect to $x_1, \ldots, x_4$. Then Stokes' theorem implies that the integrals in (7)–(10), around the boundary of a two dimensional surface in X, have the value zero.

Proof. The best proof of this theorem is obtained by repeating the proof of Stokes' theorem, beginning with the fundamental theorem of the integral calculus for surfaces [7, pp. 333–340, 346–349]. We are interested in the special case of the fundamental theorem for a two-dimensional surface in $\mathbb{C}_2$ [or $\mathbb{C}_0^4$].

Let X be a domain in $\mathbb{C}_2$, and let $h_k: X \to \mathbb{C}_0$, $k=1,2$, be two functions which have continuous derivatives. Let S be a two-dimensional surface whose trace is in X, and let ∂S denote its boundary. Introduce the following notation:

$$D_{(x_1,x_2)}(h_1,h_2) = \det\begin{bmatrix} D_{x_1}h_1 & D_{x_1}h_2 \\ D_{x_2}h_1 & D_{x_2}h_2 \end{bmatrix}, \ldots,$$

$$D_{(x_3,x_4)}(h_1,h_2) = \det\begin{bmatrix} D_{x_3}h_1 & D_{x_3}h_2 \\ D_{x_4}h_1 & D_{x_4}h_2 \end{bmatrix}.$$

Then the fundamental theorem of the integral calculus states that

$$(20) \quad \int_S D_{(x_1,x_2)}(h_1,h_2)\,d(x_1,x_2) + D_{(x_1,x_3)}(h_1,h_2)\,d(x_1,x_3) + \cdots + D_{(x_3,x_4)}(h_1,h_2)\,d(x_3,x_4)$$

$$= \int_{\partial S} h_1(x)D_{x_1}h_2(x)\,dx_1 + h_1(x)D_{x_2}h_2(x)\,dx_2 + \cdots + h_1(x)D_{x_4}h_2(x)\,dx_4.$$

Apply this theorem in turn to each of the following pairs (h_1, h_2) of functions:

(21) $\quad (g_1, x_1), \qquad (-g_2, x_2), \qquad (-g_3, x_3), \qquad (g_4, x_4).$

The four equations which result are the following:

(22) $$\int_S -D_{x_2}g_1(x)\,d(x_1, x_2) - D_{x_3}g_1(x)\,d(x_1, x_3) - D_{x_4}g_1(x)\,d(x_1, x_4) = \int_{\partial S} g_1(x)\,dx_1,$$

(23) $$\int_S -D_{x_1}g_2(x)\,d(x_1, x_2) + D_{x_3}g_2(x)\,d(x_2, x_3) + D_{x_4}g_2(x)\,d(x_2, x_4) = \int_{\partial S} -g_2(x)\,dx_2,$$

(24) $$\int_S -D_{x_1}g_3(x)\,d(x_1, x_3) - D_{x_2}g_3(x)\,d(x_2, x_3) + D_{x_4}g_3(x)\,d(x_3, x_4) = \int_{\partial S} -g_3(x)\,dx_3,$$

(25) $$\int_S D_{x_1}g_4(x)\,d(x_1, x_4) + D_{x_2}g_4(x)\,d(x_2, x_4) + D_{x_3}g_4(x)\,d(x_3, x_4) = \int_{\partial S} g_4(x)\,dx_4.$$

Add these four equations together and use elementary properties of integrals to simplify the result as follows:

(26) $$\begin{aligned}\int_S &[-D_{x_1}g_2(x) - D_{x_2}g_1(x)]\,d(x_1, x_2) + [-D_{x_1}g_3(x) - D_{x_3}g_1(x)]\,d(x_1, x_3)\\ &+ [D_{x_1}g_4(x) - D_{x_4}g_1(x)]\,d(x_1, x_4) + [-D_{x_2}g_3(x) + D_{x_3}g_2(x)]\,d(x_2, x_3)\\ &+ [D_{x_2}g_4(x) + D_{x_4}g_2(x)]\,d(x_2, x_4) + [D_{x_3}g_4(x) + D_{x_4}g_3(x)]\,d(x_3, x_4)\\ = \int_{\partial S} &g_1(x)\,dx_1 - g_2(x)\,dx_2 - g_3(x)\,dx_3 + g_4(x)\,dx_4.\end{aligned}$$

Now f is a holomorphic function by hypothesis; therefore $g_1, \ldots, g_4$ satisfy the Cauchy–Riemann differential equations in Theorem 23.1 (7)–(10), and each of the following expressions is identically zero in X and therefore on the surface S:

$$
\begin{aligned}
(27)\qquad &[-D_{x_1}g_2(x) - D_{x_2}g_1(x)], &\quad &[-D_{x_1}g_3(x) - D_{x_3}g_1(x)],\\
&[D_{x_1}g_4(x) - D_{x_4}g_1(x)], & &[-D_{x_2}g_3(x) + D_{x_3}g_2(x)],\\
&[D_{x_2}g_4(x) + D_{x_4}g_2(x)], & &[D_{x_3}g_4(x) + D_{x_4}g_3(x)].
\end{aligned}
$$

Therefore the integral over S in (26) equals zero, and (26) reduces to

$$
(28)\qquad \int_{\partial S} g_1(x)\,dx_1 - g_2(x)\,dx_2 - g_3(x)\,dx_3 + g_4(x)\,dx_4 = 0.
$$

Thus the line integral in (7) around a closed curve C which forms the boundary of S equals zero.

To prove that the line integrals in (8), (9), (10) around the boundary of S equal zero, apply the fundamental theorem of integral calculus in (20) to each pair of functions in the following three sets:

$$
\begin{aligned}
(29)\qquad &[g_2, x_1], \quad &[g_1, x_2], \quad &[-g_4, x_3], \quad &[-g_3, x_4];\\
(30)\qquad &[g_3, x_1], \quad &[-g_4, x_2], \quad &[g_1, x_3], \quad &[-g_2, x_4];\\
(31)\qquad &[g_4, x_1], \quad &[g_3, x_2], \quad &[g_2, x_3], \quad &[g_1, x_4].
\end{aligned}
$$

Add the four equations obtained for each set and use the Cauchy–Riemann equations to show that the integral over S equals zero in each case; therefore, the line integrals around the boundary of S equal zero. The proof of Theorem 37.4 is complete. □

37.5 REMARK Let X be a star-shaped domain, and let $f: X \to \mathbb{C}_2$ be a holomorphic function. Then f has a primitive F, and Theorem 37.3 has two proofs which show that the integrals (7)–(10), around a closed curve C, equal zero. By assuming that $g_1, \ldots, g_4$ have continuous derivatives, Theorem 37.4 succeeds in using Stokes' theorem to prove a similar result. Clearly, Theorem 37.3 is a better theorem than Theorem 37.4, and we have an example in which the theory of holomorphic functions of a bicomplex variable contributes a better result for real-valued functions than the real-valued function theory does. This example raises an interesting question about line integrals in spaces other than $\mathbb{R}^4$. Let Y be a domain in $\mathbb{R}^3$, and let $h_k: Y \to \mathbb{R}$, $k = 1, 2, 3$, be continuous functions. Let C be a closed curve which has a continuous derivative and whose trace is in Y. The line integral

$$
(32)\qquad \int_C h_1(y)\,dy_1 + h_2(y)\,dy_2 + h_3(y)\,dy_3, \qquad y = (y_1, y_2, y_3),
$$

is defined, but what about the methods for showing that its value is zero for some functions? Stokes' theorem is available for h_1, h_2, h_3. There is no theory of holomorphic functions in $\mathbb{R}^3$ with which to treat the problem with weaker hypotheses on h_1, h_2, h_3. From the point of view of functions of a real variable, why are the problems different in $\mathbb{R}^3$ and $\mathbb{R}^4$?

Exercises

37.1 (a) Apply the fundamental theorem of the integral calculus to the four pairs of functions in (29) and thus prove the following statement of Stokes' theorem:

$$\begin{aligned}\int_S &[D_{x_1}g_1(x) - D_{x_2}g_2(x)]\,d(x_1, x_2)\\ &+ [-D_{x_1}g_4(x) - D_{x_3}g_2(x)]\,d(x_1, x_3)\\ &+ [-D_{x_1}g_3(x) - D_{x_4}g_2(x)]\,d(x_1, x_4)\\ &+ [-D_{x_2}g_4(x) - D_{x_3}g_1(x)]\,d(x_2, x_3)\\ &+ [-D_{x_2}g_3(x) - D_{x_4}g_1(x)]\,d(x_2, x_4)\\ &+ [-D_{x_3}g_3(x) + D_{x_4}g_4(x)]\,d(x_3, x_4)\\ = \int_{\partial S} &g_2(x)\,dx_1 + g_1(x)\,dx_2 - g_4(x)\,dx_3 - g_3(x)\,dx_4.\end{aligned}$$

(b) Assume that the hypotheses of Theorem 37.4 are satisfied and use the statement of Stokes' theorem in (a) to prove that

$$\int_{\partial S} g_2(x)\,dx_1 + g_1(x)\,dx_2 - g_4(x)\,dx_3 - g_3(x)\,dx_4 = 0.$$

Compare this integral with the one in (8).

37.2 Repeat Exercise 37.1 for the set of pairs of functions in (30), and for the set of pairs of functions in (31).

37.3 The fundamental theorem of the integral calculus in (20) does not require that the boundary of S consist of a single connected piece.

(a) Prove the following theorem. Let X be a domain in $\mathbb{C}_2$, and let $f: X \to \mathbb{C}_2$ be a holomorphic function such that $g_1, \ldots, g_4$ have continuous derivatives. Let S be a surface for which (20) holds (S may be multiply connected). Then $\int_{\partial S} f(\zeta)\,d\zeta = 0$.

(b) Compare the theorem in (a) with the results in Section 36, and especially with Corollary 36.3.

37.4 Let $f: X \to \mathbb{C}_2$ be a holomorphic function in the star-shaped domain X in $\mathbb{C}_2$, and let C be a curve, with a piecewise continuous derivative, whose trace is the boundary of a Möbius strip. Prove that $\int_C f(\zeta)\,d\zeta = 0$.

38. INTEGRALS AND THE IDEMPOTENT REPRESENTATION

The definition of the integral $\int_C f(\zeta)\,d\zeta$ in Definition 32.1 employs no special representation of ζ nor of f, but on occasion (see, for example, Theorem 33.1) it has been desirable or necessary to use the following real-variable representations:

(1) $\zeta = x_1 + i_1x_2 + i_2x_3 + i_1i_2x_4, \qquad x_1, \ldots, x_4$ in $\mathbb{C}_0$,

$\zeta(t) = x_1(t) + i_1x_2(t) + i_2x_3(t) + i_1i_2x_4(t), \qquad a \leqslant t \leqslant b,$

$f(\zeta) = g_1(x) + i_1g_2(x) + i_2g_3(x) + i_1i_2g_4(x), \qquad x = x_1, \ldots, x_4.$

The idempotent representation is the following:

(2) $\zeta = z_1 + i_2z_2, \qquad z_1, z_2$ in $\mathbb{C}_1$,

(3) $\zeta = (z_1 - i_1z_2)e_1 + (z_1 + i_1z_2)e_2,$

(4) $\zeta(t) = z_1(t) + i_2z_2(t), \qquad a \leqslant t \leqslant b,$

(5) $\zeta(t) = [z_1(t) - i_1z_2(t)]e_1 + [z_1(t) + i_1z_2(t)]e_2,$

(6) $f(\zeta) = f_1(z_1 - i_1z_2)e_1 + f_2(z_1 + i_1z_2)e_2.$

If $f: X \to \mathbb{C}_2$ is a holomorphic function, then the properties of the idempotent representation show that, corresponding to each property of $\int_C f(\zeta)\,d\zeta$, there is a property in the idempotent representation. The purpose of this section is (a) to summarize the properties of the integral in the idempotent representation, and (b) to give a proof of (6). The representation (6) occurs in Theorem 21.1; it is the foundation on which much of the theory is based, and we are now able to supply the details of a proof.

Let X be a domain in $\mathbb{C}_2$, and let $f: X \to \mathbb{C}_2$ be a holomorphic function. If C is the curve in (4) whose trace is in X, then (5) shows that there are curves C_1 and C_2, with traces in X_1 and X_2 respectively, such that

(7) $C_1\colon z_1 - i_1z_2 = z_1(t) - i_1z_2(t), \qquad a \leqslant t \leqslant b,$

(8) $C_2\colon z_1 + i_1z_2 = z_1(t) + i_1z_2(t), \qquad a \leqslant t \leqslant b.$

Corresponding to the subdivision P of $[a, b]$,

(9) $a = t_0 < t_1 < \cdots < t_{i-1} < t_i < \cdots < t_n = b,$

there is a sum

$$(10)\qquad S(f,P)=\sum_{i=1}^{n} f[\zeta(t_i^*)][\zeta(t_i)-\zeta(t_{i-1})],\qquad t_{i-1}\leqslant t_i^*\leqslant t_i,$$

whose limit is the integral $\int_C f(\zeta)\,d\zeta$ (see Definition 32.1). The properties of the idempotent representation show that

$$(11)\quad S(f,P)$$
$$=\sum_{i=1}^{n} f_1[z_1(t_i^*)-i_1z_2(t_i^*)]\{[z_1(t_i)-i_1z_2(t_i)]-[z_1(t_{i-1})-i_1z_2(t_{i-1})]\}e_1$$
$$+\sum_{i=1}^{n} f_2[z_1(t_i^*)+i_1z_2(t_i^*)]\{[z_1(t_i)+i_1z_2(t_i)]-[z_1(t_{i-1})+i_1z_2(t_{i-1})]\}e_2.$$

38.1 THEOREM If the sums in (11) are formed for a sequence of subdivisions P_k, $k=1,2,\ldots$, whose norms approach zero, then the sums on the right approach the following complex integrals:

$$(12)\qquad \int_{C_1} f_1(z_1-i_1z_2)\,d(z_1-i_1z_2),$$

$$(13)\qquad \int_{C_2} f_2(z_1+i_1z_2)\,d(z_1+i_1z_2).$$

Furthermore,

$$(14)\qquad \int_C f(\zeta)\,d\zeta=\int_{C_1} f_1(z_1-i_1z_2)\,d(z_1-i_1z_2)e_1$$
$$+\int_{C_2} f_2(z_1+i_1z_2)\,d(z_1+i_1z_2)e_2.$$

Proof. The proof of this theorem follows from the definitions of the integrals, from properties of the idempotent representation, and from the idempotent representation of the norm. □

Theorem 38.1 has a converse. In Theorem 38.1 we started with a holomorphic function $f:X\to\mathbb{C}_2$ and the integral $\int_C f(\zeta)\,d\zeta$ of f on a curve C whose trace is in X. The existence of $\int_C f(\zeta)\,d\zeta$ implies the existence of the integrals (12) and (13) of the holomorphic functions f_1 and f_2 on the curves C_1 and C_2 whose traces are in X_1 and X_2, respectively. The reverse implication holds also.

38.2 THEOREM Let X_1 and X_2 be domains in $\mathbb{C}_1$, and let X be the domain in $\mathbb{C}_2$ which they generate. Let $f_1:X_1\to\mathbb{C}_1$ and $f_2:X_2\to\mathbb{C}_1$ be holomorphic functions, and let $f: X \to \mathbb{C}_2$ be the function defined as follows:

$$(15)\qquad f(z_1 + i_2z_2) = f_1(z_1 - i_1z_2)e_1 + f_2(z_1 + i_1z_2)e_2,$$
$$(z_1 - i_1z_2) \in X_1, (z_1 + i_1z_2) \in X_2.$$

Finally, let C_1 and C_2 be two curves (7) and (8) which have continuous derivatives and whose traces are in X_1 and X_2, respectively, and let C be the curve with trace in X which is defined as follows:

$$(16)\quad C: z_1 + i_2z_2 = [z_1(t) - i_1z_2(t)]e_1 + [z_1(t) + i_1z_2(t)]e_2, \qquad a \leqslant t \leqslant b.$$

Then the integrals of f, f_1, and f_2 on the curves C, C_1, and C_2 exist and the equality in (14) holds.

Proof. The integrals in (12) and (13) exist because f_1 and f_2 are holomorphic functions of a complex variable and C_1 and C_2 are curves which have continuous derivatives. The function $f: X \to \mathbb{C}_2$ defined in (15) is holomorphic by Theorem 24.3, and the curve C defined in (16) has a continuous derivative; therefore the integral $\int_C f(\zeta)\,d\zeta$ exists by Theorem 32.2. The properties of the idempotent representation show that the sums which define the three integrals satisfy (11), and (14) follows from (11). The proof of Theorem 32.2 is complete. □

Each of the properties of the integral $\int_C f(\zeta)\,d\zeta$ in Section 32 leads to a corresponding property of the idempotent integrals in (12) and (13). These properties are sufficiently obvious so that they need not be described in detail. There are also simple relations between the primitives of f and those of f_1 and f_2; they are described in the next two theorems.

38.3 THEOREM Let $f: X \to \mathbb{C}_2$ be a holomorphic function, and let $F: X \to \mathbb{C}_2$ be a primitive of f. Then

$$(17)\qquad f(z_1 + i_2z_2) = f_1(z_1 - i_1z_2)e_1 + f_2(z_1 + i_1z_2)e_2,$$

$$(18)\qquad F(z_1 + i_2z_2) = F_1(z_1 - i_1z_2)e_1 + F_2(z_1 + i_1z_2)e_2,$$

and $F_1: X_1 \to \mathbb{C}_1$ and $F_2: X_2 \to \mathbb{C}_1$ are primitives of $f_1: X_1 \to \mathbb{C}_1$ and $f_2: X_2 \to \mathbb{C}_1$, respectively.

Proof. By Theorem 21.1,

$$(19)\quad D_{z_1+i_2z_2}F(z_1 + i_2z_2) = D_{z_1-i_1z_2}F_1(z_1 - i_1z_2)e_1 + D_{z_1+i_1z_2}F_2(z_1 + i_1z_2)e_2.$$

Since F is a primitive of f, then $D_{z_1+i_2z_2}F(z_1+i_2z_2) = f(z_1+i_2z_2)$ and (17) and (19) show that

$$(20)\qquad D_{z_1-i_1z_2}F_1(z_1 - i_1z_2) = f_1(z_1 - i_1z_2),$$
$$D_{z_1+i_1z_2}F_2(z_1 + i_1z_2) = f_2(z_1 + i_1z_2).$$

Therefore, by definition, F_1 and F_2 are primitives of f_1 and f_2, respectively, and the proof of Theorem 38.3 is complete. □

38.4 THEOREM Let $f:X\to\mathbb{C}_2$ be a holomorphic function, and let $F_1:X_1\to\mathbb{C}_1$ and $F_2:X_2\to\mathbb{C}_1$ be primitives of f_1 and f_2, respectively. If $F:X\to\mathbb{C}_2$ is the function defined by (18), then F is a holomorphic function and a primitive of f.

Proof. Theorem 24.3 shows that F is a holomorphic function, and Theorem 21.1 states that (19) is true. Since F_1 and F_2 are primitives of f_1 and f_2, then the right side of the equation in (19) equals the right side of the equation in (17). Therefore the left sides of these equations are equal, and $D_{z_1+i_2z_2}F(z_1+i_2z_2)=f(z_1+i_2z_2)$ in X. Then by definition, F is a primitive of f, and the proof of Theorem 38.4 is complete. □

It is now time to give a proof of the representation (6) of a holomorphic function $f:X\to\mathbb{C}_2$ as stated in Theorem 21.1. This representation has been employed many times in the pages which follow that theorem, and care must be taken to avoid any circular arguments. As stated in Theorem 21.1, there are functions $u:X\to\mathbb{C}_1$ and $v:X\to\mathbb{C}_1$ such that

$$(21)\qquad f(\zeta)=u(\zeta)+i_2v(\zeta),\qquad \zeta\in X.$$

Also,

$$(22)\qquad \zeta=z_1+i_2z_2,\qquad z_1,\ z_2 \text{ in } \mathbb{C}_1.$$

38.5 THEOREM Let X be a domain in $\mathbb{C}_2$, and let $f:X\to\mathbb{C}_2$ in (21) be a function which satisfies the strong Stolz condition in X. Then $u(\zeta)-i_1v(\zeta)$ and $u(\zeta)+i_1v(\zeta)$ are functions of $z_1-i_1z_2$ in X_1 and $z_1+i_1z_2$ in X_2, respectively, and there are differentiable functions $f_1:X_1\to\mathbb{C}_1$ and $f_2:X_2\to\mathbb{C}_1$ such that

$$(23)\qquad f(z_1+i_2z_2)=f_1(z_1-i_1z_2)e_1+f_2(z_1+i_1z_2)e_2.$$

Proof. Let $\zeta_0:z_1^0+i_2z_2^0$ be a point in X; since X is open, there is a neighborhood $N(\zeta_0,r)=\{\zeta \text{ in } X\colon \|\zeta-\zeta_0\|<r\}$ which is in X. Now $N(\zeta_0,r)$ is star-shaped with respect to ζ_0. The proof of Cauchy's integral theorem in Section 34 and the construction of a primitive of f in Section 35 use only the fact that f satisfies the strong Stolz condition; they employ no special representations of f. Recall the construction of a primitive of f as follows. Let $\zeta':z_1'+i_2z_2'$ be a point in $N(\zeta_0,r)$, and let L be a polygonal curve in $N(\zeta_0,r)$ which joins ζ_0 to ζ'. Then the integral $\int_L f(\zeta)\,d\zeta$ has the same value for every such polygonal curve L; the proof of this statement follows from the special case of Cauchy's integral theorem in Theorem 34.1 as explained in Section 35.

Thus $\int_L f(\zeta)\,d\zeta$ defines a function $F: N(\zeta_0, r) \to \mathbb{C}_2$ as follows:

$$(24) \qquad F(\zeta') = \int_L f(\zeta)\,d\zeta, \qquad L \text{ connects } \zeta_0 \text{ to } \zeta'.$$

Now apply the idempotent representation of $\int_L f(\zeta)\,d\zeta$ as explained in this section. As indicated in (7) and (8), there are polygonal curves L_1 and L_2 which connect $z_1^0 - i_1 z_2^0$ and $z_1^0 + i_1 z_2^0$ to $z_1' - i_1 z_2'$ and $z_1' + i_1 z_2'$, respectively. Since

$$(25) \qquad f(\zeta) = u(\zeta) + i_2 v(\zeta) = [u(\zeta) - i_1 v(\zeta)]e_1 + [u(\zeta) + i_1 v(\zeta)]e_2,$$

and Theorem 38.1 shows that

$$(26) \qquad \int_L f(\zeta)\,d\zeta = \int_{L_1} [u(\zeta) - i_1 v(\zeta)]\,d(z_1 - i_1 z_2)e_1 + \int_{L_2} [u(\zeta) + i_1 v(\zeta)]\,d(z_1 + i_1 z_2)e_2.$$

Since $\int_L f(\zeta)\,d\zeta$ has the same value for every polygonal curve L which joins ζ_0 to ζ', then $\int_{L_1} [u(\zeta) - i_1 v(\zeta)]\,d(z_1 - i_1 z_2)$ has the same value for every polygonal curve L_1 which joins $z_1^0 - i_1 z_2^0$ to $z_1' - i_1 z_2'$ and it is a function of $z_1' - i_1 z_2'$. Also, $\int_{L_2} [u(\zeta) + i_1 v(\zeta)]\,d(z_1 + i_1 z_2)$ has the same value for every polygonal curve L_2 which joins $z_1^0 + i_1 z_2^0$ to $z_1' + i_1 z_2'$ and it is a function of $z_1' + i_1 z_2'$. Thus there are functions $F_1: N_1(\zeta_0, r) \to \mathbb{C}_1$ and $F_2: N_2(\zeta_0, r) \to \mathbb{C}_1$ such that

$$(27) \quad F_1(z_1' - i_1 z_2') = \int_{L_1} [u(\zeta) - i_1 v(\zeta)]\,d(z_1 - i_1 z_2), \qquad z_1' - i_1 z_2' \in N_1(\zeta_0, r),$$

$$(28) \quad F_2(z_1' + i_1 z_2') = \int_{L_2} [u(\zeta) + i_1 v(\zeta)]\,d(z_1 + i_1 z_2), \qquad z_1' + i_1 z_2' \in N_2(\zeta_0, r).$$

Here $N_1(\zeta_0, r)$ and $N_2(\zeta_0, r)$ are the sets in X_1 and X_2, respectively, which are generated by $N(\zeta_0, r)$. Equations (24), (26), (27), and (28) show that

$$(29) \qquad F(z_1' + i_2 z_2') = F_1(z_1' - i_1 z_2')e_1 + F_2(z_1' + i_1 z_2')e_2.$$

Section 35 shows that F in (24) has a derivative and that

$$(30) \qquad D_{z_1 + i_2 z_2} F(z_1' + i_2 z_2') = f(\zeta') = u(\zeta') + i_2 v(\zeta').$$

From (29) we find that

$$(31)\quad \frac{F[(z_1'+i_2z_2')+(h_1+i_2h_2)]-F(z_1'+i_2z_2')}{h_1+i_2h_2}-[u(\zeta')+i_2v(\zeta')]$$

$$=\left\{\frac{F_1[(z_1'-i_1z_2')+(h_1-i_1h_2)]-F_1(z_1'-i_1z_2')}{h_1-i_1h_2}-[u(\zeta')-i_1v(\zeta')]\right\}e_1$$

$$+\left\{\frac{F_2[(z_1'+i_1z_2')+(h_1+i_1h_2)]-F_2(z_1'+i_1z_2')}{h_1+i_1h_2}-[u(\zeta')+i_1v(\zeta')]\right\}e_2.$$

Here $h_1+i_2h_2\notin\mathcal{O}_2$ and therefore $h_1-i_1h_2$ and $h_1+i_1h_2$ are not in $\mathcal{O}_1$. By (30), the limit of the expression on the left in (31) exists as $h_1+i_2h_2\to 0$, $h_1+i_2h_2\notin\mathcal{O}_2$, and this limit is 0. Then the idempotent representation of the norm in $\mathbb{C}_2$ shows that the corresponding limits of the expressions on the right in (31) exist and are zero, and that

$$(32)\quad D_{z_1-i_1z_2}F_1(z_1'-i_1z_2')=u(\zeta')-i_1v(\zeta'),$$

$$(33)\quad D_{z_1+i_1z_2}F_2(z_1'+i_1z_2')=u(\zeta')+i_1v(\zeta').$$

Since F_1 and F_2 are functions of $z_1-i_1z_2$ and $z_1+i_1z_2$ respectively, their derivatives $D_{z_1-i_1z_2}F_1$ and $D_{z_1+i_1z_2}F_2$ are functions of the same variables. Thus (32) and (33) show that $u(\zeta)-i_1v(\zeta)$ and $u(\zeta)+i_1v(\zeta)$ also are functions of $z_1-i_1z_2$ and $z_1+i_1z_2$, respectively. Define functions $f_1:N_1(\zeta_0,r)\to\mathbb{C}_1$ and $f_2:N_2(\zeta_0,r)\to\mathbb{C}_1$ as follows:

$$(34)\quad u(\zeta)-i_1v(\zeta)=f_1(z_1-i_1z_2),\qquad u(\zeta)+i_1v(\zeta)=f_2(z_1+i_1z_2).$$

Then (21) and (34) show that

$$(35)\quad \begin{aligned} f(\zeta)&=u(\zeta)+i_2v(\zeta),\\ &=[u(\zeta)-i_1v(\zeta)]e_1+[u(\zeta)+i_1v(\zeta)]e_2,\\ &=f_1(z_1-i_1z_2)e_1+f_2(z_1+i_1z_2)e_2.\end{aligned}$$

Since f is differentiable (it satisfies the strong Stolz condition by hypothesis), the proof that f_1 and f_2 are differentiable is similar to the proof given above that F_1 and F_2 are differentiable (see also the proof of Theorem 21.1). Thus the proof of the theorem is complete in the neighborhood $N(\zeta_0,r)$. Since every point ζ_0 in X has a neighborhood in which the theorem is true, the theorem is true in X itself. □

38.6 COROLLARY Let X be a domain in $\mathbb{C}_2$, and let $f:X\to\mathbb{C}_2$ be a function which is differentiable in X. Then the conclusions of Theorem 38.5 hold as before.

Proof. Theorem 38.5 shows that, to prove this corollary, it is sufficient to prove that a differentiable function satisfies the strong Stolz condition. Theorem 21.2 contains this result, but its proof is not acceptable for present purposes because it is stated in terms of the representation (23) which we are trying to establish. However, the desired proof can be obtained from the proof of Theorem 21.1. From (24) in Section 21 it follows that there exists a function $r_1(u-i_1v;\, z_1^0-i_1z_2^0,\cdot):X_1\to\mathbb{C}_1$ such that

$$(36)\quad \begin{aligned}&[u(\zeta)-i_1v(\zeta)]-[u(\zeta_0)-i_1v(\zeta_0)]\\&\quad=[d_1(\zeta_0)-i_1d_2(\zeta_0)][(z_1-i_1z_2)-(z_1^0-i_1z_2^0)]\\&\qquad+r_1(u-i_1v;\, z_1^0-i_1z_2^0,\, z_1-i_1z_2)[(z_1-i_1z_2)-(z_1^0-i_1z_2^0)],\end{aligned}$$

$$(37)\quad \lim_{z_1-i_1z_2\to z_1^0-i_1z_2^0} r_1(u-i_1v;\, z_1^0-i_1z_2^0,\, z_1-i_1z_2)=0,$$

$$r_1(u-i_1v;\, z_1^0-i_1z_2^0,\, z_1^0-i_1z_2^0)=0.$$

In the same way, (25) in Section 21 shows that there exists a function $r_2(u+i_1v;\, z_1^0+i_1z_2^0,\cdot):X_2\to\mathbb{C}_1$ such that

$$(38)\quad \begin{aligned}&[u(\zeta)+i_1v(\zeta)]-[u(\zeta_0)+i_1v(\zeta_0)]\\&\quad=[d_1(\zeta_0)+i_1d_2(\zeta_0)][(z_1+i_1z_2)-(z_1^0+i_1z_2^0)]\\&\qquad+r_2(u+i_1v;\, z_1^0+i_1z_2^0,\, z_1+i_1z_2)[(z_1+i_1z_2)-(z_1^0+i_1z_2^0)],\end{aligned}$$

$$(39)\quad \lim_{z_1+i_1z_2\to z_1^0+i_1z_2^0} r_2(u+i_1v;\, z_1^0+i_1z_2^0,\, z_1+i_1z_2)=0,$$

$$r_2(u+i_1v;\, z_1^0+i_1z_2^0,\, z_1^0+i_1z_2^0)=0.$$

These equations and the idempotent representation of elements in $\mathbb{C}_2$ show that

$$(40)\quad \begin{aligned}&[u(\zeta)+i_2v(\zeta)]-[u(\zeta_0)+i_2v(\zeta_0)]\\&\quad=[d_1(\zeta_0)+i_2d_2(\zeta_0)][(z_1+i_2z_2)-(z_1^0+i_2z_2^0)]\\&\qquad+r(u+i_2v;\, z_1^0+i_2z_2^0,\, z_1+i_2z_2)[(z_1+i_2z_2)-(z_1^0+i_2z_2^0)],\end{aligned}$$

where

$$(41)\quad \begin{aligned}&r(u+i_2v;\, z_1^0+i_2z_2^0,\, z_1+i_2z_2)\\&\quad=r_1(u-i_1v;\, z_1^0-i_1z_2^0,\, z_1-i_1z_2)e_1\\&\qquad+r_2(u+i_1v;\, z_1^0+i_1z_2^0,\, z_1+i_1z_2)e_2.\end{aligned}$$

Equation (40) in simpler notation is

$$(42)\quad f(\zeta)-f(\zeta_0)=D_\zeta f(\zeta_0)(\zeta-\zeta_0)+r(f;\,\zeta_0,\zeta)(\zeta-\zeta_0).$$

Thus f satisfies the strong Stolz condition, and $u-i_1v$ and $u+i_1v$ are functions of $z_1-i_1z_2$ and $z_1+i_1z_2$, respectively, which satisfy the conclusions of Theorem 38.5. The proof of Corollary 38.6 is complete. □

Exercises

38.1 Use Theorem 38.4 to find a primitive of each of the following functions:
(a) $f(\zeta)=\zeta^n$;
(b) $f(\zeta)=\sin\zeta$;
(c) $f(\zeta)=\cos\zeta$;
(d) $f(\zeta)=e^\zeta$.

38.2 Let C be the curve such that $z_1(t)+i_2z_2(t)=\cos t+i_1\sin t$ for $0\leqslant t\leqslant 2\pi$. Use Theorem 33.1 to find the value of the integral $\int_C f(\zeta)\,d\zeta$ for each of the functions in Exercise 38.1.

38.3 (a) Let C be the curve with a continuous derivative whose equation in $z_1+i_2z_2=z_1(t)+i_2z_2(t)$, $a\leqslant t\leqslant b$. Find the equations of the curves C_1 and C_2.
(b) Let $f:\mathbf{C}_2\to\mathbf{C}_2$ be the function such that $f(z_1+i_2z)=(z_1+i_2z_2)^n$. Find the functions f_1 and f_2.
(c) Use (a) and (b) to find the value of each of the integrals

$$\int_{C_1} f_1(z_1-i_1z_2)\,d(z_1-i_1z_2),\qquad \int_{C_2} f_2(z_1+i_1z_2)\,d(z_1+i_1z_2).$$

(d) Use (c) to show that

$$\int_C f(z_1+i_2z_2)\,d(z_1+i_2z_2)=\frac{[z_1(b)+i_2z_2(b)]^{n+1}-[z_1(a)+i_2z_2(a)]^{n+1}}{n+1}.$$

38.4 Show that the fundamental theorem of the integral calculus can be stated in the following form. If $f:X\to\mathbf{C}_2$ is a function which has a continuous derivative, and if C is a curve which has a continuous derivative and whose trace is in X, then

$$\begin{aligned}
&\int_C D_{z_1+i_2z_2} f(z_1+i_2z_2)\,d(z_1+i_2z_2)\\
&\quad=\int_{C_1} D_{z_1-i_1z_2} f_1(z_1-i_1z_2)\,d(z_1-i_1z_2)e_1\\
&\quad+\int_{C_2} D_{z_1+i_1z_2} f_2(z_1+i_1z_2)\,d(z_1+i_1z_2)e_2\\
&\quad=\{f_1[z_1(b)-i_1z_2(b)]-f_1[z_1(a)-i_1z_2(a)]\}e_1\\
&\qquad+\{f_2[z_1(b)+i_1z_2(b)]-f_2[z_1(a)+i_1z_2(a)]\}e_2\\
&\quad=f[z_1(b)+i_2z_2(b)]-f[z_1(a)+i_2z_2(a)].
\end{aligned}$$

38.5 Let $f : \mathbb{C}_2 \to \mathbb{C}_2$ be the polynomial function such that

$$f(\zeta) = \sum_{k=0}^{n} (a_k + i_2 b_k)(z_1 + i_2 z_2)^k.$$

Give an elementary proof that $f(z_1 + i_2 z_2) = f_1(z_1 - i_1 z_2)e_1 + f_2(z_1 + i_1 z_2)e_2$ and find the functions f_1 and f_2.

39. CAUCHY'S INTEGRAL THEOREM AND THE IDEMPOTENT REPRESENTATION

Sections 34 and 36 have treated Cauchy's integral theorem, and Section 38 has treated the idempotent representation of integrals. This section establishes the relations between Cauchy's integral theorem for the integral $\int_C f(\zeta)\, d\zeta$ and Cauchy's integral theorem for the integrals

$$\int_{C_1} f_1(z_1 - i_1 z_2)\, d(z_1 - i_1 z_2), \qquad \int_{C_2} f_2(z_1 + i_1 z_2)\, d(z_1 + i_1 z_2).$$

39.1 THEOREM Let X be a domain in $\mathbb{C}_2$, let $f : X \to \mathbb{C}_2$ be a holomorphic function, and let C be a curve which has a continuous derivative and whose trace is in X. Let $f_1 : X_1 \to \mathbb{C}_1$ and $f_2 : X_2 \to \mathbb{C}_1$ be functions such that

(1) $\qquad f(z_1 + i_2 z_2) = f_1(z_1 - i_1 z_2)e_1 + f_2(z_1 + i_1 z_2)e_2.$

Then

(2) $$\int_C f(z_1 + i_2 z_2)\, d(z_1 + i_2 z_2) = 0$$

if and only if

(3) $$\int_{C_1} f_1(z_1 - i_1 z_2)\, d(z_1 - i_1 z_2) = 0, \quad \int_{C_2} f_2(z_1 + i_1 z_2)\, d(z_1 + i_1 z_2) = 0.$$

Proof. The integral in (2) exists by Theorem 32.2, and the integrals in (3) exist by Theorem 38.1. Furthermore, Theorem 38.1 shows that

(4) $$\int_C f(\zeta)\, d\zeta = \int_{C_1} f_1(z_1 - i_1 z_2)\, d(z_1 - i_1 z_2)e_1 + \int_{C_2} f_2(z_1 + i_1 z_2)\, d(z_1 + i_1 z_2)e_2.$$

Therefore, the integral in (2) equals zero if and only if the integrals in (3) equal zero. □

39.2 THEOREM Let X be a star-shaped domain in $\mathbb{C}_2$, let $f: X \to \mathbb{C}_2$ be a holomorphic function, and let C be a closed curve which has a continuous derivative and whose trace is in X. Also, let $f_1: X_1 \to \mathbb{C}_1$ and $f_2: X_2 \to \mathbb{C}_1$ be the functions such that

$$f(z_1 + i_2 z_2) = f_1(z_1 - i_1 z_2)e_1 + f_2(z_1 + i_1 z_2)e_2. \tag{5}$$

Then

$$\int_C f(z_1 + i_2 z_2)\, d(z_1 + i_2 z_2) = 0, \tag{6}$$

$$\int_{C_1} f_1(z_1 - i_1 z_2)\, d(z_1 - i_1 z_2) = 0, \tag{7}$$

$$\int_{C_2} f_2(z_1 + i_1 z_2)\, d(z_1 + i_1 z_2) = 0.$$

Proof. The hypotheses of this theorem include all of the hypotheses of Cauchy's integral theorem in Theorem 36.1; therefore, $\int_C f(z_1 + i_2 z_2)\, d(z_1 + i_2 z_2) = 0$ and (6) is true. Then (7) follows from (6) by Theorem 39.1, and the proof of Theorem 39.2 is complete. □

39.3 THEOREM Let X_1 and X_2 be star-shaped domains in $\mathbb{C}_1$, and let X be the cartesian domain generated by X_1 and X_2. Let $f_1: X_1 \to \mathbb{C}_1$ and $f_2: X_2 \to \mathbb{C}_1$ be holomorphic functions of a complex variable, and let $f: X \to \mathbb{C}_2$ be the function such that

$$f(z_1 + i_2 z_2) = f_1(z_1 - i_1 z_2)e_1 + f_2(z_1 + i_1 z_2)e_2. \tag{8}$$

Next, let

$$\begin{aligned} &C_1: z_1 - i_1 z_2 = z_1(t) - i_1 z_2(t), \\ &C_2: z_1 + i_1 z_2 = z_1(t) + i_1 z_2(t), \qquad a \leqslant t \leqslant b, \end{aligned} \tag{9}$$

be closed curves which have continuous derivatives and whose traces are in X_1 and X_2, respectively. Finally, let C be the curve

$$\begin{aligned} z_1 + i_2 z_2 &= [z_1(t) - i_1 z_2(t)]e_1 + [z_1(t) + i_1 z_2(t)]e_2 \\ &= z_1(t) + i_2 z_2(t), \qquad a \leqslant t \leqslant b. \end{aligned} \tag{10}$$

Then X is a star-shaped domain in $\mathbb{C}_2$; C is a closed curve which has a continuous derivative and whose trace is in X, and

$$\int_C f(z_1 + i_2 z_2)\, d(z_1 + i_2 z_2) = 0, \tag{11}$$

$$\int_{C_1} f_1(z_1 - i_1 z_2)\, d(z_1 - i_1 z_2) = 0, \quad \int_{C_2} f_2(z_1 + i_1 z_2)\, d(z_1 + i_1 z_2) = 0. \tag{12}$$

Proof. Observe that $f: X \to \mathbb{C}_2$ is a holomorphic function by Theorem 24.3. Also, since X is a cartesian set (see Definition 8.8) determined by X_1 and X_2, and since X_1 and X_2 are star-shaped domains, then X is a star-shaped domain by Theorems 8.9 and 8.11. Next, since C_1 and C_2 are closed curves with continuous derivatives and with traces in X_1 and X_2, respectively, then C is a closed curve with a continuous derivative and with its trace in X. Then the hypotheses of Theorem 39.2 are satisfied, and the statements in (11) and (12) follow from (6) and (7) in that theorem. □

39.4 EXAMPLE This example is an illustration of the use of (14) in Theorem 38.1 to evaluate an integral. Let f be the function such that

$$f(z_1 + i_2 z_2) = z_1 + i_2 z_2. \tag{13}$$

Then

$$f_1(z_1 - i_1 z_2) = z_1 - i_1 z_2, \qquad f_2(z_1 + i_1 z_2) = z_1 + i_1 z_2. \tag{14}$$

Let C_1 and C_2 have the following equations:

$$\begin{aligned} C_1&: z_1 - i_1 z_2 = r(\cos t + i_1 \sin t), \\ C_2&: z_1 + i_1 z_2 = r(\cos t + i_1 \sin t), \end{aligned} \qquad 0 \leqslant t \leqslant 2\pi. \tag{15}$$

Then C is the curve whose equation is

$$\begin{aligned} z_1 + i_2 z_2 &= r(\cos t + i_1 \sin t)e_1 + r(\cos t + i_1 \sin t)e_2, \\ &= r(\cos t + i_1 \sin t), \qquad 0 \leqslant t \leqslant 2\pi. \end{aligned} \tag{16}$$

Also,

$$\begin{aligned} &\int_{C_1} f_1(z_1 - i_1 z_2)\, d(z_1 - i_1 z_2) \\ &= \int_0^{2\pi} [r(\cos t + i_1 \sin t)][r(-\sin t + i_1 \cos t)]\, dt, \\ &= r^2 i_1 \int_0^{2\pi} r(\cos t + i_1 \sin t)^2\, dt, \\ &= r^2 i_1 \int_0^{2\pi} e^{2i_1 t}\, dt, \\ &= \frac{r^2 i_1 e^{2i_1 t}}{2i_1}\bigg|_0^{2\pi} = 0. \end{aligned} \tag{17}$$

A similar evaluation shows that

$$(18)\qquad \int_{C_2} f_2(z_1 + i_1 z_2)\, d(z_1 + i_1 z_2)$$

$$= \int_0^{2\pi} [r(\cos t + i_1 \sin t)][r(-\sin t + i_1 \cos t)]\, dt,$$

$$= r^2 i_1 \int_0^{2\pi} r(\cos t + i_1 \sin t)^2\, dt,$$

$$= 0.$$

Then Theorem 38.1 shows that

$$(19)\qquad \int_C f(z_1 + i_2 z_2)\, d(z_1 + i_2 z_2) = 0.$$

Since the function f such that $f(z_1 + i_1 z_2) = z_1 + i_2 z_2$ is holomorphic in all of $\mathbb{C}_2$, this result follows directly from Cauchy's integral theorem in Theorem 36.1.

Some variations on the example are instructive. If the equations of C_1 and C_2 are

$$(20)\qquad \begin{aligned} C_1&: z_1 - i_1 z_2 = r_1(\cos t + i_1 \sin t),\\ C_2&: z_1 + i_1 z_2 = r_2(\cos t + i_1 \sin t), \end{aligned} \qquad 0 \leqslant t \leqslant 2\pi,$$

then the equation of C is

$$(21)\qquad C: z_1 + i_2 z_2 = (r_1 e_1 + r_2 e_2)(\cos t + i_1 \sin t).$$

The trace of C in this case is different from that of the curve in (16), but it is still true that the values of the three integrals are zero because f is holomorphic in $\mathbb{C}_2$.

If the equations of C_1 and C_2 in (15) are replaced by the following, an even more striking change occurs in the trace of C:

$$(22)\qquad \begin{aligned} C_1&: z_1 - i_1 z_2 = r[\cos(t + c) + i_1 \sin(t + c)], \qquad 0 < c < 2\pi,\\ C_2&: z_1 + i_1 z_2 = r(\cos t + i_1 \sin t), \qquad 0 \leqslant t \leqslant 2\pi. \end{aligned}$$

The equation of C is

$$(23)\qquad \begin{aligned} z_1 + i_2 z_2 &= r[\cos(t + c) + i_1 \sin(t + c)]e_1\\ &\quad + r(\cos t + i_1 \sin t)e_2, \qquad 0 \leqslant t \leqslant 2\pi. \end{aligned}$$

Each of the curves C_1 and C_2 has the same trace in (15) and (22), but in the two cases the pairing of points on the curves to form points on C is different. Nevertheless,

$$(24)\quad \int_{C_1} f_1(z_1 - i_1 z_2)\, d(z_1 - i_1 z_2)$$

$$= \int_0^{2\pi} r[\cos(t+c) + i_1 \sin(t+c)] r[-\sin(t+c) + i_1 \cos(t+c)]\, dt$$

$$= r^2 i_1 \int_0^{2\pi} e^{2i_1(t+c)}\, dt = 0.$$

Since C_2 is unchanged from (15) and (18), then

$$(25)\quad \int_{C_2} f_2(z_1 + i_1 z_2)\, d(z_1 + i_1 z_2) = 0.$$

Thus, the traces of C in (15) and (22) are different for each c such that $0<c<2\pi$, but

$$(26)\quad \int_C f(z_1 + i_2 z_2)\, d(z_1 + i_2 z_2) = 0$$

for all of these curves as expected from Cauchy's integral theorem (Theorem 36.1). □

39.5 EXAMPLE This example describes the construction of an integral whose value is zero when f_1 and f_2 are holomorphic in ring-shaped regions;

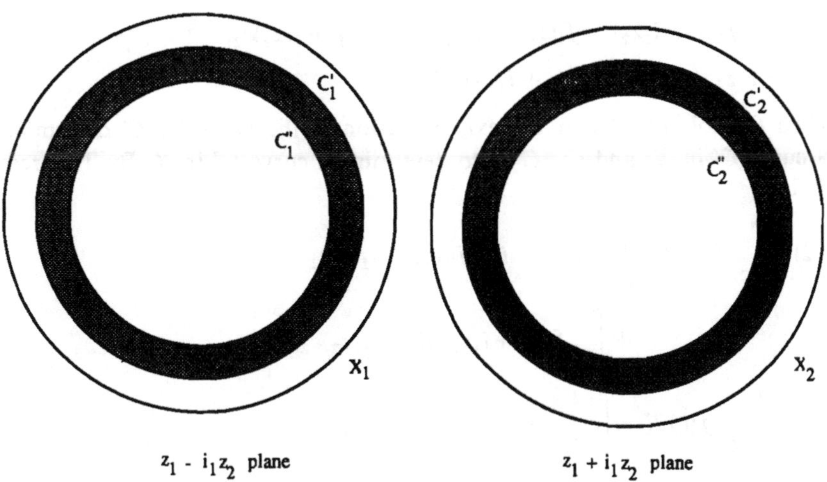

Figure 39.1. Ring regions in X_1 and X_2.

in this case the trace of C consists of more than one connected set.

Let points in the $z_1 - i_1 z_2$ and $z_1 + i_1 z_2$ planes be denoted by w_1 and w_2 respectively. Let R, r_1, and r_2 be numbers such that

(27) $$0 < r_2 < r_1 < R,$$

and let X_1 and X_2 be domains defined as follows:

(28) $$X_k = \{w_k \text{ in } \mathbb{C}_1 : 0 < |w_k| < R\}, \qquad k = 1, 2.$$

Let $f_1 : X_1 \to \mathbb{C}_1$ and $f_2 : X_2 \to \mathbb{C}_1$ be holomorphic functions. Construct circles C_1', C_2' in X_1, X_2 with centers at $w_1 = 0$ and $w_2 = 0$ and radii r_1; similarly, construct circles C_1'', C_2'' in X_1, X_2 with centers at $w_1 = 0$ and $w_2 = 0$ and radii r_2. Then (27) shows that C_1' and C_1'' form the boundary of a ring region in X_1 in which f_1 is holomorphic; also, C_2' and C_2'' bound a ring region in X_2 in which f_2 is holomorphic (see Figure 39.1). Then by the theory of functions of a complex variable,

(29) $$\int_{C_1'} f_1(z_1 - i_1 z_2)\, d(z_1 - i_1 z_2) + \int_{-C_1''} f_1(z_1 - i_1 z_2)\, d(z_1 - i_1 z_2) = 0,$$

(30) $$\int_{C_2'} f_2(z_1 + i_1 z_2)\, d(z_1 + i_1 z_2) + \int_{-C_2''} f_2(z_1 + i_1 z_2)\, d(z_1 + i_1 z_2) = 0.$$

In these integrals, C_1' and C_2' denote integration in the counterclockwise direction, and $-C_1''$ and $-C_2''$ denote integration in the clockwise direction. Let X denote the cartesian set generated by X_1 and X_2. Define $f : X \to \mathbb{C}_2$ as follows:

(31) $$f(z_1 + i_2 z_2) = f_1(z_1 - i_1 z_2) e_1 + f_2(z_1 + i_1 z_2) e_2,$$
$$z_1 - i_1 z_2 \in X_1,\ z_1 + i_1 z_2 \in X_2.$$

Then f is holomorphic in X. Next, by Section 38, the curves C_1', C_2' determine a curve C' in X; and C_1'', C_2'' also determine a curve C'' in X. By Theorem 38.1,

(32) $$\int_{C'} f(\zeta)\, d\zeta = \int_{C_1'} f_1(z_1 - i_1 z_2)\, d(z_1 - i_1 z_2) e_1 + \int_{C_2'} f_2(z_1 + i_1 z_2)\, d(z_1 + i_1 z_2) e_2,$$

(33) $$\int_{-C''} f(\zeta)\, d\zeta = \int_{-C_1''} f_1(z_1 - i_1 z_2)\, d(z_1 - i_1 z_2) e_1 + \int_{-C_2''} f_2(z_1 + i_1 z_2)\, d(z_1 + i_1 z_2) e_2.$$

Add the equations in (32) and (33); then (29) and (30) show that

$$(34) \qquad \int_{C'} f(\zeta)\,d\zeta + \int_{-C''} f(\zeta)\,d\zeta = 0.$$

This statement in (34) is the desired result; it is a form of Cauchy's integral theorem.

The proof of (34) given above has been derived from the idempotent representation of holomorphic functions of a bicomplex variable and Cauchy's integral theorem for holomorphic functions of a complex variable. A more informative proof of (34), which uses only results which have been established for holomorphic functions of a bicomplex variable, will now be given.

To begin again, the ring regions in X_1, X_2 are bounded by the curves C_1', C_1'' and C_2', C_2'', respectively, as in Figure 39.1, and f is the function defined in (31). Cut the ring regions by horizontal lines through the center, and consider the upper halves G_1 and G_2 as in Figure 39.2. The lines P_1S_1' and P_1S_1'' bound a subset Y_1 of X_1 which contains G_1 and which is star-shaped with respect to P_1 and with respect to many other points in Y_1 also. Similarly, P_2S_2' and P_2S_2'' bound a subset Y_2 of X_2 which contains G_2 and which is star-shaped with respect to P_2 and with respect to other points in Y_2. Then Y_1 and Y_2 generate a cartesian subset Y of X which is star-shaped by Theorems 8.9 and 8.11. Now it is necessary to describe curves which bound G_1 and G_2 and which generate a curve in Y. The boundary of G_1 is the trace of a continuous curve

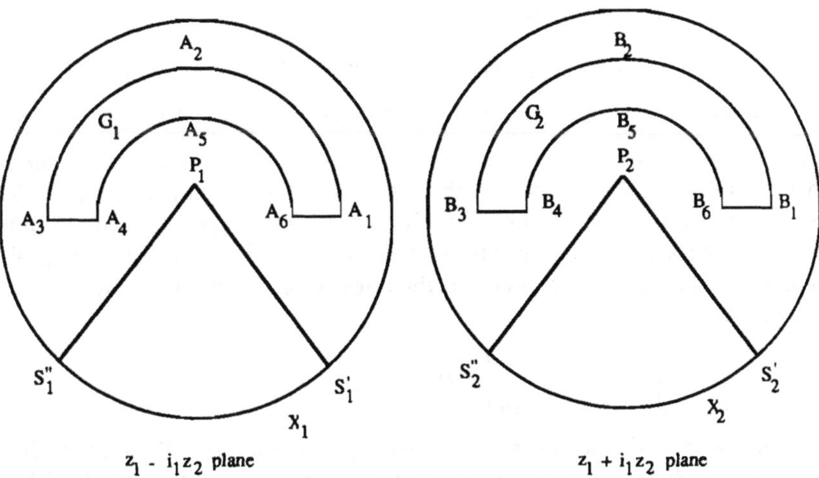

Figure 39.2. The half-rings G_1 and G_2.

which consists of two semicircles and two line segments which have the following equations:

(35) Arc $A_1A_2A_3$: $z_1-i_1z_2=r_1(\cos t+i_1 \sin t)$, $0\leqslant t\leqslant \pi$,

Seg. A_3A_4: $z_1-i_1z_2=-r_1+t(r_1-r_2)$, $0\leqslant t\leqslant 1$,

Arc $A_4A_5A_6$: $z_1-i_1z_2=r_2[\cos(-t)+i_1 \sin(-t)]$, $\pi\leqslant t\leqslant 2\pi$,

Seg. A_6A_1: $z_1-i_1z_2=r_2+t(r_1-r_2)$, $0\leqslant t\leqslant 1$.

Similarly the boundary of G_2 is the trace of a continuous curve which consists of two semicircles and two line segments which have the following equations:

(36) Arc $B_1B_2B_3$: $z_1+i_1z_2=r_1(\cos t+i_1 \sin t)$, $0\leqslant t\leqslant \pi$,

Seg. B_3B_4: $z_1+i_1z_2=-r_1+t(r_1-r_2)$, $0\leqslant t\leqslant 1$,

Arc $B_4B_5B_6$: $z_1+i_1z_2=r_2[\cos(-t)+i_1 \sin(-t)]$, $\pi\leqslant t\leqslant 2\pi$,

Seg. B_6B_1: $z_1+i_1z_2=r_2+t(r_1-r_2)$, $0\leqslant t\leqslant 1$.

The equations in (35) and (36) describe closed curves with piecewise continuous derivatives in X_1 and X_2. Pair the points for the same t on each of the four parts to form a closed curve with the following equations:

(37) (a) Arcs $A_1A_2A_3$, $B_1B_2B_3$:

$$z_1 + i_2z_2 = r_1(\cos t + i_1 \sin t), \qquad 0 \leqslant t \leqslant \pi,$$

(b) Segs. A_3A_4, B_3B_4: $z_1 + i_2z_2 = -r_1 + t(r_1 - r_2)$, $0 \leqslant t \leqslant 1$,

(c) Arcs $A_4A_5A_6$, $B_4B_5B_6$:

$$z_1 + i_2z_2 = r_2[\cos(-t) + i_1 \sin(-t)], \qquad \pi \leqslant t \leqslant 2\pi,$$

(d) Segs. A_6A_1, B_6B_1: $z_1 + i_2z_2 = r_2 + t(r_1 - r_2)$, $0\leqslant t\leqslant 1$.

Since Y_1 and Y_2 are star-shaped domains in X_1 and X_2, respectively, then Theorems 8.9 and 8.11 show that the cartesian set Y which they generate is also star-shaped. Thus f is holomorphic in the star-shaped domain Y, and (37) describes a closed curve whose trace is in Y. Then Cauchy's integral theorem in Theorem 36.1 shows that the integral of f over this curve is zero. In symbols,

$$\left[\int_{(a)} + \int_{(b)} + \int_{(c)} + \int_{(d)}\right] f(\zeta)\, d\zeta = 0. \tag{38}$$

Begin again at the beginning. The ring regions in X_1 and X_2 were cut by horizontal lines; denote the lower halves by H_1 and H_2 as in Figure 39.3. The lines Q_1T_1' and Q_1T_1'' bound a subset Z_1 of X_1 which contains H_1 and which

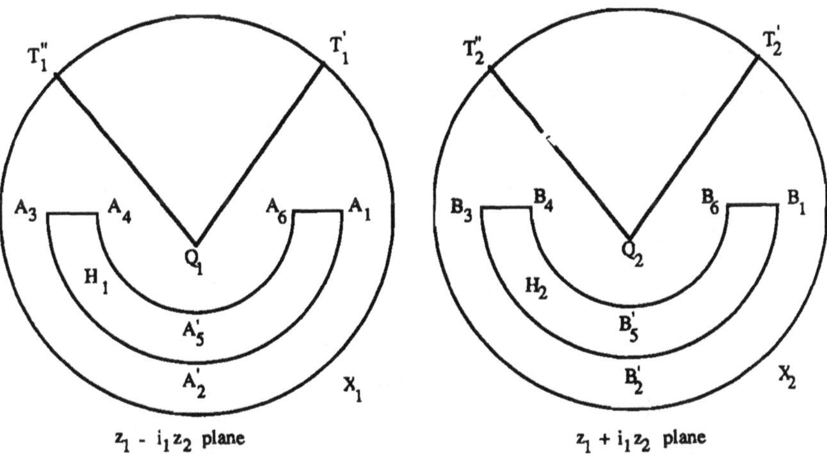

Figure 39.3. The half-rings H_1 and H_2.

is star-shaped with respect to Q_1 and other points, and Q_2T_2' and Q_2T_2'' bound a subset Z_2 of X_2 which contains H_2 and which is star-shaped with respect to Q_2 and other points. Then Z_1 and Z_2 generate a domain Z of X which is star-shaped. The boundary of H_1 is the trace of a continuous curve which consists of two semicircles and two line segments which have the following equations [compare (35)]:

$$\text{(39)}\quad \text{Arc } A_3A_2'A_1\text{: } z_1-i_1z_2=r_1(\cos t+i_1\sin t),\qquad \pi\leqslant t\leqslant 2\pi,$$

$$\text{Seg. } A_4A_3\text{: } z_1-i_1z_2=-r_2-t(r_1-r_2),\qquad 0\leqslant t\leqslant 1,$$

$$\text{Arc } A_6A_5'A_4\text{: } z_1-i_1z_2=r_2[\cos(-t)+i_1\sin(-t)],\qquad 0\leqslant t\leqslant \pi,$$

$$\text{Seg. } A_1A_6\text{: } z_1-i_1z_2=r_1-t(r_1-r_2),\qquad 0\leqslant t\leqslant 1.$$

The boundary of H_2 is the trace of a continuous curve which consists of two semicircles and two line segments which have the following equations:

$$\text{(40)}\quad \text{Arc } B_3B_2'B_1\text{: } z_1+i_1z_2=r_1(\cos t+i_1\sin t),\qquad \pi\leqslant t\leqslant 2\pi,$$

$$\text{Seg. } B_4B_3\text{: } z_1+i_1z_2=-r_2-t(r_1-r_2),\qquad 0\leqslant t\leqslant 1,$$

$$\text{Arc } B_6B_5'B_4\text{: } z_1+i_1z_2=r_2[\cos(-t)+i_1\sin(-t)],\qquad 0\leqslant t\leqslant \pi,$$

$$\text{Seg. } B_1B_6\text{: } z_1+i_1z_2=r_1-t(r_1-r_2),\qquad 0\leqslant t\leqslant 1.$$

The equations in (39) and (40) describe closed curves with piecewise continuous derivatives in X_1 and X_2. Pair the points for the same t on each of the four parts to form a closed curve with the following equations:

(41) (e) Arcs $A_3A_2'A_1$, $B_3B_2'B_1$:

$$z_1+i_2z_2=r_1(\cos t+i_1 \sin t), \qquad \pi\leqslant t\leqslant 2\pi,$$

(f) Segs. A_4A_3, B_4B_3: $z_1+i_2z_2=-r_2-t(r_1-r_2), \qquad 0\leqslant t\leqslant 1,$

(g) Arcs $A_6A_5'A_4$, $B_6B_5'B_4$:

$$z_1+i_2z_2=r_2[\cos(-t)+i_1 \sin(-t)], \qquad 0\leqslant t\leqslant \pi,$$

(h) Segs. A_1A_6, B_1B_6: $z_1+i_2z_2=r_1-t(r_1-r_2), \qquad 0\leqslant t\leqslant 1.$

Since Z_1 and Z_2 are star-shaped domains in X_1 and X_2, respectively, then Theorems 8.9 and 8.11 show that the cartesian set Z which they generate is also star-shaped. Thus f is holomorphic in the star-shaped domain Z, and (41) describes a closed curve whose trace is in Z. Then Cauchy's integral theorem in Theorem 36.1 shows that the integral of f over this curve is zero. In symbols,

$$\left[\int_{(e)}+\int_{(f)}+\int_{(g)}+\int_{(h)}\right]f(\zeta)\,d\zeta=0. \tag{42}$$

Add the equations in (38) and (42) and simplify the result. Now [see (37)(a) and (41)(e)]

$$\int_{(a)}f(\zeta)\,d\zeta+\int_{(e)}f(\zeta)\,d\zeta=\int_{C'}f(\zeta)\,d\zeta, \tag{43}$$

where C' is the curve whose equation is

$$z_1+i_2z_2=r_1(\cos t+i_1 \sin t), \qquad 0\leqslant t\leqslant 2\pi. \tag{44}$$

Here C' is the curve in X generated by the curves C_1' and C_2' in X_1 and X_2. Also,

$$\int_{(c)}f(\zeta)\,d\zeta+\int_{(g)}f(\zeta)\,d\zeta=\int_{C''}f(\zeta)\,d\zeta, \tag{45}$$

where C'' is the curve whose equation is

$$z_1+i_2z_2=r_2[\cos(-t)+i_1 \sin(-t)], \qquad 0\leqslant t\leqslant 2\pi. \tag{46}$$

This curve is generated by $-C_1''$ and $-C_2''$. Next,

$$\int_{(b)}f(\zeta)\,d\zeta+\int_{(f)}f(\zeta)\,d\zeta=0, \tag{47}$$

$$\int_{(d)}f(\zeta)\,d\zeta+\int_{(h)}f(\zeta)\,d\zeta=0. \tag{48}$$

Equations (37)(b) and (41)(f) show that (b) and (f) are the curves with equations

(49) $$z_1 + i_2 z_2 = -r_1 + t(r_1 - r_2), \qquad z_1 + i_2 z_2 = -r_2 - t(r_1 - r_2),$$
$$0 \leqslant t \leqslant 1.$$

These equations describe the same curve but with opposite orientations; thus the sum of the integrals in (47) equals zero. For similar reasons, the sum of the integrals in (48) equals zero. Thus (43), (45), (47), (48) show that the result of adding (38) and (42) can be simplified to the following form:

(50) $$\int_{C'} f(\zeta)\,d\zeta + \int_{C''} f(\zeta)\,d\zeta = 0.$$

In this case each integral may be different from zero, but their sum is zero. This statement completes Example 39.5. □

Exercises

39.1 Let f be the function such that

$$f(z_1 + i_2 z_2) = \frac{1}{(z_1 + i_2 z_2) - (a_1 + i_2 a_2)},$$
$$(z_1 + i_2 z_2) - (a_1 + i_2 a_2) \notin \mathcal{O}_2.$$

Let C be the curve defined by the equation

$$z_1 + i_2 z_2 = (a_1 + i_2 a_2) + r(\cos t + i_1 \sin t), \qquad 0 \leqslant t \leqslant 2\pi.$$

(a) Show that $\int_C f(z_1 + i_2 z_2)\, d(z_1 + i_2 z_2)$ is defined.
(b) By direct evaluation show that

$$\int_C f(\zeta)\,d\zeta = \int_0^{2\pi} \frac{r(-\sin t + i_1 \cos t)}{r(\cos t + i_1 \sin t)}\,dt = i_1 \int_0^{2\pi} dt = 2\pi i_1.$$

39.2 (a) If f is the function in Exercise 39.1, show that

$$f_1(z_1 - i_1 z_2) = \frac{1}{(z_1 - i_1 z_2) - (a_1 - i_1 a_2)},$$
$$z_1 - i_1 z_2 \neq a_1 - i_1 a_2,$$

$$f_2(z_1 + i_1 z_2) = \frac{1}{(z_1 + i_1 z_2) - (a_1 + i_1 a_2)},$$
$$z_1 + i_1 z_2 \neq a_1 + i_1 a_2.$$

(b) If C is the curve in Exercise 39.1, show that C_1 and C_2 have the following equations:

$$C_1: z_1 - i_1 z_2 = (a_1 - i_1 a_2) + r(\cos t + i_1 \sin t), \qquad 0 \leqslant t \leqslant 2\pi,$$

$$C_2: z_1 + i_1 z_2 = (a_1 + i_1 a_2) + r(\cos t + i_1 \sin t), \qquad 0 \leqslant t \leqslant 2\pi.$$

(c) Show that

$$\int_{C_1} f_1(z_1 - i_1 z_2)\, d(z_1 - i_1 z_2) = 2\pi i_1,$$

$$\int_{C_2} f_2(z_1 + i_1 z_2)\, d(z_1 + i_1 z_2) = 2\pi i_1.$$

(d) Use (c) and Theorem 38.1 to show that [compare (b) in Exercise 39.1]

$$\int_C f(z_1 + i_2 z_2)\, d(z_1 + i_2 z_2) = 2\pi i_1.$$

39.3 Let f, f_1, f_2 be the functions in Exercise 39.2, and let C_1, C_2 be the curves whose equations are the following:

$$C_1: z_1 - i_1 z_2 = (a_1 - i_1 a_2) + r(\cos t + i_1 \sin t), \qquad 0 \leqslant t \leqslant 2\pi,$$

$$C_2: z_1 + i_1 z_2 = (a_1 + i_1 a_2) + r(-\sin t + i_1 \cos t), \qquad 0 \leqslant t \leqslant 2\pi.$$

(a) Show that the traces of C_1 and C_2 in this exercise are the same as the traces of C_1 and C_2 in Exercise 39.2.

(b) Show that the curve C generated by C_1 and C_2 has the following equation:

$$C: z_1 + i_2 z_2$$
$$= (a_1 + i_2 a_2) + \tfrac{1}{2} r(1 + i_1)(1 + i_2)(\cos t + i_1 \sin t),$$
$$0 \leqslant t \leqslant 2\pi.$$

Is the trace of this curve the same as the trace of the curve C in Exercise 39.2? Explain.

(c) Verify that

$$\int_C f(z_1 + i_2 z_2)\, d(z_1 + i_2 z_2)$$
$$= \int_{C_1} f_1(z_1 - i_1 z_2)\, d(z_1 - i_1 z_2) e_1$$
$$+ \int_{C_2} f_2(z_1 + i_1 z_2)\, d(z_1 + i_1 z_2) e_2$$

by evaluating each of the integrals. Explain why the three integrals have the same values as the corresponding integrals in Exercise 39.2.

39.4 Let f be the function such that

$$f(z_1 + i_2 z_2) = \frac{1}{(z_1 + i_2 z_2) - (a_1 + i_2 a_2)},$$

$$(z_1 + i_2 z_2) - (a_1 + i_2 a_2) \notin \mathcal{O}_2,$$

and let C be the curve whose equation is

$$z_1 + i_2 z_2 = (a_1 + i_2 a_2) + r(\cos t + i_2 \sin t), \qquad 0 \leqslant t \leqslant 2\pi.$$

(a) Show that the integral of f on C is defined and that

$$\int_C f(z_1 + i_2 z_2)\, d(z_1 + i_2 z_2) = 2\pi i_2.$$

(b) Show that C_1 and C_2 have the following equations:

$$C_1\colon z_1 - i_1 z_2 = (a_1 - i_1 a_2) + r(\cos t - i_1 \sin t),$$

$$C_2\colon z_1 + i_1 z_2 = (a_1 + i_1 a_2) + r(\cos t + i_1 \sin t),$$

$$0 \leqslant t \leqslant 2\pi.$$

Show that the traces of C_1, C_2 are circles, and that C_1 and C_2 are traced in the clockwise and counterclockwise directions respectively.

(c) Show that

$$\int_{C_1} f_1(z_1 - i_1 z_2)\, d(z_1 - i_1 z_2) = -2\pi i_1,$$

$$\int_{C_2} f_2(z_1 + i_1 z_2)\, d(z_1 + i_1 z_2) = 2\pi i_1.$$

(d) Use the results in (a) and (c) to verify that

$$\int_C f(z_1 + i_2 z_2)\, d(z_1 + i_2 z_2)$$

$$= \int_{C_1} f_1(z_1 - i_1 z_2)\, d(z_1 - i_1 z_2) e_1$$

$$+ \int_{C_2} f_2(z_1 + i_1 z_2)\, d(z_1 + i_1 z_2) e_2.$$

39.5 Problem. Let X be a domain in $\mathbb{C}_2$, and let $f : X \to \mathbb{C}_2$ be a holomorphic function. Let C be a curve whose equation has a continuous derivative,

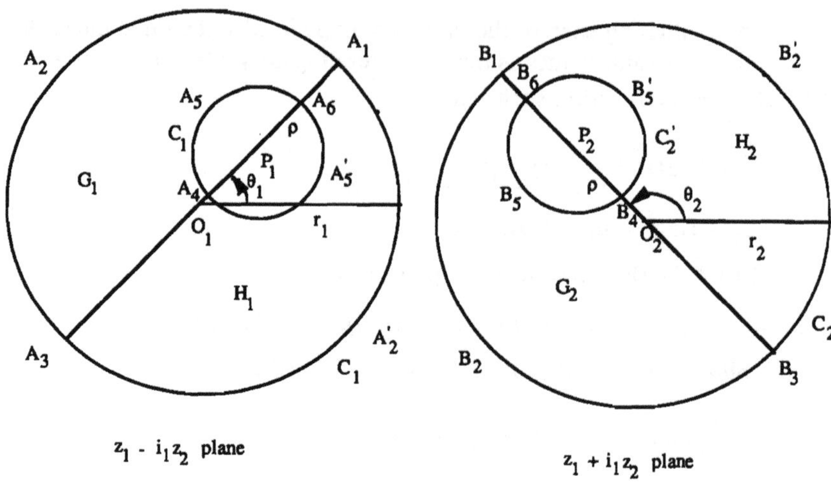

Figure 40.1. Figure for Cauchy's integral formula.

and whose trace is in X. Then $\int_C f(\zeta)\,d\zeta$ is defined; find the value of this integral. [Remarks. If X is star-shaped, then $\int_C f(\zeta)\,d\zeta=0$ by Theorem 36.1. In the general case, X is not star-shaped since f_1 and f_2 have singularities, and the value of $\int_C f(\zeta)\,d\zeta$ depends on these singularities and on the relation of C_1 and C_2 to these singularities. The complete solution of this problem is beyond the scope of this book.]

40. CAUCHY'S INTEGRAL FORMULA

The purpose of this section is to prove Cauchy's integral formula for holomorphic functions of a bicomplex variable. This formula is used in the next section to prove that a holomorphic function of a bicomplex variable has an infinite number of derivatives and that it can be represented by a Taylor series.

It is necessary to begin by describing the setting and the notation. Let X be a domain in $\mathbb{C}_2$, and let $f: X \to \mathbb{C}_2$ be a holomorphic function. Since f is holomorphic in X, then

(1) $$f(z_1+i_2z_2)=f_1(z_1-i_1z_2)e_1+f_2(z_1+i_1z_2)e_2,$$
$$z_1+i_2z_2\in X,\ z_1-i_1z_2\in X_1,\ z_1+i_1z_2\in X_2,$$

and f_1 and f_2 are holomorphic functions of a complex variable in X_1 and X_2

respectively. Let a: $a_1+i_2a_2$ be a point in X, and let r_1 and r_2 be positive constants in $\mathbb{R}$. The closed discus $\bar{D}(a;r_1,r_2)$ with center a and radii r_1 and r_2 is defined in Section 8 as follows:

$$(2)\quad \bar{D}(a;r_1,r_2)=\{z_1+i_2z_2 \text{ in } \mathbb{C}_2\colon |(z_1-i_1z_2)-(a_1-i_1a_2)| \leqslant r_1,\ |(z_1+i_1z_2)-(a_1+i_1a_2)|\leqslant r_2\}.$$

Assume that $\bar{D}(a;r_1,r_2)\subset X$; since X is an open set, $\bar{D}(a;r_1,r_2)$ is contained in X if r_1 and r_2 are sufficiently small. Let $w_1+i_2w_2$ be a fixed point in the open discus $D(a;r_1,r_2)$; then $P_1\colon w_1-i_1w_2$ is in the open circle C_1 with center O_1: $a_1-i_1a_2$ and radius r_1, and P_2: $w_1+i_1w_2$ is in the open circle C_2 with center O_2: $a_1+i_1a_2$ and radius r_2 (see Figure 40.1). The circles C_1 and C_2, together with their interiors, are in X_1 and X_2, respectively. Then C_1 and C_2 determine a curve C in X which lies in the boundary of $D(a;r_1,r_2)$.

40.1 Theorem (Cauchy's Integral Formula) If $f\colon X\to\mathbb{C}_2$, $D(a;r_1,r_2)$, and C are the holomorphic function, the discus, and the curve just described, and if $w_1+i_2w_2$ is a point in $D(a;r_1,r_2)$, then

$$(3)\qquad f(w_1+i_2w_2)=\frac{1}{2\pi i_1}\int_C \frac{f(\zeta)\,d\zeta}{\zeta-(w_1+i_2w_2)}.$$

Proof. By (1),

$$(4)\qquad f(w_1+i_2w_2)=f_1(w_1-i_1w_2)e_1+f_2(w_1+i_1w_2)e_2.$$

Now f_1 and f_2 are holomorphic functions in X_1 and X_2, respectively, and these domains contain C_1 and C_2. Then Cauchy's integral formula for holomorphic functions of a complex variable states that

$$(5)\quad f_1(w_1-i_1w_2)=\frac{1}{2\pi i_1}\int_{C_1}\frac{f_1(z_1-i_1z_2)\,d(z_1-i_1z_2)}{(z_1-i_1z_2)-(w_1-i_1w_2)},$$

$$(6)\quad f_2(w_1+i_1w_2)=\frac{1}{2\pi i_1}\int_{C_2}\frac{f_2(z_1+i_1z_2)\,d(z_1+i_1z_2)}{(z_1+i_1z_2)-(w_1+i_1w_2)}.$$

Substitute from (5) and (6) in (4); the result is

$$(7)\quad f(w_1+i_2w_2)=\frac{1}{2\pi i_1}\int_{C_1}\frac{f_1(z_1-i_1z_2)\,d(z_1-i_1z_2)}{(z_1-i_1z_2)-(w_1-i_1w_2)}e_1 +\frac{1}{2\pi i_1}\int_{C_2}\frac{f_2(z_1+i_1z_2)\,d(z_1+i_1z_2)}{(z_1+i_1z_2)-(w_1+i_1w_2)}e_2.$$

Now (1) and the idempotent representation show that

$$(8)\quad \frac{f(z_1+i_2z_2)}{(z_1+i_2z_2)-(w_1+i_2w_2)}$$

$$=\frac{f_1(z_1-i_1z_2)}{(z_1-i_1z_2)-(w_1-i_1w_2)}e_1+\frac{f_2(z_1+i_1z_2)}{(z_1+i_1z_2)-(w_1+i_1w_2)}e_2,$$

and, by construction C_1 and C_2 determine the curve C in X. Then by (14) in Section 38, equation (7) is equivalent to the following:

$$(9)\quad f(w_1+i_2w_2)=\frac{1}{2\pi i_1}\int_C \frac{f(z_1+i_2z_2)\,d(z_1+i_2z_2)}{(z_1+i_2z_2)-(w_1+i_2w_2)}.$$

This formula is (3), and the proof of Theorem 40.1 is complete. □

The proof of Theorem 40.1 given above is correct, but most of the details have been omitted and it is not very informative. Furthermore, the idempotent representation has been used to provide a proof based on the theory of functions of a complex variable. A proof of a slightly different theorem will now be given; it proceeds from first principles, employs bicomplex variable methods, and provides full details and explanations.

40.2 THEOREM (Cauchy's Integral Formula) Let $f:X\to\mathbb{C}_2$ and $D(a;r_1,r_2)$ be the holomorphic function and the discus described above. Then corresponding to each point $w_1+i_2w_2$ in $D(a;r_1,r_2)$ there is a curve C in X such that

$$(10)\qquad f(w_1+i_2w_2)=\frac{1}{2\pi i_1}\int_C \frac{f(\zeta)\,d\zeta}{\zeta-(w_1+i_2w_2)}.$$

Proof. Figure 40.1 shows the circles C_1 and C_2 with centers O_1: $a_1-i_1a_2$ and O_2: $a_1+i_1a_2$ and radii r_1 and r_2. Since $w_1+i_2w_2$ is in $D(a;r_1,r_2)$, there are points P_1: $w_1-i_1w_2$ and P_2: $w_1+i_1w_2$ in C_1 and C_2. Choose $\rho>0$ so small that the circles C_1' and C_2' with centers P_1 and P_2 and radii ρ are inside the circles C_1 and C_2. The radius from O_1 through P_1 makes an angle θ_1, $0\leqslant\theta_1<2\pi$, with the horizontal initial line; the corresponding angle for the radius from O_2 through P_2 is θ_2, $0\leqslant\theta_2<2\pi$.

Define the function g as follows:

$$(11)\quad g(z_1+i_2z_2)=\frac{f(z_1+i_2z_2)}{(z_1+i_2z_2)-(w_1+i_2w_2)},$$

$$z_1+i_2z_2\in X,\ [(z_1+i_2z_2)-(w_1+i_2w_2)]\notin\mathcal{O}_2.$$

Equation (8) shows that g is defined for every $z_1 + i_2z_2$ in X such that

$$(12)\qquad (z_1 - i_1z_2) - (w_1 - i_1w_2) \notin \mathcal{O}_1,$$
$$(z_1 + i_1z_2) - (w_1 + i_1w_2) \notin \mathcal{O}_1.$$

Let g_1 and g_2 be the functions such that

$$(13)\qquad g(z_1 + i_2z_2) = g_1(z_1 - i_1z_2)e_1 + g_2(z_1 + i_1z_2)e_2.$$

Then g_1 is defined at all points in X_1 except P_1: $w_1 - i_1w_2$, and g_2 is defined in X_2 except at P_2: $w_1 + i_1w_2$. Since f is holomorphic in X, then g is holomorphic in X where it is defined. Also, g_1 and g_2 are holomorphic functions of a complex variable where they are defined.

The next goal is to use the curves C_1 and C_2 to construct a curve C in X, and to use C_1' and C_2' to construct a curve C' in X. The construction of these curves is entirely similar to constructions carried out in Example 39.5.

Consider the regions G_1, G_2 in C_1, C_2 which are bounded as follows (see Figure 40.1):

(14) G_1: arc $A_1A_2A_3$, segment A_3A_4, arc $A_4A_5A_6$, segment A_6A_1;

(15) G_2: arc $B_1B_2B_3$, segment B_3B_4, arc $B_4B_5B_6$, segment B_6B_1.

Now G_1 is contained in a star-shaped domain Y_1 in which g_1 is holomorphic, and G_2 is contained in a star-shaped domain Y_2 in which g_2 is holomorphic. Then Y_1 and Y_2 generate a star-shaped domain Y in X in which g is holomorphic. Furthermore, the curves which bound G_1 and G_2 in (13) and (14) generate a curve in Y which can be described as follows:

(16) (a) arc$(A_1A_2A_3, B_1B_2B_3)$, (b) segment(A_3A_4, B_3B_4),

(c) arc$(A_4A_5A_6, B_4B_5B_6)$, (d) segment(A_6A_1, B_6B_1).

Since g is holomorphic in the star-shaped domain Y, and since the curve in (16) is closed, then Cauchy's integral theorem in Theorem 36.1 shows that the integral of g on this curve equals zero. Thus

$$(17)\qquad \left[\int_{(a)} + \int_{(b)} + \int_{(c)} + \int_{(d)}\right] g(z_1 + i_2z_2)\, d(z_1 + i_2z_2) = 0.$$

Again, the regions H_1, H_2 in C_1, C_2 are bounded as follows (see Figure 40.1):

(18) H_1: arc $A_3A_2'A_1$, segment A_4A_3, arc $A_6A_5'A_4$, segment A_1A_6;

(19) H_2: arc $B_3B_2'B_1$, segment B_4B_3, arc $B_6B_5'B_4$, segment B_1B_6.

Now H_1 is contained in a star-shaped domain Z_1 in which g_1 is holomorphic, and H_2 is contained in a star-shaped domain Z_2 in which g_2 is holomorphic.

Then Z_1 and Z_2 generate a star-shaped domain Z in X in which g is holomorphic. Furthermore, the curves which bound H_1 and H_2 in (18) and (19) generate a curve in Z which can be described as follows:

(20) (e) arc($A_3A'_2A_1$, $B_3B'_2B_1$), (f) segment(A_4A_3, B_4B_3),

(g) arc($A_6A'_5A_4$, $B_6B'_5B_4$), (h) segment(A_1A_6, B_1B_6).

Since g is holomorphic in the star-shaped domain Z, and since the curve in (20) is closed, then Cauchy's integral theorem shows that the integral of g on this curve equals zero. Thus

$$\left[\int_{(e)}+\int_{(f)}+\int_{(g)}+\int_{(h)}\right]g(z_1+i_2z_2)\,d(z_1+i_2z_2)=0. \tag{21}$$

Add the equations in (17) and (21) and simplify the result. Now the curves (b) and (f) are the same except that they have opposite orientations, and the same is true for (d) and (h). Thus

$$\left[\int_{(b)}+\int_{(f)}\right]g(z_1+i_2z_2)\,d(z_1+i_2z_2)=0, \tag{22}$$

$$\left[\int_{(d)}+\int_{(h)}\right]g(z_1+i_2z_2)\,d(z_1+i_2z_2)=0,$$

and the result of adding (17) and (21) can be simplified to the equation

$$\left[\int_{(a)}+\int_{(e)}\right]g(z_1+i_2z_2)\,d(z_1+i_2z_2) \tag{23}$$

$$+\left[\int_{(c)}+\int_{(g)}\right]g(z_1+i_2z_2)\,d(z_1+i_2z_2)=0.$$

It is necessary to examine the equations of the curves which occur in (23). If the arcs $A_1A_2A_3$ and $B_1B_2B_3$ are to generate a curve which has been described as arc($A_1A_2A_3$, $B_1B_2B_3$), then the equations must be chosen so that each of the arcs $A_1A_2A_3$ and $B_1B_2B_3$ corresponds to the same range of values of t. Choose equations as follows:

(24) arc $A_1A_2A_3$: $z_1-i_1z_2=(a_1-i_1a_2)$
$+r_1[\cos(\theta_1+t)+i_1\sin(\theta_1+t)], \quad 0\leqslant t\leqslant\pi,$

(25) arc $B_1B_2B_3$: $z_1+i_1z_2=(a_1+i_1a_2)$
$+r_2[\cos(\theta_2+t)+i_1\sin(\theta_2+t)], \quad 0\leqslant t\leqslant\pi,$

(26) arc $A_3A'_2A_1$: $z_1-i_1z_2=(a_1-i_1a_2)$
$+r_1[\cos(\theta_1+t)+i_1\sin(\theta_1+t)], \quad \pi\leqslant t\leqslant 2\pi,$

(27) arc $B_3B_2'B_1$: $z_1+i_1z_2=(a_1+i_1a_2)$

$$+r_2[\cos(\theta_2+t)+i_1\sin(\theta_2+t)], \qquad \pi\leqslant t\leqslant 2\pi,$$

(28) arc $A_4A_5A_6$: $z_1-i_1z_2=(w_1-i_1w_2)$

$$+\rho[\cos(\theta_1-t)+i_1\sin(\theta_1-t)], \qquad \pi\leqslant t\leqslant 2\pi,$$

(29) arc $B_4B_5B_6$: $z_1+i_1z_2=(w_1+i_1w_2)$

$$+\rho[\cos(\theta_2-t)+i_1\sin(\theta_2-t)], \qquad \pi\leqslant t\leqslant 2\pi,$$

(30) arc $A_6A_5'A_4$: $z_1-i_1z_2=(w_1-i_1w_2)$

$$+\rho[\cos(\theta_1-t)+i_1\sin(\theta_1-t)], \qquad 0\leqslant t\leqslant \pi,$$

(31) arc $B_6B_5'B_4$: $z_1+i_1z_2=(w_1+i_1w_2)$

$$+\rho[\cos(\theta_2-t)+i_1\sin(\theta_2-t)], \qquad 0\leqslant t\leqslant \pi.$$

The equation of the curve (a) in (16) can be found from (24) and (25) by using the following formula:

(32) $$z_1+i_2z_2=(z_1-i_1z_2)e_1+(z_1+i_1z_2)e_2.$$

Thus the equation of (a) is

(33) $$z_1+i_2z_2=(a_1+i_2a_2)+r_1[\cos(\theta_1+t)+i_1\sin(\theta_1+t)]e_1$$
$$+r_2[\cos(\theta_2+t)+i_1\sin(\theta_2+t)]e_2, \qquad 0\leqslant t\leqslant \pi.$$

In the same way, (26), (27), and (32) show that the equation of (e) in (20) is

(34) $$z_1+i_2z_2=(a_1+i_2a_2)+r_1[\cos(\theta_1+t)+i_1\sin(\theta_1+t)]e_1$$
$$+r_2[\cos(\theta_2+t)+i_1\sin(\theta_2+t)]e_2, \qquad \pi\leqslant t\leqslant 2\pi.$$

Equations (33) and (34) together describe a closed curve whose equation is

(35) $$z_1+i_2z_2=(a_1+i_2a_2)+r_1[\cos(\theta_1+t)+i_1\sin(\theta_1+t)]e_1$$
$$+r_2[\cos(\theta_2+t)+i_1\sin(\theta_2+t)]e_2, \qquad 0\leqslant t\leqslant 2\pi.$$

Denote this curve by C; it is the curve which corresponds to the point $w_1+i_2w_2$, and it is the curve C which occurs in the conclusion of Theorem 40.2. Similar calculations show that (c) and (g) together describe a closed curve whose equation is

(36) $$z_1+i_2z_2=(w_1+i_2w_2)+\rho[\cos(\theta_1-t)+i_1\sin(\theta_1-t)]e_1$$
$$+\rho[\cos(\theta_2-t)+i_1\sin(\theta_2-t)]e_2, \qquad 0\leqslant t\leqslant 2\pi.$$

Denote this curve by C'; it depends on the point $w_1+i_2w_2$ through the parameters θ_1 and θ_2. Then (23) can be written more simply as

$$(37)\quad \int_C g(z_1+i_2z_2)\,d(z_1+i_2z_2)+\int_{C'} g(z_1+i_2z_2)\,d(z_1+i_2z_2)=0,$$

where C and C' are the curves whose equations are given in (35) and (36) respectively. Since (11) shows that

$$(38)\quad \int_C g(z_1+i_2z_2)\,d(z_1+i_2z_2)=\int_C \frac{f(z_1+i_2z_2)\,d(z_1+i_2z_2)}{(z_1+i_2z_2)-(w_1+i_2w_2)},$$

the first integral in (37) is the integral in (10) in the conclusion of Theorem 40.2. Thus the proof of Theorem 40.2 can be completed by evaluating the second integral in (37). To simplify the notation, denote $z_1+i_2z_2$ and $w_1+i_2w_2$ by ζ and ζ_0, respectively. Then

$$(39)\quad \int_{C'} g(z_1+i_2z_2)\,d(z_1+i_2z_2)=\int_{C'} \frac{f(\zeta)\,d\zeta}{\zeta-\zeta_0},$$

and C' is the curve in (36). Since f is holomorphic at ζ_0, it satisfies the following strong Stolz condition:

$$(40)\quad f(\zeta)-f(\zeta_0)=D_\zeta f(\zeta_0)(\zeta-\zeta_0)+r(f;\zeta_0,\zeta)(\zeta-\zeta_0).$$

Substitute from (40) in (39) and simplify; equation (37) becomes

$$(41)\quad \int_C \frac{f(\zeta)\,d\zeta}{\zeta-\zeta_0}+\int_{C'} \frac{f(\zeta_0)\,d\zeta}{\zeta-\zeta_0}+\int_{C'} D_\zeta f(\zeta_0)\,d\zeta+\int_{C'} r(f;\zeta_0,\zeta)\,d\zeta=0.$$

Since $f(\zeta_0)$ and $D_\zeta f(\zeta_0)$ are constants, and since C' is a closed curve, then

$$(42)\quad \int_{C'} \frac{f(\zeta_0)\,d\zeta}{\zeta-\zeta_0}=f(\zeta_0)\int_{C'} \frac{d\zeta}{\zeta-\zeta_0},\quad \int_{C'} D_\zeta f(\zeta_0)\,d\zeta=D_\zeta f(\zeta_0)\int_{C'} d\zeta=0.$$

Use (36) to make the following evaluation:

$$(43)\quad \int_{C'} \frac{d\zeta}{\zeta-\zeta_0}$$

$$=\int_0^{2\pi} \frac{\rho[\sin(\theta_1-t)-i_1\cos(\theta_1-t)]e_1+\rho[\sin(\theta_2-t)-i_1\cos(\theta_2-t)]e_2}{\rho[\cos(\theta_1-t)+i_1\sin(\theta_1-t)]e_1+\rho[\cos(\theta_2-t)+i_1\sin(\theta_2-t)]e_2}\,dt$$

$$=-i_1\int_0^{2\pi} dt=-2\pi i_1.$$

Equations (42) and (43) show that (41) simplifies to

$$(44)\quad \int_C \frac{f(\zeta)\,d\zeta}{\zeta-\zeta_0}-2\pi i_1 f(\zeta_0)+\int_{C'} r(f;\zeta_0,\zeta)\,d\zeta=0.$$

Since the first two terms in this equation are constants, then

$$(45)\qquad \int_{C'} r(f;\zeta_0,\zeta)\,d\zeta$$

is a constant, and the proof will be completed by showing that this constant is zero. Let $\varepsilon>0$ be given. Then the Stolz condition in (40) shows that there is a $\delta>0$ such that

$$(46)\qquad \|r(f;\zeta_0,\zeta)\|<\varepsilon$$

for all ζ for which $\|\zeta-\zeta_0\|<\delta$. Equation (36) shows that $\|\zeta-\zeta_0\|=\rho$ for all ζ on C'. Choose $\rho<\delta$ in (36); then (46) shows that

$$(47)\qquad \left\|\int_{C'} r(f;\zeta_0,\zeta)\,d\zeta\right\| < \varepsilon\sqrt{2}\,L(C'),$$

where $L(C')$ is the length of C'. But (36) and Theorem 31.5 show that

$$\begin{aligned} L(C') &= \int_0^{2\pi} \|\zeta'(t)\|\,dt \\ (48)\qquad &= \int_0^{2\pi} \frac{\rho}{\sqrt{2}}\big[|\sin(\theta_1-t)-i_1\cos(\theta_1-t)|^2 \\ &\qquad\qquad + |\sin(\theta_2-t)-i_1\cos(\theta_2-t)|^2\big]^{1/2}\,dt \\ &= \int_0^{2\pi} \rho\,dt = 2\pi\rho. \end{aligned}$$

Thus if $\rho<\delta$, then (47) and (48) show that

$$(49)\qquad \left\|\int_{C'} r(f;\zeta_0,\zeta)\,d\zeta\right\| < 2^{3/2}\pi\rho\varepsilon.$$

The only constant which satisfies this inequality for every $\varepsilon>0$ and corresponding ρ is zero; therefore

$$(50)\qquad \int_{C'} r(f;\zeta_0,\zeta)\,d\zeta = 0,$$

and (44) simplifies to the statement

$$(51)\qquad \int_C \frac{f(\zeta)\,d\zeta}{\zeta-\zeta_0} - 2\pi i_1 f(\zeta_0) = 0.$$

Thus, for each point ζ_0: $w_1+i_2w_2$ in $D(a;\,r_1,\,r_2)$ there is a curve C in the boundary of the discus such that

$$(52)\qquad f(\zeta_0) = \frac{1}{2\pi i_1}\int_C \frac{f(\zeta)\,d\zeta}{\zeta-\zeta_0},$$

and the proof of Cauchy's integral formula in Theorem 40.2 is complete. □

It is interesting that many curves can be used for the representation of f as stated in Theorem 40.2; but for some applications this diversity of curves is a disadvantage, and the representation of f by means of a single curve as stated in Theorem 40.1 is to be preferred. The next theorem shows that the proof of Theorem 40.2 can be modified to yield the desired result.

40.3 THEOREM (Cauchy's Integral Formula) Let $f: X \to C_2$ and $D(a; r_1, r_2)$ be the holomorphic function and the discus described above, and let C be the curve whose equation is

$$z_1 + i_2 z_2 = (a_1 + i_2 a_2) + (r_1 e_1 + r_2 e_2)(\cos t + i_1 \sin t), \qquad 0 \leqslant t \leqslant 2\pi. \tag{53}$$

Then for every point ζ_0: $w_1 + i_2 w_2$ in $D(a; r_1, r_2)$,

$$f(\zeta_0) = \frac{1}{2\pi i_1} \int_C \frac{f(\zeta)\, d\zeta}{\zeta - \zeta_0}. \tag{54}$$

Proof. Theorem 40.3 can be proved by making some modifications in the proof of Theorem 40.2. A different evaluation of the first two integrals in (23) can be obtained as follows:

$$\begin{aligned} &\int_{(a)} g(z_1 + i_2 z_2)\, d(z_1 + i_2 z_2) \\ &= \int_{\text{arc } A_1A_2A_3} g_1(z_1 - i_1 z_2)\, d(z_1 - i_1 z_2) e_1 \\ &\quad + \int_{\text{arc } B_1B_2B_3} g_2(z_1 + i_1 z_2)\, d(z_1 + i_1 z_2) e_2, \end{aligned} \tag{55}$$

$$\begin{aligned} \text{arc } A_1A_2A_3\colon z_1 - i_1 z_2 &= (a_1 - i_1 a_2) \\ &+ r_1[\cos(\theta_1 + t) + i_1 \sin(\theta_1 + t)], \qquad 0 \leqslant t \leqslant \pi, \end{aligned} \tag{56}$$

$$\begin{aligned} \text{arc } B_1B_2B_3\colon z_1 + i_1 z_2 &= (a_1 + i_1 a_2) \\ &+ r_2[\cos(\theta_2 + t) + i_1 \sin(\theta_2 + t)], \qquad 0 \leqslant t \leqslant \pi. \end{aligned} \tag{57}$$

$$\begin{aligned} &\int_{(e)} g(z_1 + i_2 z_2)\, d(z_1 + i_2 z_2) \\ &= \int_{\text{arc } A_3A_2'A_1} g_1(z_1 - i_1 z_2)\, d(z_1 - i_1 z_2) e_1 \\ &\quad + \int_{\text{arc } B_3B_2'B_1} g_2(z_1 + i_1 z_2)\, d(z_1 + i_1 z_2) e_2, \end{aligned} \tag{58}$$

(59) arc $A_3A_2'A_1$: $z_1 - i_1z_2 = (a_1 - i_1a_2)$
$$+ r_1[\cos(\theta_1 + t) + i_1 \sin(\theta_1 + t)], \qquad \pi \leqslant t \leqslant 2\pi,$$

(60) arc $B_3B_2'B_1$: $z_1 + i_1z_2 = (a_1 + i_1a_2)$
$$+ r_2[\cos(\theta_2 + t) + i_1 \sin(\theta_2 + t)], \qquad \pi \leqslant t \leqslant 2\pi.$$

Then by an additive property of integrals,

$$(61) \quad \int_{(a)} g(z_1 + i_2z_2)\, d(z_1 + i_2z_2) + \int_{(e)} g(z_1 + i_2z_2)\, d(z_1 + i_2z_2)$$
$$= \int_{\text{arc}\, A_1A_2A_3A_2'A_1} g_1(z_1 - i_1z_2)\, d(z_1 - i_1z_2)e_1$$
$$+ \int_{\text{arc}\, B_1B_2B_3B_2'B_1} g_2(z_1 + i_1z_2)\, d(z_1 + i_1z_2)e_2,$$

(62) arc $A_1A_2A_3A_2'A_1$: $z_1 - i_1z_2 = (a_1 - i_1a_2)$
$$+ r_1[\cos(\theta_1 + t) + i_1 \sin(\theta_1 + t)], \qquad 0 \leqslant t \leqslant 2\pi,$$

(63) arc $B_1B_2B_3B_2'B_1$: $z_1 + i_1z_2 = (a_1 + i_1a_2)$
$$+ r_2[\cos(\theta_2 + t) + i_1 \sin(\theta_2 + t)], \qquad 0 \leqslant t \leqslant 2\pi.$$

The first integral on the right in (61) is an integral around a circle whose equation is given in (62); a change of variable shows that this integral is equal to

$$(64) \qquad \int_{C_1} g_1(z_1 - i_1z_2)\, d(z_1 - i_1z_2),$$

where C_1 is the curve (circle) whose equation is

$$(65) \quad C_1\colon z_1 - i_1z_2 = (a_1 - i_1a_2) + r_1(\cos t + i_1 \sin t), \qquad 0 \leqslant t \leqslant 2\pi.$$

In the same way, a change of variable shows that the second integral on the right in (61) is equal to

$$(66) \qquad \int_{C_2} g_2(z_1 + i_1z_2)\, d(z_1 + i_1z_2),$$

where C_2 is the curve (circle) whose equation is

$$(67) \quad C_2\colon z_1 + i_1z_2 = (a_1 + i_1a_2) + r_2(\cos t + i_1 \sin t), \qquad 0 \leqslant t \leqslant 2\pi.$$

Then (61), (64), and (66) show that

$$(68)\quad \left[\int_{(a)} + \int_{(e)}\right] g(z_1 + i_2 z_2)\, d(z_1 + i_2 z_2)$$

$$= \int_{C_1} g_1(z_1 - i_1 z_2)\, d(z_1 - i_1 z_2) e_1 + \int_{C_2} g_2(z_1 + i_1 z_2)\, d(z_1 + i_1 z_2) e_2,$$

where C_1 and C_2 are the curves whose equations are given in (65) and (67). Now (65), (67), and the formula

$$(69)\quad z_1 + i_2 z_2 = (z_1 - i_1 z_2) e_1 + (z_1 + i_1 z_2) e_2$$

generate a curve, denoted by C, whose equation is

$$(70)\quad C: z_1 + i_2 z_2 = (a_1 + i_2 a_2) + (r_1 e_1 + r_2 e_2)(\cos t + i_1 \sin t), \qquad 0 \leqslant t \leqslant 2\pi.$$

Then (68) above and (14) in Section 38 show that

$$(71)\quad \left[\int_{(a)} + \int_{(e)}\right] g(z_1 + i_2 z_2)\, d(z_1 + i_2 z_2) = \int_C g(z_1 + i_2 z_2)\, d(z_1 + i_2 z_2).$$

An entirely similar analysis shows that

$$(72)\quad \left[\int_{(c)} + \int_{(g)}\right] g(z_1 + i_2 z_2)\, d(z_1 + i_2 z_2) = \int_{C'} g(z_1 + i_2 z_2)\, d(z_1 + i_2 z_2),$$

where C' is the curve whose equation is

$$(73)\quad C': z_1 + i_2 z_2 = (a_1 + i_2 a_2) + \rho[\cos(-t) + i_1 \sin(-t)], \qquad 0 \leqslant t \leqslant 2\pi.$$

Then (23) and (71), (72) show that

$$(74)\quad \int_C g(z_1 + i_2 z_2)\, d(z_1 + i_2 z_2) + \int_{C'} g(z_1 + i_2 z_2)\, d(z_1 + i_2 z_2) = 0.$$

Since g is the function defined in (11), this equation is

$$(75)\quad \int_C \frac{f(\zeta)\, d\zeta}{\zeta - \zeta_0} + \int_{C'} \frac{f(\zeta)\, d\zeta}{\zeta - \zeta_0} = 0.$$

The proof of (54) can be completed by evaluating the second integral in (75). This evaluation is similar to that of the integral on the right in (39); the only difference is that the equation of C' in the former case is given in (36) and now it is given in (73). Thus

$$(76)\quad \int_{C'} \frac{f(\zeta)\, d\zeta}{\zeta - \zeta_0} = -2\pi i_1 f(\zeta_0),$$

and (75) and (76) are equivalent to

$$f(\zeta_0) = \frac{1}{2\pi i_1}\int_C \frac{f(\zeta)\,d\zeta}{\zeta - \zeta_0}. \tag{77}$$

This equation is (54), where (70) shows that C is the curve whose equation is given in (53). The proof of Theorem 40.3 is complete. □

40.4 EXAMPLE Let $f: X \to \mathbb{C}_2$ be a holomorphic function. This section has treated the evaluation of the integral

$$\frac{1}{2\pi i_1}\int_C \frac{f(\zeta)\,d\zeta}{\zeta - \zeta_0} \tag{78}$$

for certain curves C which have a continuous derivative and whose trace is in X. The special cases are interesting and important, but it is now in order to ask the following question: For what curves C does the integral exist, and what are its possible values? This example will seek to answer this question.

Let C be the curve whose equation is

$$z_1 + i_2 z_2 = z_1(t) + i_2 z_2(t), \qquad a \leqslant t \leqslant b. \tag{79}$$

The integral (78) exists if C is a curve such that $\zeta - (w_1 + i_2 w_2)$ is not in $\mathcal{O}_2$ for ζ on C. Now C generates curves C_1 and C_2 as follows:

$$C_1\colon z_1 - i_1 z_2 = z_1(t) - i_1 z_2(t), \tag{80}$$

$$C_2\colon z_1 + i_1 z_2 = z_1(t) + i_1 z_2(t), \qquad a \leqslant t \leqslant b. \tag{81}$$

For the purposes of this example, assume that C_1 and C_2 are simple closed curves; each curve may have positive orientation [counterclockwise, denoted by $C_1(+)$ and $C_2(+)$] or negative orientation [denoted by $C_1(-)$ and $C_2(-)$]. Now

$$\zeta - (w_1 + i_2 w_2) = [(z_1 - i_1 z_2) - (w_1 - i_1 w_2)]e_1 + [(z_1 + i_1 z_2) - (w_1 + i_1 w_2)]e_2. \tag{82}$$

The point P: $w_1 + i_2 w_2$ generates points P_1: $w_1 - i_1 w_2$ and P_2: $w_1 + i_1 w_2$ in the planes of C_1 and C_2, respectively. Then (82) shows that $\zeta - (w_1 + i_2 w_2)$ is not in $\mathcal{O}_2$ if and only if P_1 is not on C_1 and P_2 is not on C_2. Assume then that the condition is met so that (78) exists. Observe that P_1 may be inside or outside of C_1 [denoted by P_1(in) or P_1(out)], and that P_2 may be inside or outside of C_2 [denoted by P_2(in) or P_2(out)].

By (14) in Section 38,

$$(83)\quad \frac{1}{2\pi i_1}\int_C \frac{f(\zeta)\,d\zeta}{\zeta-(w_1+i_2w_2)}$$

$$=\frac{1}{2\pi i_1}\int_{C_1}\frac{f_1(z_1-i_1z_2)\,d(z_1-i_1z_2)}{(z_1-i_1z_2)-(w_1-i_1w_2)}e_1$$

$$+\frac{1}{2\pi i_1}\int_{C_2}\frac{f_2(z_1+i_1z_2)\,d(z_1+i_1z_2)}{(z_1+i_1z_2)-(w_1+i_1w_2)}e_2.$$

For convenience in notation, set

$$(84)\quad I=\frac{1}{2\pi i_1}\int_C\frac{f(\zeta)\,d\zeta}{\zeta-(w_1+i_2w_2)},$$

$$I_1=\frac{1}{2\pi i_1}\int_{C_1}\frac{f_1(z_1-i_1z_2)\,d(z_1-i_1z_2)}{(z_1-i_1z_2)-(w_1-i_1w_2)},$$

$$I_2=\frac{1}{2\pi i_1}\int_{C_2}\frac{f_2(z_1+i_1z_2)\,d(z_1+i_1z_2)}{(z_1+i_1z_2)-(w_1+i_1w_2)}.$$

The theory of functions of a complex variable can be used to evaluate I_1 and I_2. The possible values for I_1 are the following:

(85) $C_1(+)$, $P_1(\text{in})$: $I_1 = f_1(w_1 - i_1w_2)$,

(86) $C_1(+)$, $P_1(\text{out})$: $I_1 = 0$,

(87) $C_1(-)$, $P_1(\text{in})$: $I_1 = -f_1(w_1 - i_1w_2)$,

(88) $C_1(-)$, $P_1(\text{out})$: $I_1 = 0$.

The possible values for I_2 are the following:

(89) $C_2(+)$, $P_2(\text{in})$: $I_2 = f_2(w_1 + i_1w_2)$,

(90) $C_2(+)$, $P_2(\text{out})$: $I_2 = 0$,

(91) $C_2(-)$, $P_2(\text{in})$: $I_2 = -f_2(w_1 + i_1w_2)$,

(92) $C_2(-)$, $P_2(\text{out})$: $I_2 = 0$.

Equation (83) shows that any one of the values in (85)–(88) can be paired with each of the values in (89)–(93); therefore, the possible values of I, calculated from (83) and the possible values of I_1 and I_2, are the following:

(93) (85), (89): $I = f(w_1 + i_2w_2)$;

(94) (85), (90): $I = f_1(w_1 - i_1w_2)e_1$;

(95) (85), (91): $I = f_1(w_1 - i_1w_2)e_1 - f_2(w_1 + i_1w_2)e_2$;

(96) (85), (92): $I = f_1(w_1 - i_1w_2)e_1.$

(97) (86), (89): $I = f_2(w_1 + i_1w_2)e_2;$

(98) (86), (90): $I = 0;$

(99) (86), (91): $I = -f_2(w_1 + i_1w_2)e_2;$

(100) (86), (92): $I = 0.$

(101) (87), (89): $I = -f_1(w_1 - i_1w_2)e_1 + f_2(w_1 + i_1w_2)e_2;$

(102) (87), (90): $I = -f_1(w_1 - i_1w_2)e_1;$

(103) (87), (91): $I = -f(w_1 + i_2w_2);$

(104) (87), (92): $I = -f_1(w_1 - i_1w_2)e_1.$

(105) (88), (89): $I = f_2(w_1 + i_1w_2)e_2;$

(106) (88), (90): $I = 0;$

(107) (88), (91): $I = -f_2(w_1 + i_1w_2)e_2;$

(108) (88), (92): $I = 0.$

This tabulation shows that, if the integral

$$\frac{1}{2\pi i_1}\int_C \frac{f(\zeta)\,d\zeta}{\zeta - (w_1 + i_2w_2)} \tag{109}$$

satisfies the assumptions stated above, the integral exists and its value is the value of a holomorphic function of a bicomplex variable or of one of the singular functions

$$\pm f_1(w_1 - i_1w_2)e_1, \qquad \pm f_2(w_1 + i_1w_2)e_2. \tag{110}$$

Furthermore,

$$\frac{1}{2\pi i_1}\int_C \frac{f(\zeta)\,d\zeta}{\zeta - (w_1 + i_2w_2)} = f(w_1 + i_2w_2) \tag{111}$$

if and only if C_1 and C_2 are positively oriented and P_1 and P_2 are inside C_1 and C_2, respectively. □

Exercises

40.1 Consider Example 40.4 again. Let C be a closed curve in $\mathbb{C}_2$ such that C_1 and C_2 are positively oriented simple closed curves; assume that P_1: $w_1 - i_1w_2$ and P_2: $w_1 + i_1w_2$ are inside C_1 and C_2, respectively. For convenience in notation, let ζ_0 denote $w_1 + i_2w_2$.

(a) Use the identity $\zeta=(\zeta-\zeta_0)+\zeta_0$ to give a simple proof from first principles that

$$\frac{1}{2\pi i_1}\int_C \frac{\zeta\,d\zeta}{\zeta-\zeta_0}=\zeta_0.$$

(b) Use the identity $\zeta^2=(\zeta^2-\zeta_0^2)+\zeta_0^2=(\zeta-\zeta_0)(\zeta+\zeta_0)+\zeta_0^2$ in the same way to prove that

$$\frac{1}{2\pi i_1}\int_C \frac{\zeta^2\,d\zeta}{\zeta-\zeta_0}=\zeta_0^2.$$

(c) Give a similar proof that

$$\frac{1}{2\pi i_1}\int_C \frac{\zeta^n\,d\zeta}{\zeta-\zeta_0}=\zeta_0^n.$$

(d) Let $P(\zeta)$ be a polynomial in ζ. Use Example 40.4 to prove that

$$\frac{1}{2\pi i_1}\int_C \frac{P(\zeta)\,d\zeta}{\zeta-\zeta_0}=P(\zeta_0).$$

Use (c) in this exercise to verify this result.

40.2 Consider Example 40.4 again. Define sets $S_1,\ldots,S_4$ as follows:

$S_1=\{z_1+i_2z_2 \text{ in } \mathbb{C}_2\colon z_1-i_1z_2 \text{ is inside } C_1,$
$z_1+i_1z_2 \text{ is inside } C_2\},$

$S_2=\{z_1+i_2z_2 \text{ in } \mathbb{C}_2\colon z_1-i_1z_2 \text{ is inside } C_1,$
$z_1+i_1z_2 \text{ is outside } C_2\},$

$S_3=\{z_1+i_2z_2 \text{ in } \mathbb{C}_2\colon z_1-i_1z_2 \text{ is outside } C_1,$
$z_1+i_1z_2 \text{ is inside } C_2\},$

$S_4=\{z_1+i_2z_2 \text{ in } \mathbb{C}_2\colon z_1-i_1z_2 \text{ is outside } C_1,$
$z_1+i_1z_2 \text{ is outside } C_2\}.$

(a) Show that S_2, S_3, S_4 are unbounded sets, but that S_1 is bounded.
(b) Describe the boundaries of the sets $S_1,\ldots,S_4$.
(c) If $\zeta_0\colon w_1+i_2w_2$ is in S_4, show that

$$\frac{1}{2\pi i_1}\int_C \frac{f(\zeta)\,d\zeta}{\zeta-\zeta_0}=0.$$

(d) If $\zeta_0\in S_1$, show that the integral in (c) is equal to $f(\zeta_0)$ if C_1 and C_2 are positively oriented, and equal to $-f(\zeta_0)$ if both are negatively oriented.

(e) If ζ_0 is in S_2 or S_3, show that

$$\frac{1}{2\pi i_1}\int_C \frac{f(\zeta)\,d\zeta}{\zeta-\zeta_0}\in\mathcal{O}_2.$$

(f) Describe all of the holomorphic functions of the bicomplex variable $z_1+i_2z_2$ represented by the integral

$$\frac{1}{2\pi i_1}\int_C \frac{f(\zeta)\,d\zeta}{\zeta-\zeta_0}.$$

40.3 In Theorem 40.2, show that the same curve C corresponds to all of the points $w_1+i_2w_2$ which lie on the same ray from a: $a_1+i_2a_2$ in $D(a; r_1, r_2)$.

41. TAYLOR SERIES

The preceding section has shown that a holomorphic function $f: X\to\mathbb{C}_2$ can be represented by Cauchy's integral formula. A holomorphic function has a derivative; the present section uses Cauchy's integral formula to show that a holomorphic function has an infinite number of derivatives and that it can be represented in a discus by the usual Taylor series.

41.1 THEOREM If X is a domain in $\mathbb{C}_2$, and if $f: X\to\mathbb{C}_2$ has a derivative in X, then f has an infinite number of derivatives in X. Furthermore, if a: $a_1+i_2a_2$ is a point in X, if $\bar{D}(a; r_1, r_2)\subset X$, and if C is the curve whose equation is (see Theorem 40.3)

$$(1)\quad z_1+i_2z_2=(a_1+i_2a_2)+(r_1e_1+r_2e_2)(\cos t+i_1\sin t),\qquad 0\leqslant t\leqslant 2\pi,$$

then the nth derivative $D_\zeta^n f(\zeta_0)$ of f exists at each point ζ_0 in $D(a; r_1, r_2)$, and

$$(2)\quad D_\zeta^n f(\zeta_0)=\frac{n!}{2\pi i_1}\int_C \frac{f(\eta)}{(\eta-\zeta_0)^{n+1}}\,d\eta,\qquad n=0, 1, 2, \ldots.$$

Proof. The integral in (2) exists since f is holomorphic and Section 40 has shown that $(\eta-\zeta_0)\notin\mathcal{O}_2$ if $\zeta_0\in D(a; r_1, r_2)$ and η is on C. The proof of the theorem is by induction on n; it is similar to the proof of the corresponding theorem for holomorphic functions of a complex variable. For $n=0$, equation (2) is

$$(3)\qquad f(\zeta_0)=\frac{1}{2\pi i_1}\int_C \frac{f(\eta)}{\eta-\zeta_0}\,d\eta,\qquad \zeta_0\in D(a; r_1, r_2),$$

and this formula is true by Cauchy's integral formula in Theorem 40.3. Let ζ be a point in $D(a; r_1, r_2)$ near ζ_0 such that

$$(4) \qquad \zeta - \zeta_0 \notin \mathcal{O}_2.$$

Then by (3),

$$(5) \qquad f(\zeta) = \frac{1}{2\pi i_1} \int_C \frac{f(\eta)}{\eta - \zeta} \, d\eta,$$

and (3) and (5) show that

$$(6) \qquad f(\zeta) - f(\zeta_0) = \frac{1}{2\pi i_1} \int_C \left(\frac{1}{\eta - \zeta} - \frac{1}{\eta - \zeta_0} \right) f(\eta) \, d\eta$$

$$= \frac{1}{2\pi i_1} \int_C \frac{(\zeta - \zeta_0) f(\eta)}{(\eta - \zeta)(\eta - \zeta_0)} \, d\eta.$$

Then by (4) and (6),

$$(7) \qquad \frac{f(\zeta) - f(\zeta_0)}{\zeta - \zeta_0} = \frac{1}{2\pi i_1} \int_C \frac{f(\eta)}{(\eta - \zeta)(\eta - \zeta_0)} \, d\eta.$$

Next, subtract the expected limit of the difference quotient from each side of (7) and simplify the result to obtain the following:

$$(8) \qquad \frac{f(\zeta) - f(\zeta_0)}{\zeta - \zeta_0} - \frac{1}{2\pi i_1} \int_C \frac{f(\eta)}{(\eta - \zeta_0)^2} \, d\eta = \frac{\zeta - \zeta_0}{2\pi i_1} \int_C \frac{f(\eta)}{(\eta - \zeta)(\eta - \zeta_0)^2} \, d\eta.$$

Lemma 41.2 (below) shows that the integral on the right in (8) is bounded as ζ tends to ζ_0; therefore, because of the factor $\zeta - \zeta_0$, the limit of the right side of this equation is zero and

$$(9) \qquad D_\zeta f(\zeta_0) = \lim_{\zeta \to \zeta_0} \frac{f(\zeta) - f(\zeta_0)}{\zeta - \zeta_0} = \frac{1}{2\pi i_1} \int_C \frac{f(\eta)}{(\eta - \zeta_0)^2} \, d\eta.$$

Proceeding inductively, use this formula to show that

$$(10) \qquad \frac{D_\zeta f(\zeta) - D_\zeta f(\zeta_0)}{\zeta - \zeta_0} = \frac{1}{2\pi i_1} \int_C \frac{(\eta - \zeta_0) + (\eta - \zeta)}{(\eta - \zeta)^2 (\eta - \zeta_0)^2} f(\eta) \, d\eta.$$

Next subtract the expected limit of the difference quotient from each side of (10) and simplify the result to obtain

$$(11) \qquad \frac{D_\zeta f(\zeta) - D_\zeta f(\zeta_0)}{\zeta - \zeta_0} - \frac{2!}{2\pi i_1} \int_C \frac{f(\eta)}{(\eta - \zeta_0)^3} \, d\eta$$

$$= \frac{\zeta - \zeta_0}{2\pi i_1} \int_C \frac{(\eta - \zeta_0) + 2(\eta - \zeta)}{(\eta - \zeta)^2 (\eta - \zeta_0)^3} f(\eta) \, d\eta.$$

Lemma 41.2 shows that the integral on the right in (11) is bounded as ζ tends to ζ_0; therefore, the limit of the right side of this equation is zero and

$$(12)\qquad D_\zeta^2 f(\zeta_0)=\frac{2!}{2\pi i_1}\int_C \frac{f(\eta)}{(\eta-\zeta_0)^3}\,d\eta.$$

Next, establish the inductive step; that is, assume that

$$(13)\qquad D_\zeta^n f(\zeta_0)=\frac{n!}{2\pi i_1}\int_C \frac{f(\eta)}{(\eta-\zeta_0)^{n+1}}\,d\eta,$$

and prove that the formula obtained by replacing n by $n+1$ is also valid. Thus

$$(14)\quad D_\zeta^n f(\zeta)-D_\zeta^n f(\zeta_0)=\frac{n!}{2\pi i_1}\int_C \frac{(\eta-\zeta_0)^{n+1}-(\eta-\zeta)^{n+1}}{(\eta-\zeta)^{n+1}(\eta-\zeta_0)^{n+1}}\,f(\eta)\,d\eta.$$

The numerator of the integrand of the integral can be factored, and

$$(15)\quad (\eta-\zeta_0)^{n+1}-(\eta-\zeta)^{n+1}=[(\eta-\zeta_0)-(\eta-\zeta)]\sum_{j=0}^{n}(\eta-\zeta_0)^j(\eta-\zeta)^{n-j}$$

$$=(\zeta-\zeta_0)\sum_{j=0}^{n}(\eta-\zeta_0)^j(\eta-\zeta)^{n-j}.$$

Therefore, by (14) and (15),

$$(16)\qquad \frac{D_\zeta^n f(\zeta)-D_\zeta^n f(\zeta_0)}{\zeta-\zeta_0}=\frac{n!}{2\pi i_1}\int_C \frac{\Sigma_{j=0}^n(\eta-\zeta_0)^j(\eta-\zeta)^{n-j}}{(\eta-\zeta)^{n+1}(\eta-\zeta_0)^{n+1}}\,f(\eta)\,d\eta.$$

Next, subtract the expected limit of the difference quotient from each side of (16) and simplify the result to obtain

$$(17)\qquad \frac{D_\zeta^n f(\zeta)-D_\zeta^n f(\zeta_0)}{\zeta-\zeta_0}-\frac{(n+1)!}{2\pi i_1}\int_C \frac{f(\eta)}{(\eta-\zeta_0)^{n+2}}\,d\eta$$

$$=\frac{n!}{2\pi i_1}\int_C\left[\frac{\Sigma_{j=0}^n(\eta-\zeta_0)^j(\eta-\zeta)^{n-j}}{(\eta-\zeta)^{n+1}(\eta-\zeta_0)^{n+1}}-\frac{(n+1)}{(\eta-\zeta_0)^{n+2}}\right]f(\eta)\,d\eta.$$

The integrand of the integral on the right in (17) is

$$(18)\qquad \frac{\Sigma_{j=0}^n(\eta-\zeta_0)^{j+1}(\eta-\zeta)^{n-j}-(n+1)(\eta-\zeta)^{n+1}}{(\eta-\zeta)^{n+1}(\eta-\zeta_0)^{n+2}}.$$

The numerator of this fraction is a polynomial in $\eta-\zeta$, and this polynomial vanishes if $\eta-\zeta=\eta-\zeta_0$; therefore, by the factor theorem for polynomials, the numerator has the factor

$$(19)\qquad (\eta-\zeta_0)-(\eta-\zeta)=\zeta-\zeta_0.$$

The other factor can be found by using the division algorithm to divide the numerator of (18) by $(\eta-\zeta_0)-(\eta-\zeta)$; thus the other factor is

$$(20)\qquad P_n(\zeta,\eta)=\sum_{j=0}^{n}(n-j+1)(\eta-\zeta_0)^j(\eta-\zeta)^{n-j}.$$

Equations (18), (19), (20) show that (17) can be written in the following form:

$$(21)\qquad \frac{D_\zeta^n f(\zeta)-D_\zeta^n f(\zeta_0)}{\zeta-\zeta_0}-\frac{(n+1)!}{2\pi i_1}\int_C\frac{f(\eta)}{(\eta-\zeta_0)^{n+2}}\,d\eta$$
$$=\frac{(\zeta-\zeta_0)n!}{2\pi i_1}\int_C\frac{P_n(\zeta,\eta)f(\eta)}{(\eta-\zeta)^{n+1}(\eta-\zeta_0)^{n+2}}\,d\eta.$$

Lemma 41.2 below shows that the integral on the right is bounded as ζ tends to ζ_0; therefore, the limit of the right side of this equation is zero. The left side of the equation shows that $D_\zeta^{n+1}f(\zeta_0)$ exists and that

$$(22)\qquad D_\zeta^{n+1}f(\zeta_0)=\frac{(n+1)!}{2\pi i_1}\int_C\frac{f(\eta)}{(\eta-\zeta_0)^{n+2}}\,d\eta.$$

Thus if (2) is true for n, it is also true for $n+1$. Since (2) is true for $n=0, 1$, and 2 by (3), (9), and (12), a complete induction shows that (2) is true for all n. Thus, pending the proof of Lemma 41.2, the proof of Theorem 41.1 is complete. □

41.2 LEMMA Let $f:X\to\mathbb{C}_2$ be a holomorphic function, and let $D(a;r_1,r_2)$ be a discus such that $\bar{D}(a;r_1,r_2)\subset X$. Let C be the curve whose equation is

$$(23)\qquad z_1+i_2z_2=(a_1+i_2a_2)$$
$$+(r_1e_1+r_2e_2)(\cos t+i_1\sin t),\qquad 0\leqslant t\leqslant 2\pi.$$

Then for each ζ_0 in $D(a;r_1,r_2)$ there exists a number $\rho>0$, $\rho=\rho(\zeta_0)$, and for each ζ_0, ρ, and n there exists a number $K>0$, $K=K(\zeta_0,\rho,n)$, such that

$$(24)\qquad \left\|\int_C\frac{P_n(\zeta,\eta)f(\eta)}{(\eta-\zeta)^{n+1}(\eta-\zeta_0)^{n+2}}\,d\eta\right\|<K(\zeta_0,\rho,n),\qquad n=0,1,2,\dots,$$

for all ζ for which

$$(25)\qquad \|\zeta-\zeta_0\|\leqslant\rho(\zeta_0).$$

Proof. Begin the proof of this lemma by establishing it for $n=0$ and $n=1$. For $n=0$ the integral is found in (8) above; the problem is to show that

$$(26)\qquad \int_C\frac{f(\eta)}{(\eta-\zeta)(\eta-\zeta_0)^2}\,d\eta$$

is bounded as ζ approaches ζ_0. By Theorem 32.10,

$$(27)\qquad \left\|\int_C \frac{f(\eta)}{(\eta-\zeta)(\eta-\zeta_0)^2}\,d\eta\right\| \leqslant \sqrt{2}\,\mathrm{Bd}\left[\frac{f(\eta)}{(\eta-\zeta)(\eta-\zeta_0)^2}\right]L(C).$$

Here Bd denotes a bound, and $L(C)$ is the length of the curve C. By Theorem 31.5

$$(28)\qquad L(C) = 2\pi\|r_1e_1 + r_2e_2\| = 2\pi\left(\frac{r_1^2+r_2^2}{2}\right)^{1/2}.$$

By the inequality for the norm of a product,

$$(29)\qquad \left\|\frac{f(\eta)}{(\eta-\zeta)(\eta-\zeta_0)^2}\right\| \leqslant 2\|f(\eta)\|\left\|\frac{1}{\eta-\zeta}\right\|\left\|\frac{1}{(\eta-\zeta_0)^2}\right\|.$$

We seek bounds for the three terms on the right which are valid for all η on C and all ζ in a neighborhood of ζ_0. Now $f: X\to\mathbb{C}_2$ is a holomorphic function and therefore continuous; hence, $\|f(\eta)\|$ is a continuous function of η on the compact set C and has a maximum M. Thus

$$(30)\qquad \|f(\eta)\| \leqslant M, \qquad \forall\eta \text{ on } C.$$

Bounds for the remaining two terms on the right in (29) can be found most easily by employing the idempotent representation of bicomplex numbers. The curve C generates curves (circles) C_1 and C_2 with radii r_1 and r_2 (see Figure 41.1). The center of C is a: $a_1 + i_2a_2$; the centers of C_1 and C_2 are O_1: $a_1-i_1a_2$ and O_2: $a_1+i_1a_2$, respectively. The equations of C_1 and C_2 are found from the equation (23) of C and the equation $z_1+i_2z_2=(z_1-i_1z_2)e_1+(z_1+i_1z_2)e_2$. The point ζ_0: $w_1+i_2w_2$ in (26) is fixed; it generates points P_1: $w_1-i_1w_2$ and P_2: $w_1+i_1w_2$ in C_1 and C_2 respectively. The point ζ: $z_1+i_2z_2$ is variable; it generates points Q_1: $z_1-i_1z_2$ and Q_2: $z_1+i_1z_2$ in C_1 and C_2. Finally, the point η: $y_1+i_2y_2$ traces the curve C, and the corresponding points Y_1: $y_1-i_1y_2$ and Y_2: $y_1+i_1y_2$ trace the curves C_1 and C_2.

Let d_1 and d_2 denote the distances P_1A_1 and P_2A_2 from P_1 and P_2 to the circles C_1 and C_2. Set $\rho=\min(d_1,d_2)/2$; other choices for ρ are possible. Figure 41.1 shows the circles C' and C'' with centers P_1 and P_2 and radii ρ. The definition of ρ shows that it is a function of ζ_0. By the idempotent representation,

$$(31)\qquad \eta=(y_1-i_1y_2)e_1+(y_1+i_1y_2)e_2, \qquad \zeta=(z_1-i_1z_2)e_1+(z_1+i_1z_2)e_2,$$

$$(32)\qquad \eta-\zeta=[(y_1-i_1y_2)-(z_1-i_1z_2)]e_1+[(y_1+i_1y_2)-(z_1+i_1z_2)]e_2,$$

$$(33)\qquad \frac{1}{\eta-\zeta}=\left[\frac{1}{(y_1-i_1y_2)-(z_1-i_1z_2)}\right]e_1+\left[\frac{1}{(y_1+i_1y_2)-(z_1+i_1z_2)}\right]e_2.$$

Since $Y_1: y_1 - i_1 y_2$ and $Y_2: y_1 + i_1 y_2$ are on the circles C_1 and C_2, and since $Q_1: z_1 - i_1 z_2$ and $Q_2: z_1 + i_1 z_2$ are inside C' and C'', then $|(y_1 - i_1 y_2) - (z_1 - i_1 z_2)| > 0$ and $|(y_1 + i_1 y_2) - (z_1 + i_1 z_2)| > 0$, and the division indicated in (33) is possible. Then (33) and the idempotent representation of the norm show that

$$(34)\quad \left\|\frac{1}{\eta-\zeta}\right\| = \frac{1}{\sqrt{2}}\left\{\left|\frac{1}{(y_1 - i_1 y_2) - (z_1 - i_1 z_2)}\right|^2 + \left|\frac{1}{(y_1 + i_1 y_2) - (z_1 + i_1 z_2)}\right|^2\right\}^{1/2}.$$

Now restrict $\zeta: z_1 + i_2 z_2$ to be a point in $\bar{D}(\zeta_0; \rho, \rho)$; then $Q_1: z_1 - i_1 z_2$ and $Q_2: z_1 + i_1 z_2$ are in the closed circles C' and C'' with centers $P_1: w_1 - i_1 w_2$ and $P_2: w_1 + i_1 w_2$ and radii ρ. Figure 41.1 shows that

$$(35)\quad |(y_1 - i_1 y_2) - (z_1 - i_1 z_2)| \geqslant P_1 A_1 - \rho = d_1 - \rho$$

for all η on C and all ζ in $\bar{D}(\zeta_0; \rho, \rho)$. Similarly,

$$(36)\quad |(y_1 + i_1 y_2) - (z_1 + i_1 z_2)| \geqslant P_2 A_2 - \rho = d_2 - \rho$$

for all η on C and all ζ in $\bar{D}(\zeta_0; \rho, \rho)$. These statements and (34) show that

$$(37)\quad \left\|\frac{1}{\eta - \zeta}\right\| \leqslant \frac{1}{\sqrt{2}}\left[\frac{1}{(d_1 - \rho)^2} + \frac{1}{(d_2 - \rho)^2}\right]^{1/2}$$

for all η on C and all ζ in $\bar{D}(\zeta_0; \rho, \rho)$.

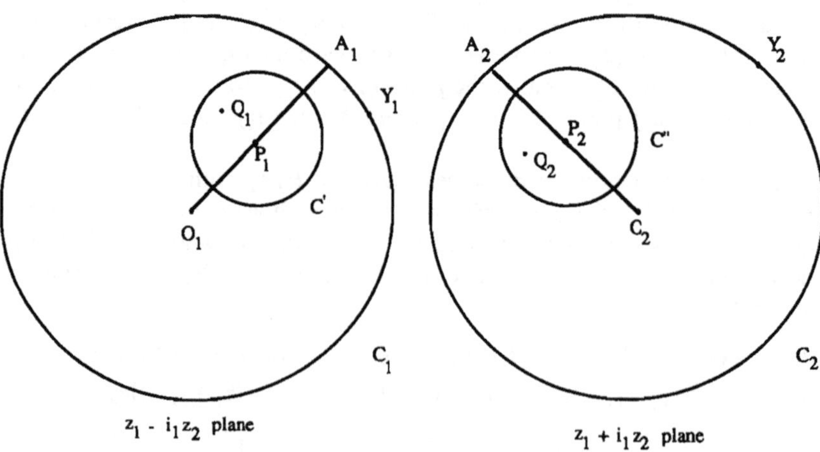

Figure 41.1. Figure for Lemma 41.2.

A bound for the last term on the right in (29) can be found in the same way. A property of the idempotent representation and (29) show that

$$(38)\quad \left\|\frac{1}{(\eta-\zeta_0)^2}\right\| = \frac{1}{\sqrt{2}}\left[\frac{1}{|(y_1-i_1y_2)-(w_1-i_1w_2)|^4} + \frac{1}{|(y_1+i_1y_2)-(w_1+i_1w_2)|^4}\right]^{1/2}.$$

Here $y_1-i_1y_2$ traces C_1 and $w_1-i_1w_2$ is P_1; then

$$(39)\quad |(y_1-i_1y_2)-(w_1-i_1w_2)| \geqslant P_1A_1 = d_1$$

for all η on C. Similarly,

$$(40)\quad |(y_1+i_1y_2)-(w_1+i_1w_2)| \geqslant P_2A_2 = d_2$$

for all η on C. Then (38), (39), and (40) show that

$$(41)\quad \left\|\frac{1}{(\eta-\zeta_0)^2}\right\| \leqslant \frac{1}{\sqrt{2}}\left(\frac{1}{d_1^4}+\frac{1}{d_2^4}\right)^{1/2}$$

for all η on C. Collect terms from (27), (29), (30), (37), and (41). Thus

$$(42)\quad \left\|\int_C \frac{f(\eta)}{(\eta-\zeta)(\eta-\zeta_0)^2}\,d\eta\right\| \leqslant \sqrt{2}L(C)M\left[\frac{1}{(d_1-\rho)^2}+\frac{1}{(d_2-\rho)^2}\right]^{1/2}\left(\frac{1}{d_1^4}+\frac{1}{d_2^4}\right)^{1/2}$$

for all ζ in $\bar{D}(\zeta_0;\rho,\rho)$. The proof of the lemma in the case $n=0$ [see the integral in (26)] is complete.

In the case $n=1$ [see (11) above] the integral is

$$(43)\quad \int_C \frac{(\eta-\zeta_0)+2(\eta-\zeta)}{(\eta-\zeta)^2(\eta-\zeta_0)^3}f(\eta)\,d\eta.$$

Here an additional term appears in the numerator; it is $P_1(\zeta,\eta)$ where

$$(44)\quad P_1(\zeta,\eta) = (\eta-\zeta_0)+2(\eta-\zeta).$$

Since $P_1(\zeta,\eta)$ is the polynomial in (44), it is clear that there exists a constant B_1 such that

$$(45)\quad \|P_1(\zeta,\eta)\| \leqslant B_1$$

for all η on C and ζ in $\bar{D}(\zeta_0; \rho, \rho)$. Next,

$$(46)\quad \left\|\int_C \frac{P_1(\zeta, \eta)f(\eta)}{(\eta-\zeta)^2(\eta-\zeta_0)^3}\,d\eta\right\| \leqslant \sqrt{2}\,\mathrm{Bd}\left[\frac{P_1(\zeta, \eta)f(\eta)}{(\eta-\zeta)^2(\eta-\zeta_0)^3}\right]L(C),$$

$$(47)\quad \left\|\frac{P_1(\zeta, \eta)f(\eta)}{(\eta-\zeta)^2(\eta-\zeta_0)^3}\right\| \leqslant 2^{3/2}\|P_1(\zeta, \eta)\|\,\|f(\eta)\|\left\|\frac{1}{(\eta-\zeta)^2}\right\|\left\|\frac{1}{(\eta-\zeta_0)^3}\right\|.$$

Then (34), (37) and (38), (41) show that

$$(48)\quad \left\|\frac{1}{(\eta-\zeta)^2}\right\| \leqslant \frac{1}{\sqrt{2}}\left[\frac{1}{(d_1-\rho)^4}+\frac{1}{(d_2-\rho)^4}\right]^{1/2},$$

$$(49)\quad \left\|\frac{1}{(\eta-\zeta_0)^3}\right\| \leqslant \frac{1}{\sqrt{2}}\left[\frac{1}{d_1^6}+\frac{1}{d_2^6}\right]^{1/2},$$

for all η on C and all ζ in $\bar{D}(\zeta_0; \rho, \rho)$. Thus (45)–(49) show that

$$(50)\quad \left\|\int_C \frac{P_1(\zeta, \eta)f(\eta)}{(\eta-\zeta)^2(\eta-\zeta_0)^3}\,d\eta\right\|$$
$$\leqslant 2L(C)B_1M\left[\frac{1}{(d_1-\rho)^4}+\frac{1}{(d_2-\rho)^4}\right]^{1/2}\left(\frac{1}{d_1^6}+\frac{1}{d_2^6}\right)^{1/2}.$$

Lemma 41.2 has now been proved in the special cases $n=0$ and $n=1$, and they show how to complete the proof in the general case. As in the special cases,

$$(51)\quad \left\|\int_C \frac{P_n(\zeta, \eta)f(\eta)}{(\eta-\zeta)^{n+1}(\eta-\zeta_0)^{n+2}}\,d\eta\right\|$$
$$\leqslant \sqrt{2}\,\mathrm{Bd}\left[\frac{P_n(\zeta, \eta)f(\eta)}{(\eta-\zeta)^{n+1}(\eta-\zeta_0)^{n+2}}\right]L(C)$$

$$(52)\quad \left\|\frac{P_n(\zeta, n)f(\eta)}{(\eta-\zeta)^{n+1}(\eta-\zeta_0)^{n+2}}\right\|$$
$$\leqslant 2^{3/2}\|P_n(\zeta, \eta)\|\,\|f(\eta)\|\left\|\frac{1}{(\eta-\zeta)^{n+1}}\right\|\left\|\frac{1}{(\eta-\zeta_0)^{n+2}}\right\|.$$

Now $P_n(\zeta, \eta)$ is the polynomial in (20), and there is constant $B_n>0$ such that

$$(53)\quad \|P_n(\zeta, \eta)\| \leqslant B_n$$

for all η on C and ζ in $\bar{D}(\zeta_0; \rho, \rho)$. The inequality in (30) holds as before. Finally, the arguments used to establish (37), (48) and (41), (49) extend easily to show that

$$(54)\quad \left\|\frac{1}{(\eta-\zeta)^{n+1}}\right\| \leqslant \frac{1}{\sqrt{2}}\left[\frac{1}{(d_1-\rho)^{2(n+1)}}+\frac{1}{(d_2-\rho)^{2(n+1)}}\right]^{1/2},$$

$$(55)\quad \left\|\frac{1}{(\eta-\zeta_0)^{n+2}}\right\| \leqslant \frac{1}{\sqrt{2}}\left[\frac{1}{d_1^{2(n+2)}}+\frac{1}{d_2^{2(n+2)}}\right]^{1/2},$$

for all η on C and all ζ in $\bar{D}(\zeta_0; \rho, \rho)$. Thus (51), (52) and the inequalities in (30), (53), (54), (55) show that

$$(56)\quad \left\|\int_C \frac{P_n(\zeta,\eta)f(\eta)}{(\eta-\zeta)^{n+1}(\eta-\zeta_0)^{n+2}}\,d\eta\right\| \leqslant 2L(C)B_nM\left[\frac{1}{(d_1-\rho)^{2(n+1)}} + \frac{1}{(d_2-\rho)^{2(n+1)}}\right]^{1/2}\left[\frac{1}{d_1^{2(n+2)}}+\frac{1}{d_2^{2(n+2)}}\right]^{1/2}$$

for all ζ in $\bar{D}(\zeta_0; \rho, \rho)$. The proof of Lemma 41.2 for $n=0,1,2,\ldots$ is now complete. □

41.3 THEOREM (Taylor's Theorem) Let X be a domain in $\mathbb{C}_2$, and let $f: X \to \mathbb{C}_2$ be a holomorphic function. Then for each a: $a_1+i_2a_2$ in X there is a discus $D(a; r_1, r_2)$, with $r_1>0$ and $r_2>0$, such that

$$(57)\quad f(\zeta) = \sum_{n=0}^{\infty} \frac{D_\zeta^n f(a)}{n!}(\zeta-a)^n, \qquad \zeta \in D(a; r_1, r_2).$$

Proof. Since X is an open set by hypothesis, then for each $a \in X$ there is a ball $B(a, r)$, with $r>0$, which is contained in X. Then Theorem 9.2 shows that $D(a; r, r) \subset B(a, r) \subset X$. Thus there is at least one discus with center a in X, and there may be many others, some of which have unequal r_1 and r_2. Assume that $D(a; r_1, r_2)$ is in X.

A first proof of Theorem 41.3 can be based on Cauchy's integral formula and Theorem 41.1. Let C be the curve in (1) as follows:

$$(58)\quad z_1+i_2z_2=(a_1+i_2a_2)+(r_1e_1+r_2e_2)(\cos t+i_1 \sin t), \qquad 0\leqslant t\leqslant 2\pi.$$

Use the proof of Taylor's theorem for functions of a complex variable as a model [7, pp. 527–530] and prove directly that

$$(59)\quad \frac{1}{2\pi i_1}\int_C \frac{f(\eta)}{\eta-\zeta}\,d\eta = \sum_{n=0}^{\infty} \frac{(\zeta-a)^n}{n!}\,\frac{n!}{2\pi i_1}\int_C \frac{f(\eta)}{(\eta-a)^{n+1}}\,d\eta,$$
$$\zeta \in D(a; r_1, r_2).$$

The details of this proof will be omitted here; they are somewhat long and tedious, and the techniques involved are well illustrated in the proof of

Lemma 41.2. By (2) in Theorem 41.1, equation (59) is equivalent to (57), and an outline of the first proof of Theorem 41.3 is complete.

A second proof of Theorem 41.3 is based directly on Taylor's theorem for functions of a complex variable. The idempotent representation of f shows that

$$f(z_1 + i_2 z_2) = f_1(z_1 - i_1 z_2)e_1 + f_2(z_1 + i_1 z_2)e_2. \tag{60}$$

Then f_1 and f_2 are holomorphic functions of $z_1 - i_1 z_2$ and $z_1 + i_1 z_2$, respectively, and they can be represented by Taylor series. Substitute these series in (60) to obtain (57). The complete details of this proof of Theorem 41.3 can be found in Section 27, and they will not be repeated here. The proof of Theorem 41.3 is complete. □

Exercises

41.1 Prove the following theorem (Cauchy's inequality). If $f: X \to \mathbb{C}_2$ is a holomorphic function, if C is a curve whose trace is in X and whose equation is

$$z_1 + i_2 z_2 = (a_1 + i_2 a_2) + r(\cos t + i_1 \sin t), \qquad 0 \leqslant t \leqslant 2\pi,$$

and if $M > 0$ is the maximum of $\|f(\zeta)\|$ for ζ on C, then

$$\|D_\zeta^n f(a)\| \leqslant \frac{n!M}{r^n}, \qquad n = 0, 1, 2, \ldots.$$

41.2 Let C in Exercise 41.1 be the circle with center a: $a_1 + i_2 a_2$ and radius $\|r_1 e_1 + r_2 e_2\|$ whose equation is

$$z_1 + i_2 z_2 = (a_1 + i_2 a_2) + (r_1 e_1 + r_2 e_2)(\cos t + i_1 \sin t), \; 0 \leqslant t \leqslant 2\pi.$$

Use this curve to find an inequality for $\|D_\zeta^n f(a)\|$. Calculate the inequality not only from

$$D_\zeta^n f(a) = \frac{n!}{2\pi i_1} \int_C \frac{f(\eta)}{(\eta - a)^{n+1}} \, d\eta,$$

but also from the idempotent representation of this integral. Compare the results. Also, when $r_1 = r_2 = r$, compare the results with each other and with the inequality in Exercise 41.1.

41.3 Prove the following theorem (Liouville's theorem). If $f: X \to \mathbb{C}_2$ is holomorphic and bounded in all of the bicomplex space $\mathbb{C}_2$, then f is a constant function. [*Hint.* Show that (2) is true for a curve C with arbitrarily large r_1 and r_2; then use Cauchy's inequality to evaluate the coefficients in the Taylor series for $n \geqslant 1$.]

41.4 The function $f:\mathbb{C}_2\to\mathbb{C}_2$ such that $f(\zeta)=\zeta^m$ is holomorphic in $\mathbb{C}_2$.
(a) Find the Taylor series of f about the point $a=0$ by using the formula in (2) to evaluate the coefficients.
(b) By the same method find the Taylor series of f about the point a. Compare this series with the expression obtained by expanding $[a+(\zeta-a)]^m$ by the binomial theorem.

41.5 Use the formula in (57) and Corollary 26.3 to expand the function

$$f(\zeta)=\frac{2}{1-\zeta^2}=\frac{1}{1-\zeta}+\frac{1}{1+\zeta}$$

in a Taylor series about the point $a=0$; find the discus in which this series converges.

42. SEQUENCES OF HOLOMORPHIC FUNCTIONS

Let X be a domain in $\mathbb{C}_2$, and let $BH(X)$ denote the set of bounded holomorphic functions $f:X\to\mathbb{C}_2$. Definition 16.3 defines the sup norm on $BH(X)$, and Section 16 shows that the set $BH(X)$ with the sup norm is a complete normed linear space (a Banach space). That proof of completeness is based on the idempotent representation of holomorphic functions of a bicomplex variable by holomorphic functions of a complex variable. Section 16 shows that the completeness of $BH(X)$ is equivalent to the completeness of the normed linear spaces $BH(X_1)$ and $BH(X_2)$ of holomorphic functions of a complex variable. In Section 16 the completeness of the complex spaces $BH(X_1)$ and $BH(X_2)$ is assumed to be known, and thus the completeness of $BH(X)$ is deduced from known complex variable theory. The purpose of the present section is to give a direct and independent proof of the completeness of $BH(X)$ using bicomplex space methods.

Let $f^n:X\to\mathbb{C}_2$, $n=1,2,\ldots$, be a Cauchy sequence of holomorphic functions in $BH(X)$; the metric is the sup norm. Then as shown in Section 16, the sequence $f^n(\zeta)$, $n=1,2,\ldots$, converges for each ζ and defines a function $f:X\to\mathbb{C}_2$. For each $\varepsilon>0$ there is an $n_0(\varepsilon)$ such that

$$(1)\qquad \|f^n(\zeta)-f(\zeta)\|\leqslant\varepsilon,\qquad n\geqslant n_0(\varepsilon),\ \forall\zeta \text{ in } X.$$

Thus the sequence of functions f^n converges uniformly to f. Since f^n is continuous for each n, then f is continuous on X; also (1) shows that f is bounded. Since f^n is holomorphic for each n, it has a derivative at each ζ in X; the problem is to show that f has a derivative at each ζ in X.

42.1 THEOREM Let $BH(X)$ be the normed linear space, with the sup norm, of bounded holomorphic functions on X. If $f^n, n=1,2,\ldots$, is a Cauchy sequence in $BH(X)$ whose limit is f, then f is a holomorphic function in $BH(X)$ and $BH(X)$ is complete and a Banach space.

Proof. As shown above, f is bounded and continuous; the proof can be completed by showing that f has a derivative at each point in X. Let a: $a_1+i_2a_2$ be a point in X, and let $\bar{D}(a; r_1, r_2)$ be a closed discus in X. Let c be the curve whose equation is

$$(2)\quad z_1+i_2z_2=(a_1+i_2a_2)+(r_1e_1+r_2e_2)(\cos t+i_1 \sin t),$$
$$0 \leqslant t \leqslant 2\pi.$$

The next step is to show that

$$(3)\qquad f(\zeta_0)=\frac{1}{2\pi i_1}\int_c \frac{f(\eta)}{\eta-\zeta_0}\,d\eta$$

for every ζ_0: $w_1+i_2w_2$ in $D(a; r_1, r_2)$. Now Theorem 40.3 establishes this result for a function f which is holomorphic, but it is not possible to appeal to this theorem in the present case since it is known only that f is continuous; in fact, the purpose of the entire theorem is to prove that f is holomorphic. Therefore, other means and methods must be found to show that f satisfies (3) at each ζ_0 in $D(a; r_1, r_2)$; the proof must be obtained from properties of the functions f^n in the sequence. Now

$$(4)\qquad \left\|f(\zeta_0)-\frac{1}{2\pi i_1}\int_c \frac{f(\eta)}{\eta-\zeta_0}\,d\eta\right\| \leqslant \|f(\zeta_0)-f^n(\zeta_0)\|$$
$$+\left\|f^n(\zeta_0)-\frac{1}{2\pi i_1}\int_c \frac{f^n(\eta)}{\eta-\zeta_0}\,d\eta\right\|$$
$$+\left\|\frac{1}{2\pi i_1}\int_c \frac{f^n(\eta)}{\eta-\zeta_0}\,d\eta-\frac{1}{2\pi i_1}\int_c \frac{f(\eta)}{\eta-\zeta_0}\,d\eta\right\|.$$

Since f^n is a holomorphic function by hypothesis, then

$$(5)\qquad f^n(\zeta_0)-\frac{1}{2\pi i_1}\int_c \frac{f^n(\eta)}{\eta-\zeta_0}\,d\eta=0$$

by Cauchy's integral formula in Theorem 40.3. Thus (4) simplifies to

$$(6)\qquad \left\|f(\zeta_0)-\frac{1}{2\pi i_1}\int_c \frac{f(\eta)}{\eta-\zeta_0}\,d\eta\right\| \leqslant \|f(\zeta_0)-f^n(\zeta_0)\|$$
$$+\left\|\frac{1}{2\pi i_1}\int_c \frac{f^n(\eta)-f(\eta)}{\eta-\zeta_0}\,d\eta\right\|.$$

Let $\varepsilon > 0$ be given. Then by (1),

(7) $\quad \|f(\zeta_0) - f^n(\zeta_0)\| \leqslant \varepsilon, \qquad n \geqslant n_0(\varepsilon),$

and

(8) $\quad \|f^n(\eta) - f(\eta)\| \leqslant \varepsilon$

for all η on C. Then Theorem 32.10 and the proof of Lemma 41.2 show that

$$(9) \quad \left\| \frac{1}{2\pi i_1} \int_C \frac{f^n(\eta) - f(\eta)}{\eta - \zeta_0} d\eta \right\| \leqslant \frac{\sqrt{2}}{2\pi} \mathrm{Bd}\left[\frac{f^n(\eta) - f(\eta)}{\eta - \zeta_0}\right] L(C)$$

$$\leqslant \frac{1}{2\pi} 2L(C)\varepsilon \frac{1}{\sqrt{2}} \left(\frac{1}{d_1^2} + \frac{1}{d_2^2}\right)^{1/2}.$$

Thus (5), (7), and (9) show that the right side of the inequality in (4) approaches zero as $n \to \infty$, and the proof of (3) is complete. Thus f is a continuous function which is represented by (3) in $D(a; r_1, r_2)$; therefore, the proof of Theorem 41.1 shows that f has a derivative. Theorem 41.1 as stated assumes that f has a derivative, but this hypothesis is needed in that theorem only to show that the function can be represented by the Cauchy integral. In the present case, we know initially only that f is continuous, but the proof given above shows that it is represented by the Cauchy integral. Then an examination of the proof of Theorem 41.1 in equations (3) through (8), and in the supporting parts of Lemma 41.2, shows that the function f in (3) has a derivative at each ζ_0 in $D(a; r_1, r_2)$. Since each point in X is contained in a discus $D(a; r_1, r_2)$, the function f has a derivative in X. The proof of Theorem 42.1 is complete. □

Definition 15.2 defines a holomorphic function as one which can be represented locally by power series. Theorems 15.3 and 15.5 show that f is holomorphic in this sense if and only if there exist holomorphic functions $f_1: X_1 \to \mathbb{C}_1$ and $f_2: X_2 \to \mathbb{C}_1$ of a complex variable such that

(10) $\quad f(z_1 + i_2 z_2) = f_1(z_1 - i_1 z_2)e_1 + f_2(z_1 + i_1 z_2)e_2.$

Thus a function f of a bicomplex variable which is holomorphic in the sense of Definition 15.2 satisfies (10), where f_1 and f_2 are holomorphic functions of a complex variable, and Theorem 24.3 shows that f has a derivative as a function of a bicomplex variable. Thus the problem has been solved once more by converting it into a problem in complex function theory. It is now possible, however, to give a direct and independent proof that a function which is represented locally by power series is holomorphic in the sense that it has a derivative.

42.2 THEOREM Let $f: X \to \mathbb{C}_2$ be a function which is represented locally by power series. More precisely, let f be a function such that for each a: $a_1 + i_1 a_2$ in X there is a discus $\bar{D}(a; r_1, r_2)$ in X and a power series with the following property:

$$(11) \qquad f(\zeta) = \sum_{k=0}^{\infty} b_k(\zeta - a)^k, \qquad \forall \zeta \text{ in } D(a; r_1, r_2).$$

Then f has a derivative $D_\zeta f(\zeta)$ at each ζ in X.

Proof. Let $D(a; r_1, r_2)$ be the discus in which the power series (11) converges. If ε is a constant such that $0 < \varepsilon < 1$, then (11) converges uniformly in

$$(12) \qquad \bar{D}[a; (1-\varepsilon)r_1, (1-\varepsilon)r_2];$$

this statement can be proved easily from the results in Section 13; see especially (29) in the proof of Theorem 13.7. Define functions $f^n: D(a; r_1, r_2) \to \mathbb{C}_2$ as follows:

$$(13) \qquad f^n(\zeta) = \sum_{k=0}^{n} b_k(\zeta - a)^k, \qquad n = 0, 1, 2, \ldots.$$

Since each f^n is a polynomial in ζ, it is a holomorphic (differentiable) function. Now the power series (11) converges uniformly to $f(\zeta)$ on the discus (12); thus the sequence f^n, $n = 0, 1, 2, \ldots$, converges in the sup norm to f on the discus in (12). Then the proof of Theorem 42.1 shows that f is a holomorphic function in the sense that it has a derivative at each point in

$$(14) \qquad D[a; (1-\varepsilon)r_1, (1-\varepsilon)r_2], \qquad 0 < \varepsilon < 1.$$

This discus contains the point a. The proof has shown that, for each a in X, there is a neighborhood of a in which f is differentiable. Thus f is differentiable at every point in X, and the proof is complete. □

Exercises

42.1 Let f be the limit in the sup norm of the sequence f^n, $n = 0, 1, 2, \ldots$, in Theorem 42.1. Show that f has an infinite number of derivatives at each point ζ_0 in X and that

$$D_\zeta^n f(\zeta_0) = \frac{n!}{2\pi i_1} \int_C \frac{f(\eta)}{(\eta - \zeta_0)^{n+1}} \, d\eta, \qquad n = 0, 1, 2, \ldots.$$

42.2 If $D_\zeta^k f^n(\zeta_0)$ and $D_\zeta^k f(\zeta_0)$ are the kth derivatives of f^n and f, respectively, in Theorem 42.1, show that

$$\lim_{n \to \infty} D_\zeta^k f^n(\zeta_0) = D_\zeta^k f(\zeta_0), \qquad k = 0, 1, 2, \ldots.$$

42.3 Let f be the function in Theorem 42.2, and let f^n be the function defined by the partial sum in (13).

(a) Use the integral formulas for $D_\zeta^k f^n(a)$ and $D_\zeta^k f(a)$ to show that

$$D_\zeta^k f^n(a) = k!b_k, \qquad n \geqslant k,$$

$$D_\zeta^k f(a) = k!b_k, \qquad k = 0, 1, 2, \ldots.$$

(b) Show that f can be expanded in a Taylor series about the point a, and that this Taylor series is the power series (11).

(c) Show that the power series which represents a holomorphic function $f: X \to \mathbb{C}_2$ is unique, and that this power series is the Taylor series.

5

Generalizations to Higher Dimensions

43. INTRODUCTION

The preceding chapters have investigated the space $\mathbb{C}_2$ and the differentiable functions $f: X \to \mathbb{C}_2$, $X \subset \mathbb{C}_2$. The space $\mathbb{C}_2$ is the first of the multicomplex spaces; it is a relatively simple special case which does not exhibit the general properties, the complexity, and the beauty of the spaces $\mathbb{C}_n$, $n > 2$, and their differentiable functions.

This chapter treats the multicomplex spaces $\mathbb{C}_n$, $n = 2, 3, \ldots$, with emphasis on the spaces $\mathbb{C}_3, \mathbb{C}_4, \ldots, \mathbb{C}_n, \ldots$, and their associated functions. Section 44 defines the multicomplex spaces $\mathbb{C}_n$ and uses $\mathbb{C}_3$ to illustrate the problems to be encountered in $\mathbb{C}_n$ and some of the differences between $\mathbb{C}_2$ and $\mathbb{C}_n$ for $n \geq 3$. The general case presents special difficulties because of the size of the expressions, formulas and equations which must be treated, and because of the complexity of the mass of details involved. For example, in $\mathbb{C}_n$ there are Cauchy–Riemann matrices with 2^n rows and columns. If n is 2 or 3, then elementary methods are adequate to prove that the determinants of these matrices can be factored into the products of determinants of two matrices whose dimensions are one-half those of the original matrix. Each of the factor matrices is a Cauchy–Riemann matrix and thus their determinants can be factored. Thus a Cauchy–Riemann matrix in $\mathbb{C}_3$ is an 8×8 matrix with 8 independent parameters, and elementary methods can be used to show that its determinant equals the product of the determinants of four 2×2 matrices.

A Cauchy–Riemann matrix in $\mathbb{C}_{10}$ has 2^{10} independent parameters; it is a matrix with 1024 rows and columns and more than a million elements. Because of their large size, elementary methods are no longer sufficient for the treatment of these matrices. Section 46 is devoted to a study of these Cauchy–Riemann matrices and their relation to the singular elements in $\mathbb{C}_n$.

Since this chapter generalizes the results in the preceding four chapters of this book, they give an indication of the nature of the topics to be treated and of the results to be obtained. However, the chapter is more than a routine replacement of 2 by n. There is a richness in the general theory; many new relationships come to light in $\mathbb{C}_n$, and many challenges arise in their treatment.

44. THE SPACES $\mathbb{C}_n$

This section defines the spaces $\mathbb{C}_n$ for $n=0,1,2,\ldots$. The spaces $\mathbb{C}_0$, $\mathbb{C}_1$, and $\mathbb{C}_2$ are either well known or they have been treated in earlier chapters of this book; therefore, the emphasis in this chapter is on $\mathbb{C}_n$ for $n\geqslant 3$. An element ζ_2 in $\mathbb{C}_2$ can be represented either as $z_1+i_2z_2$, with z_1 and z_2 in $\mathbb{C}_1$, or as $x_1+i_1x_2+i_2x_3+i_1i_2x_4$, with $x_1,\ldots,x_4$ in $\mathbb{C}_0$. Correspondingly, there are many ways in which an element ζ_n in $\mathbb{C}_n$ can be represented as a linear combination of elements in $\mathbb{C}_0,\ldots,\mathbb{C}_{n-1}$. Corresponding to each representation of the elements in $\mathbb{C}_n$, there is a linear space in which $\mathbb{C}_n$ can be imbedded. Thus $\mathbb{C}_n$ is a class of Banach algebras which are isomorphic and isometric. As in $\mathbb{C}_2$, an element in $\mathbb{C}_n$ can be represented as a linear combination of certain idempotent elements; Section 45 establishes this representation and uses it to prove some of the basic theorems about $\mathbb{C}_n$. These theorems generalize results for $\mathbb{C}_2$ in Section 6.

44.1 DEFINITION The spaces $\mathbb{C}_n$, $n=0,1,\ldots,n,\ldots$ are linear spaces whose sets of elements, norms, and operations are defined as follows:

(1) $\mathbb{C}_0=\{x: x\in\mathbb{R}\}$, ($\mathbb{C}_0$ is the space $\mathbb{R}$ of real numbers).
$\|x\|_0^2=x^2$, $x\in\mathbb{C}_0$, ($\|\ \|_0:\mathbb{C}_0\to\mathbb{R}_{\geqslant 0}$ is the absolute-value function).
Addition and multiplication are the usual operations in $\mathbb{R}$.

(2) $\mathbb{C}_1=\{z: z=x_1+i_1x_2,\ x_1 \text{ and } x_2 \text{ in } \mathbb{C}_0\}$, (space $\mathbb{C}$ of complex numbers).
$\|z\|_1^2=\|x_1\|_0^2+\|x_2\|_0^2$, ($\|\ \|_1:\mathbb{C}_1\to\mathbb{R}_{\geqslant 0}$ is the norm in $\mathbb{C}$).
Addition: $z^1+z^2=(x_1^1+i_1x_2^1)+(x_1^2+i_1x_2^2)=(x_1^1+x_1^2)+i_1(x_2^1+x_2^2)$.
Multiplication: $z^1z^2=(x_1^1x_1^2-x_2^1x_2^2)+i_1(x_1^1x_2^2+x_2^1x_1^2)$, $i_1^2=-1$.

(3) $\mathbb{C}_2 = \{\zeta_2 : \zeta_2 = z_1 + i_2 z_2,\ z_1 \text{ and } z_2 \text{ in } \mathbb{C}_1\}$.
$\|\zeta_2\|_2^2 = \|z_1\|_1^2 + \|z_2\|_1^2$.
Addition: $\zeta_2^1 + \zeta_2^2 = (z_1^1 + i_2 z_2^1) + (z_1^2 + i_2 z_2^2) = (z_1^1 + z_1^2) + i_2(z_2^1 + z_2^2)$.
Multiplication: $\zeta_2^1\zeta_2^2 = (z_1^1 z_1^2 - z_2^1 z_2^2) + i_2(z_1^1 z_2^2 + z_2^1 z_1^2)$, $i_2^2 = -1$.

(4) $\mathbb{C}_3 = \{\zeta_3 : \zeta_3 = \zeta_{21} + i_3\zeta_{22},\ \zeta_{21} \text{ and } \zeta_{22} \text{ in } \mathbb{C}_2\}$.
$\|\zeta_3\|_3^2 = \|\zeta_{21}\|_2^2 + \|\zeta_{22}\|_2^2$.
Addition: $\zeta_3^1 + \zeta_3^2 = (\zeta_{21}^1 + i_3\zeta_{22}^1) + (\zeta_{21}^2 + i_3\zeta_{22}^2)$
$= (\zeta_{21}^1 + \zeta_{21}^2) + i_3(\zeta_{22}^1 + \zeta_{22}^2)$.
Multiplication: $\zeta_3^1\zeta_3^2 = (\zeta_{21}^1\zeta_{21}^2 - \zeta_{22}^1\zeta_{22}^2) + i_3(\zeta_{21}^1\zeta_{22}^2 + \zeta_{22}^1\zeta_{21}^2)$, $i_3^2 = -1$.

If $\mathbb{C}_{n-1}$, the norm in $\mathbb{C}_{n-1}$, and the operations of addition and multiplication have been defined, then

(5) $\mathbb{C}_n = \{\zeta_n : \zeta_n = \zeta_{n-1,1} + i_n\zeta_{n-1,2},\ \zeta_{n-1,1} \text{ and } \zeta_{n-1,2} \text{ in } \mathbb{C}_{n-1}\}$.
$\|\zeta_n\|_n^2 = \|\zeta_{n-1,1}\|_{n-1}^2 + \|\zeta_{n-1,2}\|_{n-1}^2$.
Addition: $\zeta_n^1 + \zeta_n^2 = (\zeta_{n-1,1}^1 + i_n\zeta_{n-1,2}^1) + (\zeta_{n-1,1}^2 + i_n\zeta_{n-1,2}^2)$
$= (\zeta_{n-1,1}^1 + \zeta_{n-1,1}^2) + i_n(\zeta_{n-1,2}^1 + \zeta_{n-1,2}^2)$.
Multiplication: $\zeta_n^1\zeta_n^2 = (\zeta_{n-1,1}^1\zeta_{n-1,1}^2 - \zeta_{n-1,2}^1\zeta_{n-1,2}^2)$
$+ i_n(\zeta_{n-1,1}^1\zeta_{n-1,2}^2 + \zeta_{n-1,2}^1\zeta_{n-1,1}^2)$, $i_n^2 = -1$.

Many representations of $\mathbb{C}_n$ can be obtained from Definition 44.1. For example, the elements in all spaces can be represented with coefficients in $\mathbb{C}_0$. In this case, the elements in $\mathbb{C}_0, \ldots, \mathbb{C}_3$ have the following form:

(6) $\mathbb{C}_0$: x, $x \in \mathbb{R}$; $\mathbb{C}_1$: $x_1 + i_1 x_2$, x_1 and x_2 in $\mathbb{C}_0$.

(7) $\mathbb{C}_2$: $(x_1 + i_1 x_2) + i_2(x_3 + i_1 x_4)$, $x_1, \ldots, x_4$ in $\mathbb{C}_0$.

(8) $\mathbb{C}_3$: $[(x_1 + i_1 x_2) + i_2(x_3 + i_1 x_4)] + i_3[(x_5 + i_1 x_6) + i_2(x_7 + i_1 x_8)]$,
$x_1, \ldots, x_8$ in $\mathbb{C}_0$.

If elements are represented in this manner with all coefficients in $\mathbb{C}_0$, then an element in $\mathbb{C}_n$ has coefficients x_k, $k = 1, \ldots, 2^n$. Elements in the spaces $\mathbb{C}_1, \ldots, \mathbb{C}_4$ can be represented with coefficients as follows:

(9) $\mathbb{C}_1$: z (z and all z_k are in $\mathbb{C}_1$).

(10) $\mathbb{C}_2$: $z_1 + i_2 z_2$.

(11) $\mathbb{C}_3$: $(z_1 + i_2 z_2) + i_3(z_3 + i_2 z_4)$.

(12) $\mathbb{C}_4$: $[(z_1 + i_2 z_2) + i_3(z_3 + i_2 z_4)] + i_4[(z_5 + i_2 z_6) + i_3(z_7 + i_2 z_8)]$.

In this case an element ζ in $\mathbb{C}_n$, $n \geq 1$, has 2^{n-1} coefficients in $\mathbb{C}_1$. In the same way, elements in $\mathbb{C}_2, \ldots, \mathbb{C}_4$ can be represented with coefficients in $\mathbb{C}_2$ as follows:

(13) $\quad \mathbb{C}_2: \zeta \qquad (\zeta$ and all ζ_k are in $\mathbb{C}_2)$.

(14) $\quad \mathbb{C}_3: \zeta_1 + i_3\zeta_2$.

(15) $\quad \mathbb{C}_4: (\zeta_1 + i_3\zeta_2) + i_4(\zeta_3 + i_3\zeta_4)$.

In this case an element in $\mathbb{C}_n$, $n \geqslant 2$, has 2^{n-2} coefficients in $\mathbb{C}_2$.

If ζ is an element in $\mathbb{C}_n$, and if ζ is represented with coefficients in $\mathbb{C}_0$ [compare (6)–(8)], then Definition 44.1 shows that

$$(16) \qquad \|\zeta\|_n^2 = \sum_{k=1}^{2^n} |x_k|^2.$$

If ζ is an element in $\mathbb{C}_n$, $n \geqslant 1$, and if ζ is represented with coefficients in $\mathbb{C}_1$ [compare (9)–(12)], then

$$(17) \qquad \|\zeta\|_n^2 = \sum_{k=1}^{2^{n-1}} |z_k|^2.$$

In the usual way, let $\mathbb{C}_2^2$, $\mathbb{C}_1^4$, and $\mathbb{C}_0^8$ denote the linear spaces with the following elements and norm squared:

(18) $\quad (\zeta_{21}, \zeta_{22})$, ζ_{21} and ζ_{22} in $\mathbb{C}_2$, $\|\zeta_{21}\|_2^2 + \|\zeta_{22}\|_2^2$;

(19) $\quad (z_1, \ldots, z_4)$, $z_1, \ldots, z_4$ in $\mathbb{C}_1$, $\|z_1\|_1^2 + \cdots + \|z_4\|_1^2$;

(20) $\quad (x_1, \ldots, x_8)$, $x_1, \ldots, x_8$ in $\mathbb{C}_0$, $\|x_1\|_0^2 + \cdots + \|x_8\|_0^2$.

44.2 THEOREM The space $\mathbb{C}_3$ is isomorphic and isometric with each of the spaces $\mathbb{C}_2^2$, $\mathbb{C}_1^4$, and $\mathbb{C}_0^8$. Also, with the usual conventions, each of the spaces $\mathbb{C}_0$, $\mathbb{C}_1$, $\mathbb{C}_2$ is contained in $\mathbb{C}_3$, and $\mathbb{C}_3$ is contained in each of the spaces $\mathbb{C}_4$, $\mathbb{C}_5, \ldots$. Similar statements hold for each space $\mathbb{C}_k$ in $\mathbb{C}_0, \mathbb{C}_1, \ldots, \mathbb{C}_n, \ldots$.

Proof. The first statement in the theorem follows from Definition 44.1 and the definitions of the spaces $\mathbb{C}_2^2$, $\mathbb{C}_1^4$, and $\mathbb{C}_0^8$ in (18)–(20). The subspace of $\mathbb{C}_3$ consisting of the elements $[(x_1+i_10)+i_2(0+i_10)]+i_3[(0+i_10)+i_2(0+i_10)]$ is isomorphic and isometric with $\mathbb{C}_0$, and we say that $\mathbb{C}_0$ is contained in $\mathbb{C}_3$. In the same way, $\mathbb{C}_1$ is in $\mathbb{C}_3$ and $\mathbb{C}_2$ is in $\mathbb{C}_3$. The generalizations are obvious. □

44.3 THEOREM To multiply two elements in $\mathbb{C}_n$, in each representation multiply the two elements as if they were polynomials in $i_1, i_2, \ldots, i_n$, and then simplify the result by using the relations $i_1^2 = -1, \ldots, i_n^2 = -1$.

Proof. The proof of this theorem is Exercise 44.1. □

44.4 EXAMPLE The purpose of this example is to illustrate Theorem

44.3 by finding the product of two elements in $\mathbb{C}_3$ in two representations. Let the two elements x and y, with coefficients in $\mathbb{C}_0$, be

(21) $$x_1+i_1x_2+i_2x_3+i_1i_2x_4+i_3x_5+i_1i_3x_6+i_2i_3x_7+i_1i_2i_3x_8,$$

(22) $$y_1+i_1y_2+i_2y_3+i_1i_2y_4+i_3y_5+i_1i_3y_6+i_2i_3y_7+i_1i_2i_3y_8.$$

The details of the calculation can be arranged conveniently in the form of the array in (23).

(23)

	y_1	i_1y_2	i_2y_3	$\cdots$	$i_1i_2i_3y_8$
x_1	x_1y_1	$i_1x_1y_2$	$i_2x_1y_3$	$\cdots$	$i_1i_2i_3x_1y_8$
i_1x_2	$i_1x_2y_1$	$-x_2y_2$	$i_1i_2x_2y_3$	$\cdots$	$-i_2i_3x_2y_8$
i_2x_3	$i_2x_3y_1$	$i_1i_2x_3y_2$	$-x_3y_3$	$\cdots$	$-i_1i_3x_3y_8$
$i_1i_2x_4$	$i_1i_2x_4y_1$	$-i_2x_4y_2$	$-i_1x_4y_3$	$\cdots$	$i_3x_4y_8$
i_3x_5	$i_3x_5y_1$	$i_1i_3x_5y_2$	$i_2i_3x_5y_3$	$\cdots$	$-i_1i_2x_5y_8$
$i_1i_3x_6$	$i_1i_3x_6y_1$	$-i_3x_6y_2$	$i_1i_2i_3x_6y_3$	$\cdots$	$i_2x_6y_8$
$i_2i_3x_7$	$i_2i_3x_7y_1$	$i_1i_2i_3x_7y_2$	$-i_3x_7y_3$	$\cdots$	$i_1x_7y_8$
$i_1i_2i_3x_8$	$i_1i_2i_3x_8y_1$	$-i_2i_3x_8y_2$	$-i_1i_3x_8y_3$	$\cdots$	$-x_8y_8$

The product of the elements in (21) and (22) is the sum of the terms inside the array in (23). If the product of the elements in (21) and (22) is denoted by

(24) $$u_1+i_1u_2+i_2u_3+i_1i_2u_4+i_3u_5+i_1i_3u_6+i_2i_3u_7+i_1i_2i_3u_8,$$

then the following values are found from the array in (23):

(25)
$$\begin{aligned}
&x_1y_1-x_2y_2-x_3y_3+x_4y_4-x_5y_5+x_6y_6+x_7y_7-x_8y_8=u_1,\\
&x_2y_1+x_1y_2-x_4y_3-x_3y_4-x_6y_5-x_5y_6+x_8y_7+x_7y_8=u_2,\\
&x_3y_1-x_4y_2+x_1y_3-x_2y_4-x_7y_5+x_8y_6-x_5y_7+x_6y_8=u_3,\\
&x_4y_1+x_3y_2+x_2y_3+x_1y_4-x_8y_5-x_7y_6-x_6y_7-x_5y_8=u_4,\\
&x_5y_1-x_6y_2-x_7y_3+x_8y_4+x_1y_5-x_2y_6-x_3y_7+x_4y_8=u_5,\\
&x_6y_1+x_5y_2-x_8y_3-x_7y_4+x_2y_5+x_1y_6-x_4y_7-x_3y_8=u_6,\\
&x_7y_1-x_8y_2+x_5y_3-x_6y_4+x_3y_5-x_4y_6+x_1y_7-x_2y_8=u_7,\\
&x_8y_1+x_7y_2+x_6y_3+x_5y_4+x_4y_5+x_3y_6+x_2y_7+x_1y_8=u_8.
\end{aligned}$$

If (25) is considered as a system of linear equations in $y_1, \ldots, y_8$, the matrix of coefficients is the following:

(26)
$$\begin{bmatrix} x_1 & -x_2 & -x_3 & x_4 & -x_5 & x_6 & x_7 & -x_8 \\ x_2 & x_1 & -x_4 & -x_3 & -x_6 & -x_5 & x_8 & x_7 \\ x_3 & -x_4 & x_1 & -x_2 & -x_7 & x_8 & -x_5 & x_6 \\ x_4 & x_3 & x_2 & x_1 & -x_8 & -x_7 & -x_6 & -x_5 \\ x_5 & -x_6 & -x_7 & x_8 & x_1 & -x_2 & -x_3 & x_4 \\ x_6 & x_5 & -x_8 & -x_7 & x_2 & x_1 & -x_4 & -x_3 \\ x_7 & -x_8 & x_5 & -x_6 & x_3 & -x_4 & x_1 & -x_2 \\ x_8 & x_7 & x_6 & x_5 & x_4 & x_3 & x_2 & x_1 \end{bmatrix}.$$

But there is another way in which two elements in $\mathbb{C}_3$ can be multiplied. Let the two elements, with coefficients in $\mathbb{C}_1$, be

(27) $$z_1 + i_2 z_2 + i_3 z_3 + i_2 i_3 z_4,$$

(28) $$w_1 + i_2 w_2 + i_3 w_3 + i_2 i_3 w_4.$$

As before, the multiplication of these two elements can be arranged conveniently in the form of the following array.

(29)

	w_1	$i_2 w_2$	$i_3 w_3$	$i_2 i_3 w_4$
z_1	$z_1 w_1$	$i_2 z_1 w_2$	$i_3 z_1 w_3$	$i_2 i_3 z_1 w_4$
$i_2 z_2$	$i_2 z_2 w_1$	$-z_2 w_2$	$i_2 i_3 z_2 w_3$	$-i_3 z_2 w_4$
$i_3 z_3$	$i_3 z_3 w_1$	$i_2 i_3 z_3 w_2$	$-z_3 w_3$	$-i_2 z_3 w_4$
$i_2 i_3 z_4$	$i_2 i_3 z_4 w_1$	$-i_3 z_4 w_2$	$-i_2 z_4 w_3$	$z_4 w_4$

The product of the two elements in (27) and (28) is the sum of the terms inside the array in (29). If the product of the two elements is denoted by

(30) $$v_1 + i_2 v_2 + i_3 v_3 + i_2 i_3 v_4,$$

then (29) and (30) show that

(31)
$$\begin{aligned} z_1 w_1 - z_2 w_2 - z_3 w_3 + z_4 w_4 &= v_1, \\ z_2 w_1 + z_1 w_2 - z_4 w_3 - z_3 w_4 &= v_2, \\ z_3 w_1 - z_4 w_2 + z_1 w_3 - z_2 w_4 &= v_3, \\ z_4 w_1 + z_3 w_2 + z_2 w_3 + z_1 w_4 &= v_4. \end{aligned}$$

If (31) is considered as a system of linear equations in $w_1, \ldots, w_4$, the matrix of coefficients is the following:

$$(32)\qquad \begin{bmatrix} z_1 & -z_2 & -z_3 & z_4 \\ z_2 & z_1 & -z_4 & -z_3 \\ z_3 & -z_4 & z_1 & -z_2 \\ z_4 & z_3 & z_2 & z_1 \end{bmatrix}.$$

This example has carried out, in two ways, the multiplication of two elements in $\mathbb{C}_3$. The details and the results differ greatly in the two representations of elements in $\mathbb{C}_3$. The systems of equations in (25) and (31), and the matrices in (26) and (32), will be important in later sections.

Exercises

44.1 Prove Theorem 44.3 for $\mathbb{C}_4$ in the cases in which (a) the elements are represented with coefficients in $\mathbb{C}_0$, and (b) elements in $\mathbb{C}_4$ are represented with coefficients in $\mathbb{C}_1$. Prove Theorem 44.3 for $\mathbb{C}_n$.

44.2 Let ζ_1 and ζ_2 be two elements in $\mathbb{C}_2$. Recall the proof from Theorem 4.4 that $\|\zeta_1\zeta_2\|_2 \leqslant 2^{1/2}\|\zeta_1\|_2\|\zeta_2\|_2$. Use the same method to prove the following theorem. If ζ_1 and ζ_2 are two elements in $\mathbb{C}_n$, then

$$\|\zeta_1\zeta_2\|_n \leqslant 2^{(n-1)/2}\|\zeta_1\|_n\|\zeta_2\|_n.$$

44.3 Prove that every Cauchy sequence in $\mathbb{C}_n$ converges to an element in $\mathbb{C}_n$.

44.4 Let ζ be an element in $\mathbb{C}_n$ whose coefficients are the real numbers x_k, $k=1,\ldots,2^n$. If $a \in \mathbb{C}_0$, define $a \cdot \zeta$ to be the element in $\mathbb{C}_n$ whose coefficients are ax_k, $k=1,\ldots,2^n$. Prove that $\|a\cdot\zeta\|_n = |a|\,\|\zeta\|_n$. The multiplication of an element in $\mathbb{C}_n$ by a is called scalar multiplication by the (real) scalar a.

44.5 Let ζ be an element in $\mathbb{C}_n$ whose coefficients are complex numbers z_k, $k=1,\ldots,2^{n-1}$. If $c \in \mathbb{C}_1$, define $c\cdot\zeta$ to be the element in $\mathbb{C}_n$ whose coefficients are cz_k, $k=1,\ldots,2^{n-1}$. Prove that $\|c\cdot\zeta\|_n = |c|\,\|\zeta\|_n$. The multiplication of elements ζ by c is called scalar multiplication by the (complex) scalar c.

44.6 Prove that the system $\{\mathbb{C}_n, \|\ \|_n, +, \times, \cdot\}$ is a Banach algebra.

44.7 If $\zeta_1, \zeta_2, \ldots, \zeta_n, \ldots$ denote elements in $\mathbb{C}_2$, then the elements in the spaces $\mathbb{C}_2, \ldots, \mathbb{C}_n, \ldots$ can be represented as follows:

$\mathbb{C}_2$: ζ_1

$\mathbb{C}_3$: $\zeta_1 + i_3\zeta_2$

$\mathbb{C}_4$: $(\zeta_1 + i_3\zeta_2) + i_4(\zeta_3 + i_3\zeta_4)$

$\vdots$

Let ζ be an element in $\mathbb{C}_n$ whose coefficients are the elements ζ_k, $k=1,\ldots,2^{n-2}$, in $\mathbb{C}_2$. If $\eta\in\mathbb{C}_2$, define $\eta\cdot\zeta$ to be the element in $\mathbb{C}_n$ whose coefficients are $\eta\zeta_k$, $k=1,\ldots,2^{n-2}$. Prove that $\|\eta\cdot\zeta\|_n\leqslant 2^{1/2}\|\eta\|_2\|\zeta\|_n$.

44.8 Describe and establish the generalization suggested by Exercises 44.4, 44.5, and 44.7.

45. THE IDEMPOTENT REPRESENTATION

Section 6 has shown that

(1) $$\frac{1+i_1i_2}{2},\frac{1-i_1i_2}{2},$$

are idempotent elements in $\mathbb{C}_2$. Thus

(2) $$\left(\frac{1+i_1i_2}{2}\right)^2=\frac{1+i_1i_2}{2},\qquad \left(\frac{1-i_1i_2}{2}\right)^2=\frac{1-i_1i_2}{2},$$
$$\left(\frac{1+i_1i_2}{2}\right)\left(\frac{1-i_1i_2}{2}\right)=0.$$

Theorem 6.4 has shown that every element in $\mathbb{C}_2$ can be represented as a linear combination, with coefficients in $\mathbb{C}_1$, of the idempotents in (1); Theorem 6.6 shows that the algebraic operations in $\mathbb{C}_2$ are especially simple in this representation because of the properties of the idempotents in (2); and Theorem 6.8 shows that the representation provides a useful representation for the norm in $\mathbb{C}_2$. This section establishes the generalizations of all of these results for the space $\mathbb{C}_n$. Since there are many representations of an element in $\mathbb{C}_n$ as shown in Section 44, there are many forms for the generalizations.

It is easy to verify that the following are idempotent elements in $\mathbb{C}_n$:

(3) $$0,\ 1,\frac{1+i_1i_2}{2},\frac{1-i_1i_2}{2},\frac{1+i_1i_3}{2},\frac{1-i_1i_3}{2},$$
$$\frac{1+i_2i_3}{2},\frac{1-i_2i_3}{2},\ldots,\frac{1+i_{n-1}i_n}{2},\frac{1-i_{n-1}i_n}{2}.$$

It is obvious that the product of two idempotent elements is an idempotent element; thus

(4) $$\left(\frac{1+i_1i_2}{2}\right)\left(\frac{1+i_2i_3}{2}\right),\left(\frac{1-i_1i_2}{2}\right)\left(\frac{1+i_2i_3}{2}\right),$$
$$\left(\frac{1+i_1i_2}{2}\right)\left(\frac{1-i_2i_3}{2}\right),\left(\frac{1-i_1i_2}{2}\right)\left(\frac{1-i_2i_3}{2}\right)$$

are idempotents in $\mathbb{C}_3$. Furthermore, the product of each two idempotents in (4) equals zero. This example suggests how idempotents with the desired properties can be constructed in $\mathbb{C}_n$.

For convenience in notation, define the symbols $e(i_r i_s)$ and $e(-i_r i_s)$ as follows:

$$(5) \qquad e(i_r i_s) = \frac{1 + i_r i_s}{2}, \qquad e(-i_r i_s) = \frac{1 - i_r i_s}{2}.$$

Next, construct $n-1$ sets $S_1, \ldots, S_{n-1}$ of idempotents as follows:

$$(6) \qquad S_1: e(i_{n-1} i_n), \qquad e(-i_{n-1} i_n);$$

$$(7) \qquad S_2: e(i_{n-2} i_{n-1})S_1, \qquad e(-i_{n-2} i_{n-1})S_1;$$

$$\vdots$$

$$(8) \qquad S_k: e(i_{n-k} i_{n-k+1})S_{k-1}, \qquad e(-i_{n-k} i_{n-k+1})S_{k-1};$$

$$\vdots$$

$$(9) \qquad S_{n-1}: e(i_1 i_2)S_{n-2}, \qquad e(-i_1 i_2)S_{n-2}.$$

Here S_1 consists of the two elements shown. The set S_2 consists of four products of two idempotents each obtained as follows: multiply the two elements in S_1, first by $e(i_{n-2} i_{n-1})$, and then by $e(-i_{n-2} i_{n-1})$. The set S_k contains 2^k terms, each a product of k idempotents; they are obtained by multiplying each of the 2^{k-1} terms in S_{k-1}, first by $e(i_{n-k} i_{n-k+1})$, and then by $e(-i_{n-k} i_{n-k+1})$. This construction begins at $k=2$ and terminates naturally at $k=n-1$.

45.1 THEOREM For $k=1, \ldots, n-1$, the product of each two terms in S_k equals zero.

Proof. The theorem is true for $k=1$ because an easy calculation shows that $e(i_{n-1} i_n)e(-i_{n-1} i_n)=0$. For $k=2$ there are two cases: (a) one term is in $e(i_{n-2} i_{n-1})S_1$ and the other is in $e(-i_{n-2} i_{n-1})S_1$. The theorem is true in this case because the product of the two terms contains $e(i_{n-2} i_{n-1})e(-i_{n-2} i_{n-1})$, which equals zero. (b) In this case, the two terms are in $e(i_{n-2} i_{n-1})S_1$ or the two terms are in $e(-i_{n-2} i_{n-1})S_1$; the theorem is true in this case since it has already been shown to be true in S_1. If the theorem is true in S_{k-1}, similar arguments show that it is true in S_k. A complete induction shows that the theorem is true as stated. □

45.2 THEOREM Let ζ be an element in $\mathbb{C}_n$, $n \geqslant 2$, and let $\zeta = \zeta_1 + i_n \zeta_2$ with ζ_1 and ζ_2 in $\mathbb{C}_{n-1}$. Then

$$(10) \qquad \zeta = (\zeta_1 - i_{n-1}\zeta_2)e(i_{n-1} i_n) + (\zeta_1 + i_{n-1}\zeta_2)e(-i_{n-1} i_n).$$

Proof. The definitions in (5) show that the right side of (10) is

$$(11)\qquad (\zeta_1 - i_{n-1}\zeta_2)\left(\frac{1 + i_{n-1}i_n}{2}\right) + (\zeta_1 + i_{n-1}\zeta_2)\left(\frac{1 - i_{n-1}i_n}{2}\right).$$

Since $i_{n-1}^2 = -1$, simplification of this expression shows that it equals $\zeta_1 + i_n\zeta_2$, which is ζ. Thus (10) is a formal identity. □

45.3 COROLLARY Let ζ be an element in $\mathbb{C}_n$, $n \geqslant 2$, and let $\zeta = \zeta_1 + i_n\zeta_2$ with ζ_1 and ζ_2 in $\mathbb{C}_{n-1}$. Then for each k, $k = 1, \ldots, n-1$, ζ can be represented as a linear combination of the idempotents in S_k in (8), and the coefficients in this linear combination are elements in $\mathbb{C}_{n-k}$.

Proof. The corollary is true for $k=1$ [see (6)] by Theorem 45.2. Since the coefficients $\zeta_1 - i_{n-1}\zeta_2$ and $\zeta_1 + i_{n-1}\zeta_2$ in this representation are elements in $\mathbb{C}_{n-1}$, they can be represented as linear combinations of $e(i_{n-2}i_{n-1})$ and $e(-i_{n-2}i_{n-1})$ by Theorem 45.2. Substitute these representations in (10); then ζ is represented as a linear combination of the four idempotents in (7) with coefficients in $\mathbb{C}_{n-2}$. A continuation of this procedure shows that, for $k = 1, \ldots, n-1$, the element ζ can be represented as a linear combination of the 2^k idempotents in S_k and that the coefficients are elements in $\mathbb{C}_{n-k}$. If $k = n-1$, the coefficients are elements in $\mathbb{C}_1$ and the procedure terminates. The proof of Corollary 45.3 is complete. □

The next theorem and corollary, which are generalizations of Theorem 6.8, show that the norm of an element ζ in $\mathbb{C}_n$ can be represented easily in terms of the coefficients in the representations of ζ described in Theorem 45.2 and Corollary 45.3.

45.4 THEOREM Let ζ be an element in $\mathbb{C}_n$, $n \geqslant 2$, and let $\zeta = \zeta_1 + i_n\zeta_2$ with ζ_1 and ζ_2 in $\mathbb{C}_{n-1}$. Then

$$(12)\qquad \|\zeta\|_n = \left(\frac{\|\zeta_1 - i_{n-1}\zeta_2\|_{n-1}^2 + \|\zeta_1 + i_{n-1}\zeta_2\|_{n-1}^2}{2}\right)^{1/2}.$$

Proof. Let ζ be represented with coefficients in $\mathbb{C}_0$. Then ζ has the form

$$(13)\qquad (x_1 + i_1x_2 + i_2x_3 + \cdots) + i_n(y_1 + i_1y_2 + i_2y_3 + \cdots),$$

and

$$(14)\qquad \|\zeta\|_n^2 = \sum_{k=1}^{2^{n-1}} x_k^2 + \sum_{k=1}^{2^{n-1}} y_k^2.$$

Now

$$\zeta_1 - i_{n-1}\zeta_2 = (x_1 + i_1x_2 + i_2x_3 + \cdots) - i_{n-1}(y_1 + i_1y_2 + i_2y_3 + \cdots), \tag{15}$$

$$\zeta_1 + i_{n-1}\zeta_2 = (x_1 + i_1x_2 + i_2x_3 + \cdots) + i_{n-1}(y_1 + i_1y_2 + i_2y_3 + \cdots). \tag{16}$$

Carry out the indicated multiplications on the right, and then collect terms so that $\zeta_1 - i_{n-1}\zeta_2$ and $\zeta_1 + i_{n-1}\zeta_2$ are written in the standard form for an element in $\mathbb{C}_{n-1}$. If a coefficient $x_r - y_s$ occurs in (15), then the coefficient $x_r + y_s$ occurs in (16). Also, if $x_r + y_s$ occurs in (15), then $x_r - y_s$ occurs in (16). Each x_k and each y_k occur once in the coefficients in (15) and once in (16). The numerator in the formula on the right in (12) equals the sum of the squares of the coefficients in (15) and in (16). Now

$$\begin{aligned}(x_r - y_s)^2 + (x_r + y_s)^2 &= 2(x_r^2 + y_s^2),\\ (x_r + y_s)^2 + (x_r - y_s)^2 &= 2(x_r^2 + y_s^2).\end{aligned} \tag{17}$$

Thus all cross-product terms cancel in the sum of squares of the coefficients in (15) and (16); therefore,

$$\begin{aligned}&\|\zeta_1 - i_{n-1}\zeta_2\|_{n-1}^2 + \|\zeta_1 + i_{n-1}\zeta_2\|_{n-1}^2\\ &\quad = 2\left(\sum_{k=1}^{2^{n-1}} x_k^2 + \sum_{k=1}^{2^{n-1}} y_k^2\right) = 2\|\zeta\|_n^2,\end{aligned} \tag{18}$$

by (14). Formula (12) and Theorem 45.4 follow from the second equation in (18). □

As in previous cases, this theorem can be applied repeatedly to obtain a total of $n-1$ formulas for $\|\zeta\|_n$. Observe first that the coefficients in (11), which appear in (12), are elements in $\mathbb{C}_{n-1}$. Then Theorem 45.2 shows that they have the following representations:

$$\zeta_1 - i_{n-1}\zeta_2 = c_1e(i_{n-2}i_{n-1}) + c_2e(-i_{n-2}i_{n-1}), \qquad c_1, c_2 \text{ in } \mathbb{C}_{n-2}, \tag{19}$$

$$\zeta_1 + i_{n-1}\zeta_2 = c_3e(i_{n-2}i_{n-1}) + c_4e(-i_{n-2}i_{n-1}), \qquad c_3, c_4 \text{ in } \mathbb{C}_{n-2}. \tag{20}$$

45.5 COROLLARY If $\zeta_1 - i_{n-1}\zeta_2$ and $\zeta_1 + i_{n-1}\zeta_2$ have the representations shown in (19) and (20), respectively, then

$$\|\zeta\|_n = \left(\frac{\sum_{r=1}^{2^2} \|c_r\|_{n-2}^2}{2^2}\right)^{1/2}. \tag{21}$$

Theorem 45.4 can be applied repeatedly to obtain $n-1$ such formulas for $\|\zeta\|_n$; in the last formula $\|\zeta\|_n$ is represented in terms of the absolute values of coefficients in $\mathbb{C}_1$.

Proof. By Theorem 45.4,

$$\|\zeta_1 - i_{n-1}\zeta_2\|_{n-1}^2 = \frac{\|c_1\|_{n-2}^2 + \|c_2\|_{n-2}^2}{2}, \tag{22}$$

$$\|\zeta_1 + i_{n-1}\zeta_2\|_{n-1}^2 = \frac{\|c_3\|_{n-2}^2 + \|c_4\|_{n-2}^2}{2}. \tag{23}$$

Substitute from (22) and (23) in (12); the result is (21). The procedure used to obtain (21) from (12) can be applied successively to obtain all of the formulas described in the corollary. □

The next theorem is a generalization of Theorem 6.6; it describes the addition, multiplication, and division of elements in $\mathbb{C}_n$ which are represented in the form described in Corollary 45.3. Singular elements play an important role in division; although singular elements in $\mathbb{C}_n$ are not treated until the next section, Theorem 45.6 includes a statement here about division for the sake of completeness.

Let the idempotents in S_k in (8) be denoted by

$$e_j, \quad j = 1, \ldots, 2^k. \tag{24}$$

If ζ and η are two elements in $\mathbb{C}_n$, $n \geqslant 2$, then Corollary 45.3 shows that they can be represented as follows:

$$\zeta = \sum_{j=1}^{2^k} c_j e_j, \quad c_j \in \mathbb{C}_{n-k}, j = 1, \ldots, 2^k, \tag{25}$$

$$\eta = \sum_{j=1}^{2^k} d_j e_j, \quad d_j \in \mathbb{C}_{n-k}, j = 1, \ldots, 2^k. \tag{26}$$

45.6 THEOREM Let ζ and η be two elements in $\mathbb{C}_n$, $n \geqslant 2$, which are represented as shown in (25) and (26). Then

$$\zeta + \eta = \sum_{j=1}^{2^k} (c_j + d_j) e_j, \tag{27}$$

$$\zeta\eta = \sum_{j=1}^{2^k} c_j d_j e_j. \tag{28}$$

If η is nonsingular, then ζ/η is defined, and

$$\frac{\zeta}{\eta} = \sum_{j=1}^{2^k} \left(\frac{c_j}{d_j}\right) e_j. \tag{29}$$

Proof. Because multiplication is distributive with respect to addition, and because addition is associative and commutative in a Banach algebra,

$$\sum_{j=1}^{2^k} (c_j + d_j)e_j = \sum_{j=1}^{2^k} (c_je_j + d_je_j) = \sum_{j=1}^{2^k} c_je_j + \sum_{j=1}^{2^k} d_je_j = \zeta + \eta. \tag{30}$$

Therefore (27) is true. Consider (28). Now two elements in $\mathbb{C}_n$ are multiplied as if they were polynomials (see Theorem 44.3 and Exercise 44.1). Furthermore, the e_j are idempotents which have the following property by Theorem 45.1:

$$e_re_s = 0, \qquad r \neq s,\ r, s = 1, \ldots, 2^k. \tag{31}$$

Then the product of the sums in (25) and (26) can be simplified since

$$(c_je_j)(d_je_j) = c_jd_je_j^2 = c_jd_je_j, \tag{32}$$

$$(c_re_r)(d_se_s) = c_rd_se_re_s = 0, \qquad r \neq s. \tag{33}$$

Therefore (28) is true. Consider (29). By definition, ζ/η is an element ξ such that $\zeta = \eta\xi$. Represent ξ as a linear combination of the e_j. Then

$$\frac{\zeta}{\eta} = \xi = \sum_{j=1}^{2^k} h_je_j, \qquad h_j \in \mathbb{C}_{n-k},\ j = 1, \ldots, 2^k. \tag{34}$$

Now by (26), (34), and (28),

$$\eta\xi = \sum_{j=1}^{2^k} d_jh_je_j. \tag{35}$$

Also, $\zeta = \eta\xi$ if and only if

$$c_j = d_jh_j, \qquad j = 1, \ldots, 2^k. \tag{36}$$

The assumption that η is nonsingular guarantees that each of the equations (36) can be solved to obtain

$$h_j = \frac{c_j}{d_j}, \qquad j = 1, \ldots, 2^k. \tag{37}$$

Then (34) and (37) show that (29) is true. The proof of all parts of Theorem 45.6 is complete. □

45.7 EXAMPLE Let

$$\zeta_1 = x_1 + i_1x_2 + i_2x_3 + i_1i_2x_4, \qquad \zeta_2 = x_5 + i_1x_6 + i_2x_7 + i_1i_2x_8; \tag{38}$$

then ζ_1 and ζ_2 are elements in $\mathbb{C}_2$, and $\zeta_1 + i_3\zeta_2$ is an element in $\mathbb{C}_3$. Thus

$$\zeta = (x_1 + i_1x_2 + i_2x_3 + i_1i_2x_4) + i_3(x_5 + i_1x_6 + i_2x_7 + i_1i_2x_8), \tag{39}$$

$$\|\zeta\|_3^2 = \sum_{r=1}^{8} x_r^2. \tag{40}$$

A simple calculation shows that

(41) $\zeta_1 - i_2\zeta_2 = (x_1 + x_7) + i_1(x_2 + x_8) + i_2(x_3 - x_5) + i_1 i_2(x_4 - x_6),$

(42) $\zeta_1 + i_2\zeta_2 = (x_1 - x_7) + i_1(x_2 - x_8) + i_2(x_3 + x_5) + i_1 i_2(x_4 + x_6).$

Then, as stated in Theorem 45.2 (10),

(43) $$\zeta = [(x_1 + x_7) + i_1(x_2 + x_8) + i_2(x_3 - x_5) + i_1 i_2(x_4 - x_6)]e(i_2 i_3) + [(x_1 - x_7) + i_1(x_2 - x_8) + i_2(x_3 + x_5) + i_1 i_2(x_4 + x_6)]e(-i_2 i_3).$$

Furthermore, (41) and (42) show that

(44) $\|\zeta_1 - i_2\zeta_2\|_2^2 = (x_1 + x_7)^2 + (x_2 + x_8)^2 + (x_3 - x_5)^2 + (x_4 - x_6)^2,$

(45) $\|\zeta_1 + i_2\zeta_2\|_2^2 = (x_1 - x_7)^2 + (x_2 - x_8)^2 + (x_3 + x_5)^2 + (x_4 + x_6)^2.$

Therefore,

(46) $$\|\zeta_1 - i_2\zeta_2\|_2^2 + \|\zeta_1 + i_2\zeta_2\|_2^2 = 2\sum_{r=1}^{8} x_r^2 = 2\|\zeta\|_3^2,$$

and (46) and (40) show that Theorem 45.4 (12) is verified in this special case. Inspection of (41), (42) and (44), (45) shows that $\zeta_1 - i_2\zeta_2$ and $\zeta_1 + i_2\zeta_2$ have the properties employed in the proof of Theorem 45.4.

Now

(47) $$\zeta = (\zeta_1 - i_2\zeta_2)e(i_2 i_3) + (\zeta_1 + i_2\zeta_2)e(-i_2 i_3)$$

as shown in (43), and (41) and (42) show that the coefficients in (47) are elements in $\mathbb{C}_2$. Then Theorem 45.2 and Corollary 45.3 show that each of the coefficients in (47) can be represented as a linear combination of $e(i_1 i_2)$ and $e(-i_1 i_2)$. Thus

(48) $$\begin{aligned}\zeta_1 - i_2\zeta_2 &= \{[(x_1 + x_7) + i_1(x_2 + x_8)] - i_1[(x_3 - x_5) + i_1(x_4 - x_6)]\}e(i_1 i_2)\\ &\quad + \{[(x_1 + x_7) + i_1(x_2 + x_8)] + i_1[(x_3 - x_5) + i_1(x_4 - x_6)]\}e(-i_1 i_2),\end{aligned}$$

(49) $$\begin{aligned}\zeta_1 + i_2\zeta_2 &= \{[(x_1 - x_7) + i_1(x_2 - x_8)] - i_1[(x_3 + x_5) + i_1(x_4 + x_6)]\}e(i_1 i_2)\\ &\quad + \{[(x_1 - x_7) + i_1(x_2 - x_8)] + i_1[(x_3 + x_5) + i_1(x_4 + x_6)]\}e(-i_1 i_2).\end{aligned}$$

Simplify the expressions in the braces in (48) and (49), and then replace

$\zeta_1 - i_2\zeta_2$ and $\zeta_1 + i_2\zeta_2$ in (47) by their values in (48) and (49); the result is

$$(50)\quad \zeta = \{[(x_1+x_7)+(x_4-x_6)]+i_1[(x_2+x_8)-(x_3-x_5)]\}e(i_1i_2)e(i_2i_3)$$
$$+\{[(x_1+x_7)-(x_4-x_6)]+i_1[(x_2+x_8)+(x_3-x_5)]\}e(-i_1i_2)e(i_2i_3)$$
$$+\{[(x_1-x_7)+(x_4+x_6)]+i_1[(x_2-x_8)-(x_3+x_5)]\}e(i_1i_2)e(-i_2i_3)$$
$$+\{[(x_1-x_7)-(x_4+x_6)]+i_1[(x_2-x_8)+(x_3+x_5)]\}e(-i_1i_2)e(-i_2i_3).$$

This equation represents ζ as a linear combination of the four idempotents in (4); in this case these are the idempotents in S_2 in (7), which is also S_{n-1} in (9) (compare Corollary 45.3). Since the coefficients in (50) are elements in $\mathbb{C}_1$ (that is, complex numbers), the process has terminated and no further representations of ζ are possible.

Corollary 45.5 states that the coefficients in (50) can be used to represent $\|\zeta\|_3$. If $c_1, \ldots, c_4$ are the four coefficients in (50), then

$$(51)\quad \|c_1\|_1^2 + \cdots + \|c_4\|_1^2$$
$$= [(x_1 + x_7) + (x_4 - x_6)]^2 + [(x_2 + x_8) - (x_3 - x_5)]^2$$
$$+ [(x_1 + x_7) - (x_4 - x_6)]^2 + [(x_2 + x_8) + (x_3 - x_5)]^2$$
$$+ [(x_1 - x_7) + (x_4 + x_6)]^2 + [(x_2 - x_8) - (x_3 + x_5)]^2$$
$$+ [(x_1 - x_7) - (x_4 + x_6)]^2 + [(x_2 - x_8) + (x_3 + x_5)]^2$$
$$= 4\sum_{r=1}^{8} x_r^2 = 4\|\zeta\|_3^2.$$

This equation verifies the formula in Corollary 45.5 (21) in this case. But there is another way to use (50) to represent $\|\zeta\|_3^2$. Since

$$(52)\qquad |a + ib|^2 = a^2 + b^2 = \det\begin{bmatrix} a & -b \\ b & a \end{bmatrix},$$

the sum of the squares of the absolute values of the four coefficients in (50) can be written as four determinants. Define matrices $C_1, \ldots, C_4$ as follows:

$$(53)\qquad C_1 = \begin{bmatrix} (x_1+x_7)+(x_4-x_6) & -[(x_2+x_8)-(x_3-x_5)] \\ (x_2+x_8)-(x_3-x_5) & (x_1+x_7)+(x_4-x_6) \end{bmatrix}$$

$$(54)\qquad C_2 = \begin{bmatrix} (x_1+x_7)-(x_4-x_6) & -[(x_2+x_8)+(x_3-x_5)] \\ (x_2+x_8)+(x_3-x_5) & (x_1+x_7)-(x_4-x_6) \end{bmatrix}$$

$$(55)\qquad C_3=\begin{bmatrix}(x-x_7)+(x_4+x_6) & -[(x_2-x_8)-(x_3+x_5)]\\(x_2-x_8)-(x_3+x_5) & (x_1-x_7)+(x_4+x_6)\end{bmatrix}$$

$$(56)\qquad C_4=\begin{bmatrix}(x_1-x_7)-(x_4+x_6) & -[(x_2-x_8)+(x_3+x_5)]\\(x_2-x_8)+(x_3+x_5) & (x_1-x_7)-(x_4+x_6)\end{bmatrix}.$$

Then (50) and (51) show that

$$(57)\qquad \|\zeta\|_3^2=\frac{\det C_1+\det C_2+\det C_3+\det C_4}{4}.$$

Exercises

45.1 Prove the following theorem by induction: if ζ and η are elements in $\mathbb{C}_n$, $n\geqslant 2$, then

$$(58)\qquad \|\zeta\eta\|_n\leqslant 2^{n-1/2}\|\zeta\|_n\|\eta\|_n.$$

[*Hints.* The theorem is true for $n=2$ by Theorem 4.4. Next let ζ and η be elements in $\mathbb{C}_3$. Then

$$\zeta=a_1e(i_2i_3)+a_2e(-i_2i_3),\qquad a_1,\,a_2\text{ in }\mathbb{C}_2,$$
$$\eta=b_1e(i_2i_3)+b_2e(-i_2i_3),\qquad b_1,\,b_2\text{ in }\mathbb{C}_2,$$
$$\zeta\eta=a_1b_1e(i_2i_3)+a_2b_2e(-i_2i_3).$$

By Theorem 45.4,

$$\|\zeta\|_3=\left(\frac{\|a_1\|_2^2+\|a_2\|_2^2}{2}\right)^{1/2},\qquad \|\eta\|_3=\left(\frac{\|b_1\|_2^2+\|b_2\|_2^2}{2}\right)^{1/2},$$

$$\|\zeta\eta\|_3=\left(\frac{\|a_1b_1\|_2^2+\|a_2b_2\|_2^2}{2}\right)^{1/2}\leqslant\frac{\|a_1b_1\|_2+\|a_2b_2\|_2}{2^{1/2}}$$
$$\leqslant\|a_1\|_2\|b_1\|_2+\|a_2\|_2\|b_2\|_2$$
$$\leqslant 2\left(\frac{\|a_1\|_2^2+\|a_2\|_2^2}{2}\right)^{1/2}\left(\frac{\|b_1\|_2^2+\|b_2\|_2^2}{2}\right)^{1/2}.$$

Then $\|\zeta\eta\|_3\leqslant 2^{3-1/2}\|\zeta\|_3\,\|\eta\|_3$, and the theorem is true for $n=3$. Proceed by induction.]

45.2 Give a second proof of the theorem in Exercise 45.1. [*Hint.* Use Corollary 45.3 to represent ζ and η as linear combinations of the idempotents in S_{n-1} in (9) with coefficients in $\mathbb{C}_1$. Use Corollary 45.5 to find $\|\zeta\eta\|_n$, and then use Theorem 45.6 and Schwarz's inequality to complete the proof as in Exercise 45.1.]

45.3 Define N be the following statement:

$$N = e(i_1 i_2)e(i_2 i_3) \cdots e(i_{n-1} i_n).$$

(a) Show that $\|N\|_n = \dfrac{1}{2^{n-1/2}}$.

(b) Prove that the inequality in Exercise 45.1 (58) is the best possible in $\mathbb{C}_n$ by showing that

$$\|NN\|_n = 2^{n-1/2}\|N\|_n\|N\|_n.$$

45.4 Show that $\|e(i_1 i_2)e(i_2 i_3)\|_3 = \|e(i_1 i_2)\|_3 \|e(i_2 i_3)\|_3$.

46. SINGULAR ELEMENTS; CAUCHY–RIEMANN MATRICES

The condition that an element in $\mathbb{C}_n$ be a nonsingular element can be expressed in terms of a certain matrix called a Cauchy–Riemann matrix (compare Section 28). This section investigates the properties of the singular and nonsingular elements in $\mathbb{C}_n$ and of the associated Cauchy–Riemann matrices. Many of the results of the section are consequences of the following fundamental theorem for Cauchy–Riemann matrices: for $n \geqslant 2$ the determinant of a $2^n \times 2^n$ Cauchy–Riemann matrix can be factored into the product of determinants of two $2^{n-1} \times 2^{n-1}$ Cauchy–Riemann matrices. For example, this theorem can be used to prove that, for $n \geqslant 2$, the determinant of a $2^n \times 2^n$ Cauchy–Riemann matrix with elements in $\mathbb{C}_0$ is nonnegative (this statement is trivially true also for $n = 1$). Elementary properties of determinants can be used to prove the fundamental theorem for small values of n (say $n \leqslant 4$), but, although the method applies in all cases, as a practical matter similar proofs cannot be carried out for large values of n because of the size of the matrices and the complexity of the great mass of details.

Table 46.1 helps to organize and understand the work of this section. First, singular elements and Cauchy–Riemann matrices of elements will be investigated in the representation shown in the $\mathbb{C}_0$ column of the table. Next, the same study will be carried out for the representations in column $\mathbb{C}_1$ and later columns. Finally, the relations between the results for the various columns, and especially those for columns $\mathbb{C}_0$ and $\mathbb{C}_1$, will be investigated.

By definition (compare Definition 4.7), an element ζ in $\mathbb{C}_n$ is nonsingular if and only if there exists a unique η in $\mathbb{C}_n$ such that $\zeta\eta = 1$, and ζ is singular if and only if it is not nonsingular. The set of singular elements in $\mathbb{C}_n$ is denoted by $\mathcal{O}_n$ for $n = 0, 1, 2, \ldots$.

Consider then an element x in $\mathbb{C}_n$ represented with coefficients x_k,

46.1 Table Representations of Elements

Elements in space	Coefficients in						
	$\mathbb{C}_0$	$\mathbb{C}_1$	$\mathbb{C}_2$	$\mathbb{C}_3$	$\mathbb{C}_4$	$\mathbb{C}_5$	$\cdots$
$\mathbb{C}_0$	x						
$\mathbb{C}_1$	x_1, x_2	z					
$\mathbb{C}_2$	$x_1,\ldots,x_{2^2}$	z_1, z_2	ζ				
$\mathbb{C}_3$	$x_1,\ldots,x_{2^3}$	$z_1,\ldots,z_{2^2}$	ζ_1, ζ_2	η			
$\mathbb{C}_4$	$x_1,\ldots,x_{2^4}$	$z_1,\ldots,z_{2^3}$	$\zeta_1,\ldots,\zeta_{2^2}$	η_1, η_2	ξ		
$\mathbb{C}_5$	$x_1,\ldots,x_{2^5}$	$z_1,\ldots,z_{2^4}$	$\zeta_1,\ldots,\zeta_{2^3}$	$\eta_1,\ldots,\eta_{2^2}$	ξ_1, ξ_2	ω	
$\cdots$	$\cdots$	$\cdots$	$\cdots$	$\cdots$	$\cdots$	$\cdots$	$\cdots$

$k=1,\ldots,2^n$, as in column $\mathbb{C}_0$ in Table 46.1. To determine whether or not x is nonsingular, it is necessary to determine whether or not there exists a unique element y in $\mathbb{C}_n$ such that $xy=1$. Multiply x and y to obtain an element in $\mathbb{C}_n$; equate the coefficients of $1, i_1, i_2, \ldots, i_1 \cdots i_n$ to those of 1 in $\mathbb{C}_n$ represented as in column $\mathbb{C}_0$ of Table 46.1. The result is a system of 2^n linear equations in the unknowns y_k, $k=1,\ldots,2^n$; the coefficients of the unknowns in this system are formed from the coefficients x_k, $k=1,\ldots,2^n$, of x.

46.2 DEFINITION The matrix of coefficients of the system of equations just described is called the ($\mathbb{C}_0$) Cauchy–Riemann matrix $M(x)$ determined by the element x in $\mathbb{C}_n$, and the set of ($\mathbb{C}_0$) Cauchy–Riemann matrices is the set of matrices determined in this manner by all of the elements x in $\mathbb{C}_n$. More generally, there are Cauchy–Riemann matrices corresponding to each of the representations of elements in $\mathbb{C}_n$.

For an example, see (25) in Section 44. In that system of equations, set $u_1=1$ and $u_2=\cdots=u_8=0$. The resulting equations form the system for determining whether $[(x_1 + i_1x_2) + i_2(x_3 + i_1x_4)] + i_3[(x_5 + i_1x_6) + i_2(x_7 + i_1x_8)]$ in $\mathbb{C}_3$ is singular or nonsingular. The matrix in (26) in Section 44 is the Cauchy–Riemann matrix determined by this x. Earlier sections of this book have defined and treated 2×2 and 4×4 Cauchy–Riemann matrices (see especially Section 28).

46.3 THEOREM Let x be an element in $\mathbb{C}_n$, $n\geqslant 1$, and let $M(x)$ be the matrix which it determines. Then x is nonsingular if and only if $\det M(x)\notin \mathcal{O}_0$, and it is singular if and only if $\det M(x)\in \mathcal{O}_0$.

Proof. Cramer's rule states that a system of linear equations has a unique solution if and only if the determinant of its matrix of coefficients is not zero. □

46.4 COROLLARY For $n \geqslant 1$, $\mathcal{O}_n = \{x \text{ in } \mathbb{C}_n\colon \det M(x) \in \mathcal{O}_0\}$.

46.5 THEOREM The set of matrices $M(x)$ for x in $\mathbb{C}_n$, $n \geqslant 1$, has the following properties:

(1) It is closed under scalar multiplication by scalars in $\mathbb{C}_0$.

(2) $M(xy) = M(x)M(y)$.

(3) It is closed under matrix addition and multiplication, and both operations are commutative.

Proof. It is easy to verify that $aM(x) = M(ax)$ if $a \in \mathbb{C}_0$ and $x \in \mathbb{C}_n$; therefore (1) is true. Consider (2). The matrices $M(x)$ and $M(y)$ are given, and thus there are corresponding elements x and y in $\mathbb{C}_n$. Multiply all the elements in $\mathbb{C}_n$ by the element y in $\mathbb{C}_n$. Because $\mathbb{C}_n$ is a Banach algebra, this operation corresponds to performing a linear transformation with matrix $M(y)$ on the space $\mathbb{C}_0^n$. Multiplication by x also corresponds to a linear transformation on $\mathbb{C}_0^n$, and the composition of the two transformations has matrix $M(xy)$. It is known in the theory of linear transformations that a transformation with matrix $M(y)$ followed by a transformation with matrix $M(x)$ is a linear transformation whose matrix is the matrix product $M(x)M(y)$. Therefore, $M(xy) = M(x)M(y)$, and (2) is true. Since $M(x) + M(y) = M(x+y) = M(y+x) = M(y) + M(x)$, and since, by (2), $M(x)M(y) = M(xy) = M(yx) = M(y)M(x)$, the set of Cauchy–Riemann matrices is closed under matrix addition and multiplication, and both operations are commutative. The proof of all parts of Theorem 46.5 is complete. □

46.6 EXAMPLE The following example illustrates parts of Theorem 46.5. Let $x\colon x_1 + i_1x_2 + i_2x_3 + i_1i_2x_4$ and $y\colon y_1 + i_1y_2 + i_2y_3 + i_1i_2y_4$ be two elements in $\mathbb{C}_2$. Then xy is an element $u\colon u_1 + i_1u_2 + i_2u_3 + i_1i_2u_4$ where

(4)
$$\begin{aligned}
u_1 &= x_1y_1 - x_2y_2 - x_3y_3 + x_4y_4,\\
u_2 &= x_2y_1 + x_1y_2 - x_4y_3 - x_3y_4,\\
u_3 &= x_3y_1 - x_4y_2 + x_1y_3 - x_2y_4,\\
u_4 &= x_4y_1 + x_3y_2 + x_2y_3 + x_1y_4.
\end{aligned}$$

It is easy to verify the following matrix products and to show that the three matrices are Cauchy–Riemann matrices.

(5)
$$\begin{bmatrix} x_1 & -x_2 & -x_3 & x_4 \\ x_2 & x_1 & -x_4 & -x_3 \\ x_3 & -x_4 & x_1 & -x_2 \\ x_4 & x_3 & x_2 & x_1 \end{bmatrix}\begin{bmatrix} y_1 & -y_2 & -y_3 & y_4 \\ y_2 & y_1 & -y_4 & -y_3 \\ y_3 & -y_4 & y_1 & -y_2 \\ y_4 & y_3 & y_2 & y_1 \end{bmatrix}$$

$$
= \begin{bmatrix} u_1 & -u_2 & -u_3 & u_4 \\ u_2 & u_1 & -u_4 & -u_3 \\ u_3 & -u_4 & u_1 & -u_2 \\ u_4 & u_3 & u_2 & u_1 \end{bmatrix}
$$

$$
= \begin{bmatrix} y_1 & -y_2 & -y_3 & y_4 \\ y_2 & y_1 & -y_4 & -y_3 \\ y_3 & -y_4 & y_1 & -y_2 \\ y_4 & y_3 & y_2 & y_1 \end{bmatrix} \begin{bmatrix} x_1 & -x_2 & -x_3 & x_4 \\ x_2 & x_1 & -x_4 & -x_3 \\ x_3 & -x_4 & x_1 & -x_2 \\ x_4 & x_3 & x_2 & x_1 \end{bmatrix}.
$$

46.7 THEOREM The space $\mathbb{C}_n$, $n \geqslant 1$, and its associated Cauchy–Riemann matrices have the following properties:

(6) $\det M(xy) = \det M(x) \det M(y)$.

(7) $\mathcal{O}_n$ is the solution set of the equation $\det M(x) = 0$.

(8) If $x \in \mathbb{C}_n$ and $y \in \mathcal{O}_n$, then $xy \in \mathcal{O}_n$.

(9) If x and y are nonsingular, then xy is nonsingular.

(10) If x and y are in $\mathbb{C}_n$ and $y \notin \mathcal{O}_n$, then x/y is defined and is an element in $\mathbb{C}_n$.

Proof. The determinant of the product of two matrices equals the product of the determinants of these matrices; thus (6) follows from (2). The statement in (7) follows from Corollary 46.4. If $y \in \mathcal{O}_n$, then $\det M(y) = 0$ and $\det M(xy) = 0$ by (6). Therefore, $xy \in \mathcal{O}_n$ by Corollary 46.4 and (8) is true. Next, if x and y are nonsingular, then $\det M(x) \notin \mathcal{O}_0$ and $\det M(y) \notin \mathcal{O}_0$ by Theorem 46.3. Then $\det M(xy) \notin \mathcal{O}_0$ by (6), and therefore xy is nonsingular by Theorem 46.3. Thus (9) is true. Consider (10). By definition, x/y is a number u such that $x = yu$ if this equation has a unique solution for u. The equation $x = yu$ is equivalent to a system of linear equations in u_k, $k = 1, \ldots, 2^n$, and $M(y)$ is the matrix of coefficients. Since $y \notin \mathcal{O}_n$, then $\det M(y) \notin \mathcal{O}_0$ and the system of linear equations has a unique solution for the u_k. Thus x/y is defined as a unique element in $\mathbb{C}_n$, and (10) is true. The proof of all parts of Theorem 46.7 is complete. □

The next steps in this treatment of $\mathbb{C}_n$ are the following: (a) state a certain theorem concerning the determinant of a $2^n \times 2^n$ Cauchy–Riemann matrix for $n \geqslant 2$; (b) prove the theorem, first for small values of n by elementary methods, and then for all values n; and (c) prove certain results which follow from the theorem. To begin these developments, let $\zeta_1 + i_n \zeta_2$ denote an element in $\mathbb{C}_n$ as in Theorem 45.2. Then ζ_1 and ζ_2 are elements in $\mathbb{C}_{n-1}$. The notation is for convenience only, since elements in $\mathbb{C}_n$ continue to be represented with coefficients in $\mathbb{C}_0$ as in the $\mathbb{C}_0$ column of Table 46.1. Then $M(\zeta_1+i_n\zeta_2)$ denotes the $2^n \times 2^n$ Cauchy–Riemann matrix, with elements in $\mathbb{C}_0$, which is associated with $\zeta_1+i_n\zeta_2$ in $\mathbb{C}_n$; also $M(\zeta_1-i_{n-1}\zeta_2)$ and $M(\zeta_1+i_{n-1}\zeta_2)$ are $2^{n-1}\times 2^{n-1}$ Cauchy–Riemann matrices, with elements in $\mathbb{C}_0$, associated with the elements $\zeta_1-i_{n-1}\zeta_2$ and $\zeta_1+i_{n-1}\zeta_2$ in $\mathbb{C}_{n-1}$.

46.8 THEOREM If $M(\zeta_1+i_n\zeta_2)$, $M(\zeta_1-i_{n-1}\zeta_2)$, and $M(\zeta_1+i_{n-1}\zeta_2)$ are the Cauchy–Riemann matrices just described, then

$$(11)\quad \det M(\zeta_1+i_n\zeta_2)=\det M(\zeta_1-i_{n-1}\zeta_2)\det M(\zeta_1+i_{n-1}\zeta_2),\qquad n\geqslant 2.$$

The first evidence in support of this theorem is the fact that it can be proved by elementary methods for small values of n. If $n=2$, then

$$(12)\quad \begin{aligned} \zeta_1 + i_2\zeta_2 &= (x_1 + i_1x_2) + i_2(x_3 + i_1x_4),\\ \zeta_1 - i_1\zeta_2 &= (x_1 + i_1x_2) - i_1(x_3 + i_1x_4) = (x_1 + x_4) + i_1(x_2 - x_3),\\ \zeta_1 + i_1\zeta_2 &= (x_1 + i_1x_2) + i_1(x_3 + i_1x_4) = (x_1 - x_4) + i_1(x_2 + x_3). \end{aligned}$$

Then as in (5) above and in Theorem 28.6,

$$(13)\quad \det M(\zeta_1 + i_2\zeta_2) = \det\begin{bmatrix} x_1 & -x_2 & -x_3 & x_4\\ x_2 & x_1 & -x_4 & -x_3\\ x_3 & -x_4 & x_1 & -x_2\\ x_4 & x_3 & x_2 & x_1 \end{bmatrix}.$$

Add column 1 to column 4, and subtract column 2 from column 3; then subtract row 4 from row 1, and add row 3 to row 2. These operations do not change the value of the determinant, but (13) becomes

$$(14)\quad \det M(\zeta_1+i_2\zeta_2)=\det\begin{bmatrix} x_1-x_4 & -(x_2+x_3) & 0 & 0\\ x_2+x_3 & x_1-x_4 & 0 & 0\\ x_3 & -x_4 & x_1+x_4 & -(x_2-x_3)\\ x_4 & x_3 & x_2-x_3 & x_1+x_4 \end{bmatrix}.$$

Now Laplace's expansion of the determinant shows that

$$(15)\quad \det M(\zeta_1+i_2\zeta_2)=\det\begin{bmatrix} x_1-x_4 & -(x_2+x_3)\\ x_2+x_3 & x_1-x_4 \end{bmatrix}$$

$$\det\begin{bmatrix} x_1+x_4 & -(x_2-x_3)\\ x_2-x_3 & x_1+x_4 \end{bmatrix}.$$

From the equations in (12) it is easily verified that

$$(16)\quad M(\zeta_1 + i_{n-1}\zeta_2) = \begin{bmatrix} x_1 - x_4 & -(x_2 + x_3) \\ x_2 + x_3 & x_1 - x_4 \end{bmatrix},$$

$$M(\zeta_1 - i_{n-1}\zeta_2) = \begin{bmatrix} x_1 + x_4 & -(x_2 - x_3) \\ x_2 - x_3 & x_1 + x_4 \end{bmatrix}.$$

Thus (15) is

$$(17)\quad \det M(\zeta_1 + i_2\zeta_2) = \det M(\zeta_1 - i_1\zeta_2) \det M(\zeta_1 + i_1\zeta_2),$$

and (11) in Theorem 46.8 is true for $n=2$.

The same procedure will now be used to establish (11) for $n=3$. The Cauchy–Riemann matrix $M(\zeta_1 + i_3\zeta_2)$ is shown in (26) in Section 44; it is

$$(18)\quad \begin{bmatrix} x_1 & -x_2 & -x_3 & x_4 & -x_5 & x_6 & x_7 & -x_8 \\ x_2 & x_1 & -x_4 & -x_3 & -x_6 & -x_5 & x_8 & x_7 \\ x_3 & -x_4 & x_1 & -x_2 & -x_7 & x_8 & -x_5 & x_6 \\ x_4 & x_3 & x_2 & x_1 & -x_8 & -x_7 & -x_6 & -x_5 \\ x_5 & -x_6 & -x_7 & x_8 & x_1 & -x_2 & -x_3 & x_4 \\ x_6 & x_5 & -x_8 & -x_7 & x_2 & x_1 & -x_4 & -x_3 \\ x_7 & -x_8 & x_5 & -x_6 & x_3 & -x_4 & x_1 & -x_2 \\ x_8 & x_7 & x_6 & x_5 & x_4 & x_3 & x_2 & x_1 \end{bmatrix}.$$

First perform the following operations on columns: subtract column 1 from column 7 and column 2 from column 8; then add column 3 to column 5 and column 4 to column 6. Next perform the following operations on rows: add row 7 to row 1 and row 8 to row 2; then subtract row 5 from row 3 and row 6 from row 4. The result is a 4×4 block of zeros in the upper right corner. Then Laplace's expansion shows that the determinant of the matrix in (18) equals

$$(19)\quad \det \begin{bmatrix} x_1+x_7 & -(x_2+x_8) & -(x_3-x_5) & x_4-x_6 \\ x_2+x_8 & x_1+x_7 & -(x_4-x_6) & -(x_3-x_5) \\ x_3-x_5 & -(x_4-x_6) & x_1+x_7 & -(x_2+x_8) \\ x_4-x_6 & x_3-x_5 & x_2+x_8 & x_1+x_7 \end{bmatrix}$$

$$\times \det \begin{bmatrix} x_1-x_7 & -(x_2-x_8) & -(x_3+x_5) & x_4+x_6 \\ x_2-x_8 & x_1-x_7 & -(x_4+x_6) & -(x_3+x_5) \\ x_3+x_5 & -(x_4+x_6) & x_1-x_7 & -(x_2-x_8) \\ x_4+x_6 & x_3+x_5 & x_2-x_8 & x_1-x_7 \end{bmatrix}.$$

In this case

$$(20)\qquad \zeta_1+i_3\zeta_2=[(x_1+i_1x_2)+i_2(x_3+i_1x_4)]+i_3[(x_5+i_1x_6)+i_2(x_7+i_1x_8)],$$

$$(21)\qquad \zeta_1-i_2\zeta_2=(x_1+x_7)+i_1(x_2+x_8)+i_2(x_3-x_5)+i_1i_2(x_4-x_6),$$

$$(22)\qquad \zeta_1+i_2\zeta_2=(x_1-x_7)+i_1(x_2-x_8)+i_2(x_3+x_5)+i_1i_2(x_4+x_6).$$

Now the $2^2\times 2^2$ Cauchy–Riemann matrix is shown in (13); this pattern and (21) and (22) show that the expression in (19) is $\det M(\zeta_1-i_2\zeta_2)$ $\det M(\zeta_1+i_2\zeta_2)$. Since the determinant of (18), which is $\det(\zeta_1+i_3\zeta_2)$ in this case, equals (19), the proof is complete that (11) in Theorem 46.8 is true for $n=3$.

Now each of the matrices in (19) is a Cauchy–Riemann matrix. Therefore each of the determinants can be factored by the part of the theorem already proved [see (13)–(17)]. Use (13) and (15) with (19) to show that the determinant of the matrix in (18) equals the product of the following four determinants:

$$(23)\quad \det\begin{bmatrix}(x_1+x_7)-(x_4-x_6) & -[(x_2+x_8)+(x_3-x_5)]\\ (x_2+x_8)+(x_3-x_5) & (x_1+x_7)-(x_4-x_6)\end{bmatrix},$$

$$(24)\quad \det\begin{bmatrix}(x_1+x_7)+(x_4-x_6) & -[(x_2+x_8)-(x_3-x_5)]\\ (x_2+x_8)-(x_3-x_5) & (x_1+x_7)+(x_4-x_6)\end{bmatrix},$$

$$(25)\quad \det\begin{bmatrix}(x_1-x_7)-(x_4+x_6) & -[(x_2-x_8)+(x_3+x_5)]\\ (x_2-x_8)+(x_3+x_5) & (x_1-x_7)-(x_4+x_6)\end{bmatrix},$$

$$(26)\quad \det\begin{bmatrix}(x_1-x_7)+(x_4+x_6) & -[(x_2-x_8)-(x_3+x_5)]\\ (x_2-x_8)-(x_3+x_5) & (x_1-x_7)+(x_4+x_6)\end{bmatrix}.$$

Although elementary methods have been successful in proving Theorem 46.8 for $n=2$ and $n=3$, and although the same methods could prove the theorem for $n=4$ and $n=5$, they are not capable of proving it for large values of n because of the size of the matrix $M(\zeta_1+i_n\zeta_2)$ and the complexity of the great mass of details. Therefore other methods must be developed. The following theorem and corollaries will provide a foundation for a proof of the theorem for all $n\geqslant 2$.

46.9 THEOREM Let $\zeta_1+i_n\zeta_2$ be an element in $\mathbb{C}_n$, $n\geqslant 2$, which is represented with coefficients in $\mathbb{C}_0$. A necessary and sufficient condition that $\eta_1+i_n\eta_2$ be the inverse of $\zeta_1+i_n\zeta_2$ is that $\eta_1-i_{n-1}\eta_2$ be the inverse of $\zeta_1-i_{n-1}\zeta_2$, and that $\eta_1+i_n\eta_2$ be the inverse of $\zeta_1+i_{n-1}\zeta_2$.

Proof. Assume first that $\eta_1+i_n\eta_2$ is the inverse of $\zeta_1+i_n\zeta_2$. Then $\eta_1+i_n\eta_2$ is the unique element such that

(27) $(\zeta_1 + i_n\zeta_2)(\eta_1 + i_n\eta_2) = 1.$

By Theorems 45.2 (10) and 45.6 (28),

(28)
$$\begin{aligned}(\zeta_1 + i_n\zeta_2)(\eta_1 + i_n\eta_2)\\ &= (\zeta_1 - i_{n-1}\zeta_2)(\eta_1 - i_{n-1}\eta_2)e(i_{n-1}i_n)\\ &\quad + (\zeta_1 + i_{n-1}\zeta_2)(\eta_1 + i_{n-1}\eta_2)e(-i_{n-1}i_n).\end{aligned}$$

By (27), equation (28) is equivalent to the following two equations:

(29)
$$\begin{aligned}&1/2(\zeta_1 - i_{n-1}\zeta_2)(\eta_1 - i_{n-1}\eta_2) + 1/2(\zeta_1 + i_{n-1}\zeta_2)(\eta_1 + i_{n-1}\eta_2) = 1,\\ &1/2(\zeta_1 - i_{n-1}\zeta_2)(\eta_1 - i_{n-1}\eta_2) - 1/2(\zeta_1 + i_{n-1}\zeta_2)(\eta_1 + i_{n-1}\eta_2) = 0.\end{aligned}$$

Therefore,

(30) $(\zeta_1 - i_{n-1}\zeta_2)(\eta_1 - i_{n-1}\eta_2) = 1, \qquad (\zeta_1 + i_{n-1}\zeta_2)(\eta_1 + i_{n-1}\eta_2) = 1,$

and $\eta_1 - i_{n-1}\eta_2$ and $\eta_1 + i_{n-1}\eta_2$ are the inverses of $\zeta_1 - i_{n-1}\zeta_2$ and $\zeta_1 + i_{n-1}\zeta_2$, respectively. Therefore, (27) implies (30). Conversely, (30) implies (27). More precisely, if $\eta_1 - i_{n-1}\eta_2$ and $\eta_1 + i_{n-1}\eta_2$ are unique elements such that (30) is true, then (30) and (28) show that $\eta_1 + i_n\eta_2$ is the unique element such that (27) is true. Thus $\eta_1 + i_n\eta_2$ is the inverse of $\zeta_1 + i_n\zeta_2$. The proof of Theorem 46.9 is complete. □

As a result of Theorem 46.9, it is possible to determine whether $\zeta_1 + i_n\zeta_2$ is nonsingular or singular by determining whether $\zeta_1 - i_{n-1}\zeta_2$ and $\zeta_1 + i_{n-1}\zeta_2$ are nonsingular or singular.

46.10 COROLLARY The element $\zeta_1 + i_n\zeta_2$ in $\mathbb{C}_n$, $n \geq 2$, is nonsingular if and only if $\zeta_1 - i_{n-1}\zeta_2$ and $\zeta_1 + i_{n-1}\zeta_2$ are nonsingular; the element $\zeta_1 + i_n\zeta_2$ is singular if at least one of the elements $\zeta_1 - i_{n-1}\zeta_2$ and $\zeta_1 + i_{n-1}\zeta_2$ is singular.

Proof. If $\zeta_1 + i_n\zeta_2$ is nonsingular, it has an inverse. Then by Theorem 46.9, both $\zeta_1 - i_{n-1}\zeta_2$ and $\zeta_1 + i_{n-1}\zeta_2$ have inverses, and they are nonsingular by definition. The converse is true by Theorem 46.9. If $\zeta_1 + i_n\zeta_2$ is singular, then at least one of the elements $\zeta_1 - i_{n-1}\zeta_2$ and $\zeta_1 + i_{n-1}\zeta_2$ is singular; otherwise $\zeta_1 + i_n\zeta_2$ would be nonsingular by Theorem 46.9. If at least one of the elements $\zeta_1 - i_{n-1}\zeta_2$ and $\zeta_1 + i_{n-1}\zeta_2$ is singular, then $\zeta_1 + i_n\zeta_2$ is singular, for otherwise a contradiction would result from Theorem 46.9. □

Now Theorem 46.3 shows that the determinant of the Cauchy–Riemann matrix of an element determines whether or not it is singular or nonsingular. That theorem and Corollary 46.10 supply the proof for the following corollary.

46.11 COROLLARY Let $\zeta_1+i_n\zeta_2$ be an element in $\mathbb{C}_n$, $n\geqslant 2$. Then $\zeta_1+i_n\zeta_2$ is nonsingular if and only if

(31) $\det M(\zeta_1 - i_{n-1}\zeta_2)\det(\zeta_1 + i_{n-1}\zeta_2) \neq 0.$

Also, $\zeta_1+i_n\zeta_2$ is singular if and only if

(32) $\det M(\zeta_1 - i_{n-1}\zeta_2)\det M(\zeta_1 + i_{n-1}\zeta_2) = 0.$

Proof of Theorem 46.8. To establish the theorem, it is necessary to prove that

(33) $\det M(\zeta_1 + i_n\zeta_2) = \det M(\zeta_1 - i_{n-1}\zeta_2)\det M(\zeta_1 + i_{n-1}\zeta_2)$

for all elements $\zeta_1+i_n\zeta_2$ in $\mathbb{C}_n$, $n\geqslant 2$. Equation (32) shows that the expressions on the two sides of (33) vanish on the same set $\mathcal{O}_n$ in $\mathbb{C}_n$. This fact alone is not sufficient to prove that the two expressions are equal for all $\zeta_1+i_n\zeta_2$ in $\mathbb{C}_n$, but other information is available. Since $\zeta_1+i_n\zeta_2$ is in $\mathbb{C}_n$, then $M(\zeta_1+i_n\zeta_2)$ is a $2^n\times 2^n$ matrix, and $M(\zeta_1-i_{n-1}\zeta_2)$ and $M(\zeta_1+i_{n-1}\zeta_2)$ are $2^{n-1}\times 2^{n-1}$ matrices. The elements in these matrices are formed from the elements x_k, $k=1,\ldots,2^n$, in $\mathbb{C}_0$. Then, by the definition of the determinant, $\det M(\zeta_1+i_n\zeta_2)$ is a homogeneous polynomial of degree 2^n in the variables x_k, $k=1,\ldots,2^n$, and each of the determinants on the right in (33) is a homogeneous polynomial of degree 2^{n-1} in the same variables. Thus the product of determinants on the right in (33) is a homogeneous polynomial of degree $2^{n-1}+2^{n-1}=2^n$. Thus the expressions on the two sides of (33) are homogeneous polynomials of degree 2^n, and Corollary 46.11 shows that they vanish on the same sets in $\mathbb{C}_0^{2^n}$. With only this information, it would still be possible for one of the polynomials to be a multiple of the other. However, by examining the definition (or construction) of the Cauchy–Riemann matrices, it is easy to see that the polynomial on both the right and on the left contain the leading term $x_1^{2^n}$. We conclude that the two polynomials are identical, and that the proof of Theorem 46.8 is complete. □

46.12 COROLLARY Let $\zeta_1+i_n\zeta_2$ be an element in $\mathbb{C}_n$, $n\geqslant 2$. Then $\det M(\zeta_1+i_n\zeta_2)$ can be factored into a product of 2^{n-1} determinants of 2×2 Cauchy–Riemann matrices. Stated in terms of polynomials, the polynomial $\det M(\zeta_1+i_n\zeta_2)$ can be factored into a product of 2^{n-1} polynomials of the second degree in the variables x_k, $k=1,\ldots,2^n$.

Proof. Theorem 46.8 states [see (11) and (33)] that $\det M(\zeta_1+i_n\zeta_2)$ can be factored into the product of the determinants of two Cauchy–Riemann matrices. By the same theorem, each of the latter determinants can be factored into the product of two determinants. This process can be repeated until $\det M(\zeta_1+i_n\zeta_2)$ is represented as the product of determinants of 2×2 matrices. □

46.13 COROLLARY Let $\zeta_1+i_n\zeta_2$ be an element in $\mathbf{C}_n$, $n\geqslant 1$, which is represented with coefficients in $\mathbf{C}_0$, and let $M(\zeta_1+i_n\zeta_2)$ be the associated Cauchy–Riemann matrix. Then

$$\text{(34)} \qquad \det M(\zeta_1 + i_n\zeta_2) \geqslant 0, \qquad n \geqslant 1.$$

Proof. A 2×2 Cauchy–Riemann matrix with elements in $\mathbf{C}_0$ has the form

$$\text{(35)} \qquad \begin{bmatrix} a & -b \\ b & a \end{bmatrix}.$$

The determinant of (35) is the nonnegative number a^2+b^2. Since Corollary 46.12 shows that every $\det M(\zeta_1+i_n\zeta_2)$, $n\geqslant 2$, can be represented as the product of determinants of matrices of the form (35), then (34) and Corollary 46.13 are true. □

46.14 COROLLARY If $\mathcal{O}_n$, $n\geqslant 0$, is the set of singular elements in $\mathbf{C}_n$ (see Definition 46.4), then

$$\text{(36)} \qquad \mathcal{O}_0 = \mathcal{O}_1 \subset \mathcal{O}_2 \subset \cdots \subset \mathcal{O}_n \subset \cdots.$$

Proof. In proving this corollary, it is necessary to remember the convention in Theorem 44.2 under which $\mathbf{C}_{n-1}$ is considered to be contained in $\mathbf{C}_n$. With this convention, it is clear that $\mathcal{O}_0=\mathcal{O}_1\subset\mathcal{O}_2$. The proof, by induction, will be completed by showing that $\mathcal{O}_{n-1}\subset\mathcal{O}_n$ for $n\geqslant 3$. Let ζ_1 be an element in $\mathbf{C}_{n-1}$, $n\geqslant 3$, which is in $\mathcal{O}_{n-1}$. If $M(\zeta_1)$ is the associated Cauchy–Riemann matrix, then $\det M(\zeta_1)\in\mathcal{O}_0$. Let 0 denote the zero element in $\mathbf{C}_{n-1}$. Then ζ_1+i_n0 is an element in $\mathbf{C}_n$; it is also ζ_1 represented as an element in $\mathbf{C}_n$. By Theorem 46.8,

$$\text{(37)} \qquad \begin{aligned} \det M(\zeta_1 + i_n0) &= \det M(\zeta_1 - i_{n-1}0) \det M(\zeta_1 + i_{n-1}0) \\ &= \det M(\zeta_1) \det M(\zeta_1). \end{aligned}$$

This equation shows that, since $\det M(\zeta_1)\in\mathcal{O}_0$, then $\det M(\zeta_1+i_n0)\in\mathcal{O}_0$ also. Therefore, by Definition 46.4, the element ζ_1+i_n0, which is the element ζ_1 in $\mathcal{O}_{n-1}$, is also in $\mathcal{O}_n$. Since this statement is true for every element ζ_1 in $\mathcal{O}_{n-1}$, then $\mathcal{O}_{n-1}\subset\mathcal{O}_n$ and the proof of Corollary 46.14 is complete. □

46.15 COROLLARY The set $\mathcal{O}_n$ in $\mathbf{C}_n$, $n\geqslant 2$, is the union of 2^{n-1} linear subspaces of $\mathbf{C}_0^{2^n}$, each of which has dimension 2^n-2.

Proof. First, prove this corollary for $\mathcal{O}_3$ in $\mathbf{C}_3$. In the setting of the $\mathbf{C}_0$ column of Table 46.1, the space $\mathbf{C}_3$ is imbedded in $\mathbf{C}_0^8$. Also, $\mathcal{O}_3$ is the set of points at which the determinant of the matrix in (18) vanishes. Since this determinant equals the product of the determinants in (23)–(26) as shown above and proved also in Corollary 46.12, the set $\mathcal{O}_3$ is the union of the sets on which the four determinants in (23)–(26) vanish. Consider the determinant in (23); since it equals

$$(38) \qquad [(x_1 + x_7) - (x_4 - x_6)]^2 + [(x_2 + x_8) + (x_3 - x_5)]^2,$$

it vanishes on the intersection of the following planes through the origin:

$$(39) \qquad (x_1 + x_7) - (x_4 - x_6) = 0, \qquad (x_2 + x_8) + (x_3 - x_5) = 0.$$

Each plane in (39) has dimension 7 (that is, $2^3 - 1$), and their intersection is a linear subspace of $\mathbb{C}_0^8$ which has dimension 6 (that is, $2^3 - 2$). There are 4 (that is, 2^{3-1}) such linear subspaces of $\mathbb{C}_0^8$ derived from the determinants in (23)–(26). Therefore, Corollary 46.15 is true for $\mathcal{O}_3$ in $\mathbb{C}_3$. Next, Corollary 46.12, together with the methods used in proving the corollary for $\mathcal{O}_3$ in $\mathbb{C}_3$, can be used to prove Corollary 46.15 for all $n \geqslant 2$. □

Thus far this section has emphasized the problems and methods which arise in the $\mathbb{C}_0$ column in Table 46.1. This means that all elements in $\mathbb{C}_1, \ldots, \mathbb{C}_n, \ldots$ have been represented with coefficients in $\mathbb{C}_0$, and that the zero in $\mathcal{O}_0$ in $\mathbb{C}_0$ is the fundamental element used to characterize all singular elements. But as Table 46.1 indicates, all elements in $\mathbb{C}_2, \ldots, \mathbb{C}_n, \ldots$ can be represented in terms of z, the element in $\mathbb{C}_1$. In this treatment, the zero in $\mathbb{C}_1$ is used to characterize the singular elements. It is in order at this time to examine the similarities and differences in the two treatments.

First examine the singular elements in the $\mathbb{C}_1$ column setting. An element ζ is nonsingular if and only if there exists a unique element η such that $\zeta\eta = 1$. This condition leads to a system of linear equations. For example, if ζ is $z_1 + i_2 z_2$ in $\mathbb{C}_2$, then it is nonsingular if and only if there exists a unique element $w_1 + i_2 w_2$ such that

$$(40) \qquad (z_1 + i_2 z_2)(w_1 + i_2 w_2) = 1.$$

This equation is equivalent to the following system:

$$(41) \qquad z_1 w_1 - z_2 w_2 = 1, \qquad z_2 w_1 + z_1 w_2 = 0.$$

This system of equations has a unique solution if and only if

$$(42) \qquad \det M(\zeta) \notin \mathcal{O}_1, \qquad \zeta = z_1 + i_2 z_2, \qquad M(\zeta) = \begin{bmatrix} z_1 & -z_2 \\ z_2 & z_1 \end{bmatrix}.$$

Here $M(\zeta)$ is a Cauchy–Riemann matrix with elements in $\mathbb{C}_1$. If $\zeta = (z_1 + i_2 z_2) + i_3(z_3 + i_2 z_4)$, then the calculations for finding the inverse have been carried out in (27)–(32) in Section 44. The Cauchy–Riemann matrix of ζ in this case is shown in (32) of Section 44, and $(z_1 + i_2 z_2) + i_3(z_3 + i_2 z_4)$ is nonsingular if and only if $M(\zeta)\notin \mathcal{O}_1$, where

$$\det M(\zeta) = \det\begin{bmatrix} z_1 & -z_2 & -z_3 & z_4 \\ z_2 & z_1 & -z_4 & -z_3 \\ z_3 & -z_4 & z_1 & -z_2 \\ z_4 & z_3 & z_2 & z_1 \end{bmatrix}. \tag{43}$$

The Cauchy–Riemann matrix in (42) has the same form as the 2×2 Cauchy–Riemann matrix

$$\begin{bmatrix} x_1 & -x_2 \\ x_2 & x_1 \end{bmatrix} \tag{44}$$

with elements in $\mathbb{C}_0$. Also, $M(\zeta)$ in (43) has the same form as the Cauchy–Riemann matrix in (13) with elements in $\mathbb{C}_0$. The next theorem summarizes some of the similarities and differences in the results in the $\mathbb{C}_0$ and $\mathbb{C}_1$ columns of Table 46.1.

46.16 THEOREM A $2^n\times 2^n$ Cauchy–Riemann matrix (with elements x_k in $\mathbb{C}_0$) determines whether an element in the space

$$\mathbb{C}_n\colon i_1,\ldots,i_n;\ \text{coefficients } x_k,\ k=1,\ldots,2^n, \tag{45}$$

is nonsingular or singular; and a $2^{n-1}\times 2^{n-1}$ Cauchy–Riemann matrix (with elements z_k in $\mathbb{C}_1$) determines whether an element in the space

$$\mathbb{C}_n\colon i_2,\ldots,i_n;\ \text{coefficients } z_k,\ k=1,\ldots,2^{n-1}, \tag{46}$$

is nonsingular or singular. The system

$$\mathbb{C}_{n-1}\colon i_1,\ldots,i_{n-1};\ \text{coefficients } x_k,\ k=1,\ldots,2^{n-1}, \tag{47}$$

is isomorphic with respect to multiplication to the system in (46), and the Cauchy–Riemann matrices in the two systems are $2^{n-1}\times 2^{n-1}$ matrices which are identical in form.

Proof. The statements about (45) and (46) have been established above. As an example, the $2^3\times 2^3$ matrix in (18) is the Cauchy–Riemann matrix determined by $[(x_1 + i_1x_2) + i_2(x_3 + i_1x_4)] + i_3[(x_5 + i_1x_6) + i_2(x_7 + i_1x_8)]$ in $\mathbb{C}_3$, and the $2^2\times 2^2$ matrix in (43) is the Cauchy–Riemann matrix associated with the element $(z_1+i_2z_2)+i_3(z_3+i_2z_4)$ in $\mathbb{C}_3$. Since $i_2^2=-1,\ldots, i_n^2=-1$ in (46) and $i_1^2=-1,\ldots,i_{n-1}^2=-1$ in (47), and since the z_k in (46) and the x_k in (47) obey the same formal laws of operation, the results of multiplication in the two systems are isomorphic and the Cauchy–Riemann matrices have the same form and size. For example, (5) contains the Cauchy–Riemann matrix for $(x_1+i_1x_2)+i_2(x_3+i_1x_4)$ in $\mathbb{C}_2$, and (43) contains the

Cauchy–Riemann matrix for $(z_1+i_2z_2)+i_3(z_3+i_2z_4)$ in $\mathbb{C}_3$. The two matrices have the same size and form although the elements are called x in the former and z in the latter. □

Because of the similarities and isomorphisms described in Theorem 46.16, many of the theorems already established for column $\mathbb{C}_0$ representation can be converted into theorems for column $\mathbb{C}_1$ representation (see Table 46.1) by changing $i_1,\ldots,i_{n-1}$ to $i_2,\ldots,i_n$ and x_k to z_k, $k=1,\ldots,2^{n-1}$. For example, the following theorem is an automatic consequence of Theorem 46.8.

46.17 THEOREM If $M(\zeta_1+i_n\zeta_2)$, $M(\zeta_1-i_{n-1}\zeta_2)$, and $M(\zeta_1+i_{n-1}\zeta_2)$ are the Cauchy–Riemann matrices of $\zeta_1+i_n\zeta_2$ in $\mathbb{C}_n$, $n\geqslant 3$, and $\zeta_1-i_{n-1}\zeta_2$, $\zeta_1+i_{n-1}\zeta_2$ in $\mathbb{C}_{n-1}$, respectively, in the $\mathbb{C}_1$-column representation of Table 46.1, then

$$\det M(\zeta_1+i_n\zeta_2)=\det M(\zeta_1-i_{n-1}\zeta_2)\det M(\zeta_1+i_{n-1}\zeta_2). \tag{48}$$

Examples of the derivation of Theorem 46.17 from Theorem 46.8 are easy to provide. Equations (13) and (16) above contain the following example of Theorem 46.8 for the element $(x_1+i_1x_2)+i_2(x_3+i_1x_4)$ in $\mathbb{C}_2$:

$$\det\begin{bmatrix} x_1 & -x_2 & -x_3 & x_4\\ x_2 & x_1 & -x_4 & -x_3\\ x_3 & -x_4 & x_1 & -x_2\\ x_4 & x_3 & x_2 & x_1\end{bmatrix} =\det\begin{bmatrix} x_1+x_4 & -(x_2-x_3)\\ x_2-x_3 & x_1+x_4\end{bmatrix}\det\begin{bmatrix} x_1-x_4 & -(x_2+x_3)\\ x_2+x_3 & x_1-x_4\end{bmatrix}. \tag{49}$$

Then by Theorem 46.17 the following identity holds for the Cauchy–Riemann matrix of $(z_1+i_2z_2)+i_3(z_3+i_2z_4)$ in $\mathbb{C}_3$:

$$\det\begin{bmatrix} z_1 & -z_2 & -z_3 & z_4\\ z_2 & z_1 & -z_4 & -z_3\\ z_3 & -z_4 & z_1 & -z_2\\ z_4 & z_3 & z_2 & z_1\end{bmatrix} =\det\begin{bmatrix} z_1+z_4 & -(z_2-z_3)\\ z_2-z_3 & z_1+z_4\end{bmatrix}\det\begin{bmatrix} z_1-z_4 & -(z_2+z_3)\\ z_2+z_3 & z_1-z_4\end{bmatrix}. \tag{50}$$

Repeated application of Theorem 46.8 proved Corollary 46.12. In the same way, Theorem 46.17 can be applied repeatedly to prove that the determinant of a Cauchy–Riemann matrix of an element in $\mathbb{C}_n$, $n\geqslant 3$, can be factored into a product of determinants of 2×2 Cauchy–Riemann matrices of elements in $\mathbb{C}_1$. We shall not undertake a systematic restatement, for column $\mathbb{C}_1$

representation, of all of the results given above for column $\mathbb{C}_0$ representation since the reader, with the help of the examples should find it easy to fill in the missing details.

There are an infinite number of columns in Table 46.1, but an examination of one more column should be a sufficient introduction to all of them. In the $\mathbb{C}_2$ column, all elements in $\mathbb{C}_n$, $n \geqslant 3$, are represented in terms of elements in $\mathbb{C}_2$. Thus an element in $\mathbb{C}_4$ is represented as $(\zeta_1 + i_3\zeta_2) + i_4(\zeta_3 + i_3\zeta_4)$, where $\zeta_1, \dots, \zeta_4$ are elements in $\mathbb{C}_2$. To determine whether an element ζ in $\mathbb{C}_n$ is nonsingular or singular, proceed to determine its inverse. The calculation leads to a system of linear equations as before. The matrix $M(\zeta)$ of coefficients has elements in $\mathbb{C}_2$; it is the Cauchy–Riemann matrix for the element in $\mathbb{C}_n$. If $\det M(\zeta) \notin \mathcal{O}_2$, then ζ is nonsingular; if $\det M(\zeta) \in \mathcal{O}_2$, then ζ is singular. Most of the results established above in this section can be generalized to obtain similar results for the $\mathbb{C}_2$ column of Table 46.1; these generalizations depend on the following theorem, which is similar to Theorem 46.16.

46.18 THEOREM A $2^{n-1} \times 2^{n-1}$ Cauchy–Riemann matrix (with elements z_k in $\mathbb{C}_1$) determines whether an element in the space

(51) $\quad \mathbb{C}_n\colon i_2, \dots, i_n;$ coefficients z_k, $k = 1, \dots, 2^{n-1}$,

is nonsingular or singular; and a $2^{n-2} \times 2^{n-2}$ Cauchy–Riemann matrix (with elements ζ_k in $\mathbb{C}_2$) determines whether an element in the space

(52) $\quad \mathbb{C}_n\colon i_3, \dots, i_n;$ coefficients ζ_k, $k = 1, \dots, 2^{n-2}$,

is nonsingular or singular. The system

(53) $\quad \mathbb{C}_{n-1}\colon i_2, \dots, i_{n-1};$ coefficients z_k, $k = 1, \dots, 2^{n-2}$,

is isomorphic with respect to multiplication to the system in (52), and the Cauchy–Riemann matrices in the two systems are $2^{n-2} \times 2^{n-2}$ matrices which are identical in form.

The generalization, to the column $\mathbb{C}_2$ of Table 46.1, of the results established above in this section is left to the reader. The following identity, which is derived from (50), is an example of the generalization of Theorems 46.8 and 46.17; the ζ_k denote elements in $\mathbb{C}_2$ and $(\zeta_1 + i_3\zeta_2) + i_4(\zeta_3 + i_3\zeta_4)$ is an element in $\mathbb{C}_4$.

(54)
$$\det \begin{bmatrix} \zeta_1 & -\zeta_2 & -\zeta_3 & \zeta_4 \\ \zeta_2 & \zeta_1 & -\zeta_4 & -\zeta_3 \\ \zeta_3 & -\zeta_4 & \zeta_1 & -\zeta_2 \\ \zeta_4 & \zeta_3 & \zeta_2 & \zeta_1 \end{bmatrix}$$
$$= \det \begin{bmatrix} \zeta_1 + \zeta_4 & -(\zeta_2 - \zeta_3) \\ \zeta_2 - \zeta_3 & \zeta_1 + \zeta_4 \end{bmatrix} \det \begin{bmatrix} \zeta_1 - \zeta_4 & -(\zeta_2 + \zeta_3) \\ \zeta_2 + \zeta_3 & \zeta_1 - \zeta_4 \end{bmatrix}.$$

We turn now to the final topic to be investigated in this section. Thus far the section has emphasized two topics as follows: (a) proofs of certain properties of the space $\mathbb{C}_n$; and (b) demonstration that the treatment of $\mathbb{C}_n$ can be carried out (1) with elements in $\mathbb{C}_n$ represented with coefficients in $\mathbb{C}_0$ ($\mathbb{C}_0$-column representation in Table 46.1), or (2) with elements in $\mathbb{C}_n$ represented with coefficients in $\mathbb{C}_1$ ($\mathbb{C}_1$-column representation), ..., or $(n-1)$ with elements in $\mathbb{C}_n$ represented with coefficients in $\mathbb{C}_{n-1}$ ($\mathbb{C}_{n-1}$-column representation). The remainder of the section is concerned with an investigation of the relations among some of the results on different representations of an element. We begin with some examples.

An element in $\mathbb{C}_2$ is denoted either by x: $(x_1+i_1x_2)+i_2(x_3+i_1x_4)$ or by z: $z_1+i_2z_2$. The element x is nonsingular if and only if

(55) $$\det\begin{bmatrix} x_1 & -x_2 & -x_3 & x_4 \\ x_2 & x_1 & -x_4 & -x_3 \\ x_3 & -x_4 & x_1 & -x_2 \\ x_4 & x_3 & x_2 & x_1 \end{bmatrix} \notin \mathcal{O}_0.$$

The element z is nonsingular if and only if

(56) $$\det\begin{bmatrix} z_1 & -z_2 \\ z_2 & z_1 \end{bmatrix} \notin \mathcal{O}_1.$$

Equations (55) and (56) are two forms of the same condition, namely, the condition that the element x or z be nonsingular in $\mathbb{C}_2$. The conditions, although not the same, are nevertheless equivalent. What is the relation between them; more precisely, what is the relation between the determinants in (55) and (56)? Similarly, there are two forms for the condition that an element in $\mathbb{C}_3$ be nonsingular. The element $[(x_1+i_1x_2)+i_2(x_3+i_1x_4)] + i_3[(x_5+i_1x_6)+i_2(x_7+i_1x_8)]$ in $\mathbb{C}_3$ is nonsingular if and only if the determinant of the matrix in (18) is not an element in $\mathcal{O}_0$, and the element z: $(z_1+i_2z_2)+i_3(z_3+i_2z_4)$ in $\mathbb{C}_3$ is nonsingular if and only if the determinant in (43) is not in $\mathcal{O}_1$. The conditions, although equivalent, are very different. What relation between the two determinants explains the equivalence?

Let $\det M(x)$ denote the determinant in (55) and also the determinant of the matrix in (18); also, let $\det M(z)$ denote the determinants in (56) and in (43).

46.19 THEOREM For each of the pairs $\det M(x)$ and $\det M(z)$ just described, the following relation holds:

(57) $$|\det M(z)|^2 = \det M(x).$$

Proof. First, prove that the square of the absolute value of (56) equals (55). Now

(58) $$|\det M(z)|^2 = |z_1^2 + z_2^2|^2 = |z_1 - i_1 z_2|^2 |z_1 + i_1 z_2|^2.$$

Since

(59) $$\begin{aligned} z_1 - i_1 z_2 &= (x_1 + i_1 x_2) - i_1(x_3 + i_1 x_4) = (x_1 + x_4) + i_1(x_2 - x_3), \\ z_1 + i_1 z_2 &= (x_1 + i_1 x_2) + i_1(x_3 + i_1 x_4) = (x_1 - x_4) + i_1(x_2 + x_3), \end{aligned}$$

then

(60) $$\begin{aligned} |z_1 - i_1 z_2|^2 &= (x_1 + x_4)^2 + (x_2 - x_3)^2 = \det\begin{bmatrix} x_1 + x_4 & -(x_2 - x_3) \\ x_2 - x_3 & x_1 + x_4 \end{bmatrix}, \\ |z_1 + i_1 z_2|^2 &= (x_1 - x_4)^2 + (x_2 + x_3)^2 = \det\begin{bmatrix} x_1 - x_4 & -(x_2 + x_3) \\ x_2 + x_3 & x_1 - x_4 \end{bmatrix}. \end{aligned}$$

Substitute from (60) in (58) to obtain

(61) $$|\det M(z)|^2 = \det\begin{bmatrix} x_1 + x_4 & -(x_2 - x_3) \\ x_2 - x_3 & x_1 + x_4 \end{bmatrix} \det\begin{bmatrix} x_1 - x_4 & -(x_2 + x_3) \\ x_2 + x_3 & x_1 - x_4 \end{bmatrix}.$$

By (13)–(17) above, the product of the two determinants on the right equals the determinant in (55). The proof of (57) is complete for a pair x and z in $\mathbb{C}_2$.

Next, consider (57) for a pair x and z in $\mathbb{C}_3$. The problem is to prove that the determinant of (18) equals the square of the absolute value of (43). By the identity in (50),

(62) $$|\det M(z)|^2 = \left|\det\begin{bmatrix} z_1 + z_4 & -(z_2 - z_3) \\ z_2 - z_3 & z_1 + z_4 \end{bmatrix}\right|^2 \left|\det\begin{bmatrix} (z_1 - z_4) & -(z_2 + z_3) \\ z_2 + z_3 & z_1 - z_4 \end{bmatrix}\right|^2.$$

Now (57), in the case already proved, provides an evaluation for each of the terms on the right. We have

(63) $$\begin{aligned} &\left|\det\begin{bmatrix} z_1 + z_4 & -(z_2 - z_3) \\ z_2 - z_3 & z_1 + z_4 \end{bmatrix}\right|^2 \\ &= \left|\det\begin{bmatrix} (x_1 + x_7) + i_1(x_2 + x_8) & -[(x_3 - x_5) + i_1(x_4 - x_6)] \\ (x_3 - x_5) + i_1(x_4 - x_6) & (x_1 + x_7) + i_1(x_2 + x_8) \end{bmatrix}\right|^2 \\ &= \det\begin{bmatrix} x_1 + x_7 & -(x_2 + x_8) & -(x_3 - x_5) & x_4 - x_6 \\ x_2 + x_8 & x_1 + x_7 & -(x_4 - x_6) & -(x_3 - x_5) \\ x_3 - x_5 & -(x_4 - x_6) & x_1 + x_7 & -(x_2 + x_8) \\ x_4 - x_6 & x_3 - x_5 & x_2 + x_8 & x_1 + x_7 \end{bmatrix}; \end{aligned}$$

(64)
$$\left|\det\begin{bmatrix} z_1-z_4 & -(z_2+z_3) \\ z_2+z_3 & z_1-z_4 \end{bmatrix}\right|^2$$
$$=\left|\det\begin{bmatrix} (x_1-x_7)+i_1(x_2-x_8) & -[(x_3+x_5)+i_1(x_4+x_6)] \\ (x_3+x_5)+i_1(x_4+x_6) & (x_1-x_7)+i_1(x_2-x_8) \end{bmatrix}\right|^2$$
$$=\det\begin{bmatrix} x_1-x_7 & -(x_2-x_8) & -(x_3+x_5) & x_4+x_6 \\ x_2-x_8 & x_1-x_7 & -(x_4+x_6) & -(x_3+x_5) \\ x_3+x_5 & -(x_4+x_6) & x_1-x_7 & -(x_2-x_8) \\ x_4+x_6 & x_3+x_5 & x_2-x_8 & x_1-x_7 \end{bmatrix}.$$

Substitute from (63) and (64) in (62). The product of the two 4×4 determinants on the right equals, by (19), the determinant of the matrix in (18). The latter determinant is $M(x)$ in the present case. The proof of (57) for the pair x and z in $\mathbb{C}_3$ is complete. Thus the proof of Theorem 46.19 is complete. □

The questions that led to Theorem 46.19 can now be repeated for column $\mathbb{C}_1$ and column $\mathbb{C}_2$ representations. Because of Theorem 46.18, it is to be expected that Theorem 46.19 implies a corresponding theorem for the new pair of representations. Before this new theorem can be stated and proved, it is necessary to define a new function on $\mathbb{C}_2$.

46.20 DEFINITION The $\mathbb{C}_2$-absolute value squared function is the function $|\ |_2^2:\mathbb{C}_2\to\mathbb{C}_1$ which is defined as follows: for every $z_1+i_2z_2$ in $\mathbb{C}_2$,

(65)
$$|z_1+i_2z_2|_2^2=z_1^2+z_2^2=\det\begin{bmatrix} z_1 & -z_2 \\ z_2 & z_1 \end{bmatrix}.$$

46.21 LEMMA If $z_1+i_2z_2$ and $w_1+i_2w_2$ are two elements in $\mathbb{C}_2$, then

(66)
$$|(z_1+i_2z_2)(w_1+i_2w_2)|_2^2=|z_1+i_2z_2|_2^2|w_1+i_2w_2|_2^2.$$

Proof. Since $(z_1+i_2z_2)(w_1+i_2w_2)=(z_1w_1-z_2w_2)+i_2(z_2w_1+z_1w_2)$, then by Definition 46.20,

(67)
$$|(z_1+i_2z_2)(w_1+i_2w_2)|_2^2$$
$$=\det\begin{bmatrix} z_1w_1-z_2w_2 & -(z_2w_1+z_1w_2) \\ z_2w_1+z_1w_2 & z_1w_1-z_2w_2 \end{bmatrix}$$
$$=\det\begin{bmatrix} z_1 & -z_2 \\ z_2 & z_1 \end{bmatrix}\det\begin{bmatrix} w_1 & -w_2 \\ w_2 & w_1 \end{bmatrix}$$
$$=|z_1+i_2z_2|_2^2|w_1+i_2w_2|_2^2.$$

□

The element $(z_1+i_2z_2)+i_3(z_3+i_2z_4)$ in $\mathbb{C}_3$ is nonsingular if and only if

$$(68)\qquad \det\begin{bmatrix} z_1 & -z_2 & -z_3 & z_4 \\ z_2 & z_1 & -z_4 & -z_3 \\ z_3 & -z_4 & z_1 & -z_2 \\ z_4 & z_3 & z_2 & z_1 \end{bmatrix}$$

is not in $\mathcal{O}_1$. Denote this determinant by $\det M(z)$. Let $\zeta_1+i_3\zeta_2$, with ζ_1 and ζ_2 in $\mathbb{C}_2$, denote the same element in $\mathbb{C}_3$; it is nonsingular if and only if

$$(69)\qquad \det\begin{bmatrix} \zeta_1 & -\zeta_2 \\ \zeta_2 & \zeta_1 \end{bmatrix}$$

is not in $\mathcal{O}_2$. Denote this determinant by $\det M(\zeta)$.

46.22 THEOREM If $(z_1+i_2z_2)+i_3(z_3+i_2z_4)$ and $\zeta_1+i_3\zeta_2$ denote the same element in $\mathbb{C}_3$, in the representations of columns $\mathbb{C}_1$ and $\mathbb{C}_2$, respectively, of Table 46.1, and if $\det M(z)$ and $\det M(\zeta)$ are the determinants in (68) and (69), respectively, then

$$(70)\qquad |\det M(\zeta)|_2^2 = \det M(z).$$

Proof. The proof is isomorphic [see Theorem 48.18] to the proof of the first part of Theorem 46.19 in (58)–(61). Thus by (69), (65), and Lemma 46.21,

$$(71)\qquad \begin{aligned} |\det M(\zeta)|_2^2 &= |\zeta_1^2+\zeta_2^2|_2^2 = |(\zeta_1-i_2\zeta_2)(\zeta_1+i_2\zeta_2)|_2^2 \\ &= |(\zeta_1-i_2\zeta_2)|_2^2|(\zeta_1+i_2\zeta_2)|_2^2. \end{aligned}$$

Since

$$(72)\qquad \begin{aligned} \zeta_1-i_2\zeta_2 &= (z_1+i_2z_2)-i_2(z_3+i_2z_4) = (z_1+z_4)+i_2(z_2-z_3), \\ \zeta_1+i_2\zeta_2 &= (z_1+i_2z_2)+i_2(z_3+i_2z_4) = (z_1-z_4)+i_2(z_2+z_3), \end{aligned}$$

then

$$(73)\qquad \begin{aligned} |\zeta_1-i_2\zeta_2|_2^2 &= (z_1+z_4)^2+(z_2-z_3)^2 = \det\begin{bmatrix} z_1+z_4 & -(z_2-z_3) \\ z_2-z_3 & z_1+z_4 \end{bmatrix}, \\ |\zeta_1+i_2\zeta_2|_2^2 &= (z_1-z_4)^2+(z_2+z_3)^2 = \det\begin{bmatrix} z_1-z_4 & -(z_2+z_3) \\ z_2+z_3 & z_1-z_4 \end{bmatrix}. \end{aligned}$$

Substitute from (73) in (71); by the identity in (50), the result shows that

$$(74)\qquad |\det M(\zeta)|_2^2 = \det\begin{bmatrix} z_1 & -z_2 & -z_3 & z_4 \\ z_2 & z_1 & -z_4 & -z_3 \\ z_3 & -z_4 & z_1 & -z_2 \\ z_4 & z_3 & z_2 & z_1 \end{bmatrix}.$$

The right side of this equation is the determinant which has been denoted by $\det M(z)$, and thus the proof of (70) and of Theorem 46.22 is complete. □

Theorems 46.19 and 46.22 suggest several conjectures. The relation in (57) has been established for the representations x and z (corresponding to columns $\mathbb{C}_0$ and $\mathbb{C}_1$ in Table 46.1) of an element in $\mathbb{C}_2$ and also for an element in $\mathbb{C}_3$, and similar proofs of (57) can certainly be given for representations x and z of an element in $\mathbb{C}_4$. The conjecture states that (57) is true for the representations x and z for an element in $\mathbb{C}_n$ for $n \geqslant 2$. Some new, general methods are needed to prove the generalization of Theorem 46.19 for large values of n.

Again, the relation in (70) has been established for the representations z and ζ corresponding to the columns $\mathbb{C}_1$ and $\mathbb{C}_2$ of Table 46.1) of an element in $\mathbb{C}_3$. The conjecture states that (70) is true for the representations z and ζ of an element in $\mathbb{C}_n$ for $n \geqslant 3$. New methods are needed for the proof for large values of n.

Finally, a conjecture states that Theorems 46.19 and 46.22 are special cases of a more general theorem. Define the $\mathbb{C}_k$-absolute value squared function $|\ |_k^2 : \mathbb{C}_k \to \mathbb{C}_{k-1}$ as follows: if $\zeta_1 + i_k\zeta_2$ is in $\mathbb{C}_k$, then ζ_1 and ζ_2 are in $\mathbb{C}_{k-1}$, and

$$(75) \qquad |\zeta_1 + i_k\zeta_2|_k^2 = \zeta_1^2 + \zeta_2^2.$$

Let η_{k-1} and η_k denote an element in $\mathbb{C}_n$ in the representations of columns $\mathbb{C}_{k-1}$ and $\mathbb{C}_k$ of Table 46.1, and let $M(\eta_{k-1})$ and $M(\eta_k)$ denote the corresponding Cauchy–Riemann matrices. Then the conjecture states that

$$(76) \qquad |\det M(\eta_k)|_k^2 = \det M(\eta_{k-1})$$

for all $k \geqslant 1$ and $n \geqslant k+1$. Theorem 46.19 is this conjecture for $k=1$ and $n=2$ and $n=3$; Theorem 46.22 is this conjecture for $k=2$ and $n=3$. The proofs of these theorems show that elementary methods are sufficient to prove the conjecture for small values of k and n, but they also emphasize the inadequacy of those methods for large k and n.

Exercises

46.1 Show that, for $k=1$, the $\mathbb{C}_k$-absolute value squared function in (75) is the traditional absolute value squared function for complex numbers.

46.2 Let x and z denote, in the representations

$$\begin{aligned} x\colon &\{[(x_1+i_1x_2)+i_2(x_3+i_1x_4)]+i_3[(x_5+i_1x_6)+i_2(x_7+i_1x_8)]\} \\ &+i_4\{[(x_9+i_1x_{10})+i_2(x_{11}+i_1x_{12})]+i_3[(x_{13}+i_1x_{14}) \\ &\qquad +i_2(x_{15}+i_1x_{16})]\}, \end{aligned}$$

$$z:\ [(z_1+i_2z_2)+i_3(z_3+i_2z_4)]+i_4[(z_5+i_2z_6)+i_3(z_7+i_2z_8)],$$

of the columns $\mathbb{C}_0$ and $\mathbb{C}_1$, respectively, of Table 46.1, the same element in $\mathbb{C}_4$.

(a) Find the Cauchy–Riemann matrices $M(x)$ and $M(z)$.

(b) Factor $\det M(x)$ and $\det M(z)$ into the product of two determinants by the method used to show that the determinant of the matrix in (18) equals the product of the determinants in (19).

(c) Factor $\det M(x)$ into the product of (i) two determinants, (ii) four determinants, and (iii) eight determinants.

(d) Factor $\det M(z)$ into the product of two determinants and of four determinants.

(e) Prove that $|\det M(z)|^2=\det M(x)$.

(f) Give a complete description of the solution set of the equation $\det M(x)=0$. What is this solution set called?

(g) Give two proofs that $\det M(x)\geqslant 0$.

46.3 Let η_2 and η_3 denote, in the representations of columns $\mathbb{C}_2$ and $\mathbb{C}_3$ of Table 46.1, the same element in $\mathbb{C}_4$.

(a) Find the Cauchy–Riemann matrices $M(\eta_2)$ and $M(\eta_3)$.

(b) If ζ_1 and ζ_2 are two elements in $\mathbb{C}_3$, prove that $|\zeta_1\zeta_2|_3^2=|\zeta_1|_3^2|\zeta_2|_3^2$.

(c) Prove that $|\det M(\eta_3)|_3^2=\det M(\eta_2)$.

46.4 Prove that the zero element in $\mathbb{C}_n$ is a singular element, and that every neighborhood of the zero element contains singular elements in addition to the zero element.

46.5 Prove that the element 1 in $\mathbb{C}_n$ is a nonsingular element, and that every sufficiently small neighborhood of 1 contains no singular elements.

46.6 Prove the following: (a) the nonsingular elements in $\mathbb{C}_n$ form an open set in $\mathbb{C}_n$; and (b) the set of singular elements contains no neighborhood of a point.

46.7 (a) Show that

$$e(i_1i_2)e(i_2i_3)=\frac{1+i_1i_2+i_2i_3-i_1i_3}{4}$$

and that, in the $\mathbb{C}_0$-column representation of Table 46.1, $x_1=\frac{1}{4}$, $x_2=0$, $x_3=0$, $x_4=\frac{1}{4}$, $x_5=0$, $x_6=-\frac{1}{4}$, $x_7=\frac{1}{4}$, $x_8=0$.

(b) Show that, in the representation of part (a) the Cauchy–Riemann matrix of $e(i_1i_2)e(i_2i_3)$ is

$$\begin{bmatrix} \frac{1}{4} & 0 & 0 & \frac{1}{4} & 0 & -\frac{1}{4} & \frac{1}{4} & 0 \\ 0 & \frac{1}{4} & -\frac{1}{4} & 0 & \frac{1}{4} & 0 & 0 & \frac{1}{4} \\ 0 & -\frac{1}{4} & \frac{1}{4} & 0 & -\frac{1}{4} & 0 & 0 & -\frac{1}{4} \\ \frac{1}{4} & 0 & 0 & \frac{1}{4} & 0 & -\frac{1}{4} & \frac{1}{4} & 0 \\ 0 & \frac{1}{4} & -\frac{1}{4} & 0 & \frac{1}{4} & 0 & 0 & \frac{1}{4} \\ -\frac{1}{4} & 0 & 0 & -\frac{1}{4} & 0 & \frac{1}{4} & -\frac{1}{4} & 0 \\ \frac{1}{4} & 0 & 0 & \frac{1}{4} & 0 & -\frac{1}{4} & \frac{1}{4} & 0 \\ 0 & \frac{1}{4} & -\frac{1}{4} & 0 & \frac{1}{4} & 0 & 0 & \frac{1}{4} \end{bmatrix}.$$

(c) Find the Cauchy–Riemann matrices of the idempotent elements $e(-i_1i_2)e(i_2i_3)$, $e(i_1i_2)e(-i_2i_3)$, $e(-i_1i_2)e(-i_2i_3)$.

(d) Use the matrices found in (b) and (c) to show that the four idempotent elements $e(i_1i_2)e(i_2i_3), \ldots, e(-i_1i_2)e(-i_2i_3)$ are singular elements in $\mathbb{C}_3$.

(e) Show that the matrices in (b) and (c) are idempotent matrices.

46.8 (a) Show that $e(i_1i_2)e(i_2i_3)$, in the $\mathbb{C}_1$-column representation of Table 46.1 is $[(1/4)+i_2(i_1/4)]+i_3[(-i_1/4)+i_2(1/4)]$, and that the associated Cauchy–Riemann matrix is

$$\begin{bmatrix} \frac{1}{4} & -\frac{i_1}{4} & \frac{i_1}{4} & \frac{1}{4} \\ \frac{i_1}{4} & \frac{1}{4} & -\frac{1}{4} & \frac{i_1}{4} \\ -\frac{i_1}{4} & -\frac{1}{4} & \frac{1}{4} & -\frac{i_1}{4} \\ \frac{1}{4} & -\frac{i_1}{4} & \frac{i_1}{4} & \frac{1}{4} \end{bmatrix}.$$

(b) Show that the square of the absolute value of the determinant of the matrix in (a) equals the determinant of the matrix in Exercise 46.7 (b).

46.9 Show that $z_1+i_2z_2 \in \mathcal{O}_2$ if and only if $|z_1+i_2z_2|_2^2 \in \mathcal{O}_1$. [*Hint.* $z_1+i_2z_2=(z_1-i_1z_2)e(i_1i_2)+(z_1+i_1z_2)e(-i_1i_2)$.]

46.10 Consider equation (70) in Theorem 46.22. Show that $\det M(\zeta) \in \mathcal{O}_2$ if and only if $\det M(z) \in \mathcal{O}_1$.

47. POWER SERIES AND HOLOMORPHIC FUNCTIONS IN $\mathbb{C}_n$

This book has treated holomorphic functions on $\mathbb{C}_2$ both by power series (Chapter 2) and by integrals (Chapter 4). The remainder of this chapter treats holomorphic functions on $\mathbb{C}_n$; the methods employed are an extension of those used with functions of a bicomplex variable. The present section defines holomorphic functions on $\mathbb{C}_n$, $n \geqslant 1$, and uses power series to establish their existence and basic properties. Proofs by induction and multiple representations are characteristic features of the subject.

47.1 DEFINITION Let γ_k, $k=0, 1, \ldots$, be constants in $\mathbb{C}_n$, and let ζ and ζ_0 denote elements in $\mathbb{C}_n$. Then a power series about ζ_0 in $\mathbb{C}_n$ is an infinite series of the form

(1) $$\sum_{k=0}^{\infty} \gamma_k(\zeta - \zeta_0)^k.$$

The study of the convergence of (1) is based on the idempotent representation of the infinite series. In order to state the relevant theorem, it is necessary to represent (1) in the following notation:

(2) $$\sum_{k=0}^{\infty} (c_k + i_n d_k)[(\zeta_1 + i_n\zeta_2) - (a_1 + i_n a_2)]^k.$$

Now by Theorem 45.2,

(3) $$c_k + i_n d_k = (c_k - i_{n-1}d_k)e(i_{n-1}i_n) + (c_k + i_{n-1}d_k)e(-i_{n-1}i_n),$$

(4) $$\zeta_1 + i_n\zeta_2 = (\zeta_1 - i_{n-1}\zeta_2)e(i_{n-1}i_n) + (\zeta_1 + i_{n-1}\zeta_2)e(-i_{n-1}i_n),$$

(5) $$a_1 + i_n a_2 = (a_1 - i_{n-1}a_2)e(i_{n-1}i_n) + (a_1 + i_{n-1}a_2)e(-i_{n-1}i_n).$$

Then Theorem 45.6 shows that the infinite series in (2) has the following idempotent representation:

(6) $$\sum_{k=0}^{\infty} (c_k - i_{n-1}d_k)[(\zeta_1 - i_{n-1}\zeta_2) - (a_1 - i_{n-1}a_2)]^k e(i_{n-1}i_n)$$

(7) $$+ \sum_{k=0}^{\infty} (c_k + i_{n-1}d_k)[(\zeta_1 + i_{n-1}\zeta_2) - (a_1 + i_{n-1}a_2)]^k e(-i_{n-1}i_n).$$

47.2 THEOREM If the infinite series (2) converges at $\zeta_1 + i_n\zeta_2$, $n \geqslant 2$, then the infinite series in (6) and (7) converge at $\zeta_1 - i_{n-1}\zeta_2$ and $\zeta_1 + i_{n-1}\zeta_2$, respectively, and (2) equals the expression in (6) and (7). If the series in (6) and (7) converge at $\zeta_1 - i_{n-1}\zeta_2$ and $\zeta_1 + i_{n-1}\zeta_2$, respectively, then the series in (2) converges at $\zeta_1 + i_n\zeta_2$ and equals the expression in (6) and (7).

Proof. First, assume that (2) converges, and let S, in $\mathbb{C}_n$, denote its sum. Let S_r, T_r, and U_r denote the partial sums of the series in (2), (6), and (7), respectively. Since

$$(8)\qquad \sum_{k=0}^{r} (c_k + i_n d_k)[(\zeta_1 + i_n\zeta_2) - (a_1 + i_n a_2)]^k$$

$$= \sum_{k=0}^{r} (c_k - i_{n-1} d_k)[(\zeta_1 - i_{n-1}\zeta_2) - (a_1 - i_{n-1} a_2)]^k e(i_{n-1} i_n)$$

$$+ \sum_{k=0}^{r} (c_k + i_{n-1} d_k)[(\zeta_1 + i_{n-1}\zeta_2) - (a_1 + i_{n-1} a_2)]^k e(-i_{n-1} i_n)$$

by Theorem 45.6, then

$$(9)\qquad S_r = T_r e(i_{n-1} i_n) + U_r e(-i_{n-1} i_n).$$

Define elements T and U by the following equation:

$$(10)\qquad S = Te(i_{n-1} i_n) + Ue(-i_{n-1} i_n).$$

Then Theorem 45.4 (12) shows that

$$(11)\qquad \|S_r - S\|_n = \left[\frac{\|T_r - T\|_{n-1}^2 + \|U_r - U\|_{n-1}^2}{2}\right]^{1/2}.$$

Since by hypothesis,

$$(12)\qquad \lim_{r\to\infty} S_r = S,$$

then (11) shows that

$$(13)\qquad \lim_{r\to\infty} T_r = T, \qquad \lim_{r\to\infty} U_r = U.$$

Therefore, the infinite series in (6) and (7) converge at $\zeta_1 - i_{n-1}\zeta_2$ and $\zeta_1 + i_{n-1}\zeta_2$, respectively, and (2) equals the expression in (6) and (7). Thus the proof of the first part of the theorem is complete. Next, assume that the series in (6) and (7) converge; that is, assume that (13) holds. Define S by (10). Then (11) shows that (12) is true; that is, the series in (2) converges, and (10) holds as before. □

47.3 COROLLARY The series in (6) and (7) are infinite series in $\zeta_1 - i_{n-1}\zeta_2$ and $\zeta_1 + i_{n-1}\zeta_2$ in $\mathbb{C}_{n-1}$, $n \geq 2$. By Theorem 47.2 each of these series has a representation in terms of power series in $\mathbb{C}_{n-2}$ if $n > 2$. This process can be repeated until finally the series in (2) is represented by means of 2^{n-1} infinite series in $\mathbb{C}_1$.

Earlier chapters of this book have treated holomorphic functions of a complex variable z and holomorphic functions of a bicomplex variable

$z_1+i_2z_2$ (see Section 15). The following definition extends the earlier definition (see Definition 15.2) to include holomorphic functions on $\mathbb{C}_n$, $n\geqslant 3$.

47.4 DEFINITION A holomorphic function f of $\zeta_1+i_n\zeta_2$ in $\mathbb{C}_n$, $n\geqslant 1$, is a function with the following properties: (a) f is defined on a domain X in $\mathbb{C}_n$, and $f(\zeta_1+i_n\zeta_2)\in\mathbb{C}_n$ for each $\zeta_1+i_n\zeta_2$ in X; and (b) for each $a_1+i_na_2$ in X there exists a domain D, which contains $a_1+i_na_2$, and a power series such that

$$f(\zeta_1+i_n\zeta_2)=\sum_{k=0}^{\infty}(c_k+i_nd_k)[(\zeta_1+i_n\zeta_2)-(a_1+i_na_2)]^k \tag{14}$$

for all $\zeta_1+i_n\zeta_2$ in D.

The next goal is to establish the existence of holomorphic functions in $\mathbb{C}_n$ for all $n\geqslant 2$, but a lemma is required first. Let X be a domain in $\mathbb{C}_n$, that is, a set which is open and connected (see Section 15). Define sets X_1 and X_2 as follows:

$$X_1=\{\zeta_1-i_{n-1}\zeta_2 \text{ in } \mathbb{C}_{n-1}\colon \zeta_1+i_n\zeta_2\in X\}, \tag{15}$$

$$X_2=\{\zeta_1+i_{n-1}\zeta_2 \text{ in } \mathbb{C}_{n-1}\colon \zeta_1+i_n\zeta_2\in X\}. \tag{16}$$

47.5 LEMMA If X is a domain in $\mathbb{C}_n$, $n\geqslant 2$, then the sets X_1 and X_2 are domains.

Proof. The proof is similar to the proofs of Theorems 8.7 and 8.11. To prove the lemma, it is necessary to show that X_1 and X_2 are open and connected. First, show that each point in X_1 has a neighborhood in X_1, and that each point in X_2 has a neighborhood in X_2. Let w_1^0 be a point in X_1; then there is some point $a+i_nb$ in X such that $w_1^0=a-i_{n-1}b$. Also, $a+i_{n-1}b$ is a point w_2^0 in X_2. Since X is open, there is a neighborhood $N(a+i_nb,\varepsilon)$ which is contained in X. The proof will show that $N(a-i_{n-1}b,\varepsilon)\subset X_1$ and $N(a+i_{n-1}b,\varepsilon)\subset X_2$. Let w_1 and w_2 be arbitrary points such that

$$w_1\in N(a-i_{n-1}b,\varepsilon),\qquad w_2\in N(a+i_{n-1}b,\varepsilon). \tag{17}$$

The proof will show that $w_1\in X_1$ and $w_2\in X_2$. Because of (17),

$$\|w_1-(a-i_{n-1}b)\|_{n-1}<\varepsilon,\qquad \|w_2-(a+i_{n-1}b)\|_{n-1}<\varepsilon. \tag{18}$$

The points w_1 and w_2 determine a unique point $\zeta_1+i_n\zeta_2$ such that

$$\begin{gathered}\zeta_1+i_n\zeta_2=w_1e(i_{n-1}i_n)+w_2e(-i_{n-1}i_n),\\ w_1=\zeta_1-i_{n-1}\zeta_2,\qquad w_2=\zeta_1+i_{n-1}\zeta_2.\end{gathered} \tag{19}$$

The proof will show that $\zeta_1+i_n\zeta_2$, a point in $\mathbb{C}_n$, is also in X. By Theorem 45.4 (12),

$$(20)\quad \|(\zeta_1+i_n\zeta_2)-(a+i_nb)\|_n$$
$$=(1/\sqrt{2})[\|(\zeta_1-i_{n-1}\zeta_2)-(a-i_{n-1}b)\|_{n-1}^2$$
$$+\|(\zeta_1+i_{n-1}\zeta_2)-(a+i_{n-1}b)\|_{n-1}^2]^{1/2}$$
$$<\left(\frac{\varepsilon^2+\varepsilon^2}{2}\right)^{1/2}=\varepsilon.$$

Therefore, $\zeta_1+i_n\zeta_2$ is in $N(a+i_nb,\varepsilon)$, which is in X. Thus w_1 (which is $\zeta_1-i_{n-1}\zeta_2$) is in X_1, and w_2 (which is $\zeta_1+i_{n-1}\zeta_2$) is in X_2. Therefore, $N(a-i_{n-1}b,\varepsilon)\subset X_1$, and X_1 is an open set. A similar proof can be given to show that there is a neighborhood $N(c+i_{n-1}d,\delta)$ of an arbitrary point $c+i_{n-1}d$ in X_2 which is in X_2. Thus the proof that X_1 and X_2 are open sets is complete.

The proof of Lemma 47.4 will be completed by showing that X_1 and X_2 are connected (compare the proof of Theorem 8.11). If u and v are two points in X_1, then there are points $u_1+i_nu_2$ and $v_1+i_nv_2$ in X such that $u=u_1-i_{n-1}u_2$ and $v=v_1-i_{n-1}v_2$. Since X is connected by hypothesis, there is a polygonal curve in X which connects $u_1+i_nu_2$ and $v_1+i_nv_2$. Each line segment of this polygonal curve corresponds to a line segment in X_1 and a line segment in X_2. For example, if the segment $t(u_1+i_nu_2)+(1-t)(v_1+i_nv_2)$, $0\leqslant t\leqslant 1$, is in X, then the segment $t(u_1-i_{n-1}u_2)+(1-t)(v_1-i_{n-1}v_2)$, $0\leqslant t\leqslant 1$, which connects $u_1-i_{n-1}u_2$ and $v_1-i_{n-1}v_2$, is in X_1; and the line segment $t(u_1+i_{n-1}u_2)+(1-t)(v_1+i_{n-1}v_2)$, $0\leqslant t\leqslant 1$, which connects $u_1+i_{n-1}u_2$ and $v_1+i_{n-1}v_2$, is in X_2.

Since X_1 and X_2 are both open and connected, they are domains by definition. The proof of Lemma 47.5 is complete. □

47.6 THEOREM Let X be a domain in $\mathbb{C}_n$, $n\geqslant 1$. Then there exists a function $f:X\to\mathbb{C}_n$ which is holomorphic on X.

Proof. The proof is by induction on n. If $n=1$, then the theory of functions of a complex variable shows that there exists a function $f:X\to\mathbb{C}_1$ which is holomorphic on the domain X in $\mathbb{C}_1$. Thus the theorem is true for $n=1$. Also, the theorem is true for $n=2$ by Theorem 15.3. Next, assume that the theorem is true for $n-1$; that is, assume that there exists a holomorphic function on every domain in $\mathbb{C}_{n-1}$. The domain X in $\mathbb{C}_n$ is given. As shown in Lemma 47.5, the sets X_1 and X_2 are domains in $\mathbb{C}_{n-1}$. By the induction hypothesis, there exist holomorphic functions $f_1:X_1\to\mathbb{C}_{n-1}$ and $f_2:X_2\to\mathbb{C}_{n-1}$. Define a function $f:X\to\mathbb{C}_n$ as follows:

$$(21)\quad f(\zeta_1+i_n\zeta_2)=f_1(\zeta_1-i_{n-1}\zeta_2)e(i_{n-1}i_n)+f_2(\zeta_1+i_{n-1}\zeta_2)e(-i_{n-1}i_n).$$

In this equation, $\zeta_1+i_n\zeta_2$ is an arbitrary point in X, and $\zeta_1-i_{n-1}\zeta_2$ and $\zeta_1+i_{n-1}\zeta_2$ are the corresponding points in X_1 and X_2, respectively. Thus the function $f:X\to\mathbb{C}_n$ is effectively defined by (21). The proof will show that $f:X\to\mathbb{C}_n$ is holomorphic on X. In order to do so, it is necessary to show that f can be represented by a power series in a neighborhood of each point in X.

Let $a_1+i_nb_2$ be a point in X. Then $a_1-i_{n-1}a_2$ is in X_1 and $a_1+i_{n-1}a_2$ is in X_2. Since $f_1:X_1\to\mathbb{C}_{n-1}$ is holomorphic in X_1, there is a neighborhood $N(a_1-i_{n-1}a_2,r_1)$ in X_1 in which f_1 is represented by a power series

$$(22)\quad f_1(\eta)=\sum_{k=0}^{\infty}\alpha_k[\eta-(a_1-i_{n-1}a_2)]^k,\qquad \eta\in N(a_1-i_{n-1}a_2,r_1).$$

Likewise, since $f_2:X_2\to\mathbb{C}_{n-1}$ is holomorphic in X_2, there is a neighborhood $N(a_1+i_{n-1}a_2,r_2)$ in X_2 in which f_2 is represented by a power series

$$(23)\quad f_2(\eta)=\sum_{k=0}^{\infty}\beta_k[\eta-(a_1+i_{n-1}a_2)]^k,\qquad \eta\in N(a_1+i_{n-1}a_2,r_2).$$

Let

$$(24)\qquad r=\min(r_1,r_2).$$

Now since X is an open set, there is a neighborhood $N(a_1+i_na_2,\delta)$ of $a_1+i_na_2$ which is contained in X. Let

$$(25)\qquad R=\min(\delta,r/\sqrt{2}).$$

Next, let ζ: $\zeta_1+i_n\zeta_2$ be a point in $N(a_1+i_na_2,R)$; then

$$(26)\qquad \|\zeta-(a_1+i_na_2)\|_n<R.$$

Since

$$(27)\qquad \begin{aligned}&\|\zeta-(a_1+i_na_2)\|_n\\&\quad=(1/\sqrt{2})[\|(\zeta_1-i_{n-1}\zeta_2)-(a_1-i_{n-1}a_2)\|_{n-1}^2\\&\qquad+\|(\zeta_1+i_{n-1}\zeta_2)-(a_1+i_{n-1}a_2)\|_{n-1}^2]^{1/2},\end{aligned}$$

then

$$(28)\qquad \begin{aligned}&\|(\zeta_1-i_{n-1}\zeta_2)-(a_1-i_{n-1}a_2)\|_{n-1}<r,\\&\|(\zeta_1+i_{n-1}\zeta_2)-(a_1+i_{n-1}a_2)\|_{n-1}<r,\end{aligned}$$

and $\zeta_1-i_{n-1}\zeta_2$ and $\zeta_1+i_{n-1}\zeta_2$ are in $N(a_1-i_{n-1}a_2,r)$ and $N(a_1+i_{n-1}a_2,r)$, respectively. Thus by (22) and (23),

$$(29)\qquad f_1(\zeta_1-i_{n-1}\zeta_2)=\sum_{k=0}^{\infty}\alpha_k[(\zeta_1-i_{n-1}\zeta_2)-(a_1-i_{n-1}a_2)]^k,$$

$$(30)\qquad f_2(\zeta_1+i_{n-1}\zeta_2)=\sum_{k=0}^{\infty}\beta_k[(\zeta_1+i_{n-1}\zeta_2)-(a_1+i_{n-1}a_2)]^k.$$

Therefore, by (21),

$$(31) \qquad f(\zeta_1+i_n\zeta_2)=\sum_{k=0}^{\infty}\alpha_k[(\zeta_1-i_{n-1}\zeta_2)-(a_1-i_{n-1}a_2)]^k e(i_{n-1}i_n)$$
$$+\sum_{k=0}^{\infty}\beta_k[(\zeta_1+i_{n-1}\zeta_2)-(a_1+i_{n-1}a_2)]^k e(-i_{n-1}i_n),$$

for every $\zeta_1+i_n\zeta_2$ in $N(a_1+i_na_2, R)$. The proof will be complete by showing that the expression on the right in (31) represents a power series in $[(\zeta_1+i_n\zeta_2)-(a_1-i_na_2)]$. Now α_k and β_k are elements in $\mathbb{C}_{n-1}$; define new elements c_k and d_k in $\mathbb{C}_{n-1}$ by the following equations:

$$(32) \qquad c_k - i_{n-1}d_k = \alpha_k, \qquad c_k + i_{n-1}d_k = \beta_k, \qquad k = 0, 1, 2, \ldots.$$

These equations have the following solution:

$$(33) \qquad c_k = \frac{\alpha_k+\beta_k}{2}, \qquad d_k = \frac{i_{n-1}(\alpha_k-\beta_k)}{2}, \qquad k = 0, 1, 2, \ldots.$$

Then (32) shows that (31) can be written in the following form:

$$(34) \qquad f(\zeta_1+i_n\zeta_2)$$
$$=\sum_{k=0}^{\infty}(c_k-i_{n-1}d_k)[(\zeta_1-i_{n-1}\zeta_2)-(a_1-i_{n-1}a_2)]^k e(i_{n-1}i_n)$$
$$+\sum_{k=0}^{\infty}(c_k+i_{n-1}d_k)[(\zeta_1+i_{n-1}\zeta_2)-(a_1+i_{n-1}a_2)]^k e(-i_{n-1}i_n).$$

By Theorem 47.2, the expression on the right in (34) is equal to a single power series, and

$$(35) \qquad f(\zeta_1 + i_n\zeta_2) = \sum_{k=0}^{\infty}(c_k + i_nd_k)[(\zeta_1 + i_n\zeta_2) - (a_1 + i_na_2)]^k$$

for every $\zeta_1+i_n\zeta_2$ in $N(a_1+i_na_2, R)$. The following summarizes the proof: f_1 and f_2 are holomorphic functions which are represented in $N(a_1-i_{n-1}a_2, r)$ and $N(a_1+i_{n-1}a_2, r)$ by the series in (29) and (30). If $\zeta_1+i_n\zeta_2$ is in $N(a_1+i_na_2, R)$, then $\zeta_1-i_{n-1}\zeta_2$ is in $N(a_1-i_{n-1}a_2, r)$ and $\zeta_1+i_{n-1}\zeta_2$ is in $N(a_1+i_{n-1}a_2, r)$. Therefore, (29) and (30) are true, and the function f defined by (21) is represented as in (34) and, by Theorem 47.2, as in (35). Finally, $f: X\to\mathbb{C}_n$ is a holomorphic function by Definition 47.4. As stated, Theorem 47.6 is true for $n=1$ and $n=2$; also, if it is true for $n-1$, it is true for n. A complete induction shows that Theorem 47.6 is true for all values of n, and the proof is complete. □

Since there are holomorphic functions on domains in $\mathbb{C}_1$, the following corollary suggests the great abundance of holomorphic functions in $\mathbb{C}_n$ (compare Theorem 15.3).

47.7 COROLLARY Let X_1 and X_2 be domains in $\mathbb{C}_{n-1}$, and let $f_1 : X_1 \to \mathbb{C}_{n-1}$ and $f_2 : X_2 \to \mathbb{C}_{n-1}$ be holomorphic functions. If X is the set of points $\zeta_1 + i_n\zeta_2$ such that

$$(36) \qquad \begin{aligned} &\zeta_1 + i_n\zeta_2 = (\zeta_1 - i_{n-1}\zeta_2)e(i_{n-1}i_n) + (\zeta_1 + i_{n-1}\zeta_2)e(-i_{n-1}i_n), \\ &\zeta_1 - i_{n-1}\zeta_2 \in X_1,\ \zeta_1 + i_{n-1}\zeta_2 \in X_2, \end{aligned}$$

and if f is the function defined by the statement

$$(37) \qquad \begin{aligned} &f(\zeta_1 + i_n\zeta_2) = f_1(\zeta_1 - i_{n-1}\zeta_2)e(i_{n-1}i_n) + f_2(\zeta_1 + i_{n-1}\zeta_2)e(-i_{n-1}i_n), \\ &\zeta_1 - i_{n-1}\zeta_2 \in X_1,\ \zeta_1 + i_{n-1}\zeta_2 \in X_2, \end{aligned}$$

then $f : X \to \mathbb{C}_n$ is a holomorphic function.

Proof. The following lemma contains the proof that the set X defined by (36) is a domain. Then the proof that the function $f : X \to \mathbb{C}_n$ defined by (37) is holomorphic is contained in the proof of Theorem 47.6. □

47.8 LEMMA Let X_1 and X_2 be domains in $\mathbb{C}_{n-1}$, $n \geqslant 2$, and let X be the set of points $\zeta_1 + i_n\zeta_2$ defined in (36). Then X is a domain in $\mathbb{C}_n$.

Proof. The proof is similar to the proofs of Theorems 8.9 and 8.12. To establish the lemma, it is necessary to show that X is open and connected. Let $a + i_nb$ be a point in X. Then there are points $a - i_{n-1}b$ in X_1 and $a + i_{n-1}b$ in X_2 such that

$$(38) \qquad a + i_nb = (a - i_{n-1}b)e(i_{n-1}i_n) + (a + i_{n-1}b)e(-i_{n-1}i_n).$$

Since X_1 and X_2 are open by hypothesis, there are neighborhoods $N(a - i_{n-1}b, r)$ in X_1 and $N(a + i_{n-1}b, r)$ in X_2. The proof will show that $N(a + i_nb, r/\sqrt{2})$ is in X, and that X is therefore open. Let $\zeta_1 + i_n\zeta_2$ be a point in $N(a + i_nb, r/\sqrt{2})$. Then there are points $\zeta_1 - i_{n-1}\zeta_2$ and $\zeta_1 + i_{n-1}\zeta_2$ in $\mathbb{C}_{n-1}$ such that

$$(39) \qquad \zeta_1 + i_n\zeta_2 = (\zeta_1 - i_{n-1}\zeta_2)e(i_{n-1}i_n) + (\zeta_1 + i_{n-1}\zeta_2)e(-i_{n-1}i_n).$$

Now by Theorem 45.4 (12),

$$(40) \qquad \begin{aligned} &\|(\zeta_1 + i_n\zeta_2) - (a + i_nb)\|_n \\ &\quad = (1/\sqrt{2})[\|(\zeta_1 - i_{n-1}\zeta_2) - (a - i_{n-1}b)\|_{n-1}^2 + \|(\zeta_1 + i_{n-1}\zeta_2) \\ &\qquad - (a + i_{n-1}b)\|_{n-1}^2]^{1/2}. \end{aligned}$$

Since the left side of this equation is less than $r/\sqrt{2}$, then

$$(41) \qquad \begin{aligned} &\|(\zeta_1 - i_{n-1}\zeta_2) - (a - i_{n-1}b)\|_{n-1} < r, \\ &\|(\zeta_1 + i_{n-1}\zeta_2) - (a + i_{n-1}b)\|_{n-1} < r. \end{aligned}$$

Thus $(\zeta_1 - i_{n-1}\zeta_2) \in X_1$ and $(\zeta_1 + i_{n-1}\zeta_2) \in X_2$, and (36) and (39) show that $(\zeta_1 + i_n\zeta_2)$ is in X. Therefore X is an open set. The proof that X is connected is similar to the proof of Theorem 8.12. Therefore, X is a domain since it is open and connected. □

Corollary 47.7 shows that a holomorphic function $f: X \to \mathbb{C}_n$, $X \subset \mathbb{C}_n$, can be constructed from each pair of holomorphic functions $f_1: X_1 \to \mathbb{C}_{n-1}$, $X_1 \subset \mathbb{C}_{n-1}$, and $f_2: X_2 \to \mathbb{C}_{n-1}$, $X_2 \subset \mathbb{C}_{n-1}$. Furthermore,

$$\text{(42)} \quad f(\zeta_1 + i_n\zeta_2) = f_1(\zeta_1 - i_{n-1}\zeta_2)e(i_{n-1}i_n) + f_2(\zeta_1 + i_{n-1}\zeta_2)e(-i_{n-1}i_n),$$
$$\zeta_1 - i_{n-1}\zeta_2 \in X_1,\ \zeta_1 + i_{n-1}\zeta_2 \in X_2.$$

The next theorem shows that every holomorphic function can be constructed in this manner.

47.9 THEOREM Let X be a domain in $\mathbb{C}_n$, $n \geqslant 2$, and let $f: X \to \mathbb{C}_n$ be a holomorphic function. Then there exist domains X_1 and X_2 in $\mathbb{C}_{n-1}$ and holomorphic functions $f_1: X_1 \to \mathbb{C}_{n-1}$, $f_2: X_2 \to \mathbb{C}_{n-1}$ such that (42) is true.

Proof. Define sets X_1 and X_2 as in (15) and (16). Then Lemma 47.5 shows that X_1 and X_2 are domains. Let $a_1 + i_n a_2$ be a point in X. Then $a_1 - i_{n-1}a_2$ and $a_1 + i_{n-1}a_2$ are in X_1 and X_2, respectively. Since f is holomorphic on X, there is a power series such that

$$\text{(43)} \quad f(\zeta_1 + i_n\zeta_2) = \sum_{k=0}^{\infty} (c_k + i_n d_k)[(\zeta_1 + i_n\zeta_2) - (a_1 + i_n a_2)]^k$$

for all $\zeta_1 + i_n\zeta_2$ in some neighborhood of $a_1 + i_n a_2$. Now $f(\zeta_1 + i_n\zeta_2)$ is an element $h_1(\zeta_1 + i_n\zeta_2) + i_n h_2(\zeta_1 + i_n\zeta_2)$ in $\mathbb{C}_n$ for each $\zeta_1 + i_n\zeta_2$ in X; therefore,

$$\text{(44)} \quad f(\zeta_1 + i_n\zeta_2) = [h_1(\zeta_1 + i_n\zeta_2) - i_{n-1}h_2(\zeta_1 + i_n\zeta_2)]e(i_{n-1}i_n)$$
$$+ [h_1(\zeta_1 + i_n\zeta_2) + i_{n-1}h_2(\zeta_1 + i_n\zeta_2)]e(-i_{n-1}i_n).$$

The proof will be completed by showing that there are holomorphic functions $f_1: X_1 \to \mathbb{C}_{n-1}$ and $f_2: X_2 \to \mathbb{C}_{n-1}$ such that

$$\text{(45)} \quad f_1(\zeta_1 - i_{n-1}\zeta_2) = h_1(\zeta_1 + i_n\zeta_2) - i_{n-1}h_2(\zeta_1 + i_n\zeta_2),$$
$$\zeta_1 + i_n\zeta_2 \in X,$$
$$f_2(\zeta_1 + i_{n-1}\zeta_2) = h_1(\zeta_1 + i_n\zeta_2) + i_{n-1}h_2(\zeta_1 + i_n\zeta_2),$$
$$\zeta_1 + i_n\zeta_2 \in X.$$

Now Theorem 47.2 shows that, if the series in (43) and (2) converge, then the series in (6) and (7) converge; therefore, in the region of convergence of (43),

(46) $$h_1(\zeta_1 + i_n\zeta_2) - i_{n-1}h_2(\zeta_1 + i_n\zeta_2)$$
$$= \sum_{k=0}^{\infty} (c_k - i_{n-1}d_k)[(\zeta_1 - i_{n-1}\zeta_2) - (a_1 - i_{n-1}a_2)]^k,$$

(47) $$h_1(\zeta_1 + i_n\zeta_2) + i_{n-1}h_2(\zeta_1 + i_n\zeta_2)$$
$$= \sum_{k=0}^{\infty} (c_k + i_{n-1}d_k)[(\zeta_1 + i_{n-1}\zeta_2) - (a_1 + i_{n-1}a_2)]^k.$$

There is a series such as the one in (43) for each point $a_1 + i_na_2$ in X, and there is a series such as the one in (46) for each point $a_1 - i_{n-1}a_2$ in X_1 and a series such as the one in (47) for each point $a_1 + i_{n-1}a_2$ in X_2. At first glance it is not clear that the collection of series (46) defines a function $f_1: X_1 \to \mathbb{C}_{n-1}$ and that the collection of series (47) defines a function $f_2: X_2 \to \mathbb{C}_{n-1}$. However, all series (43) which converge at $\zeta_1 + i_n\zeta_2$ converge to the value $f(\zeta_1 + i_n\zeta_2)$, which is $h_1(\zeta_1 + i_n\zeta_2) + i_nh_2(\zeta_1 + i_n\zeta_2)$. Correspondingly, all series (46) which converge at $\zeta_1 - i_{n-1}\zeta_2$ converge to the same value $h_1(\zeta_1 + i_n\zeta_2) - i_{n-1}h_2(\zeta_1 + i_n\zeta_2)$. Thus the infinite series (46) define a function $f_1: X_1 \to \mathbb{C}_{n-1}$ whose value at $\zeta_1 - i_{n-1}\zeta_2$ is given by (45). The value of f_1 at any point is given by the sum of any one of the series (46) which converges at the point; thus, in the region of convergence of the infinite series,

(48) $$f_1(\zeta_1 - i_{n-1}\zeta_2) = \sum_{k=0}^{\infty} (c_k - i_{n-1}d_k)[(\zeta_1 - i_{n-1}\zeta_2) - (a_1 - i_{n-1}a_2)]^k.$$

Since f_1 is defined on the domain X_1, and since f_1 is represented by a power series in the neighborhood of each $a_1 - i_{n-1}a_2$ in X_1, f_1 is a holomorphic function of $\zeta_1 - i_{n-1}\zeta_2$ by Definition 47.4. In the same way, the infinite series (47) define a function $f_2: X_2 \to \mathbb{C}_{n-1}$ whose value at $\zeta_1 + i_{n-1}\zeta_2$ is given by (45), and

(49) $$f_2(\zeta_1 + i_{n-1}\zeta_2) = \sum_{k=0}^{\infty} (c_k + i_{n-1}d_k)[(\zeta_1 + i_{n-1}\zeta_2) - (a_1 + i_{n-1}a_2)]^k$$

in a neighborhood of $a_1 + i_{n-1}a_2$. Thus f_2 is a holomorphic function which is defined on the domain X_2 in $\mathbb{C}_{n-1}$. Finally, (44) and (45) show that the functions $f_1: X_1 \to \mathbb{C}_{n-1}$ and $f_2: X_2 \to \mathbb{C}_{n-1}$ satisfy (42), and the proof of Theorem 47.9 is complete. □

47.10 COROLLARY Let X be a domain in $\mathbb{C}_n$, $n \geqslant 2$, and let $f: X \to \mathbb{C}_n$ be holomorphic on X. Then there exist domains X_1 and X_2 in $\mathbb{C}_{n-1}$ and holomorphic functions $f_1: X_1 \to \mathbb{C}_{n-1}$ and $f_2: X_2 \to \mathbb{C}_{n-1}$ such that

(50) $$f(\zeta_1 + i_n\zeta_2) = f_1(\zeta_1 - i_{n-1}\zeta_2)e(i_{n-1}i_n) + f_2(\zeta_1 + i_{n-1}\zeta_2)e(-i_{n-1}i_n),$$
$$\zeta_1 + i_n\zeta_2 \in X,\ \zeta_1 - i_{n-1}\zeta_2 \in X_1,\ \zeta_1 + i_{n-1}\zeta_2 \in X_2.$$

If $n-1\geq 2$, then Theorem 47.9 can be used again to give similar representations for each of the functions f_1 and f_2. This series of representations of f can be continued to obtain a total of $n-1$ representations of f. In the last of these, f is represented by means of 2^{n-1} holomorphic functions $g_t: W_t\to\mathbb{C}_1$, $t=1,\ldots,2^{n-1}$, where each W_t is a domain in $\mathbb{C}_1$.

47.11 EXAMPLE The purpose of this example is to obtain the two representations of the holomorphic function $f: X\to\mathbb{C}_3$, $X\subset\mathbb{C}_3$, which exist by Corollary 47.10. The first of these representations is described in the corollary itself. The sets X_1 and X_2, defined as in (15) and (16), are domains by Lemma 47.5. Then by Theorem 47.9 there exist holomorphic functions $f_1: X_1\to\mathbb{C}_2$ and $f_2: X_2\to\mathbb{C}_2$ such that

$$(51)\qquad \begin{aligned} &f(\zeta_1+i_3\zeta_2)=f_1(\zeta_1-i_2\zeta_2)e(i_2i_3)+f_2(\zeta_1+i_2\zeta_2)e(-i_2i_3),\\ &\zeta_1+i_3\zeta_2\in X\subset\mathbb{C}_3,\ \zeta_1-i_2\zeta_2\in X_1\subset\mathbb{C}_2,\ \zeta_1+i_2\zeta_2\in X_2\subset\mathbb{C}_2.\end{aligned}$$

The second representation of f is obtained by using Theorem 47.9 to obtain similar representations of f_1 and f_2. Since $\zeta_1+i_3\zeta_2=(z_1+i_2z_2)+i_3(z_3+i_2z_4)$, then (51) can be written as follows:

$$(52)\qquad \begin{aligned} &f[(z_1+i_2z_2)+i_3(z_3+i_2z_4)]\\ &\quad=f_1[(z_1+z_4)+i_2(z_2-z_3)]e(i_2i_3)\\ &\qquad+f_2[(z_1-z_4)+i_2(z_2+z_3)]e(-i_2i_3),\ (z_1+i_2z_2)+i_3(z_3+i_2z_4)\in X,\\ &(z_1+z_4)+i_2(z_2-z_3)\in X_1,\ (z_1-z_4)+i_2(z_2+z_3)\in X_2.\end{aligned}$$

Define sets $Y_1,\ldots,Y_4$ as follows:

$$(53)\qquad \begin{aligned} &Y_1=\{(z_1+z_4)-i_1(z_2-z_3) \text{ in } \mathbb{C}_1: (z_1+z_4)+i_2(z_2-z_3)\in X_1\},\\ &Y_2=\{(z_1+z_4)+i_1(z_2-z_3) \text{ in } \mathbb{C}_1: (z_1+z_4)+i_2(z_2-z_3)\in X_1\},\\ &Y_3=\{(z_1-z_4)-i_1(z_2+z_3) \text{ in } \mathbb{C}_1: (z_1-z_4)+i_2(z_2+z_3)\in X_2\},\\ &Y_4=\{(z_1-z_4)+i_1(z_2+z_3) \text{ in } \mathbb{C}_1: (z_1-z_4)+i_2(z_2+z_3)\in X_2\}.\end{aligned}$$

Then by Theorem 47.9 there are functions $g_1: Y_1\to\mathbb{C}_1,\ldots,g_4: Y_4\to\mathbb{C}_1$ such that

$$(54)\qquad \begin{aligned} &f_1[(z_1+z_4)+i_2(z_2-z_3)]\\ &\quad=g_1[(z_1+z_4)-i_1(z_2-z_3)]e(i_1i_2)\\ &\qquad+g_2[(z_1+z_4)+i_1(z_2-z_3)]e(-i_1i_2),\end{aligned}$$

$$(55)\qquad \begin{aligned} &f_2[(z_1-z_4)+i_2(z_2+z_3)]\\ &\quad=g_3[(z_1-z_4)-i_1(z_2+z_3)]e(i_1i_2)\\ &\qquad+g_4[(z_1-z_4)+i_1(z_2+z_3)]e(-i_1i_2).\end{aligned}$$

Substitute from (54) and (55) in (52) to obtain the following second representation of $f: X \to \mathbb{C}_3$:

(56) $$f[(z_1 + i_2 z_2) + i_3(z_3 + i_2 z_4)]$$
$$= g_1[(z_1 + z_4) - i_1(z_2 - z_3)]e(i_1 i_2)e(i_2 i_3)$$
$$+ g_2[(z_1 + z_4) + i_1(z_2 - z_3)]e(-i_1 i_2)e(i_2 i_3)$$
$$+ g_3[(z_1 - z_4) - i_1(z_2 + z_3)]e(i_1 i_2)e(-i_2 i_3)$$
$$+ g_4[(z_1 - z_4) + i_1(z_2 + z_3)]e(-i_1 i_2)e(-i_2 i_3).$$

Observe that (56) represents $f: X \to \mathbb{C}_3$, $X \subset \mathbb{C}_3$, in terms of holomorphic functions of a complex variable. In the same way, repeated application of Theorem 47.9 can be used, as stated in Corollary 47.10, to represent every holomorphic function $f: X \to \mathbb{C}_n$, $X \subset \mathbb{C}_n$, $n \geqslant 2$, in terms of holomorphic functions of a complex variable.

Exercises

47.1 (a) Does every power series $\Sigma_{k=0}^{\infty} \gamma_k(\zeta - \zeta_0)^k$ in $\mathbb{C}_n$, $n \geqslant 1$, represent a holomorphic function? Explain your answer.
(b) Determine the exact set in which a power series such as the one in (a) converges. Does this set always contain a neighborhood of ζ_0? Explain your answer.

47.2 Show that there exist domains of holomorphism in $\mathbb{C}_n$, $n \geqslant 1$; that is, show that there exist domains X and holomorphic functions $f: X \to \mathbb{C}_n$, $X \subset \mathbb{C}_n$, which cannot be continued analytically into a larger domain. [*Hint.* Theorem 15.11.]

47.3 (a) Define the function $\exp: \mathbb{C}_n \to \mathbb{C}_n$ for $n \geqslant 1$ (see Section 17).
(b) Show that exp is a holomorphic function in $\mathbb{C}_n$.
(c) Show that exp is an entire function in $\mathbb{C}_n$; that is, show that the power series which represents exp at any point ζ_0 in $\mathbb{C}_n$ converges and represents exp for every ζ in $\mathbb{C}_n$.
(d) Show that $\exp(\zeta) \notin \mathcal{O}_n$ for every ζ in $\mathbb{C}_n$.

47.4 Define the sine and cosine functions in $\mathbb{C}_n$. Show that these functions are periodic, and find all of their periods. [*Hint.* Theorem 17.19.]

47.5 Prove the following form of Euler's theorem:

$$\exp(i_n \zeta) = \cos \zeta + i_n \sin \zeta, \qquad \zeta \in \mathbb{C}_n.$$

47.6 Prove the following theorem:

$$\exp(\zeta_1 + \zeta_2) = \exp(\zeta_1) \exp(\zeta_2), \qquad \zeta_1 \text{ and } \zeta_2 \text{ in } \mathbb{C}_n.$$

47.7 Investigate the definition, existence, and properties of $\log \zeta$ for ζ in $\mathbb{C}_n$, $n \geqslant 1$.

47.8 (a) Show that the function $f:\mathbb{C}_n\to\mathbb{C}_n$ such that $f(\zeta)=\zeta^m$ is a holomorphic function on $\mathbb{C}_n$ for every integer $m\geqslant 1$.
(b) Obtain the power series representation of f about the point a in $\mathbb{C}_n$ by setting $\zeta=(\zeta-a)+a$.
(c) For $n=2$, 3, and 4 obtain the representations of f described in Corollary 47.10.

47.9 If ζ is a variable in $\mathbb{C}_n$, $n\geqslant 1$, show that every polynomial in ζ with coefficients in $\mathbb{C}_n$ is a holomorphic function.

48. DERIVATIVES OF FUNCTIONS IN $\mathbb{C}_n$

Chapter 3 has treated the derivatives of functions defined on domains in $\mathbb{C}_2$; the purpose of the present section is to extend these results to functions defined on domains in $\mathbb{C}_n$, $n>2$. Thus the section treats derivatives, the strong and weak Stolz conditions and their relation to the existence of the derivative, necessary conditions for the existence of the derivative, and sufficient conditions for differentiability. In spite of important differences, the theorems and their proofs are surprisingly similar to those for functions in $\mathbb{C}_2$.

The definitions of derivatives, strong Stolz condition, and weak Stolz condition for functions $f:X\to\mathbb{C}_n$, $X\subset\mathbb{C}_n$, can be obtained by changing $\mathbb{C}_2$ to $\mathbb{C}_n$ in Definitions 20.13 and 20.14. Because there are no other changes, these definitions will not be repeated here.

48.1 THEOREM If $f:X\to\mathbb{C}_n$, $X\subset\mathbb{C}_n$, $n\geqslant 1$, satisfies the strong Stolz condition at ζ_0 in X, then f is differentiable at ζ_0, and it satisfies the weak Stolz condition at ζ_0.

Proof. If $n=1$, the theorem is true by Theorem 20.12; if $n=2$, it is true by Theorem 20.15. Since f satisfies the strong Stolz condition, then for all $n\geqslant 1$ there exists a constant d (which depends on ζ_0) in $\mathbb{C}_n$ and a function $r(f;\zeta_0,\cdot)$, defined in a neighborhood of ζ_0 and with values in $\mathbb{C}_n$, such that

$$(1)\qquad f(\zeta)-f(\zeta_0)=d(\zeta-\zeta_0)+r(f;\zeta_0,\zeta)(\zeta-\zeta_0),$$

$$(2)\qquad \lim_{\zeta\to\zeta_0} r(f;\zeta_0,\zeta)=0,\qquad r(f;\zeta_0,\zeta_0)=0.$$

Then

$$(3)\qquad \frac{f(\zeta)-f(\zeta_0)}{\zeta-\zeta_0}=d+r(f;\zeta_0,\zeta),\qquad \zeta-\zeta_0\notin\mathcal{O}_n,$$

$$(4)\qquad \lim_{\zeta\to\zeta_0}\frac{f(\zeta)-f(\zeta_0)}{\zeta-\zeta_0}=d,$$

and f has a derivative $D_\zeta f(\zeta_0)$ which is equal to d. By (1),

$$(5)\qquad f(\zeta)-f(\zeta_0)=d(\zeta-\zeta_0)+\left[\frac{r(f;\zeta_0,\zeta)(\zeta-\zeta_0)}{\|\zeta-\zeta_0\|_n}\right]\|\zeta-\zeta_0\|_n.$$

Set $d'=d$ and define $r'(f;\zeta_0,\cdot)$ as follows:

$$(6)\qquad r'(f;\zeta_0,\zeta)=\frac{r(f;\zeta_0,\zeta)(\zeta-\zeta_0)}{\|\zeta-\zeta_0\|_n},\qquad \zeta\neq\zeta_0,$$

$$(7)\qquad r'(f;\zeta_0,\zeta_0)=0.$$

Then (6) and Exercise 44.2 show that

$$(8)\qquad \|r'(f;\zeta_0,\zeta_0)\|_n\leqslant\frac{2^{n-1/2}\|r(f;\zeta_0,\zeta)\|_n\|\zeta-\zeta_0\|_n}{\|\zeta-\zeta_0\|_n}=2^{n-1/2}\|r(f;\zeta_0,\zeta)\|_n.$$

Then by (2), (7), and (8),

$$(9)\qquad \lim_{\zeta\to\zeta_0} r'(f;\zeta_0,\zeta)=0,\qquad r'(f;\zeta_0,\zeta_0)=0.$$

By (5) and (6),

$$(10)\qquad f(\zeta)-f(\zeta_0)=d'(\zeta-\zeta_0)+r'(f;\zeta_0,\zeta)\|\zeta-\zeta_0\|_n,$$

and (9) completes the proof that f satisfies the weak Stolz condition at ζ_0. The proofs of the two conclusions in Theorem 48.1 are complete. □

The purpose of the next theorem is to prove that a function $f:X\to\mathbb{C}_n$, $X\subset\mathbb{C}_n$, which is differentiable in X satisfies the strong Stolz condition in X. This theorem has been proved already for $n=2$ in Theorems 21.1 and 21.2. The proof will be given now for $n=3$. The method employed in this proof is valid in the general case for $n\geqslant 2$, and the notational difficulties are the only reason for not presenting the proof in the general case. Before stating the theorem it is necessary to describe the notation to be employed. Let X be a domain in $\mathbb{C}_3$. If ζ is in X, then

$$(11)\qquad \zeta=(z_1+i_2z_2)+i_3(z_3+i_2z_4),\qquad z_k\in\mathbb{C}_1.$$

Let $f:X\to\mathbb{C}_3$ be a function with derivative $D_\zeta f(\zeta)$ such that

$$(12)\qquad f(\zeta)=[u_1(\zeta)+i_2u_2(\zeta)]+i_3[u_3(\zeta)+i_2u_4(\zeta)],\qquad u_k(\zeta)\in\mathbb{C}_1,$$

$$(13)\qquad D_\zeta f(\zeta)=[d_1(\zeta)+i_2d_2(\zeta)]+i_3[d_3(\zeta)+i_2d_4(\zeta)],\qquad d_k(\zeta)\in\mathbb{C}_1.$$

Set

$$(14)\qquad e_1=e(i_1i_2)e(i_2i_3),\qquad e_2=e(-i_1i_2)e(i_2i_3),$$
$$e_3=e(i_1i_2)e(-i_2i_3),\qquad e_4=e(-i_1i_2)e(-i_2i_3).$$

Then (compare Example 45.7),

$$\zeta = [(z_1 + z_4) - i_1(z_2 - z_3)]e_1 + [(z_1 + z_4) + i_1(z_2 - z_3)]e_2 + [(z_1 - z_4) - i_1(z_2 + z_3)]e_3 + [(z_1 - z_4) + i_1(z_2 + z_3)]e_4. \tag{15}$$

Set

$$Z_1 = (z_1 + z_4) - i_1(z_2 - z_3), \tag{16}$$

$$Z_2 = (z_1 + z_4) + i_1(z_2 - z_3), \tag{17}$$

$$Z_3 = (z_1 - z_4) - i_1(z_2 + z_3), \tag{18}$$

$$Z_4 = (z_1 - z_4) + i_1(z_2 + z_3). \tag{19}$$

Then

$$\zeta = \sum_{k=1}^{4} Z_k e_k. \tag{20}$$

Likewise,

$$f(\zeta) = \sum_{k=1}^{4} U_k(\zeta)e_k, \tag{21}$$

where

$$U_1(\zeta) = [u_1(\zeta) + u_4(\zeta)] - i_1[u_2(\zeta) - u_3(\zeta)], \tag{22}$$

$$U_2(\zeta) = [u_1(\zeta) + u_4(\zeta)] + i_1[u_2(\zeta) - u_3(\zeta)], \tag{23}$$

$$U_3(\zeta) = [u_1(\zeta) - u_4(\zeta)] - i_1[u_2(\zeta) + u_3(\zeta)], \tag{24}$$

$$U_4(\zeta) = [u_1(\zeta) - u_4(\zeta)] + i_1[u_2(\zeta) + u_3(\zeta)]. \tag{25}$$

Next,

$$D_\zeta f(\zeta) = \sum_{k=1}^{4} D_k(\zeta)e_k, \tag{26}$$

where

$$D_1(\zeta) = [d_1(\zeta) + d_4(\zeta)] - i_1[d_2(\zeta) - d_3(\zeta)], \tag{27}$$

$$D_2(\zeta) = [d_1(\zeta) + d_4(\zeta)] + i_1[d_2(\zeta) - d_3(\zeta)], \tag{28}$$

$$D_3(\zeta) = [d_1(\zeta) - d_4(\zeta)] - i_1[d_2(\zeta) + d_3(\zeta)], \tag{29}$$

$$D_4(\zeta) = [d_1(\zeta) - d_4(\zeta)] + i_1[d_2(\zeta) + d_3(\zeta)]. \tag{30}$$

Finally, define sets $X_1, \ldots, X_4$ in $\mathbb{C}_1$ as follows:

$$X_k = \{Z_k \text{ in } \mathbb{C}_1 : \zeta \in X\}, \qquad k = 1, \ldots, 4. \tag{31}$$

48.2 THEOREM Let $f: X \to \mathbb{C}_3$, $X \subset \mathbb{C}_3$, be a function as described above which has a derivative $D_\zeta f(\zeta)$ at each point ζ in X. Then:

(32) $U_k: X \to \mathbb{C}_1$, for $k=1, \ldots, 4$, is a function which has a derivative with respect to Z_k;

(33) $$f(\zeta) = \sum_{k=1}^{4} U_k(\zeta) e_k;$$

(34) $$D_{Z_k} U_k(\zeta) = D_k(\zeta), \qquad \zeta \text{ in } X,\ k = 1, \ldots, 4;$$

(35) $$D_\zeta f(\zeta) = \sum_{k=1}^{4} D_{Z_k} U_k(\zeta) e_k;$$

(36) f satisfies the strong Stolz condition in X.

Proof. Since f has a derivative at ζ_0,

(37) $$\lim_{\zeta \to \zeta_0} \left[\frac{f(\zeta) - f(\zeta_0)}{\zeta - \zeta_0} - D_\zeta f(\zeta_0) \right] = 0, \qquad \zeta - \zeta_0 \notin \mathcal{O}_3.$$

By Corollary 45.3, the expression inside the braces in (37) can be represented as a linear combination of $e_1, \ldots, e_4$. This corollary and the explanation of the notation given above show that

(38) $$\frac{f(\zeta) - f(\zeta_0)}{\zeta - \zeta_0} - D_\zeta f(\zeta_0) = \sum_{k=1}^{4} \left[\frac{U_k(\zeta) - U_k(\zeta_0)}{Z_k - Z_k^0} - D_k(\zeta_0) \right] e_k.$$

Since $\zeta - \zeta_0 \notin \mathcal{O}_3$ by (37), then $Z_k - Z_k^0 \notin \mathcal{O}_1$ by (15)–(19), and the expressions in the braces in (38) are well defined. Next, Corollary 45.5 shows that

(39) $$\left\| \frac{f(\zeta) - f(\zeta_0)}{\zeta - \zeta_0} - D_\zeta f(\zeta_0) \right\|_3 = \left[\frac{1}{2^2} \sum_{k=1}^{4} \left| \frac{U_k(\zeta) - U_k(\zeta_0)}{Z_k - Z_k^0} - D_k(\zeta_0) \right|^2 \right]^{1/2}.$$

Now, with ζ restricted so that $\zeta - \zeta_0 \notin \mathcal{O}_3$, let ζ tend to ζ_0; then (37) shows that

(40) $$\lim_{\zeta \to \zeta_0} \left\| \frac{f(\zeta) - f(\zeta_0)}{\zeta - \zeta_0} - D_\zeta f(\zeta_0) \right\|_3 = 0.$$

Furthermore, $\zeta - \zeta_0 \notin \mathcal{O}_3$ implies that $Z_k - Z_k^0 \notin \mathcal{O}_1$ by (15). Also,

(41) $$\|\zeta - \zeta_0\|_3 = \left(\frac{1}{2^2} \sum_{k=1}^{4} |Z_k - Z_k^0|^2 \right)^{1/2}$$

by (21) in Section 45, and this identity shows that Z_k tends to Z_k^0 as ζ tends to ζ_0. Then (39) and (40) show that

(42) $$\lim_{Z_k \to Z_k^0} \left| \frac{U_k(\zeta) - U_k(\zeta_0)}{Z_k - Z_k^0} - D_k(\zeta_0) \right| = 0, \qquad k = 1, \ldots, 4,$$

or

$$(43)\qquad \lim_{Z_k\to Z_k^0} \frac{U_k(\zeta)-U_k(\zeta_0)}{Z_k-Z_k^0} = D_k(\zeta_0).$$

Now (12), (22)–(25) show that $U_k(\zeta)\in\mathbb{C}_1$ for ζ in X, and (11), (16)–(19) show that $Z_k\in\mathbb{C}_1$; thus the difference quotient in (43) equals a complex number in $\mathbb{C}_1$. Furthermore, (43) suggests that U_k, which is patently a function of ζ, is actually a function of Z_k which has a derivative with respect to Z_k as stated in (43). Formal proofs of this fact will be given later, but it is not assumed in the statement of Theorem 48.2 and in its proof (see Corollary 48.3, however). Thus (43) defines a derivative to be denoted by $D_{Z_k}U_k(\zeta)$, and

$$(44)\qquad D_{Z_k}U_k(\zeta_0) = D_k(\zeta_0), \qquad k=1,\ldots,4.$$

Hence, the conclusion in (32) has been established; (33) is the statement in (21); (34) is (44); and (26) and (44) show that (35) is true.

The proof of Theorem 48.2 can be completed by showing that f satisfies the strong Stolz condition. Since $U_1,\ldots,U_4$ are differentiable with respect to $Z_1,\ldots,Z_4$, respectively, as stated in (44), then (43) shows that there exist functions $r(U_k; Z_k^0,\cdot)$, $k=1,\ldots,4$, such that

$$(45)\qquad U_k(\zeta)-U_k(\zeta_0) = D_{Z_k}U_k(\zeta_0)(Z_k-Z_k^0) + r(U_k;\,Z_k^0,\,Z_k)(Z_k-Z_k^0),$$

$$(46)\qquad \lim_{Z_k\to Z_k^0} r(U_k;\,Z_k^0,\,Z_k)=0, \qquad r(U_k;\,Z_k^0,\,Z_k^0)=0.$$

From (21) we see that

$$(47)\qquad f(\zeta)-f(\zeta_0) = \sum_{k=1}^{4} [U_k(\zeta)-U_k(\zeta_0)]e_k.$$

Substitute from (45) in (47). In order to simplify the result, we observe first from (20) that

$$(48)\qquad \zeta-\zeta_0 = \sum_{k=1}^{4} (Z_k-Z_k^0)e_k.$$

Next, recall (35). Finally, define a new function $r(f;\zeta_0,\cdot)$ by the following statement:

$$(49)\qquad r(f;\,\zeta_0,\,\zeta) = \sum_{k=1}^{4} r(U_k;\,Z_k^0,\,Z_k)e_k, \qquad \zeta\in X.$$

Then

$$(50)\qquad \|r(f;\,\zeta_0,\,\zeta)\|_3 = \left(\frac{1}{2^2}\sum_{k=1}^{4} |r(U_k;\,Z_k^0,\,Z_k)|^2\right)^{1/2},$$

and this formula and (46) show that

$$\lim_{\zeta\to\zeta_0} r(f;\zeta_0,\zeta)=0, \qquad r(f;\zeta_0,\zeta_0)=0. \tag{51}$$

Using all of these facts and also the properties of the idempotent representation of elements in $\mathbb{C}_3$ (see Section 45), the result of substituting from (45) in (47) and simplifying is the following:

$$f(\zeta)-f(\zeta_0)=D_\zeta f(\zeta_0)(\zeta-\zeta_0)+r(f;\zeta_0,\zeta)(\zeta-\zeta_0). \tag{52}$$

Finally, (52) and (51) show that f satisfies the strong Stolz condition in X (see Definition 20.14). Thus the statement in (36) is true, and the proof of all parts of Theorem 48.2 is complete. □

48.3 COROLLARY If it is known that $U_k: X\to\mathbb{C}_1$ in Theorem 48.2 is a function $f_k: X_k\to\mathbb{C}_1$ of the variable Z_k in X_k, then Theorem 48.2 can be stated as follows: Let $f: X\to\mathbb{C}_3$, $X\subset\mathbb{C}_3$, be a function as described above which has a derivative $D_\zeta f(\zeta)$ at each point ζ in X. Then:

(53) $U_k: X\to\mathbb{C}_1$, for $k=1,\ldots,4$, is a differentiable function $f_k: X_k\to\mathbb{C}_1$ of the variable Z_k;

$$f(\zeta)=\sum_{k=1}^{4} f_k(Z_k)e_k; \tag{54}$$

$$D_{Z_k} f_k(Z_k)=D_k(\zeta), \qquad \zeta \text{ in } X,\ k=1,\ldots,4; \tag{55}$$

$$D_\zeta f(\zeta)=\sum_{k=1}^{4} D_{Z_k} f_k(Z_k)e_k; \tag{56}$$

(57) f satisfies the strong Stolz condition in X.

Theorem 48.2 and Corollary 48.3 have been proved for a function $f: X\to\mathbb{C}_3$, $X\subset\mathbb{C}_3$. The methods, however, are perfectly general, and they can be used to show that every differentiable function $f: X\to\mathbb{C}_n$, $X\subset\mathbb{C}_n$, $n\geqslant 2$, has an idempotent representation in terms of 2^{n-1} differentiable functions of certain complex variables, and that it satisfies the strong Stolz condition.

The next step is to prove that a function $f: X\to\mathbb{C}_n$, $X\subset\mathbb{C}_n$, is differentiable in X and satisfies the strong Stolz condition in X if it satisfies the weak Stolz condition. This result will complete the proof that differentiability and the strong and weak Stolz conditions are equivalent conditions.

Let $f: X\to\mathbb{C}_n$ be a function which satisfies the weak Stolz condition at each point ζ_0 in X. Then there exists a constant $d(\zeta_0)$ in $\mathbb{C}_n$ and a function $r(f;\zeta_0,\cdot)$ such that

$$f(\zeta)-f(\zeta_0)=d(\zeta_0)(\zeta-\zeta_0)+r(f;\zeta_0,\zeta)\|\zeta-\zeta_0\|_n \tag{58}$$

$$\lim_{\zeta\to\zeta_0} r(f;\zeta_0,\zeta)=0, \qquad r(f;\zeta_0,\zeta_0)=0. \tag{59}$$

Then

$$(60) \qquad \frac{f(\zeta)-f(\zeta_0)}{\zeta-\zeta_0} = d(\zeta_0) + r(f;\zeta_0,\zeta)\frac{\|\zeta-\zeta_0\|_n}{\zeta-\zeta_0}, \qquad \zeta-\zeta_0 \notin \mathcal{O}_n.$$

Let ρ be a constant in $\mathbb{C}_0$ such that $\rho > 1$. If the approach of ζ to ζ_0 is restricted so that $\zeta-\zeta_0 \notin \mathcal{O}_n$ and

$$(61) \qquad \left\| \frac{\|\zeta-\zeta_0\|_n}{\zeta-\zeta_0} \right\|_n < \rho,$$

then

$$(62) \qquad \lim_{\zeta\to\zeta_0} r(f;\zeta_0,\zeta)\frac{\|\zeta-\zeta_0\|_n}{\zeta-\zeta_0} = 0,$$

$$(63) \qquad \lim_{\zeta\to\zeta_0} \frac{f(\zeta)-f(\zeta_0)}{\zeta-\zeta_0} = d(\zeta_0).$$

The proof of the next theorem is similar to the proof of Theorem 48.2, and it also will be proved in the special case $n=3$. The proof in this special case illustrates everything involved in the proof in the general case, and it is presented because it is notationally simpler. Thus we assume once more the situation and the notation described in equations (11)–(31) above.

48.4 LEMMA Let $\zeta = (z_1 + i_2 z_2) + i_3(z_3 + i_2 z_4)$, $z_k \in \mathbb{C}_1$. If

$$(64) \qquad |Z_1 - Z_1^0| = |Z_2 - Z_2^0| = \cdots = |Z_4 - Z_4^0| > 0,$$

then $\zeta-\zeta_0 \notin \mathcal{O}_3$ and

$$(65) \qquad \left\| \frac{\|\zeta-\zeta_0\|_3}{\zeta-\zeta_0} \right\|_3 = 1.$$

Proof. Now

$$(66) \qquad \left\| \frac{\|\zeta-\zeta_0\|_3}{\zeta-\zeta_0} \right\|_3 = \|\zeta-\zeta_0\|_3 \left\| \frac{1}{\zeta-\zeta_0} \right\|_3.$$

Let r be the value of the four equal quantities in (64); then $r = |Z_k - Z_k^0|$, $k = 1, \ldots, 4$, and, by (41),

$$(67) \qquad \|\zeta-\zeta_0\|_3 = r.$$

Also, since

$$(68) \qquad \frac{1}{\zeta-\zeta_0} = \sum_{k=1}^{4} \frac{e_k}{Z_k - Z_k^0},$$

then

(69) $$\left\|\frac{1}{\zeta-\zeta_0}\right\|_3=\left(\frac{1}{2^2}\sum_{k=1}^{4}\frac{1}{|Z_k-Z_k^0|^2}\right)^{1/2}=\frac{1}{r}.$$

Finally, (66), (67), and (69) show that (65) is true. □

Lemma 48.4 shows that it is possible for ζ to approach ζ_0 so that $\zeta-\zeta_0\notin\mathcal{O}_3$ and so that the regularity condition is satisfied. In this case, (63) is true. Let

(70) $$d(\zeta_0)=[d_1(\zeta_0)+i_2d_2(\zeta_0)]+i_3[d_3(\zeta_0)+i_2d_4(\zeta_0)],$$

and define $D_1(\zeta_0),\ldots,D_4(\zeta_0)$ as in (27)–(30). Thus, if (64) is satisfied, then

(71) $$\lim_{\zeta\to\zeta_0}\frac{f(\zeta)-f(\zeta_0)}{\zeta-\zeta_0}=\sum_{k=1}^{4}D_k(\zeta_0)e_k.$$

The next theorem proves the following: if (71) holds when ζ tends to ζ_0 so that (64) is satisfied, then it also holds when ζ tends to ζ_0 so that the weaker restriction $\zeta-\zeta_0\notin\mathcal{O}_3$ is satisfied; that is, the weak Stolz condition implies the strong Stolz condition. If the hypothesis in Theorem 48.2 that $f:X\to\mathbb{C}_3$, $X\subset\mathbb{C}_3$, is differentiable in X is replaced by the hypothesis that f satisfies the weak Stolz condition in X, the result is the following Theorem 48.5; the proofs of the two theorems are similar.

48.5 THEOREM Let $f:X\to\mathbb{C}_3$, $X\subset\mathbb{C}_3$, be a function which satisfies the weak Stolz condition in X. Then:

(72) $U_k:X\to\mathbb{C}_1$, for $k=1,\ldots,4$, is a function which has a derivative with respect to Z_k;

(73) $$f(\zeta)=\sum_{k=1}^{4}U_k(\zeta)e_k;$$

(74) $$D_{Z_k}U_k(\zeta)=D_k(\zeta),\qquad \zeta \text{ in } X,\ k=1,\ldots,4;$$

(75) $$D_\zeta f(\zeta)=\sum_{k=1}^{4}D_{Z_k}U_k(\zeta)e_k;$$

(76) f satisfies the strong Stolz condition in X.

Proof. Since f satisfies the weak Stolz condition,

(77) $$\lim_{\zeta\to\zeta_0}\left[\frac{f(\zeta)-f(\zeta_0)}{\zeta-\zeta_0}-d(\zeta_0)\right]$$

$$=0 \text{ provided } \zeta-\zeta_0\notin\mathcal{O}_3,\ \left\|\frac{\|\zeta-\zeta_0\|_n}{\zeta-\zeta_0}\right\|_n<\rho.$$

By Corollary 45.3, the expression inside the square brackets in (77) can be represented as a linear combination of $e_1, \ldots, e_4$. This corollary and the explanation of the notation show that

$$\frac{f(\zeta) - f(\zeta_0)}{\zeta - \zeta_0} - d(\zeta_0) = \sum_{k=1}^{4} \left[\frac{U_k(\zeta) - U_k(\zeta_0)}{Z_k - Z_k^0} - D_k(\zeta_0) \right] e_k. \tag{78}$$

By Corollary 45.5,

$$\left\| \frac{f(\zeta) - f(\zeta_0)}{\zeta - \zeta_0} - d(\zeta_0) \right\|_3 = \left[\frac{1}{2^2} \sum_{k=1}^{4} \left| \frac{U_k(\zeta) - U_k(\zeta_0)}{Z_k - Z_k^0} - D_k(\zeta_0) \right|^2 \right]^{1/2}. \tag{79}$$

Let $\varepsilon > 0$ be given. Then by (77) there exists a $\delta > 0$ such that

$$\left\| \frac{f(\zeta) - f(\zeta_0)}{\zeta - \zeta_0} - d(\zeta_0) \right\|_3 < \varepsilon \tag{80}$$

provided

$$\|\zeta - \zeta_0\|_3 < \delta, \qquad \zeta - \zeta_0 \notin \mathcal{O}_3, \left\| \frac{\|\zeta - \zeta_0\|_3}{\zeta - \zeta_0} \right\|_3 < \rho. \tag{81}$$

The next step in the proof is to show that

$$\lim_{Z_k \to Z_k^0} \left\{ \frac{U_k(\zeta) - U_k(\zeta_0)}{Z_k - Z_k^0} - D_k(\zeta_0) \right\} = 0, \qquad k = 1, \ldots, 4. \tag{82}$$

First prove (82) for $k = 1$. Choose r so that $0 < r < \delta$, and let Z_1 be an arbitrary point in X_1 such that $|Z_1 - Z_1^0| = r$. Next, choose points $Z_2, \ldots, Z_4$ in $X_2, \ldots, X_4$ so that

$$|Z_k - Z_k^0| = r, \qquad k = 1, \ldots, 4. \tag{83}$$

The points $Z_1, \ldots, Z_4$ thus chosen determine a point ζ as in (20), and

$$\zeta - \zeta_0 = \sum_{k=1}^{4} (Z_k - Z_k^0) e_k, \tag{84}$$

$$\|\zeta - \zeta_0\|_3 = \left(\frac{1}{2^2} \sum_{k=1}^{4} |Z_k - Z_k^0|^2 \right)^{1/2} = r < \delta. \tag{85}$$

Next, (83) and Lemma 48.4 show that

$$\left\| \frac{\|\zeta - \zeta_0\|_3}{\zeta - \zeta_0} \right\|_3 = 1. \tag{86}$$

Then (85), (83), and (86) show that ζ satisfies each of the restrictions in (81); therefore, the inequality in (80) is valid for the special ζ just constructed.

Therefore, (79) shows that

$$\left|\frac{U_k(\zeta)-U_k(\zeta_0)}{Z_k-Z_k^0}-D_k(\zeta_0)\right|<2\varepsilon, \qquad k=1,\ldots,4. \tag{87}$$

In these inequalities, Z_1 is an arbitrary point in X_1 such that $|Z_1-Z_1^0|=r$, and $Z_2,\ldots,Z_4$ are conveniently chosen auxiliary points such that (83) is true. Thus this statement and (87) show that (82) is true for $k=1$. Similar arguments show that (82) is true also for $k=2,\ldots,4$. Thus

$$\lim_{Z_k\to Z_k^0}\frac{U_k(\zeta)-U_k(\zeta_0)}{Z_k-Z_k^0}=D_k(\zeta_0), \qquad k=1,\ldots,4. \tag{88}$$

Now (12), (22)–(25) show that $U_k(\zeta)\in\mathbb{C}_1$ for ζ in X, and (11), (16)–(19) show that $Z_k\in\mathbb{C}_1$; thus the difference quotient in (88) equals a complex number in $\mathbb{C}_1$. Furthermore, (88) suggests that U_k, which appears as a function of ζ, is actually a function of Z_k in X_k which has a derivative with respect to Z_k as shown in (88). As stated in the proof of Theorem 48.2, a proof of this statement will be given in a later section; however, at this time it will not be assumed to be true. Thus (88) defines a derivative to be denoted by $D_{Z_k}U_k(\zeta_0)$, and

$$D_{Z_k}U_k(\zeta_0)=D_k(\zeta_0), \qquad k=1,\ldots,4. \tag{89}$$

Hence, the conclusion in (72) has been established; (73) is the statement in (21); (74) is (89); and (26) and (89) show that (75) is true.

The proof of Theorem 48.5 can be completed by showing that f satisfies the strong Stolz condition. Since $U_1,\ldots,U_4$ are differentiable with respect to $Z_1,\ldots,Z_4$, respectively, as stated in (89), then (88) shows that there exist functions $r(U_k;Z_k^0,\cdot)$, $k=1,\ldots,4$, such that

$$U_k(\zeta)-U_k(\zeta_0)=D_{Z_k}U_k(\zeta_0)(Z_k-Z_k^0)+r(U_k;Z_k^0,Z_k)(Z_k-Z_k^0), \tag{90}$$

$$\lim_{Z_k\to Z_k^0} r(U_k;Z_k^0,Z_k)=0, \qquad r(U_k;Z_k^0,Z_k^0)=0. \tag{91}$$

Then, as shown in the proof of Theorem 48.2, the function f satisfies the following strong Stolz condition:

$$f(\zeta)-f(\zeta_0)=\left[\sum_{k=1}^{4}D_{Z_k}U_k(\zeta_0)e_k\right](\zeta-\zeta_0)+r(f;\zeta_0,\zeta)(\zeta-\zeta_0) \tag{92}$$

$$\lim_{\zeta\to\zeta_0} r(f;\zeta_0,\zeta)=0, \qquad r(f;\zeta_0,\zeta_0)=0. \tag{93}$$

Thus (76) is true. Also, (92), (93), and Theorem 48.1 show that f has a derivative $D_\zeta f(\zeta)$ for all ζ in X, and that

$$D_\zeta f(\zeta_0)=\sum_{k=1}^{4}D_{Z_k}U_k(\zeta_0)e_k. \tag{94}$$

Thus (75) is true, and all parts of Theorem 48.5 have been proved. □

Section 48 has, thus far, established the equivalence of the following three conditions:

(95) f satisfies the strong Stolz condition in X;

(96) f is differentiable in X;

(97) f satisfies the weak Stolz condition in X.

Theorem 48.1 shows that (95) implies (96) and (97); Theorem 48.2 shows that (96) implies (95) which implies (97); and Theorem 48.5 shows that (97) implies (95) and (96).

The next part of this section will be used to prove some theorems which establish necessary conditions for differentiability. The first of these theorems is a generalization of Theorem 23.1; the latter theorem states that if $f:X\to\mathbb{C}_2$ is a function

$$f(\zeta)=[g_1(x)+i_1g_2(x)]+i_2[g_3(x)+i_1g_4(x)],\qquad x\in X,\tag{98}$$

which satisfies the strong Stolz condition, then the functions $g_k:X\to\mathbb{C}_0$, $k=1,\ldots,4$, satisfy the strong Stolz condition and the Cauchy–Riemann differential equations. There is a similar theorem in $\mathbb{C}_n$, and the only difficulty which arises in its statement and proof is the complexity of the details rather than any difficulty in the method. The theorem and its proof will be given for a function $f:X\to\mathbb{C}_3$, $X\subset\mathbb{C}_3$. The notation is similar to the notation in Theorem 23.1. Thus in the next theorem, $f:X\to\mathbb{C}_3$, $X\subset\mathbb{C}_3$, is a function which satisfies the strong Stolz condition as follows: for each ζ_0 in X, there is a derivative $D_\zeta f(\zeta_0)$ and a function $r(f;\zeta_0,\cdot)$ such that

$$f(\zeta)-f(\zeta_0)=D_\zeta f(\zeta_0)(\zeta-\zeta_0)+r(f;\zeta_0,\zeta)(\zeta-\zeta_0)\tag{99}$$

$$\lim_{\zeta\to\zeta_0}r(f;\zeta_0,\zeta)=0,\qquad r(f;\zeta_0,\zeta_0)=0.\tag{100}$$

Then let x denote the point $(x_1,\ldots,x_8)$ in $\mathbb{C}_0^8$, and let ζ denote the corresponding point in $\mathbb{C}_3$; thus

$$\zeta=[(x_1+i_1x_2)+i_2(x_3+i_1x_4)]+i_3[(x_5+i_1x_6)+i_2(x_7+i_1x_8)].\tag{101}$$

Let X denote a domain in $\mathbb{C}_3$ and also the corresponding domain in $\mathbb{C}_0^8$. Since $f(\zeta)\in\mathbb{C}_3$, there are functions $g_k:X\to\mathbb{C}_0$, $k=1,\ldots,8$, such that, for ζ in X,

$$\begin{aligned}f(\zeta)&=\{[g_1(x)+i_1g_2(x)]+i_2[g_3(x)+i_1g_4(x)]\}\\&\quad+i_3\{[g_5(x)+i_1g_6(x)]+i_2[g_7(x)+i_1g_8(x)]\}.\end{aligned}\tag{102}$$

Also, since $D_\zeta f(\zeta)\in\mathbb{C}_3$, there are functions $d_k:X\to\mathbb{C}_0$, $k=1,\ldots,8$, such that

$$\begin{aligned}D_\zeta f(\zeta)&=\{[d_1(\zeta)+i_1d_2(\zeta)]+i_2[d_3(\zeta)+i_1d_4(\zeta)]\}\\&\quad+i_3\{[d_5(\zeta)+i_1d_6(\zeta)]+i_2[d_7(\zeta)+i_1d_8(\zeta)]\}.\end{aligned}\tag{103}$$

Finally, since $r(f;\zeta_0,\zeta)\in\mathbb{C}_3$, there are real-valued functions $r_k(f;x^0,\cdot)$, $k=1,\ldots,8$, such that

$$(104)\quad r(f;\zeta_0,\zeta) = \{[r_1(f;x^0,x)+i_1r_2(f;x^0,x)]+i_2[r_3(f;x^0,x)+i_1r_4(f;x^0,x)]\} + i_3\{[r_5(f;x^0,x)+i_1r_6(f;x^0,x)]+i_2[r_7(f;x^0,x)+i_1r_8(f;x^0,x)]\}.$$

$$(105)\quad \lim_{x\to x^0} r_k(f;x^0,x)=0,\qquad r_k(f;x^0,x^0)=0,\qquad k=1,\ldots,8.$$

48.6 THEOREM If $f:X\to\mathbb{C}_3$ satisfies the strong Stolz condition in (99) and (100), then the functions $g_1,\ldots,g_8$ satisfy the following strong Stolz conditions:

(106)

$$g_1(x)-g_1(x^0)=d_1(x^0)(x_1-x_1^0)-d_2(x^0)(x_2-x_2^0)-\cdots-d_8(x^0)(x_8-x_8^0)+R_1,$$
$$\vdots$$
$$g_8(x)-g_8(x^0)=d_8(x^0)(x_1-x_1^0)+d_7(x^0)(x_2-x_2^0)+\cdots+d_1(x^0)(x_8-x_8^0)+R_8.$$

$$(107)\quad R_1=r_1(f;x^0,x)(x_1-x_1^0)-\cdots-r_8(f;x^0,x)(x_8-x_8^0),$$

and $R_2,\ldots,R_8$ are similar expressions in the $x_k-x_k^0$ and $r_k(f;x^0,x)$ [compare (6) in Section 23]. If d_k denotes $d_k(x^0)$, the matrix of coefficients of $(x_1-x_1^0),\ldots,(x_8-x_8^0)$ in (106) is shown in (108).

$$(108)\quad \begin{bmatrix} d_1 & -d_2 & -d_3 & d_4 & -d_5 & d_6 & d_7 & -d_8\\ d_2 & d_1 & -d_4 & -d_3 & -d_6 & -d_5 & d_8 & d_7\\ d_3 & -d_4 & d_1 & -d_2 & -d_7 & d_8 & -d_5 & d_6\\ d_4 & d_3 & d_2 & d_1 & -d_8 & -d_7 & -d_6 & -d_5\\ d_5 & -d_6 & -d_7 & d_8 & d_1 & -d_2 & -d_3 & d_4\\ d_6 & d_5 & -d_8 & -d_7 & d_2 & d_1 & -d_4 & -d_3\\ d_7 & -d_8 & d_5 & -d_6 & d_3 & -d_4 & d_1 & -d_2\\ d_8 & d_7 & d_6 & d_5 & d_4 & d_3 & d_2 & d_1 \end{bmatrix}.$$

The functions $g_1,\ldots,g_8$ are differentiable in X, and their jacobian matrix is

$$(109)\quad \begin{bmatrix} D_{x_1}g_1 & D_{x_2}g_1 & \cdots & D_{x_8}g_1\\ D_{x_1}g_2 & D_{x_2}g_2 & \cdots & D_{x_8}g_2\\ \vdots & \vdots & \vdots & \vdots\\ D_{x_1}g_8 & D_{x_2}g_8 & \cdots & D_{x_8}g_8 \end{bmatrix}.$$

The jacobian matrix (109) is the same matrix as the Cauchy–Riemann matrix in (108), and the functions $g_1, \ldots, g_8$ satisfy the Cauchy–Riemann differential equations obtained by equating all derivatives in (109) which are equal, in turn, to $d_1, \ldots, d_8$. Finally,

$$(110) \quad D_\zeta f(\zeta_0) = \{[D_{x_1}g_1(x_0) + i_1 D_{x_1}g_2(x_0)] + i_2[D_{x_1}g_3(x_0) + i_1 D_{x_1}g_4(x_0)]\} + i_3\{[D_{x_1}g_5(x_0) + i_1 D_{x_1}g_6(x_0)] + i_2[D_{x_1}g_7(x_0) + i_1 D_{x_1}g_8(x_0)]\}.$$

Proof. Substitute from (101)–(104) in (99), carry out the indicated multiplications and additions, and then equate the coefficients of 1, i_1, i_2, i_1i_2, i_3, i_1i_3, i_2i_3, $i_1i_2i_3$, respectively on the two sides of the resulting equation to obtain the eight equations in (106). These equations show that each of the functions $g_1, \ldots, g_8$ satisfies the strong Stolz condition for functions of several real variables as stated in Definition 20.5, and Theorem 20.6 shows that these functions are differentiable in X. The derivatives $D_{x_1}g_k(x^0), \ldots,$ $D_{x_8}g_k$, $k = 1, \ldots, 8$, are the coefficients of $(x_1 - x_1^0), \ldots, (x_8 - x_8^0)$ respectively in the Stolz condition (106) which g_k satisfies. Thus (106) shows that the Cauchy–Riemann matrix (108) is equal to the jacobian matrix (109). Each of the derivatives on the principal diagonal of (109) is equal to d_1 on the principal diagonal in (108); the first of the Cauchy–Riemann differential equations is obtained by equating the eight derivatives on the principal diagonal of (109). The matrix in (108) shows that each derivative in (109) has one of eight values; the eight Cauchy–Riemann equations are obtained by equating the eight derivatives which have the same value. The matrices (108) and (109) show that

$$(111) \quad d_k(\zeta_0) = d_k(x^0) = D_{x_1}g_k(x^0), \qquad k = 1, \ldots, 8.$$

Thus (110) follows from (103) and (111). Observe that $g_1, \ldots, g_8$ have derivatives [7, pp. 19–20] and not merely the weaker partial derivatives. The proof of Theorem 48.6 is complete. □

Another necessary condition for the differentiability of $f: X \to \mathbb{C}_n$, $X \subset \mathbb{C}_n$, has been established already in Theorem 48.2. The theorem was stated and proved there only for the special case $n = 3$. Nevertheless, the methods used there can be employed to prove the theorem in the general case. The special case $n = 3$, stated in Corollary 48.3, is partially restated here for emphasis. The notation is explained in equations (11)–(31). A proof of Theorem 48.7 is given in Section 49.

48.7 THEOREM Let $f: X \to \mathbb{C}_3$, $X \subset \mathbb{C}_3$, be a function which has a derivative at each point ζ in the domain X. Then there exist differentiable functions $f_k : X_k \to \mathbb{C}_1$, $k=1,\ldots,4$ of the complex variable Z_k such that

$$(112)\qquad f(\zeta) = f_1(Z_1)e(i_1 i_2)e(i_2 i_3) + f_2(Z_2)e(-i_1 i_2)e(i_2 i_3) + f_3(Z_3)e(i_1 i_2)e(-i_2 i_3) + f_4(Z_4)e(-i_1 i_2)e(-i_2 i_3).$$

Section 44 has emphasized the great variety of representations of elements in $\mathbb{C}_n$; they range from representations in terms of coordinates in $\mathbb{C}_0$ to coordinates in $\mathbb{C}_{n-1}$. Theorem 48.6 has given necessary conditions for differentiability in terms of functions with values in $\mathbb{C}_0$ (the lower end of the range of representations), and the next theorem gives similar necessary conditions in terms of functions with values in $\mathbb{C}_{n-1}$ (the upper end of the range of representations). The following statements explain the notations in the next theorem:

$$(113)\qquad \zeta = \zeta_1 + i_n\zeta_2, \qquad \zeta \in \mathbb{C}_n,\ n \geqslant 2;\ \zeta_1, \zeta_2 \in \mathbb{C}_{n-1};$$

$$(114)\qquad f(\zeta) = u(\zeta) + i_n v(\zeta), \qquad u(\zeta), v(\zeta) \in \mathbb{C}_{n-1};$$

$$(115)\qquad D_\zeta f(\zeta) = d_1(\zeta) + i_n d_2(\zeta), \qquad d_1(\zeta), d_2(\zeta) \in \mathbb{C}_{n-1};$$

$$(116)\qquad X_1 = \{\zeta_1 - i_{n-1}\zeta_2 \text{ in } \mathbb{C}_{n-1} : \zeta_1 + i_n\zeta_2 \in X\},$$
$$X_2 = \{\zeta_1 + i_{n-1}\zeta_2 \text{ in } \mathbb{C}_{n-1} : \zeta_1 + i_n\zeta_2 \in X\}.$$

48.8 THEOREM Let X be a domain in $\mathbb{C}_n$, and let $f : X \to \mathbb{C}_n$ be a function which satisfies the strong Stolz condition in X. Then:

(117) $u - i_{n-1}v$ and $u + i_{n-1}v$ are functions $f_1 : X_1 \to \mathbb{C}_{n-1}$ and $f_2 : X_2 \to \mathbb{C}_{n-1}$ of $\zeta_1 - i_{n-1}\zeta_2$ in X_1 and $\zeta_1 + i_{n-1}\zeta_2$ in X_2, respectively, which satisfy the strong Stolz condition;

$$(118)\qquad f(\zeta) = f_1(\zeta_1 - i_{n-1}\zeta_2)e(i_{n-1}i_n) + f_2(\zeta_1 + i_{n-1}\zeta_2)e(-i_{n-1}i_n), \qquad \zeta \in X;$$

$$(119)\qquad D_{\zeta_1 - i_{n-1}\zeta_2} f_1(\zeta_1 - i_{n-1}\zeta_2) = d_1(\zeta) - i_{n-1}d_2(\zeta),$$
$$D_{\zeta_1 + i_{n-1}\zeta_2} f_2(\zeta_1 + i_{n-1}\zeta_2) = d_1(\zeta) + i_{n-1}d_2(\zeta), \qquad \zeta \in X;$$

$$(120)\qquad D_\zeta f(\zeta) = D_{\zeta_1 - i_{n-1}\zeta_2} f_1(\zeta_1 - i_{n-1}\zeta_2)e(i_{n-1}i_n) + D_{\zeta_1 + i_{n-1}\zeta_2} f_2(\zeta_1 + i_{n-1}\zeta_2)e(-i_{n-1}i_n).$$

Proof. Since f satisfies the strong Stolz condition, it has a derivative $D_\zeta f(\zeta)$ in X, and there is a function $r(f;\zeta^0,\cdot)$ such that

$$(121)\qquad f(\zeta) - f(\zeta^0) = D_\zeta f(\zeta)(\zeta - \zeta^0) + r(f;\zeta^0,\zeta)(\zeta - \zeta^0).$$

By (113)–(115) and the idempotent representation,

$$\zeta = (\zeta_1 - i_{n-1}\zeta_2)e(i_{n-1}i_n) + (\zeta_1 + i_{n-1}\zeta_2)e(-i_{n-1}i_n), \tag{122}$$

$$f(\zeta) = [u(\zeta) - i_{n-1}v(\zeta)]e(i_{n-1}i_n) + [u(\zeta) + i_{n-1}v(\zeta)]e(-i_{n-1}i_n), \tag{123}$$

$$D_\zeta f(\zeta) = [d_1(\zeta) - i_{n-1}d_2(\zeta)]e(i_{n-1}i_n) + [d_1(\zeta) + i_{n-1}d_2(\zeta)]e(-i_{n-1}i_n), \tag{124}$$

$$r(f; \zeta^0, \zeta) = r_1(f; \zeta^0, \zeta)e(i_{n-1}i_n) + r_2(f; \zeta^0, \zeta)e(-i_{n-1}i_n). \tag{125}$$

Substitute from (122)–(125) in (121) and then equate the coefficients of $e(i_{n-1}i_n)$ and $e(-i_{n-1}i_n)$ on the two sides of the equation. The result is the following two equations:

$$\begin{aligned}&[u(\zeta) - i_{n-1}v(\zeta)] - [u(\zeta^0) - i_{n-1}v(\zeta^0)]\\ &\quad= [d_1(\zeta^0) - i_{n-1}d_2(\zeta^0)][(\zeta_1 - i_{n-1}\zeta_2) - (\zeta_1^0 - i_{n-1}\zeta_2^0)]\\ &\qquad + r_1(f; \zeta^0, \zeta)[(\zeta_1 - i_{n-1}\zeta_2) - (\zeta_1^0 - i_{n-1}\zeta_2^0)],\end{aligned} \tag{126}$$

$$\begin{aligned}&[u(\zeta) + i_{n-1}v(\zeta)] - [u(\zeta^0) + i_{n-1}v(\zeta^0)]\\ &\quad= [d_1(\zeta^0) + i_{n-1}d_2(\zeta^0)][(\zeta_1 + i_{n-1}\zeta_2) - (\zeta_1^0 + i_{n-1}\zeta_2^0)]\\ &\qquad + r_2(f; \zeta^0, \zeta)[(\zeta_1 + i_{n-1}\zeta_2) - (\zeta_1^0 + i_{n-1}\zeta_2^0)].\end{aligned} \tag{127}$$

We show, as follows, that $r_1(f;\zeta^0,\cdot)$ and $r_2(f;\zeta^0,\cdot)$ are equal to functions of $\zeta_1 - i_{n-1}\zeta_2$ and $\zeta_1 + i_{n-1}\zeta_2$, respectively. Equation (122) shows that ζ, which is $\zeta_1 + i_n\zeta_2$, determines a unique $\zeta_1 - i_{n-1}\zeta_2$ and $\zeta_1 + i_{n-1}\zeta_2$. Then define new functions $r(f_1; \zeta_1^0 - i_{n-1}\zeta_2^0, \cdot)\colon X_1 \to \mathbb{C}_{n-1}$ and $r(f_2; \zeta_1^0 + i_{n-1}\zeta_2^0, \cdot)\colon X_2 \to \mathbb{C}_{n-1}$ as follows:

$$\begin{aligned}&r(f_1; \zeta_1^0 - i_{n-1}\zeta_2^0, \zeta_1 - i_{n-1}\zeta_2) = r_1(f; \zeta^0, \zeta),\\ &r(f_2; \zeta_1^0 + i_{n-1}\zeta_2^0, \zeta_1 + i_{n-1}\zeta_2) = r_2(f; \zeta^0, \zeta).\end{aligned} \tag{128}$$

Now

$$\lim_{\zeta\to\zeta_0} r(f; \zeta^0, \zeta) = 0, \qquad r(f; \zeta^0, \zeta^0) = 0, \tag{129}$$

and since

$$\|r(f; \zeta^0, \zeta)\|_n = \left(\frac{1}{2}\sum_{k=1}^{2} \|r_k(f; \zeta^0, \zeta)\|_{n-1}^2\right)^{1/2} \tag{130}$$

by (125), then

$$\lim_{\zeta\to\zeta_0} r_k(f; \zeta^0, \zeta) = 0, \qquad r_k(f; \zeta^0, \zeta^0) = 0, \qquad k = 1, 2. \tag{131}$$

Then (131) and (128) show that

$$\text{(132)}\qquad \lim_{\zeta_1 - i_{n-1}\zeta_2 \to \zeta_1^0 - i_{n-1}\zeta_2^0} r(f_1;\zeta_1^0 - i_{n-1}\zeta_2^0, \zeta_1 - i_{n-1}\zeta_2) = 0,$$

$$r(f_1;\zeta_1^0 - i_{n-1}\zeta_2^0, \zeta_1^0 - i_{n-1}\zeta_2^0) = 0,$$

$$\text{(133)}\qquad \lim_{\zeta_1 + i_{n-1}\zeta_2 \to \zeta_1^0 + i_{n-1}\zeta_2^0} r(f_2;\zeta_1^0 + i_{n-1}\zeta_2^0, \zeta_1 + i_{n-1}\zeta_2) = 0,$$

$$r(f_2;\zeta_1^0 + i_{n-1}\zeta_2^0, \zeta_1^0 + i_{n-1}\zeta_2^0) = 0.$$

Replace $r_1(f;\zeta^0,\zeta)$ in (126) by its value in (128); the resulting equation shows that $u - i_{n-1}v$ is a function $f_1 : X_1 \to \mathbb{C}_{n-1}$ of $\zeta_1 - i_{n-1}\zeta_2$ as stated in (117), that it satisfies the strong Stolz condition in X_1, and that the first equation in (119) is true. Likewise, replace $r_2(f;\zeta^0,\zeta)$ in (127) by its value in (128); the resulting equation shows that $u + i_{n-1}v$ is a function $f_2 : X_2 \to \mathbb{C}_{n-1}$ of $\zeta_1 + i_{n-1}\zeta_2$ as stated in (117), that it satisfies the strong Stolz condition in X_2, and that the second equation in (119) is true. The statement in (118) follows from (123) and the definition of f_1 and f_2. The statement in (120) follows from (124) and the two statements in (119). The first proof of Theorem 48.8 is complete; a second proof is given in Section 49. □

This section will be completed by proving (a) two theorems which state sufficient conditions that a function be differentiable, and (b) a corollary which contains an existence theorem.

48.9 THEOREM Let $f_1 : X_1 \to \mathbb{C}_{n-1}$, $X_1 \subset \mathbb{C}_{n-1}$, and $f_2 : X_2 \to \mathbb{C}_{n-1}$, $X_2 \subset \mathbb{C}_{n-1}$, be functions which are differentiable in the domains X_1 and X_2, respectively. If X is the domain in $\mathbb{C}_n$ determined by X_1 and X_2, and if $f : X \to \mathbb{C}_n$ is the function such that

$$\text{(134)}\qquad f(\zeta_1 + i_n\zeta_2) = f_1(\zeta_1 - i_{n-1}\zeta_2)e(i_{n-1}i_n) + f_2(\zeta_1 + i_{n-1}\zeta_2)e(-i_{n-1}i_n),$$

$$\zeta_1 + i_n\zeta_2 \in X, \qquad \zeta_1 - i_{n-1}\zeta_2 \in X_1, \qquad \zeta_1 + i_{n-1}\zeta_2 \in X_2,$$

then f is differentiable in X.

Proof. Since f_1 and f_2 are differentiable in X_1 and X_2, respectively, they satisfy the strong Stolz condition by the general version of Theorem 48.2 (36). Thus

$$\text{(135)}\quad f_1(\zeta_1 - i_{n-1}\zeta_2) - f_1(\zeta_1^0 - i_{n-1}\zeta_2^0)$$

$$= D_{\zeta_1 - i_{n-1}\zeta_2} f_1(\zeta_1^0 - i_{n-1}\zeta_2^0)[(\zeta_1 - i_{n-1}\zeta_2) - (\zeta_1^0 - i_{n-1}\zeta_2^0)]$$

$$+ r(f_1;\zeta_1^0 - i_{n-1}\zeta_2^0, \zeta_1 - i_{n-1}\zeta_2)[(\zeta_1 - i_{n-1}\zeta_2) - (\zeta_1^0 - i_{n-1}\zeta_2^0)],$$

$$(136)\quad f_2(\zeta_1 + i_{n-1}\zeta_2) - f_2(\zeta_1^0 + i_{n-1}\zeta_2^0)$$
$$= D_{\zeta_1 + i_{n-1}\zeta_2} f_2(\zeta_1^0 + i_{n-1}\zeta_2^0)[(\zeta_1 + i_{n-1}\zeta_2) - (\zeta_1^0 + i_{n-1}\zeta_2^0)]$$
$$+ r(f_2; \zeta_1^0 + i_{n-1}\zeta_2^0, \zeta_1 + i_{n-1}\zeta_2)[(\zeta_1 + i_{n-1}\zeta_2) - (\zeta_1^0 + i_{n-1}\zeta_2^0)].$$

From (134),

$$(137)\quad f(\zeta_1 + i_n\zeta_2) - f_2(\zeta_1^0 + i_n\zeta_2^0)$$
$$= [f_1(\zeta_1 - i_{n-1}\zeta_2) - f_1(\zeta_1^0 - i_{n-1}\zeta_2^0)]e(i_{n-1}i_n)$$
$$+ [f_2(\zeta_1 + i_{n-1}\zeta_2) - f_2(\zeta_1^0 + i_{n-1}\zeta_2^0)]e(-i_{n-1}i_n).$$

Substitute from (135) and (136) in (137); then simplify the right side to show that f satisfies the strong Stolz condition. Therefore, f is differentiable by Theorem 48.1. This theorem is a generalization of Theorem 24.3, and the details which have been omitted here are similar to those in the proof of that theorem. □

Theorems 48.8 and 48.9 completely characterize the class of differentiable functions $f: X \to \mathbb{C}_n$, $X \subset \mathbb{C}_n$, $n \geqslant 2$. If f is defined by differentiable functions f_1 and f_2 as stated in (134), then Theorem 48.9 shows that f is differentiable. Conversely, if f is differentiable, Theorem 48.8 shows that there exist differentiable functions $f_1: X_1 \to \mathbb{C}_{n-1}$ and $f_2: X_2 \to \mathbb{C}_{n-1}$ such that (134) is true.

48.10 COROLLARY There exists a differentiable function $f: X \to \mathbb{C}_n$ in every domain X in every space $\mathbb{C}_n$ for $n = 1, 2, \ldots$.

Proof. The proof is by induction on n. By the theory of functions of a complex variable, there is a differentiable function $f: X \to \mathbb{C}_1$ in every domain X in $\mathbb{C}_1$. Let X be a domain in $\mathbb{C}_2$. Then X generates sets X_1 and X_2 in $\mathbb{C}_1$, and there are holomorphic functions $f_1: X_1 \to \mathbb{C}_1$ and $f_2: X_2 \to \mathbb{C}_1$ by the first case. Then as shown by the proof of Theorem 48.9, the functions f_1 and f_2 can be used to define a function f which is differentiable at least in X, and perhaps in a larger domain. Thus the corollary is true for $n = 2$. Assume then that it is true for $n - 1$, and let X be a domain in $\mathbb{C}_n$. If X_1 and X_2 are the sets determined by X, they are domains in $\mathbb{C}_{n-1}$ and there are differentiable functions $f_1: X_1 \to \mathbb{C}_{n-1}$ and $f_2: X_2 \to \mathbb{C}_{n-1}$ by the induction hypothesis. Then Theorem 48.9 shows how to use f_1 and f_2 to construct a function f which is differentiable at least in X, and perhaps in a larger set. Thus the corollary is true for n. Since the corollary is true for $n = 1$ and $n = 2$, and since it is true for n whenever it is true for $n - 1$, a complete induction shows that the corollary is true for all $n = 1, 2, \ldots$. □

Let X be a domain in $\mathbb{C}_n$, $n \geqslant 1$. Then X can be considered also as a domain in $\mathbb{C}_0^{2n}$. The coordinates of a point in $\mathbb{C}_0^{2n}$, denoted by $(x_k\colon k=1,\ldots,2^k)$ and also by x, determine a point ζ in $\mathbb{C}_n$ as in (101). The functions $g_k\colon X\to\mathbb{C}_0$, $k=1,\ldots,2^n$, with values denoted by $g_k(x)$, determine a function $f\colon X\to\mathbb{C}_n$ as in (102). The next theorem is a generalization of Theorem 24.1 (see also Theorem 48.6).

48.11 THEOREM If the functions $g_k\colon X\to\mathbb{C}_0$, $k=1,\ldots,2^n$, are differentiable and satisfy the Cauchy–Riemann differentiable equations in X, then $f\colon X\to\mathbb{C}_n$ satisfies the strong Stolz condition and is differentiable in X.

Proof. Since each g_k is differentiable in X, it satisfies the following weak Stolz condition (see Definition 20.5) by Theorem 20.6.

$$g_k(x) - g_k(x^0) = \sum_{j=1}^{2^n} D_{x_j} g_k(x^0)(x_j - x_j^0) + r(g_k; x^0, x)|x - x^0|, \tag{138}$$

$$\lim_{x\to x^0} r(g_k; x^0, x) = 0, \qquad r(g_k; x^0, x^0) = 0. \tag{139}$$

Now $f(\zeta)$ equals an expression similar to that in (102), and $f(\zeta)-f(\zeta^0)$ is a linear combination of the $g_k(x)-g_k(x^0)$, $k=1,\ldots,2^n$. In this linear combination, replace $g_k(x)-g_k(x^0)$ by its value in (138). Then use the Cauchy–Riemann differential equations to simplify the result to the following:

$$f(\zeta) - f(\zeta^0) = D_\zeta f(\zeta^0)(\zeta - \zeta_0) + r(f; \zeta^0, \zeta)\|\zeta - \zeta^0\|_n. \tag{140}$$

Here $\|\zeta-\zeta^0\|=|x-x^0|$, and $r(f;\zeta^0,\zeta)$ is a linear combination of the $r(g_k; x^0, x)$ in (139). Also, (139) shows that

$$\lim_{\zeta\to\zeta^0} r(f; \zeta^0, \zeta) = 0, \qquad r(f; \zeta^0, \zeta^0) = 0. \tag{141}$$

Then (140) and (141) show that $f\colon X\to\mathbb{C}_n$ satisfies the weak Stolz condition (compare Definition 20.14). Finally, f satisfies the strong Stolz condition by Theorem 48.5 (76), and it is differentiable by Theorem 48.1. The proof is complete. □

Exercises

48.1 Write out the complete details of the proof of Theorem 48.9. In particular, give the details of the proof that equations (135)–(137) imply that f satisfies the strong Stolz condition.

48.2 Use (108) and (109) to write out explicitly the Cauchy–Riemann differential equations satisfied by the functions $g_1, \ldots, g_8$ in Theorem 48.6.

48.3 Write out the proof of Theorem 48.11 in the special case $n=3$. The proof of Theorem 24.1 contains these details in the special case $n=2$ and can serve as a guide in the case $n=3$.

48.4 Write out the Cauchy–Riemann differential equations in the case $n=4$. Then write out the proof of Theorem 48.11 in detail for $n=4$. In order to complete this exercise satisfactorily, it will be necessary to devise some special scheme for organizing and managing a large mass of details or to construct an induction procedure.

48.5 Explain why you believe that Theorem 48.11 can, or can not, be proved in the special case $n=1000$. Do you believe that Theorem 48.11 is true for every $n \geqslant 1$? For what values of n do you consider that Theorem 48.11 has been, or can be, proved? Explain your answer.

48.6 Let X be a domain in $\mathbb{C}_3$, and let $f: X \to \mathbb{C}_3$ be a function which is differentiable in X. Then f defines a mapping $\eta = f(\zeta)$ of X into $\mathbb{C}_3$. Also, $\eta = f(\zeta)$ can be considered as a mapping of X (considered as a domain in $\mathbb{C}_0^8$) into $\mathbb{C}_0^8$.

(a) If $M(\zeta)$ denotes the jacobian matrix (109) at ζ, prove that $\det M(\zeta) \geqslant 0$ for every ζ in X.

(b) Assume that the derivative $D_\zeta f$ of f is continuous in X. Prove that each derivative in the jacobian matrix (109) is continuous in X.

(c) Prove the following theorem. If $D_\zeta f$ is continuous in X and if $D_\zeta f(\zeta_0) \notin \mathcal{O}_3$ at a point ζ_0 in X, then in a sufficiently small neighborhood of ζ_0 the mapping $\eta = f(\zeta)$ is one-to-one [7, pp. 97–98].

48.7 Attempt to establish the following generalization of Exercise 48.6; if you encounter difficulties, explain why they arise and what is needed to overcome them. Let X be a domain in $\mathbb{C}_n$, and let $f: X \to \mathbb{C}_n$ be a differentiable function in X. Then f defines a mapping $\eta = f(\zeta)$ of X into $\mathbb{C}_n$ and a mapping of X (considered as a domain in $\mathbb{C}_0^{2^n}$) into $\mathbb{C}_0^{2^n}$.

(a) If $M(\zeta)$ denotes the jacobian matrix which corresponds to (109), prove that $\det M(\zeta) \geqslant 0$ for every ζ in X.

(b) Assume that the derivative $D_\zeta f$ is continuous in X. Prove that each of the derivatives in $M(\zeta)$ is continuous in X.

(c) Prove the following theorem. If $D_\zeta f$ is continuous in X, and if $D_\zeta f(\zeta_0) \notin \mathcal{O}_n$ at a point ζ_0 in X, then in a sufficiently small neighborhood of ζ_0 the mapping $\eta = f(\zeta)$ is one-to-one.

48.8 Problem for investigation. Let $f: X \to \mathbb{C}_n$, $X \subset \mathbb{C}_n$, be a function with a continuous derivative as described in Exercise 48.7. Investigate the nature of the mapping $\eta = f(\zeta)$ in the neighborhood of a point ζ_0 at which $D_\zeta f(\zeta) \in \mathcal{O}_n$.

49. INTEGRALS AND THEIR APPLICATIONS

Chapter 4 of this book treats integrals of holomorphic functions and uses these integrals to establish some of the fundamental properties of holomorphic functions in $\mathbb{C}_2$. This section establishes, for functions $f: X \to \mathbb{C}_n$, $X \subset \mathbb{C}_n$, $n > 2$, results which are generalizations of those in Chapter 4 for functions in $\mathbb{C}_2$. In some cases the generalization is obvious; then the treatment is brief or it is omitted altogether. The purpose of the section is to round out the treatment of holomorphic functions $f: X \to \mathbb{C}_n$, $X \subset \mathbb{C}_n$, $n \geqslant 1$, by providing proofs, or alternative proofs, based on derivatives and integrals, for the basic properties of these functions. One of these results is the representation

$$(1) \quad f(\zeta_1 + i_n\zeta_2) = f_1(\zeta_1 - i_{n-1}\zeta_2)e(i_{n-1}i_n) + f_2(\zeta_1 + i_{n-1}\zeta_2)e(-i_{n-1}i_n)$$

of a holomorphic function, on which much of the theory is based. For $n = 2$, a simple proof of this representation is given in Theorem 38.5. For $n \geqslant 2$, the representation (1) is contained in Theorem 48.8, but the present section contains two proofs which are more formal than the one offered there. The first of these proofs is similar to the proof in Theorem 38.5 for the case $n = 2$, and the second is obtained from Cauchy's integral formula. In both proofs the treatment includes the sequence of generalizations of (1) obtained by applying the formula (1) to each of the functions f_1 and f_2. The other result in this section to be mentioned in this introduction is the use of Cauchy's integral formula to prove that a holomorphic function can be represented by its Taylor series. Since $n = 1$ is the case of the classical holomorphic functions of a complex variable, and since Chapter 4 treats functions $f: X \to \mathbb{C}_n$ for $n = 2$, this section emphasizes the cases $n > 2$.

Sections 31 and 32 present the elementary properties of integrals with values in $\mathbb{C}_2$, and the treatment can be extended with little difficulty to include the integrals of functions with values in $\mathbb{C}_n$. Thus, only a brief introduction is needed here to provide a review and to point out a few changes.

Let X be a domain in $\mathbb{C}_n$, $n \geqslant 1$. A curve C in X is a mapping $\zeta: [a, b] \to X$, $t \mapsto \zeta(t)$, which has a continuous derivative $\zeta': [a, b] \to \mathbb{C}_n$. Then C has finite length $L(C)$, which is defined and calculated as follows: Let $P_1, P_2, \ldots$ be a sequence of subdivisions of $[a, b]$ whose norms approach zero. If P_m is the subdivision $a = t_0 < t_1 < \cdots < t_{k-1} < t_k < \cdots < t_m = b$, then

$$(2) \qquad L(C) = \lim_{m \to \infty} \sum_{k=1}^{m} \|\zeta(t_k) - \zeta(t_{k-1})\|_n,$$

$$(3) \qquad L(C) = \int_a^b \|\zeta'(t)\|_n \, dt.$$

Let $f: X \to \mathbb{C}_n$ be a holomorphic function; then f has a derivative, f is continuous, and the integral $\int_C f(\zeta)\,d\zeta$ is defined as follows:

$$\text{(4)} \qquad \int_C f(\zeta)\,d\zeta = \lim_{m\to\infty} \sum_{k=1}^{m} f[\zeta(t_{k-1})][\zeta(t_k) - \zeta(t_{k-1})].$$

Let

$$\text{(5)} \qquad M = \max\{\,\|f[\zeta(t)]\|_n : a \leqslant t \leqslant b\}.$$

Then (2) and Exercise 44.2 show that for every subdivision P_m,

$$\text{(6)} \qquad \left\| \sum_{k=1}^{m} f[\zeta(t_{k-1})][\zeta(t_k) - \zeta(t_{k-1})] \right\|_n$$

$$\leqslant \sum_{k=1}^{m} \|f[\zeta(t_{k-1})][\zeta(t_k) - \zeta(t_{k-1})]\|_n$$

$$\leqslant \sum_{k=1}^{m} 2^{n-1/2} \|f[\zeta(t_{k-1})]\|_n \,\|[\zeta(t_k) - \zeta(t_{k-1})]\|_n$$

$$\leqslant 2^{n-1/2} M \sum_{k=1}^{m} \|\zeta(t_k) - \zeta(t_{k-1})\|_n$$

$$\leqslant 2^{n-1/2} M L(C).$$

Therefore,

$$\text{(7)} \qquad \left\| \int_C f(\zeta)\,d\zeta \right\|_n \leqslant 2^{n-1/2} M L(C).$$

If t is introduced as the variable of integration (compare Theorem 32.9), then

$$\text{(8)} \qquad \int_C f(\zeta)\,d\zeta = \int_a^b f[\zeta(t)]\zeta'(t)\,dt.$$

The next step is to prove, for holomorphic functions $f: X \to \mathbb{C}_n$, $n \geqslant 1$, the special case of Cauchy's integral theorem which was proved in Section 34 for functions $f: X \to \mathbb{C}_2$. Both the notation and the method will be the same as in Section 34. Let T denote the oriented curve whose trace consists of the segments S_0: $[\zeta_0, \zeta_1]$, S_1: $[\zeta_1, \zeta_2]$, S_2: $[\zeta_2, \zeta_0]$ of the triangle with vertices $\zeta_0, \zeta_1, \zeta_2$. Then $T = S_0 + S_1 + S_2$ [see (1) in Section 34]. Let $c(T)$ denote the convex extension of $\{\zeta_0, \zeta_1, \zeta_2\}$. By the $\mathbb{C}_n$ analog of Theorem 32.7,

$$\text{(9)} \qquad \int_T f(\zeta)\,d\zeta = \int_{S_0} f(\zeta)\,d\zeta + \int_{S_1} f(\zeta)\,d\zeta + \int_{S_2} f(\zeta)\,d\zeta.$$

49.1 THEOREM (Cauchy's Integral Theorem, Special Case) If $f: X \to$

$\mathbb{C}_n$, $X \subset \mathbb{C}_n$, $n \geq 1$, satisfies the strong Stolz condition, and if T is a curve such that $c(T) \subset X$, then

$$\int_T f(\zeta)\, d\zeta = 0. \tag{10}$$

Proof. Assume that the theorem is false. Denote the triangle T by T_0; then

$$\left\| \int_{T_0} f(\zeta)\, d\zeta \right\|_n > 0. \tag{11}$$

Set

$$\zeta_{00} = \frac{\zeta_0 + \zeta_1}{2}, \qquad \zeta_{01} = \frac{\zeta_1 + \zeta_2}{2}, \qquad \zeta_{02} = \frac{\zeta_2 + \zeta_0}{2}. \tag{12}$$

Then the triangles

$$\begin{aligned} &T_{01}: [\zeta_{02}, \zeta_0, \zeta_{00}], \qquad &&T_{02}: [\zeta_{00}, \zeta_1, \zeta_{01}],\\ &T_{03}: [\zeta_{01}, \zeta_2, \zeta_{02}], \qquad &&T_{04}: [\zeta_{00}, \zeta_{01}, \zeta_{02}] \end{aligned} \tag{13}$$

have the same orientation as $[\zeta_0, \zeta_1, \zeta_2]$, and

$$\int_{T_0} f(\zeta)\, d\zeta = \sum_{k=1}^{4} \int_{T_{0k}} f(\zeta)\, d\zeta. \tag{14}$$

A side which belongs to two of the triangles in (13) has opposite orientations in the two triangles. Next, by the triangle inequality,

$$\left\| \int_{T_0} f(\zeta)\, d\zeta \right\|_n \leqslant \sum_{k=1}^{4} \left\| \int_{T_{0k}} f(\zeta)\, d\zeta \right\|_n. \tag{15}$$

Because of (11), at least one of the terms on the right is not zero; let the term on T_{0r} be the maximum term, or one of the maximum terms, on the right in (15). Let T_1 be the new notation for T_{0r}; then

$$\left\| \int_{T_1} f(\zeta)\, d\zeta \right\|_n > 0, \tag{16}$$

$$\left\| \int_{T_0} f(\zeta)\, d\zeta \right\|_n \leqslant 4 \left\| \int_{T_1} f(\zeta)\, d\zeta \right\|_n. \tag{17}$$

Use the midpoints of the sides of T_1 to divide T_1 into four triangles. A repetition of the analysis used to find T_1 shows that there is a triangle T_2 such that

$$\left\| \int_{T_2} f(\zeta)\, d\zeta \right\|_n > 0, \tag{18}$$

$$(19)\qquad \left\| \int_{T_1} f(\zeta)\,d\zeta \right\|_n \leqslant 4 \left\| \int_{T_2} f(\zeta)\,d\zeta \right\|_n .$$

Then (17) and (19) show that

$$(20)\qquad \left\| \int_{T_0} f(\zeta)\,d\zeta \right\|_n \leqslant 4^2 \left\| \int_{T_2} f(\zeta)\,d\zeta \right\|_n .$$

A continuation of this process shows that

$$(21)\qquad \left\| \int_{T_0} f(\zeta)\,d\zeta \right\|_n \leqslant 4^m \left\| \int_{T_m} f(\zeta)\,d\zeta \right\|_n .$$

If $L(T_m)$ is the length of T_m (that is, the sum of the lengths of the sides of T_m), then $L(T_1)=(1/2)L(T_0)$, $L(T_2)=(1/2)L(T_1)=(1/2)^2L(T_0)$, and

$$(22)\qquad L(T_m) = (1/2)^m L(T_0), \qquad m = 1, 2, \ldots .$$

Also,

$$(23)\qquad c(T_0) \supset c(T_1) \supset \cdots \supset c(T_m) \supset \cdots .$$

Since the diameter of $c(T_m)$ tends to zero as $m \to \infty$, there is a single point ζ^* which belongs to all of the sets $c(T_m)$.

Now ζ^* is in X, and f satisfies the strong Stolz condition in X by hypothesis. Let $\varepsilon > 0$ be given; then there is a $\delta(\varepsilon) > 0$ such that

$$(24)\qquad f(\zeta) - f(\zeta^*) = D_\zeta f(\zeta^*)(\zeta - \zeta^*) + r(f; \zeta^*, \zeta)(\zeta - \zeta^*),$$

$$(25)\qquad \|r(f; \zeta^*, \zeta)\|_n < \varepsilon, \qquad \|\zeta - \zeta^*\|_n < \delta(\varepsilon).$$

Choose m so large that the diameter of $c(T_m)$ is less than $\delta(\varepsilon)$. Now ζ^* is in $c(T_m)$; then $\|\zeta - \zeta^*\|_n < \delta(\varepsilon)$ for every ζ in $c(T_m)$. Use (24) to evaluate $\int_{T_m} f(\zeta)\,d\zeta$ in (21) as follows:

$$(26)\qquad \int_{T_m} f(\zeta)\,d\zeta$$

$$= \int_{T_m} f(\zeta^*)\,d\zeta + \int_{T_m} D_\zeta f(\zeta^*)(\zeta - \zeta^*)\,d\zeta + \int_{T_m} r(f; \zeta^*, \zeta)(\zeta - \zeta^*)\,d\zeta$$

$$= [f(\zeta^*) - D_\zeta f(\zeta^*)\zeta^*] \int_{T_m} d\zeta + D_\zeta f(\zeta^*) \int_{T_m} \zeta\,d\zeta + \int_{T_m} r(f; \zeta^*, \zeta)(\zeta - \zeta^*)\,d\zeta.$$

Now T_m is a closed curve and $\int_{T_m} d\zeta = 0$ by the definition of the integral; furthermore, $\int_{T_m} \zeta\,d\zeta = 0$ (compare Exercises 32.4 and 32.5). Thus

$$(27)\qquad \int_{T_m} f(\zeta)\,d\zeta = \int_{T_m} r(f; \zeta^*, \zeta)(\zeta - \zeta^*)\,d\zeta.$$

Use (7) to estimate the norm of the integral on the right. Now (25) shows that $\|r(f;\zeta^*,\zeta)\|_n<\varepsilon$ for all ζ on T_m. Also, $\|\zeta-\zeta^*\|_n$ is equal to or less than the diameter of $c(T_m)$, which is less than $1/2L(T_m)$, for all ζ on T_m; then (22) shows that

(28) $\|\zeta-\zeta^*\|_n < 1/2(1/2)^m L(T_0), \qquad \zeta \text{ on } T_m.$

Therefore, by (25), (28), (7), and (22),

(29) $\|r(f;\zeta^*,\zeta)(\zeta-\zeta^*)\|_n < 2^{n-1/2}\varepsilon(1/2)^{m+1}L(T_0),$

(30) $$\left\|\int_{T_m} r(f;\zeta^*,\zeta)(\zeta-\zeta^*)\,d\zeta\right\|_n$$
$$< 2^{n-1/2}[2^{n-1/2}\varepsilon(1/2)^{m+1}L(T_0)]L(T_m),$$
$$< 2^{n-1/2}[2^{n-1/2}\varepsilon(1/2)^{m+1}L(T_0)](1/2)^m L(T_0),$$
$$< 2^{n-2}\varepsilon(1/4)^m[L(T_0)]^2.$$

Therefore, by (21),

(31) $$\left\|\int_{T_0} f(\zeta)\,d\zeta\right\|_n < 2^{n-2}\varepsilon[L(T_0)]^2.$$

Since this inequality holds for every $\varepsilon>0$, then

(32) $$\left\|\int_{T_0} f(\zeta)\,d\zeta\right\|_n = 0.$$

This conclusion contradicts the assumption in (11); therefore, (11) is impossible, and

(33) $$\left\|\int_{T_0} f(\zeta)\,d\zeta\right\|_n = 0, \qquad \int_{T_0} f(\zeta)\,d\zeta = 0.$$

The proof of (10) and of Theorem 49.1 is complete. □

The next theorem is one of the consequences of Theorem 49.1. Observe that the proof does not employ any of the special representations for elements in $\mathbb{C}_n$; furthermore, a single proof establishes the theorem for all values of n. The proof is similar to the proofs of Theorems 35.2 and 33.4.

49.2 THEOREM Let X be a domain in $\mathbb{C}_n$, $n\geqslant 1$, which is star-shaped with respect to a point ζ^*, and let $f: X\to\mathbb{C}_n$ be a function which satisfies the strong Stolz condition in X. Then there exists a function $F: X\to\mathbb{C}_n$ which has the following properties:

(34) F has a continuous derivative, and $D_\zeta F(\zeta)=f(\zeta)$ in X.

(35) F satisfies the uniform strong Stolz condition in every compact set S in X; that is, if $\varepsilon>0$ is given, there exists a $\delta(\varepsilon)$ such that $\|r(F;\zeta_1,\zeta_2)\|_n<\varepsilon$ for every segment $[\zeta_1,\zeta_2]$ in S such that $\|\zeta_2-\zeta_1\|_n<\delta(\varepsilon)$.

Proof. Let ζ_0 be a fixed point in X, and let ζ_1 be an arbitrary point. Connect ζ_0 to ζ_1 by a polygonal curve P in X with segments

(36) $$p_0p_1,\ p_1p_2,\dots,\ p_{k-1}p_k,\dots,\ p_{m-1}p_m,\qquad p_0=\zeta_0,\ p_m=\zeta_1.$$

Such polygonal curves exist; since X is star-shaped with respect to ζ^*, one such curve consists of the segments $\zeta_0\zeta^*$ and $\zeta^*\zeta_1$. Then f has an integral on each segment in (36), and

(37) $$\int_P f(\eta)\,d\eta=\sum_{k=1}^{m}\int_{p_{k-1}p_k} f(\eta)\,d\eta.$$

The value of the integral on the left in (37) appears to depend on the polygonal curve P as well as on ζ_1, but the proof which follows shows that $\int_P f(\eta)\,d\eta$ has the same value for every polygonal curve P in X which connects ζ_0 to ζ_1; therefore $\int_P f(\eta)\,d\eta$ is the value at ζ_1 of a function $F:X\to\mathbb{C}_n$. Let Q with segments

(38) $$q_0q_1,\ q_1q_2,\dots,\ q_{k-1}q_k,\dots,\ q_{r-1}q_r,\ q_0=\zeta_0,\qquad q_r=\zeta_1,$$

be a second polygonal curve in X which connects ζ_0 to ζ_1. Reverse the orientation of each segment $q_{k-1}q_k$ in Q to form a closed curve $P-Q$. Connect ζ^* to each point p_k and q_k to form oriented triangles (curves) $\zeta^*p_{k-1}p_k$ and $\zeta^*q_kq_{k-1}$. Since X is star-shaped by hypothesis, then the convex extensions $c(\zeta^*p_{k-1}p_k)$ and $c(\zeta^*q_kq_{k-1})$ are contained in X. Then the special case of Cauchy's integral theorem in Theorem 49.1 shows that

(39) $$\int_{\zeta^*p_{k-1}p_k} f(\eta)\,d\eta=0,\qquad \int_{\zeta^*q_kq_{k-1}} f(\eta)\,d\eta=0,$$

(40) $$\sum_{k=1}^{m}\int_{\zeta^*p_{k-1}p_k} f(\eta)\,d\eta+\sum_{k=1}^{r}\int_{\zeta^*q_kq_{k-1}} f(\eta)\,d\eta=0.$$

Now the segment ζ^*p_k belongs to $\zeta^*p_{k-1}p_k$ and also to $\zeta^*p_kp_{k+1}$, and it has opposite orientations in these triangles; then the corresponding integrals on these segments cancel (compare Theorem 32.6). Also, ζ^*q_k belongs to $\zeta^*q_kq_{k-1}$ and $\zeta^*q_{k+1}q_k$, and it has opposite orientation in these triangles; again the integrals cancel. The result of making all such cancellations is to reduce (40) to the following:

(41) $$\sum_{k=1}^{m}\int_{p_{k-1}p_k} f(\eta)\,d\eta+\sum_{k=1}^{r}\int_{q_kq_{k-1}} f(\eta)\,d\eta=0.$$

Reverse the orientation of the segments $q_k q_{k-1}$ in the second sum; then

$$(42)\qquad \sum_{k=1}^{m}\int_{p_{k-1}p_k} f(\eta)\,d\eta - \sum_{k=1}^{r}\int_{q_{k-1}q_k} f(\eta)\,d\eta = 0,$$

$$(43)\qquad \sum_{k=1}^{m}\int_{p_{k-1}p_k} f(\eta)\,d\eta = \sum_{k=1}^{r}\int_{q_{k-1}q_k} f(\eta)\,d\eta.$$

This equation shows that $\int_P f(\eta)\,d\eta$ has the same value for every polygonal curve P which connects ζ_0 to ζ_1, and the value of the integral depends on ζ_1 alone. Define the function $F: X \to \mathbb{C}_n$ as follows:

$$(44)\qquad F(\zeta_1) = \int_P f(\eta)\,d\eta, \qquad P \text{ connects } \zeta_0 \text{ to } \zeta_1.$$

The next step is to show that F has a derivative and to find its value. Both are accomplished by showing that

$$(45)\qquad \lim_{h\to 0}\frac{F(\zeta_1+h)-F(\zeta_1)}{h} = f(\zeta_1), \qquad h \notin \mathcal{O}_n.$$

Recall (44); then

$$(46)\qquad F(\zeta_1+h) = \int_P f(\eta)\,d\eta + \int_{\zeta_1(\zeta_1+h)} f(\eta)\,d\eta = F(\zeta_1) + \int_{\zeta_1(\zeta_1+h)} f(\eta)\,d\eta.$$

Make the change of variable $\eta=\zeta_1+th$, $d\eta = h\,dt$, $0\leqslant t\leqslant 1$; the result is

$$(47)\qquad F(\zeta_1+h) = F(\zeta_1) + \int_0^1 f(\zeta_1+th)h\,dt = F(\zeta_1) + h\int_0^1 f(\zeta_1+th)\,dt,$$

$$(48)\qquad \frac{F(\zeta_1+h)-F(\zeta_1)}{h} = \int_0^1 f(\zeta_1+th)\,dt, \qquad h\notin \mathcal{O}_n,$$

$$(49)\qquad \frac{F(\zeta_1+h)-F(\zeta_1)}{h} - f(\zeta_1) = \int_0^1 [f(\zeta_1+th)-f(\zeta_1)]\,dt.$$

Now f satisfies the strong Stolz condition in X; it is therefore continuous in X. Let $\varepsilon>0$ be given. Then there exists a $\delta(\varepsilon,\zeta_1)$ such that $\|f(\eta)-f(\zeta_1)\|_n<\varepsilon$ for every η for which $\|\eta-\zeta_1\|_n<\delta(\varepsilon,\zeta_1)$. Next

$$(50)\qquad \left\|\frac{F(\zeta_1+h)-F(\zeta_1)}{h} - f(\zeta_1)\right\|_n \leqslant \int_0^1 \|f(\zeta_1+th)-f(\zeta_1)\|_n\,dt$$
$$< \varepsilon, \qquad \|h\|_n < \delta(\varepsilon,\zeta_1),\ h\notin\mathcal{O}_n.$$

Therefore, $D_\zeta F(\zeta)$ exists and

$$(51)\qquad D_\zeta F(\zeta) = f(\zeta).$$

Since f is continuous, then (51) shows that $D_\zeta F$ is continuous, and the proof of (34) is complete.

To prove (35), begin with (47). Set $\zeta_1+h=\zeta$; then $h=\zeta-\zeta_1$, and (47) can be written as follows:

$$(52)\qquad F(\zeta)-F(\zeta_1)=\int_0^1 f(\zeta_1+th)\,dt(\zeta-\zeta_1).$$

In the integral on the right, use (51) to replace $f(\zeta_1+th)$ by $D_\zeta F(\zeta_1+th)$; then add and subtract $D_\zeta F(\zeta_1)(\zeta-\zeta_1)$ on the right in (52). The result is

$$(53)\qquad F(\zeta)-F(\zeta_1)=D_\zeta F(\zeta_1)(\zeta-\zeta_1)+\int_0^1 [D_\zeta F(\zeta_1+th)-D_\zeta F(\zeta_1)]\,dt(\zeta-\zeta_1).$$

Set

$$(54)\qquad r(F;\zeta_1,\zeta)=\int_0^1 [D_\zeta F(\zeta_1+th)-D_\zeta F(\zeta_1)]\,dt.$$

Since $D_\zeta F$ is continuous in X by (51) and $h\to 0$ as $\zeta\to\zeta_1$, then (54) shows that

$$(55)\qquad \lim_{\zeta\to\zeta_1} r(F;\zeta_1,\zeta)=\lim_{h\to 0}\int_0^1 [D_\zeta F(\zeta_1+th)-D_\zeta F(\zeta_1)]\,dt=0,$$

$$(56)\qquad r(F;\zeta_1,\zeta_1)=0.$$

Then (53) and (54) show that

$$(57)\qquad F(\zeta)-F(\zeta_1)=D_\zeta F(\zeta_1)(\zeta-\zeta_1)+r(F;\zeta_1,\zeta)(\zeta-\zeta_1).$$

Finally, F satisfies the strong Stolz condition by (57), (55), and (56). But it is necessary to prove even more. Since $D_\zeta F$ is continuous in X, it is uniformly continuous in a compact set S in X. Let $\varepsilon>0$ be given. Then there exists a $\delta(\varepsilon)$, which depends on ε but not on ζ_1, such that

$$(58)\qquad \|D_\zeta F(\zeta_1+th)-D_\zeta F(\zeta_1)\|_n<\varepsilon,\qquad 0\leqslant t\leqslant 1,$$

for every pair of points ζ_1 and ζ provided the segment $[\zeta_1,\zeta]$ is in S and $\|\zeta-\zeta_1\|_n<\delta(\varepsilon)$. Then (54) and (58) show that

$$(59)\qquad \|r(F;\zeta_1,\zeta)\|_n<\varepsilon,\qquad \|\zeta-\zeta_1\|_n<\delta(\varepsilon)\text{ and }[\zeta_1,\zeta]\subset S.$$

Finally, (57) and (59) show that F satisfies the uniform strong Stolz condition in S, and the proof of (35) and of all parts of Theorem 49.2 is complete. □

Theorem 49.2 and the idempotent representation of integrals can be used as in Section 38 to establish the set of fundamental representations of a differentiable function $f:X\to\mathbb{C}_n$ for $n\geqslant 2$. The methods are not difficult, but the notational problems are formidable. Begin by describing the notation and the idempotent representation of integrals. Let

(60) $\zeta = Z_1 + i_n Z_2, \qquad \zeta \in \mathbb{C}_n,\ Z_1 \text{ and } Z_2 \text{ in } \mathbb{C}_{n-1};$

(61) $\zeta = (Z_1 - i_{n-1}Z_2)e(i_{n-1}i_n) + (Z_1 + i_{n-1}Z_2)e(-i_{n-1}i_n);$

(62) $f(\zeta) = u_1(\zeta) + i_n u_2(\zeta), \qquad u_1(\zeta) \text{ and } u_2(\zeta) \text{ in } \mathbb{C}_{n-1}.$

Then f has the following idempotent representation:

(63)
$$\begin{aligned} f(\zeta) = {} & [u_1(\zeta) - i_{n-1}u_2(\zeta)]e(i_{n-1}i_n) \\ & + [u_1(\zeta) + i_{n-1}u_2(\zeta)]e(-i_{n-1}i_n). \end{aligned}$$

The problem is to determine the nature of the functions $u_1(\zeta) - i_{n-1}u_2(\zeta)$ and $u_1(\zeta) + i_{n-1}u_2(\zeta)$. A curve C in $\mathbb{C}_n$ has an equation

(64) $Z_1 + i_n Z_2 = Z_1(t) + i_n Z_2(t), \qquad a \leqslant t \leqslant b.$

Then C has the following idempotent representation:

(65)
$$\begin{aligned} Z_1 + i_n Z_2 = {} & [Z_1(t) - i_{n-1}Z_2(t)]e(i_{n-1}i_n) \\ & + [Z_1(t) + i_{n-1}Z_2(t)]e(-i_{n-1}i_n). \end{aligned}$$

Thus associated with C there are two curves C_1 and C_2 in $\mathbb{C}_{n-1}$ with the following equations:

(66) $Z_1 - i_{n-1}Z_2 = Z_1(t) - i_{n-1}Z_2(t), \qquad a \leqslant t \leqslant b,$

(67) $Z_1 + i_{n-1}Z_2 = Z_1(t) + i_{n-1}Z_2(t), \qquad a \leqslant t \leqslant b.$

An integral is the limit of a sum as shown in (4) above. Represent all terms in the sum by the idempotent representations and use the properties of limits in $\mathbb{C}_n$ to show that

(68)
$$\begin{aligned} \int_C f(\eta)\,d\eta = {} & \int_{C_1} [u_1(\zeta) - i_{n-1}u_2(\zeta)]\,d(Z_1 - i_{n-1}Z_2)e(i_{n-1}i_n) \\ & + \int_{C_2} [u_1(\zeta) + i_{n-1}u_2(\zeta)]\,d(Z_1 + i_{n-1}Z_2)e(-i_{n-1}i_n). \end{aligned}$$

49.3 THEOREM Let X be a domain in $\mathbb{C}_n$, and let $f: X \to \mathbb{C}_n$, $n \geqslant 2$, be a differentiable function. Then there exist differentiable functions $f_1: X_1 \to \mathbb{C}_{n-1}$ and $f_2: X_2 \to \mathbb{C}_{n-1}$ such that

(69)
$$\begin{gathered} f(Z_1 + i_n Z_2) = f_1(Z_1 - i_{n-1}Z_2)e(i_{n-1}i_n) + f_2(Z_1 + i_{n-1}Z_2)e(-i_{n-1}i_n), \\ Z_1 + i_n Z_2 \in X, \qquad Z_1 - i_{n-1}Z_2 \in X_1, \qquad Z_1 + i_{n-1}Z_2 \in X_2. \end{gathered}$$

Proof. Let ζ_0 be a point in X. Since X is open, there is a neighborhood $N(\zeta_0, r)$ of ζ_0 which is contained in X. Now $N(\zeta_0, r)$ is star-shaped with respect to ζ_0, and the construction of the function F described in the proof of Theorem 49.2 can be carried out in $N(\zeta_0, r)$. Thus

$$(70) \qquad F(\zeta_1) = \int_P f(\eta)\,d\eta,$$

and P is a polygonal curve which connects ζ_0: $Z_1^0 + i_n Z_2^0$ to ζ_1: $Z_1' + i_n Z_2'$ in $N(\zeta_0, r)$. Then by the representation of the integral in (68),

$$(71) \quad \int_P f(\eta)\,d\eta = \int_{P_1} [u_1(\eta) - i_{n-1}u_2(\eta)]\,d(Z_1 - i_{n-1}Z_2)e(i_{n-1}i_n)$$
$$+ \int_{P_2} [u_1(\eta) + i_{n-1}u_2(\eta)]\,d(Z_1 + i_{n-1}Z_2)e(-i_{n-1}i_n).$$

Since P connects ζ_0 to ζ_1, then P_1 connects $Z_1^0 - i_{n-1}Z_2^0$ to $Z_1' - i_{n-1}Z_2'$, and P_2 connects $Z_1^0 + i_{n-1}Z_2^0$ to $Z_1' + i_{n-1}Z_2'$ (P_1 and P_2 correspond to C_1 and C_2 above). Since the integral in (70) is independent of the path, then the value of

$$(72) \qquad \int_{P_1} [u_1(\eta) - i_{n-1}u_2(\eta)]\,d(Z_1 - i_{n-1}Z_2)$$

is independent of the path and depends only on $Z_1' - i_{n-1}Z_2'$. Thus the integral in (72) is the value of a function $F_1 : [N(\zeta_0, r)]_1 \to \mathbb{C}_{n-1}$ at $Z_1' - i_{n-1}Z_2'$, and

$$(73) \qquad F_1(Z_1' - i_{n-1}Z_2') = \int_{P_1} [u_1(\eta) - i_{n-1}u_2(\eta)]\,d(Z_1 - i_{n-1}Z_2).$$

Similarly, the integral

$$(74) \qquad \int_{P_2} [u_1(\eta) + i_{n-1}u_2(\eta)]\,d(Z_1 + i_{n-1}Z_2)$$

is independent of the path, and its value depends only on $Z_1' + i_{n-1}Z_2'$. Thus there is a function $F_2 : [N(\zeta_0, r)]_2 \to \mathbb{C}_{n-1}$ such that

$$(75) \qquad F_2(Z_1' + i_{n-1}Z_2') = \int_{P_2} [u_1(\eta) + i_{n-1}u_2(\eta)]\,d(Z_1 + i_{n-1}Z_2).$$

Now (70), (71), (73), and (75) show that

$$(76) \qquad F(Z_1' + i_n Z_2') = F_1(Z_1' - i_{n-1}Z_2')e(i_{n-1}i_n) + F_2(Z_1' + i_{n-1}Z_2')e(-i_{n-1}i_n).$$

The proof of Theorem 49.2 has shown that F has a derivative $D_\zeta F$, and (51) and (62) show that

$$(77) \qquad D_\zeta F(\zeta) = f(\zeta) = u_1(\zeta) + i_n u_2(\zeta).$$

Because F has a derivative with the value shown in (77), the idempotent representation of the norm in $\mathbb{C}_n$ shows that F_1 and F_2 have derivatives, and that

$$(78) \qquad D_{Z_1-i_{n-1}Z_2}F_1(Z_1 - i_{n-1}Z_2) = u_1(\zeta) - i_{n-1}u_2(\zeta),$$

$$(79) \qquad D_{Z_1+i_{n-1}Z_2}F_2(Z_1 + i_{n-1}Z_2) = u_1(\zeta) + i_{n-1}u_2(\zeta).$$

Now F_1 and F_2 are functions of $Z_1 - i_{n-1}Z_2$ and $Z_1 + i_{n-1}Z_2$, respectively, and their derivatives are functions of the same variables. Thus (78) and (79) show that there are functions f_1 and f_2 such that

$$(80) \qquad f_1(Z_1 - i_{n-1}Z_2) = D_{Z_1-i_{n-1}Z_2}F_1(Z_1 - i_{n-1}Z_2) = u_1(\zeta) - i_{n-1}u_2(\zeta),$$

$$(81) \qquad f_2(Z_1 + i_{n-1}Z_2) = D_{Z_1+i_{n-1}Z_2}F_2(Z_1 + i_{n-1}Z_2) = u_1(\zeta) + i_{n-1}u_2(\zeta), \qquad \zeta \in N(\zeta_0, r).$$

Then (63), (80), and (81) show that

$$(82) \qquad f(Z_1 + i_nZ_2) = f_1(Z_1 - i_{n-1}Z_2)e(i_{n-1}i_n) + f_2(Z_1 + i_{n-1}Z_2)e(-i_{n-1}i_n)$$

for $Z_1 + i_nZ_2$ in $N(\zeta_0, r)$. Since f is differentiable in X by hypothesis, then the idempotent representation of the norm in $\mathbb{C}_n$ shows that f_1 and f_2 are differentiable in $[N(\zeta_0, r)]_1$ and $[N(\zeta_0, r)]_2$, respectively. Thus the equation in (69) and Theorem 49.3 have been proved in $N(\zeta_0, r)$. Since every point ζ_0 in X has a neighborhood in which Theorem 49.3 is true, the theorem is true in X. □

49.4 COROLLARY Let X be a domain in $\mathbb{C}_n$, and let $f: X \to \mathbb{C}_n$, $n \geqslant 2$, be a differentiable function. Then there are $n-1$ idempotent representations of f with coefficients which are differentiable functions. These representations can be described as follows. The first representation is shown in (69); here f has an idempotent representation by two differentiable functions f_1 and f_2 of variables $Z_1 - i_{n-1}Z_2$ and $Z_1 + i_{n-1}Z_2$ in $\mathbb{C}_{n-1}$ and with values in $\mathbb{C}_{n-1}$. In representation number 2, f has an idempotent representation by 2^2 differentiable functions $f_1, \ldots, f_4$ of variables in $\mathbb{C}_{n-2}$ and with values in $\mathbb{C}_{n-2}$. Representations number $3, \ldots, n-1$ are similar. In representation number $n-1$, f has an idempotent representation by 2^{n-1} differentiable functions f_k, $k = 1, \ldots, 2^{n-1}$, of variables in $\mathbb{C}_1$ and with values in $\mathbb{C}_1$.

Proof. There are two proofs of this corollary. The first is a proof by induction beginning with the representation in (69). Apply Theorem 49.3 to each of the functions f_1 and f_2 in (69) to obtain the second representation. Continue in this fashion until the process terminates in a representation of f by means of 2^{n-1} functions of complex variables. The second proof of

Corollary 49.4 is similar to the proof of Theorem 49.3. The latter process uses Theorem 49.2 and the representation

$$(83)\qquad Z_1 + i_n Z_2 = (Z_1 - i_{n-1}Z_2)e(i_{n-1}i_n) + (Z_1 + i_{n-1}Z_2)e(-i_{n-1}i_n),$$
$$Z_1 + i_n Z_2 \in \mathbb{C}_n, \quad Z_1 - i_{n-1}Z_2 \in \mathbb{C}_{n-1}, \quad Z_1 + i_{n-1}Z_2 \in \mathbb{C}_{n-1}.$$

Use this formula to obtain an idempotent representation of $Z_1 - i_{n-1}Z_2$ and $Z_1 + i_{n-1}Z_2$; the result is an idempotent representation of $Z_1 + i_n Z_2$ with four coefficients in $\mathbb{C}_{n-2}$. Use this idempotent representation of elements in $\mathbb{C}_n$, with four coefficients in $\mathbb{C}_{n-2}$, to repeat the process described in the proof of Theorem 49.3. The result is representation number 2 of f as described in the corollary. Thus the method used in proving Theorem 49.3 can be employed to show that to each idempotent representation of $Z_1 + i_n Z_2$ in $\mathbb{C}_n$ there corresponds an idempotent representation of f with differentiable coefficients; the number of functions in each representation is the same as the number of coefficients in the corresponding representation of an element in $\mathbb{C}_n$. The description of the two proofs of Corollary 49.4 is complete. □

The proofs of some theorems have employed special representations for differentiable functions $f: X \to \mathbb{C}_n$. For example, the fundamental theorem of the integral calculus in Theorem 33.1 uses the representation

$$(84)\qquad F(\zeta) = g_1(x) + i_1 g_2(x) + i_2 g_3(x) + i_1 i_2 g_4(x).$$

Here each g_k is a real-valued function of the real variables $x_1, \ldots, x_4$. This real-variable representation makes it possible to take advantage of the methods and properties of functions of several real variables. Although this resort to real variables achieves numerous successes, it encounters difficulties in treating functions $f: X \to \mathbb{C}_n$ when n is large. The mass of detail grows so rapidly with n that meaningful proofs become impossible. Furthermore, proofs based on the real-variable representation are aesthetically not pleasing, and they are not in keeping with the spirit of multicomplex analysis. For all of these reasons every effort has been made to provide proofs which depend on ζ alone in $\mathbb{C}_n$ and are independent of n. There have been some successes, but it is the opinion of this author that further new methods are needed in order to bring the subject of holomorphic functions of a multicomplex variable to a satisfactory state of development. The proof of Cauchy's integral theorem which follows employs no special representations of f or of ζ and it is independent of n since it is valid without change for all $\mathbb{C}_n$, $n \geqslant 1$; therefore this theorem must be considered to have a proof of the desirable type. However, the proof of the fundamental theorem of the integral calculus is less than satisfactory. Two proofs are suggested below. The first employs the real-variable representation of the function of the type shown in (84) and (87). Although the same method applies for all, a separate proof is required for each n, and there is doubt that the proof can be effectively carried out for large values of n. The second method of proof is by induction, and it employs the special representation of the function f shown in (1).

49.5 THEOREM (Fundamental Theorem of the Integral Calculus) Let X be a domain in $\mathbb{C}_n$, $n \geqslant 1$; let $f: X \to \mathbb{C}_n$ be a function whose derivative $D_\zeta f$ is continuous in X; and let C be a curve $\zeta: [a, b] \to X$ which has a continuous derivative. Then

$$\int_C D_\zeta f(\zeta)\,d\zeta = f[\zeta(b)] - f[\zeta(a)]. \tag{85}$$

Proof. This theorem has been proved already for $n=2$ in Theorem 33.1, and the same method can be used for larger values of n. The proof employs the real-valued representation of ζ and of f. For example, if $n=3$, then

$$\zeta = \{[x_1 + i_1x_2] + i_2[x_3 + i_1x_4]\} + i_3\{[x_5 + i_1x_6] + i_2[x_7 + i_1x_8]\}, \tag{86}$$

$$f(\zeta) = \{[g_1(x) + i_1g_2(x)] + i_2[g_3(x) + i_1g_4(x)]\} + i_3\{[g_5(x) + i_1g_6(x)] + i_2[g_7(x) + i_1g_8(x)]\}. \tag{87}$$

Also,

$$\zeta(t) = \{[x_1(t) + i_1x_2(t)] + i_2[x_3(t) + i_1x_4(t)]\} + i_3\{[x_5(t) + i_1x_6(t)] + i_2[x_7(t) + i_1x_8(t)]\}, \tag{88}$$

and $\zeta'(t)$ is obtained by replacing each $x_k(t)$ in (88) by $x_k'(t)$. As in (9) in Section 33,

$$\int_C D_\zeta f(\zeta)\,d\zeta = \int_a^b D_\zeta f[\zeta(t)]\zeta'(t)\,dt. \tag{89}$$

Next, $D_\zeta f(\zeta)$ is given in (110) in Section 48. Substitute these values in the integral on the right in (89) and then multiply out the product $D_\zeta f[\zeta(t)]\zeta'(t)$. Finally, use the Cauchy–Riemann differential equations in Theorem 48.6 to write the integrand of the integral on the right in (89) as the derivative of a function of t. Then the fundamental theorem of the integral calculus for functions of a real variable can be used to evaluate the integral and obtain the result stated in (85). The details are entirely similar to those in the proof of Theorem 33.1. □

49.6 REMARK Theoretically, the method used to prove Theorem 49.5 for $n=2$ (in Theorem 33.1) and $n=3$ (above) can be used to prove the theorem for all $n \geqslant 1$. In reality, this method of proof has severe limitations: the proof cannot be exhibited in a convincing manner because of the great mass of detail which lacks an easy and obvious pattern. Nevertheless, these proofs in the special cases $n=2$ and $n=3$ are important because they provide a beginning for a complete proof of the theorem by induction. The inductive step can be proved as follows: Let $f: X \to \mathbb{C}_n$, $n>2$, be a function which has a continuous derivative in X. If ζ is in $\mathbb{C}_n$, then $\zeta=\zeta_1+i_n\zeta_2$ and ζ_1 and ζ_2 are in $\mathbb{C}_{n-1}$. Then by (1),

$$
\begin{aligned}
(90)\qquad f(\zeta_1+i_n\zeta_2) &= f_1(\zeta_1-i_{n-1}\zeta_2)e(i_{n-1}i_n) \\
&\quad + f_2(\zeta_1+i_{n-1}\zeta_2)e(-i_{n-1}i_n),
\end{aligned}
$$

$$
\begin{aligned}
(91)\qquad & D_{\zeta_1+i_n\zeta_2} f(\zeta_1+i_n\zeta_2) \\
&= D_{\zeta_1-i_{n-1}\zeta_2} f_1(\zeta_1-i_{n-1}\zeta_2)e(i_{n-1}i_n) \\
&\quad + D_{\zeta_1+i_{n-1}\zeta_2} f_2(\zeta_1+i_{n-1}\zeta_2)e(-i_{n-1}i_n).
\end{aligned}
$$

Since $D_{\zeta_1+i_n\zeta_2} f$ is continuous by hypothesis, then $D_{\zeta_1-i_{n-1}\zeta_2} f_1$ and $D_{\zeta_1+i_{n-1}\zeta_2} f_2$ are also continuous, and

$$
\begin{aligned}
(92)\qquad & \int_C D_{\zeta_1+i_n\zeta_2} f(\zeta_1+i_n\zeta_2)\,d(\zeta_1+i_n\zeta_2) \\
&= \int_{C_1} D_{\zeta_1-i_{n-1}\zeta_2} f_1(\zeta_1-i_{n-1}\zeta_2)\,d(\zeta_1-i_{n-1}\zeta_2)e(i_{n-1}i_n) \\
&\quad + \int_{C_2} D_{\zeta_1+i_{n-1}\zeta_2} f_2(\zeta_1+i_{n-1}\zeta_2)\,d(\zeta_1+i_{n-1}\zeta_2)e(-i_{n-1}i_n).
\end{aligned}
$$

Since the theorem is assumed to be true in $\mathbb{C}_{n-1}$ by the induction hypothesis, then (92) shows that it is true in $\mathbb{C}_n$. This outline explains how the proof of Theorem 49.5 can be completed for all values of n. □

49.7 COROLLARY If the curve C in Theorem 49.5 is a closed curve, then

$$(93)\qquad \int_C D_\zeta f(\zeta)\,d\zeta = 0.$$

49.8 COROLLARY Let $f: X \to \mathbb{C}_n$ and $h: X \to \mathbb{C}_n$, $n \geqslant 1$, be two functions which have continuous derivatives $D_\zeta f$ and $D_\zeta h$ in X such that

$$(94)\qquad D_\zeta f(\zeta) = D_\zeta h(\zeta), \qquad \zeta \text{ in } X.$$

Then there is a constant c in $\mathbb{C}_n$ such that

$$(95)\qquad f(\zeta) = h(\zeta) + c, \qquad \zeta \in X.$$

Proof. The proof of this corollary is similar to the proof of Theorem 33.2. □

49.9 COROLLARY Let $f: X \to \mathbb{C}_n$, $n \geqslant 1$, be a function which has a continuous derivative in X, and let S be a compact set in X. Then f satisfies the uniform strong Stolz condition in S.

Proof. The proof of this corollary is similar to the proof of Theorem 33.4. □

49.10 REMARK Let $f: X \to \mathbb{C}_n$, $n \geqslant 1$, be a function which has a continuous derivative in X. Then f satisfies the fundamental theorem of the integral calculus (Theorem 49.5), and Corollary 49.9 shows that this theorem can be used to prove that f satisfies the uniform strong Stolz condition in a compact set S in X. Conversely, if $f: X \to \mathbb{C}_n$ has a continuous derivative and it satisfies the uniform strong Stolz condition in every compact set S in X, then f satisfies the fundamental theorem of the integral calculus; the proof is similar to the proof of Theorem 33.8. This proof of the fundamental theorem is of the desirable type: a single proof establishes the theorem for all values of n, and the proof does not require any preliminary results such as the special representation of f in (1).

Theorem 49.1 contains the special case of Cauchy's integral theorem, and this account of differentiable functions $f: X \to \mathbb{C}_n$ will be completed by establishing the following: (a) the general case of Cauchy's integral theorem; (b) Cauchy's integral formula; and (c) the Taylor series representation of f.

49.11 THEOREM (Cauchy's Integral Theorem) Let X be a domain in $\mathbb{C}_n$, $n \geqslant 1$, which is star-shaped with respect to a point ζ^*, and let $f: X \to \mathbb{C}_n$ be a function which satisfies the strong Stolz condition in X. Let C be a closed curve $\zeta: [a, b] \to X$, $t \mapsto \zeta(t)$, which has a continuous derivative ζ'. Then

(96) $$\int_C f(\zeta)\, d\zeta = 0.$$

Proof. The function $f: X \to \mathbb{C}_n$ satisfies the hypotheses of Theorem 49.2; therefore, there exists a function $F: X \to \mathbb{C}_n$ which is a primitive of f:

(97) $$D_\zeta F(\zeta) = f(\zeta), \qquad \zeta \in X.$$

Furthermore, by Theorem 49.2 (35), F satisfies the uniform strong Stolz condition in compact sets S in X; therefore, by Remark 49.10, F satisfies the fundamental theorem of the integral calculus. Then

(98) $$\int_C f(\zeta)\, d\zeta = \int_C D_\zeta F(\zeta)\, d\zeta = F[\zeta(b)] - F[\zeta(a)] = 0.$$

The proof of (96) and of Theorem 49.11 is complete. □

This proof of Cauchy's integral theorem must be considered highly satisfactory for the following reasons: (a) the proof of the theorem, and the proof Theorem 49.2 on which it is based, do not employ any of the special representations of f or F; and (b) one short proof establishes Theorem 49.11 for all values of n, and the same statement is true for the preliminary results in Theorem 49.2 and Remark 49.10.

The final theorem to be treated in this section is Cauchy's integral formula. The theorem to be proved is not the definitive result, but rather a special case which throws light on the general theorem. Before stating the theorem, a description of the setting and notation will be helpful. Let X be a domain in $\mathbb{C}_n$, $n \geqslant 1$, and let $f: X \to \mathbb{C}_n$ be a function which satisfies the strong Stolz condition in X. Let a be a point in X; since X is open, a has a neighborhood $N(a, d)$ in X. Define a subset Y of $N(a, d)$ as follows:

(99) $$Y = \{\zeta \text{ in } N(a, d)\colon \zeta = a + (x_1 + i_1 x_2), (x_1 + i_1 x_2) \in \mathbb{C}_1\}.$$

Then Y is the part of a two-dimensional plane in $\mathbb{C}_n$ which is contained in $N(a, d)$. If $0 < r < d$, then the circle C, whose equation is

(100) $$\zeta = a + r(\cos\theta + i_1 \sin\theta), \qquad 0 \leqslant \theta \leqslant 2\pi,$$

is contained in Y. Let ζ_0 be a point in the interior of C.

49.12 THEOREM (Cauchy's Integral Formula) If $f: X \to \mathbb{C}_n$ satisfies the strong Stolz condition in X, and if ζ_0 is a point in the interior of C, then

(101) $$f(\zeta_0) = \frac{1}{2\pi i_1} \int_C \frac{f(\zeta)\, d\zeta}{\zeta - \zeta_0}.$$

Proof. Let C' be a circle in Y whose center is ζ_0; its equation is

(102) $$\zeta = \zeta_0 + \rho(\cos\theta + i_1 \sin\theta), \qquad 0 \leqslant \theta \leqslant 2\pi.$$

Choose ρ small enough so that C' lies inside C as shown in Figure 49.1; eventually a limit will be taken in which ρ tends to zero. Consider the function

(103) $$\frac{f(\zeta)}{\zeta - \zeta_0}, \qquad \zeta \neq \zeta_0, \zeta \in Y.$$

Since $\zeta - \zeta_0 \in \mathbb{C}_1$ and $\zeta \neq \zeta_0$, then $\zeta - \zeta_0 \notin \mathcal{O}_n$ and the function in (103) is defined. Furthermore, since f satisfies the strong Stolz condition in X by hypothesis, and since $\zeta - \zeta_0 \notin \mathcal{O}_n$, then the function in (103) satisfies the strong Stolz condition in $Y - \{\zeta_0\}$ [compare Corollary 26.3 (17)]. Then a proof, the details of which are given below, shows that

(104) $$\int_C \frac{f(\zeta)\, d\zeta}{\zeta - \zeta_0} - \int_{C'} \frac{f(\zeta)\, d(\zeta)}{\zeta - \zeta_0} = 0.$$

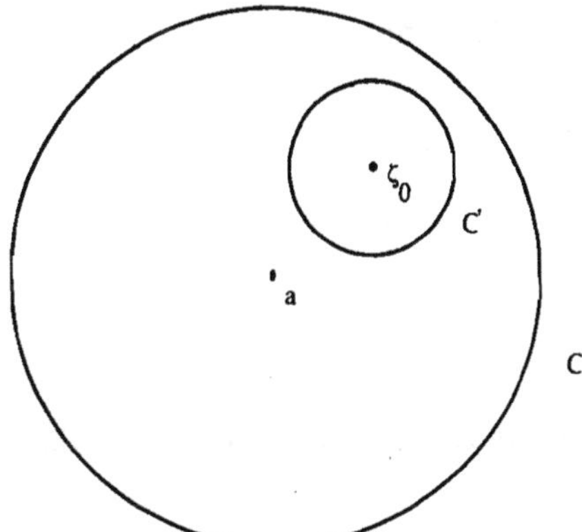

Figure 49.1. Figure for Cauchy's integral formula.

Here the integrals around C and C' are taken in the positive or counter-clockwise direction. The next step is to evaluate the second integral in (104). Since f satisfies the strong Stolz condition in X by hypothesis, then

$$f(\zeta) - f(\zeta_0) = D_\zeta f(\zeta_0)(\zeta - \zeta_0) + r(f; \zeta_0, \zeta)(\zeta - \zeta_0) \tag{105}$$
$$\lim_{\zeta \to \zeta_0} r(f; \zeta_0, \zeta) = 0, \qquad r(f; \zeta_0, \zeta_0) = 0.$$

Substitute from (105) in the second integral in (104) to obtain

$$\int_{C'} \frac{f(\zeta)\,d\zeta}{\zeta - \zeta_0} = \int_{C'} \frac{f(\zeta_0)\,d\zeta}{\zeta - \zeta_0} + \int_{C'} D_\zeta f(\zeta_0)\,d\zeta + \int_{C'} r(f; \zeta_0, \zeta)\,d\zeta. \tag{106}$$

Now by (102),

$$\int_{C'} \frac{f(\zeta_0)\,d\zeta}{\zeta - \zeta_0} = f(\zeta_0) \int_{C'} \frac{d\zeta}{\zeta - \zeta_0} = f(\zeta_0) \int_0^{2\pi} \frac{\rho(-\sin\theta + i_1 \cos\theta)}{\rho(\cos\theta + i_1 \sin\theta)}\,d\theta \tag{107}$$
$$= 2\pi i_1 f(\zeta_0);$$

$$\int_{C'} D_\zeta f(\zeta_0)\,d\zeta = D_\zeta f(\zeta_0) \int_{C'} d\zeta = 0. \tag{108}$$

Equation (106) shows that the value of the third integral on the right is a constant which is independent of ρ; a proof will now show that the value of this integral is zero. Let $\varepsilon > 0$ be given; then (105) shows that there is a $\delta(\varepsilon, \zeta_0)$ such that

$$\|r(f; \zeta_0, \zeta)\|_n < \varepsilon, \qquad \|\zeta - \zeta_0\|_n < \delta(\varepsilon, \zeta_0). \tag{109}$$

Choose ρ in (102) so that $\rho < \delta(\varepsilon, \zeta_0)$; then (109) shows that

$$\left\|\int_{C'} r(f; \zeta_0, \zeta)\, d\zeta\right\|_n \leqslant 2^{n-1/2} 2\pi\rho\varepsilon. \tag{110}$$

The only constant which satisfies this inequality for every positive ε is zero; therefore,

$$\int_{C'} r(f; \zeta_0, \zeta)\, d\zeta = 0. \tag{111}$$

Collect from (106), (107), (108), and (111) and substitute in (104) to show that

$$\int_C \frac{f(\zeta)\, d\zeta}{\zeta - \zeta_0} - 2\pi i_1 f(\zeta_0) = 0. \tag{112}$$

This equation is equivalent to (101), and the proof of the theorem can be completed by establishing the equation in (104).

Consider the proof of (104). Use a diameter of Y through a and ζ_0 to construct two closed curves C_1 and C_2 as shown in Figure 49.2. The curve C_1 consists of semicircles from C and C' which lie on the same side of the diameter, and segments of the diameter to form a closed curve. The curve C_2 consists of semicircles of C and C' on the other side of the diameter, together with the same segments as before of the diameter to form a closed curve. Choose points ζ_1^* and ζ_2^* in the two figures in Figure 49.2 and draw rays out from ζ_1^* so that C_1 lies in a region S_1 of Y which is star-shaped with respect to ζ_1^*, and so that C_2 lies in a region S_2 of Y which is star-shaped with respect to ζ_2^*. Observe that ζ_0 does not belong to S_1, and that ζ_0 does not belong to S_2.

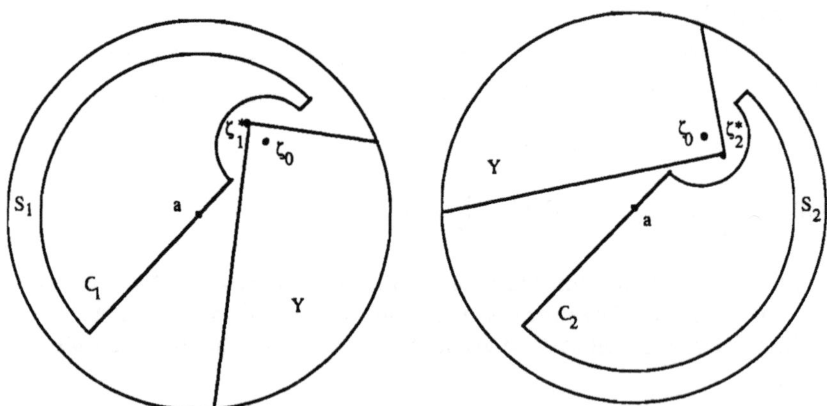

Figure 49.2. Star-shaped regions which contain C_1 and C_2.

Then the function in (103) satisfies the strong Stolz condition in S_1 and also in S_2.

The next step in the proof is to show that

$$\int_{C_1} \frac{f(\zeta)\,d\zeta}{\zeta-\zeta_0} = 0, \qquad \int_{C_2} \frac{f(\zeta)\,d\zeta}{\zeta-\zeta_0} = 0. \tag{113}$$

For convenience let $g(\zeta)$ denote $f(\zeta)/(\zeta-\zeta_0)$. The method used in proving Theorem 49.2 can be used again to construct a function $F_1 : S_1 \to \mathbb{C}_n$ which has the following properties:

(114) F_1 has a derivative $D_\zeta F_1$, and $D_\zeta F_1(\zeta) = g(\zeta)$ for all ζ in S_1.

(115) F_1 satisfies the uniform strong Stolz condition in compact sets in S_1.

(116) F_1 satisfies the fundamental theorem of the integral calculus in S_1.

Then

$$\int_{C_1} \frac{f(\zeta)\,d\zeta}{\zeta-\zeta_0} = \int_{C_1} g(\zeta)\,d\zeta = \int_{C_1} D_\zeta F_1(\zeta)\,d\zeta = 0. \tag{117}$$

In the same way, a function $F_2 : S_2 \to \mathbb{C}_n$ can be constructed which has the properties in (114)–(116). Then

$$\int_{C_2} \frac{f(\zeta)\,d\zeta}{\zeta-\zeta_0} = \int_{C_2} g(\zeta)\,d\zeta = \int_{C_2} D_\zeta F_2(\zeta)\,d\zeta = 0. \tag{118}$$

Equations (117) and (118) show that

$$\int_{C_1} \frac{f(\zeta)\,d\zeta}{\zeta-\zeta_0} + \int_{C_2} \frac{f(\zeta)\,d\zeta}{\zeta-\zeta_0} = 0. \tag{119}$$

The integration in each of these integrals is in the counterclockwise direction. Now the segments of the diameter belong to both C_1 and C_2, but the directions of integration are opposite in the two curves. Thus the integrals on the segments of the diameter cancel, and only integrals on C and C' remain. Figure 49.2 shows that the integration on C' is in the clockwise or negative direction; therefore (119) simplifies to

$$\int_{C} \frac{f(\zeta)\,d\zeta}{\zeta-\zeta_0} + \int_{-C'} \frac{f(\zeta)\,d\zeta}{\zeta-\zeta_0} = 0. \tag{120}$$

Since the sign of an integral is reversed when the direction of integration is reversed, the equation in (120) can be written as follows:

$$\int_{C} \frac{f(\zeta)\,d\zeta}{\zeta-\zeta_0} - \int_{C'} \frac{f(\zeta)\,d\zeta}{\zeta-\zeta_0} = 0. \tag{121}$$

Thus the proof of (104) and of Theorem 49.12 is complete. □

49.13 COROLLARY Let $f: X \to \mathbb{C}_n$ be the function in Theorem 49.12, and let $\zeta = \zeta_1 + i_n \zeta_2$ and $f(\zeta) = u_1(\zeta) + i_n u_2(\zeta)$. Then there are functions f_1 of $\zeta_1 - i_{n-1}\zeta_2$ and f_2 of $\zeta_1 + i_{n-1}\zeta_2$ such that, for all ζ in X,

$$(122) \qquad f_1(\zeta_1 - i_{n-1}\zeta_2) = u_1(\zeta) - i_{n-1}u_2(\zeta),$$

$$f_2(\zeta_1 + i_{n-1}\zeta_2) = u_1(\zeta) + i_{n-1}u_2(\zeta),$$

$$(123) \qquad f(\zeta_1 + i_n\zeta_2) = f_1(\zeta_1 - i_{n-1}\zeta_2)e(i_{n-1}i_n) + f_2(\zeta_1 + i_{n-1}\zeta_2)e(-i_{n-1}i_n).$$

Proof. Let C^1 and C^2 be the idempotent representations of the curve C in Theorem 49.12. Then by (101) and the idempotent representation of elements in $\mathbb{C}_n$,

$$(124) \qquad [u_1(\zeta_0) - i_{n-1}u_2(\zeta_0)]e(i_{n-1}i_n) + [u_1(\zeta_0) + i_{n-1}u_2(\zeta_0)]e(-i_{n-1}i_n)$$
$$= \frac{1}{2\pi i_1}\int_{C^1} \frac{[u_1(\zeta) - i_{n-1}u_2(\zeta)]\, d(\zeta_1 - i_{n-1}\zeta_2)}{(\zeta_1 - i_{n-1}\zeta_2) - (\zeta_1^0 - i_{n-1}\zeta_2^0)}\, e(i_{n-1}i_n)$$
$$+ \frac{1}{2\pi i_1}\int_{C^2} \frac{[u_1(\zeta) + i_{n-1}u_2(\zeta)]\, d(\zeta_1 + i_{n-1}\zeta_2)}{(\zeta_1 + i_{n-1}\zeta_2) - (\zeta_1^0 + i_{n-1}\zeta_2^0)}\, e(-i_{n-1}i_n).$$

This equation shows that

$$(125) \qquad u_1(\zeta_0) - i_{n-1}u_2(\zeta_0) = \frac{1}{2\pi i_1}\int_{C^1} \frac{[u_1(\zeta) - i_{n-1}u_2(\zeta)]\, d(\zeta_1 - i_{n-1}\zeta_2)}{(\zeta_1 - i_{n-1}\zeta_2) - (\zeta_1^0 - i_{n-1}\zeta_2^0)},$$

$$(126) \qquad u_1(\zeta_0) + i_{n-1}u_2(\zeta_0) = \frac{1}{2\pi i_1}\int_{C^2} \frac{[u_1(\zeta) + i_{n-1}u_2(\zeta)]\, d(\zeta_1 + i_{n-1}\zeta_2)}{(\zeta_1 + i_{n-1}\zeta_2) - (\zeta_1^0 + i_{n-1}\zeta_2^0)}.$$

Since there is a circle C which contains in its interior an arbitrary point ζ_0 in X, formula (101) can be applied to derive equations (125) and (126), from which the statements in Corollary 49.13 follow. □

The representation (123) is the first in a sequence of 2^{n-1} such idempotent representations. In the second, f is represented by 2^2 functions of variables in $\mathbb{C}_{n-2}$. The last in the sequence is especially important in the present context; in it, f is represented by 2^{n-1} functions of variables which are in $\mathbb{C}_1$. If e_k, $k = 1, \ldots, 2^{n-1}$, are the standard idempotents in $\mathbb{C}_n$, and if

$$(127) \qquad \zeta = \sum_{k=1}^{2^{n-1}} Z_k e_k, \qquad \zeta \in \mathbb{C}_n,\ Z_k \in \mathbb{C}_1,$$

then for each k there is a function f_k of Z_k such that

(128) $$f(\zeta) = \sum_{k=1}^{2^{n-1}} f_k(Z_k)e_k, \qquad f_k(Z_k) \in \mathbb{C}_1.$$

The entire sequence of representations can be obtained from the sequence of idempotent representations of the integral in (101) in the same way that (123) was obtained from this integral. Other proofs of this fundamental sequence of representations have been given; see, for example, Theorem 49.3. The sequence of idempotent representations of ζ and of f constitutes the fundamental theorem of the multicomplex calculus.

Cauchy's integral formula in Theorem 49.12 establishes formula (101) only for points ζ_0 in Y which are inside the circle C. Thus only in $\mathbb{C}_1$ does Theorem 49.12 establish formula (101) in a neighborhood (or ball). Nevertheless, the integral

(129) $$\int_C \frac{f(\zeta)\,d\zeta}{\zeta - \zeta_0}$$

is obviously defined for points ζ_0 which are not in Y. If the statement $\zeta - \zeta_0 \notin \mathcal{O}_n$ is true for every ζ on C, then the integrand in (129) is a differentiable function and the integral exists. This fact shows that (129) is defined for some points ζ_0, not in Y, but sufficiently near a point in the interior of C. Some method is needed to provide a description of the set of points ζ_0 for which (129) is defined and for which the equality in (101) holds. A method is available, but it depends on n and on the idempotent representation of the integral in (101). Define the points a_k by the following equation:

(130) $$a = \sum_{k=1}^{2^{n-1}} a_k e_k, \qquad a_k \in \mathbb{C}_1.$$

The idempotent representation of the circle C is a set of circles C_k, $k = 1, \ldots, 2^{n-1}$, whose equations are

(131) $$C_k\colon Z_k = a_k + r(\cos\theta + i_1 \sin\theta), \qquad k = 1, \ldots, 2^{n-1}.$$

The formula (101) has the following representation:

(132) $$f(\zeta_0) = \sum_{k=1}^{2^{n-1}} \frac{1}{2\pi i_1} \int_{C_k} \frac{f_k(Z_k)\,dZ_k}{Z_k - Z_k^0} e_k, \qquad \zeta_0 = \sum_{k=1}^{2^{n-1}} Z_k^0 e_k.$$

Now by the theory of functions of a complex variable, it is known that, if Z_k^0 is inside C_k, then

(133) $$f_k(Z_k^0) = \frac{1}{2\pi i_1} \int_{C_k} \frac{f_k(Z_k)\,dZ_k}{Z_k - Z_k^0}, \qquad k = 1, \ldots, 2^{n-1}.$$

Thus if ζ_0 is a point in $\mathbb{C}_n$ such that Z_k^0 is inside C_k for $k=1,\dots,2^{n-1}$, then

$$(134)\qquad \frac{1}{2\pi i_1}\int_C \frac{f(\zeta)\,d\zeta}{\zeta-\zeta_0} = \sum_{k=1}^{2^{n-1}} \frac{1}{2\pi i_1}\int_{C_k} \frac{f_k(Z_k)\,dZ_k}{Z_k - Z_k^0} e_k,$$
$$= \sum_{k=1}^{2^{n-1}} f_k(Z_k^0)e_k,$$
$$= f(\zeta_0).$$

Thus formula (101) in Theorem 49.12 is valid for every point ζ_0 in $\mathbb{C}_n$ such that Z_k is in C_k for $k=1,\dots,2^{n-1}$. This set of points ζ_0 is the generalization of the set which, in $\mathbb{C}_2$ in Section 9, was called a discus.

The theory of functions of a complex variable shows that the Cauchy integral in (133) can be used to obtain the Taylor series for f_k and for f. From (133),

$$(135)\qquad f_k(Z_k) = \sum_{m=0}^{\infty} b_m^k (Z_k - a_k)^m, \qquad k = 1,\dots,2^{n-1}.$$

Then

$$(136)\qquad f(\zeta) = \sum_{k=1}^{2^{n-1}} f_k(Z_k)e_k,$$
$$= \sum_{k=1}^{2^{n-1}} \left[\sum_{m=0}^{\infty} b_m^k (Z_k - a_k)^m\right] e_k,$$
$$= \sum_{m=0}^{\infty} b_m(\zeta - a)^m, \qquad b_m = \sum_{k=1}^{2^{n-1}} b_m^k e_k.$$

Exercises

49.1 Show that all points in Y and inside C are contained in the set D defined as follows:

$$D = \left\{\zeta \text{ in } \mathbb{C}_n : \zeta = \sum_{k=1}^{2^{n-1}} Z_k e_k,\ Z_k \text{ is a point inside } C_k\right\}.$$

[*Hint.* Find how the points in Y and inside C are mapped by the idempotent representation.]

49.2 In (136) show that

$$b_m^k = \frac{D_{Z_k}^m f_k(a_k)}{m!}, \qquad k = 1,\dots,2^{n-1},\ m = 0, 1,\dots;$$

$$b_m = \frac{D_\zeta^m f(a)}{m!}, \qquad m = 0, 1, \dots.$$

6

Epilogue

This book has presented an introduction to the theory of functions of a bicomplex, and of a multicomplex, variable. This theory combines important parts of the theory of functions of a real variable and also of a complex variable to construct a theory of a new class of functions known as holomorphic functions of a (bicomplex or) multicomplex variable. This theory seems to be an altogether natural continuation of the theory of functions of a complex variable. The theory of functions of a complex variable gains much relative simplicity from the fact that it is a two-dimensional theory; the theory of functions in $\mathbb{C}_n$ gains much interest and difficulty from the fact that it is a 2^n-dimensional theory. But $\mathbb{C}_n$ has a structure which provides some special tools which facilitate the study of certain high-dimensional spaces where geometric intuition is largely wanting.

But the full scope of the multicomplex spaces is much greater than the book has suggested thus far. Recall that $\mathbb{C}_0$ has been used to denote the real numbers $\mathbb{R}$, and that $\mathbb{C}_1$ has denoted the complex numbers $\mathbb{C}$. Consider the following array of spaces:

$$\begin{array}{lccccc}
(1) & \mathbb{C}_0 & \mathbb{C}_0^2 & \cdots & \mathbb{C}_0^m & \cdots \\
 & \mathbb{C}_1 & \mathbb{C}_1^2 & \cdots & \mathbb{C}_1^m & \cdots \\
 & \mathbb{C}_2 & \mathbb{C}_2^2 & \cdots & \mathbb{C}_2^m & \cdots \\
 & \vdots & & & & \\
 & \mathbb{C}_n & \mathbb{C}_n^2 & \cdots & \mathbb{C}_n^m & \cdots \\
 & \vdots & & & &
\end{array}$$

The first row in this array contains the real euclidean spaces usually denoted by $\mathbb{R}^m$, $m=1,2,\ldots$. The second row contains the complex spaces usually denoted by $\mathbb{C}^m$. These spaces $\mathbb{C}_0^m$ and $\mathbb{C}_1^m$, and especially $\mathbb{C}_0^{2^n}$ and $\mathbb{C}_1^{2^{n-1}}$, have played an auxiliary role in studying holomorphic functions in $\mathbb{C}_n$. This book has emphasized the study of the spaces in the first column of the array (1) and of the functions defined on them. Thus the book has dealt with the spaces on the borders of the array (1), and it has ignored the spaces $\mathbb{C}_n^m$, $m, n=2,3,\ldots$, in the interior. Our book leaves a vast territory still to be surveyed. For example, what are the special properties of the linear space $\mathbb{C}_n^m$ whose elements are

$$(2) \qquad (\zeta_1, \zeta_2, \ldots, \zeta_m), \qquad \zeta_k \in \mathbb{C}_n,\ k = 1, 2, \ldots, m?$$

There are all the usual topics to investigate: bases, linear transformations, solution of systems of linear equations, determinants and their properties, etc. Another topic concerns the study of functions of the form

$$(3) \qquad f: X \to \mathbb{C}_n, \qquad X \subset \mathbb{C}_n^m;$$

these are the functions of m variables $\zeta_1, \ldots, \zeta_m$ in $\mathbb{C}_n$ with values $f(\zeta_1, \ldots, \zeta_m)$ in $\mathbb{C}_n$. More simply, the functions in (3) are the functions of several multicomplex variables. These functions can be represented as follows:

$$(4) \qquad f(\zeta_1, \ldots, \zeta_m) = u_1(\zeta_1, \ldots, \zeta_m) + i_n u_2(\zeta_1, \ldots, \zeta_m),$$

$$u_1(\zeta_1, \ldots, \zeta_m) \in \mathbb{C}_{n-1},\ u_2(\zeta_1, \ldots, \zeta_m) \in \mathbb{C}_{n-1};$$

$$(5) \qquad f(\zeta_1, \ldots, \zeta_m) = [u_1(\zeta_1, \ldots, \zeta_m) - i_{n-1} u_2(\zeta_1, \ldots, \zeta_m)] e(i_{n-1} i_n)$$

$$+ [u_1(\zeta_1, \ldots, \zeta_m) + i_{n-1} u_2(\zeta_1, \ldots, \zeta_m)] e(-i_{n-1} i_n).$$

If $m=1$, the fundamental theorem of holomorphic functions in $\mathbb{C}_n$ states that

$$(6) \qquad u_1(\zeta_1, \ldots, \zeta_m) - i_{n-1} u_2(\zeta_1, \ldots, \zeta_m),$$

$$u_1(\zeta_1, \ldots, \zeta_m) + i_{n-1} u_2(\zeta_1, \ldots, \zeta_m),$$

have an especially simple form (see Theorem 49.3 and Corollary 49.4) which unlocks much of the theory. Is there a corresponding theorem when $m > 1$? If so, what is it? But the answers to these questions and the complete development of the theory of such functions are beyond the scope of this book.

The purpose of this book is to provide an introduction to the multicomplex spaces and their functions. We have examined the borders; the vast territory of these spaces and their functions lies largely unexplored before us.

Bibliography

[1] G. Birkhoff and S. Mac Lane, *A Survey of Modern Algebra*, Revised Edition, Macmillan, New York, 1953, xi + 472 pp.

[2] G. Scorza Dragoni, Sulle funzioni olomorfe di una variabile bicomplessa, Reale Accad. d'Italia, Mem. Classe Sci. Fis., Mat. Nat. *5* (1934), 597–665.

[3] Michiji Futagawa, On the theory of functions of a quaternary variable, Tôhoku Math. J. *29* (1928), 175–222; *35* (1932), 69–120.

[4] E. Hille, *Functional Analysis and Semi-Groups*, Vol. 31, American Mathematical Society Colloquium Publications, New York, 1948.

[5] E. Hille, *Analytic Function Theory*, Chelsea Publishing, New York, Vol. I, 1982, xi + 308 pp.; Vol. II, 1977, xii + 496 pp.

[6] U. Morin, Richerche sull'algebra bicomplessa, Mem. Acad. Ital. *6* (1935), 1241–1265.

[7] G. Baley Price, *Multivariable Analysis*, Springer-Verlag, New York, Berlin, Heidelberg, Tokyo, 1984, xiv + 655 pp.

[8] James D. Riley, Contributions to the theory of functions of a bicomplex variable, Tôhoku Math. J. 2nd series *5* (1953), 132–165.

[9] Friedrich Ringleb, Beiträge zur Funktionentheorie in hyperkomplexen Systemen, I, Rend. Circ. Mat. Palermo *57* (1933), 311–340.

[10] G. Battista Rizza, Teoremi e formule integrali nelle algebre ipercomplesse, Atti IV Cong. Un. mat. Ital. *2* (1953), 204–209.

[11] G. Scheffers, Verallgemeinerung der Grundlagen der gewöhnlich complexen Funktionen, I, II, Berichte über die Verhandlungen der Sächsischen Akademie der Wissenschaften zu Leipzig, Math.-Phys. Klasse *45* (1893), 828–848; *46* (1894), 120–134.

[12] C. Segre, Le Rappresentazioni Reali delle Forme Complesse e Gli Enti Iperalgebrici, Math. Ann. *40* (1892), 413–467.

[13] N. Spampinato, Estensione nel Campo Bicomplesso di Due Teoremi, del Levi-Civita e del Severi, per le Funzione Olomorfe di Due Variabili Complesse, I, II, Atti Reale Accad. Naz. Lincei, Rend. (6) *22* (1935), 38–43, 96–102.

[14] N. Spampinato, Sulla Rappresentazione delle Funzioni di Variabile Bicomplessa Totalmente Derivabili, Ann. Mat. Pura Appl. (4) *14* (1936), 305–325.

[15] N. Spampinato, Sulle Funzione di Variabili bicomplessa o Biduale, Scritte Matematici Offerto a Luigi Berzolari, published by Zanichelli, 1936, pp. 595–611.

[16] Tsurusaburo Takasu, Theorie der Funktionen einer allgemeinen bikomplexen Veranderlichen, I, Tôhoku Sci. Rep. *32* (1943), 1.

[17] V. A. Tret′jakov, On the properties of some elementary functions that are defined on the algebra of bicomplex numbers (Russian), pp. 99–106 in *Mathematical Analysis and the Theory of Functions* (Russian), edited by I. I. Bavrin, Moskov. Oblast. Ped. Inst., Moscow, 1980, 164 pp.

[18] J. A. Ward, Theory of analytic functions in linear associative algebras, Duke Math. J. *7* (1940), 233–248.

[19] M. Spivak, *Calculus on Manifolds*, W. A. Benjamin, New York and Amsterdam, 1965, xii + 146 pp.

[20] R. Fueter and E. Bareiss, *Functions of a Hyper Complex Variable*, v + 318 pp. The title page of this book contains the following information:

Lectures by Rudolf Fueter, Written and supplemented by Erwin Bareiss, Fall Semester 1948/49.

Following recent requests, this monograph is reproduced by Argonne National Laboratory with permission of the University of Zurich, Switzerland.

The manuscript, written and supplemented from lecture notes by Erwin Bareiss, was approved by Rudolf Fueter (1879–1950), and typed by Wilfred Bauert. It summarizes Professor Fueter's research in hyper complex function theory, and points out where more research would lead to new insight in number theory, function theory and its applications.

Index

Milton Keynes UK
Ingram Content Group UK Ltd.
UKHW021832071024
449327UK00021B/1483

9 780367 403119